Studies in Choice and Welfare

Peter C. Fishburn

Thanks to Diane Brown for this photograph.

Editors

Professor Steven J. Brams
Department of Politics
New York University
New York, NY 10012
USA
steven.brams@nyu.edu

Professor Fred S. Roberts
DIMACS
Rutgers University
96 Frelinghuysen Rd
Piscataway, NJ 08854
USA
froberts@dimacs.rutgters.edu

Professor William V. Gehrlein
Department of Business Administration
University of Delaware
Newark, DE 19716
USA
wvg@udel.edu

ISBN 978-3-642-09800-0 e ISBN 978-3-540-79128-7

Studies in Choice and Welfare ISSN 1614-0311

Softcover reprint of the hardcover 1st edition 2009

Cover design: WMX Design GmbH, Heidelberg

Printed on acid-free paper

9 8 7 6 5 4 3 2 1

springer.com

Steven J. Brams · William V. Gehrlein
Fred S. Roberts
Editors

The Mathematics of Preference, Choice and Order

Essays in Honor of Peter C. Fishburn

Springer

Preface

Peter Fishburn has had a splendidly productive career that led to path-breaking contributions in a remarkable variety of areas of research. His contributions have been published in a vast literature, ranging through journals of social choice and welfare, decision theory, operations research, economic theory, political science, mathematical psychology, and discrete mathematics. This work was done both on an individual basis and with a very long list of coauthors.

The contributions that Fishburn made can roughly be divided into three major topical areas, and contributions to each of these areas are identified by sections of this monograph. Section 1 deals with topics that are included in the general areas of utility, preference, individual choice, subjective probability, and measurement theory. Section 2 covers social choice theory, voting models, and social welfare. Section 3 deals with more purely mathematical topics that are related to combinatorics, graph theory, and ordered sets. The common theme of Fishburn's contributions to all of these areas is his ability to bring rigorous mathematical analysis to bear on a wide range of difficult problems.

Part 1 covers a variety of topics stemming from several of Fishburn's books: *Decision and Value Theory* [Fishburn (1964)], *Utility Theory for Decision Making* [Fishburn (1970)], *Mathematics of Decision Theory* [Fishburn (1973a)], *The Foundations of Expected Utility Theory* [Fishburn (1982)], and *Nonlinear Preference and Utility Theory* [Fishburn (1988)]. Fishburn has made cutting-edge contributions to the theory of utility, including work on nontransitive preference, stochastic utility, and decision theory, broadly speaking. He has contributed greatly to the theory of expected utility, including important work on axioms for expected utility, the study of multiattribute expected utility, behavioral models of risk taking, and the study of dominance relations, as well as fundamental contributions to the understanding of subjective expected utility. He has also contributed to nonlinear utility theory, with contributions dealing with risk and with uncertainty. Fishburn's work on choice has dealt with choice probability, choice functions, and nonprobabilistic preference and utility. His work on measurement theory has concentrated on uniqueness of representations, as well as on additive and on nondecomposable representations.

The contributions in Part 1 reflect different facets of the aforementioned research. They start with three papers that are related to the general topic of utility theory. Luce, Marley and Ng (Entropy-Related Measures of the Utility of Gambling) develop utility models to explain individual behavior in gambling situations by adding an additional 'entropy' term to the individual's utility function to account for an individual's preference, or aversion, for gambling in specific situations. Bell and Keeney (Altruistic Utility Functions for Joint Decisions) consider situations in which two or more individuals are involved in selecting some alternative in a decision making situation. The specific situation that is considered describes scenarios in which each of the decision makers has an underlying interest in selecting an alternative that will please the other decision makers. Nakamura (SSB Preferences: Nonseparable Utilities or Nonseparable Beliefs) extends aspects of Fishburn's Skew-symmetric Bilinear Utility model to the case of decision making under uncertainty.

Four contributions are on the general topic of decision theory. The paper by Jia and Dyer (Decision Making Based on Risk-Value Tradeoffs) starts this section with a survey of risk-value decision models that have been developed in the last decade. This paper merges two streams of research, the modeling of individual preference and the modeling of risk judgment, in an effort to develop a more descriptively powerful risk-value model. Bodurtha and Shen (Normally Distributed Admissible Choices are Optimal) consider one particular aspect of risk-value models by examining mean-variance analysis to determine the characteristics of optimal solutions for decisions involving both mutually exclusive investments and financial portfolios of investments with normally distributed returns. Their analysis shows that these optimal solutions meet the conditions that are described by Fishburn's definitions of stochastic dominance of admissible choices. Bouyssou, Marchant and Pirlot (A Conjoint Measurement Approach to the Discrete Suengo Integral) extend Fishburn's work on subjective expected utility in multiple criteria decision making by showing conditions under which a noncompensatory multiple-criteria decision-making model is equivalent to a model that is based on the discrete Suengo integral. Slinko (Additive Representability of Finite Measurement Structures) presents a survey of recent developments that are related to Fishburn's work on the additive representation of finite measurement structures, work relating closely to the classical measurement-theoretic topic of additive conjoint measurement. The paper highlights the remaining open problems that Fishburn formulated in this area.

Part 2 mirrors Fishburn's interest in voting and social-choice theory that he developed in two major books: *The Theory of Social Choice* [Fishburn (1973b)] and *Approval Voting* [Brams and Fishburn (1983)]. He has made pioneering contributions to the understanding of social choice functions, which includes work on anonymity conditions, paradoxes of preferential voting, and Borda's rule and Condorcet's principle. His research on scoring-rule sensitivity and scoring vectors is also significant. Fishburn has been a leader in developing and analyzing new voting rules, with his analysis of approval voting being an important case in point. His contributions to the comparison of voting methods are also noteworthy, including work on two-stage voting systems, single transferable vote, and positional voting rules. He has also undertaken important studies of majority choice, including finding conditions on preferences that guarantee a simple majority winner and a location theorem for single-peaked preferences. He has also studied Condorcet proportions and probabilities, social-choice lotteries, and impossibility theorems. Finally, his work includes beautiful results on social welfare and equity, including equity axioms for public risks and fair-cost allocation schemes.

Four contributions in this part are connected to Fishburn's work on identifying conditions that require the existence of a Condorcet winner in an election and the associated probabilities of observing events in election outcomes. Monjardet (Acyclic Domains of Linear Orders) presents a survey of work that has focused on the identification of domains of voters' preferences that require an 'acyclic set' or transitive majority rule relationship; he develops intriguing connections between some of these domains. Saari (Condorcet Domains: A Geometric Perspective) addresses the same topic, taking a geometric approach that offers intuitive insight to the problem. Gehrlein (Condorcet's Paradox with Three Candidates) analyzes the probability that a Condorcet winner will exist and shows that this probability is quite large for a small number of candidates when voters have preferences that are at all close to being mutually coherent (according to any of several possible measures of mutual coherence in group preferences). Feix, Lepelley, Merlin, and Rouet (On the Probability to Act in the European Union) extend some of Fishburn's work on probabilities of election outcomes to analyze the probability that the voting rules used by the European Union will produce deadlock.

Two contributions consider properties of voting rules. Brams and Sanver (Voting Systems that Combine Approval and Preference) provide an extensive analysis of two hybrid voting systems that combine approval voting with voting procedures that require either a complete ranking of candidates or a partial ranking of only the candidates in the approved subset. Zwicker (Anonymous Voting with Abstention: Weighted Voting) considers an extension of the standard case of yes-no legislative voting in which abstention is viewed as being a voter preference position somewhere between a yes and a no vote. Characterizations are provided in which a specified set of weighted scores are linked to voter responses of yes, abstain or no.

Two contributions address the general topic of social choice. Campbell and Kelly (Social Welfare Functions that Satisfy Pareto, Anonymity and Neutrality, but not IIA: Countably Many Alternatives) extend earlier work that showed that in the presence of the conditions of Pareto, non-dictatorship, full domain, and transitivity, an extremely weak independence condition is incompatible with anonymity and neutrality for a finite number of alternatives; here they consider the case of countably many alternatives.

Hopkins and Jones (Bruhat Orders and the Sequential Selection of Indivisible Items) extend some of Fishburn's work on fair division by considering the case in which two players sequentially make selections from a set of indivisible items. Necessary and sufficient conditions are found under which players receive their most preferred and least preferred outcomes.

Part 3 explores fundamental mathematical constructs that arise in the more applied work, described in Parts 1 and 2, through the study of binary relations, partial orders, graphs and networks, combinatorics, number theory, linear programming, inequalities, and coding theory. Fishburn's partial order work includes foundational introductions to the theory of partial-order dimension, linear extensions of partial orders, the FKG property, and so on, as well as research on geometric partial orders such as angle orders and circle orders. Interval orders and semiorders are important special classes of partial orders that arise in problems in economics, psychology, biology, scheduling, and so on. Fishburn has made both theoretical and applied contributions to the understanding of such orders, highlighted in his book, *Interval Graphs and Interval Orders* [Fishburn (1985)].

Graph theory topics are the subject of a wide variety of Fishburn's papers. His research in that area includes important contributions to such topics as niche overlap graphs (arising in ecology), tolerance graphs (arising in psychology and operations research), and L(2,1)-colorings (arising in communications) as well as the design of various kinds of communication and other networks. Combinatorial geometry involves the study of various configurations, and Fishburn's work here has included the study of convex n-gons, planar sets, partial set covering, and a wide variety of related topics.

In addition, coding problems often can be analyzed using combinatorial and related algebraic methods. Fishburn's contributions to a variety of coding problems have included important results on sequence-based methods for data transmission and source compression, binary convolutional codes, and related lattice concepts.

Much of Fishburn's work involves counting, enumeration, and asymptotic behavior of structures, including posets and graphs, but also sequences arising in number theory, solutions to inequalities, and types of geometries. This work falls at the interface among combinatorics, probability, number theory, and a number of other subdisciplines and often intersects ideas of convexity, linear programming, and so on.

Seven contributions tie into Fishburn's work on posets, graphs, and networks. There are two different representations for interval orders and semiorders. The basic definitions of interval orders and semiorders both relate a poset to a set of intervals on the number line. A second representation describes interval orders as the subset of posets that do not include a $2+2$ configuration; it describes semiorders as the subset of interval orders that do not include a $3+1$ configuration. Shuchat, Shull and Trenk (Fractional Weak Discrepancy of Posets and Certain Forbidden Configurations) find the range of possible values of fractional weak discrepancy for the subset of posets that contain a $3+1$ but no $2+2$. Isaak (Interval Order Representation via Shortest Paths) develops an alternative proof of the second representation for interval orders and semiorders by showing that they are special instances of existence results that are related to the measure of potentials in digraphs. Brown and Langley (Probe Interval Orders) investigate probe interval graphs that arise in molecular biology and are obtained with a variation of the model by which interval orders are determined by intervals on the number line; they also consider restrictions that must be placed on these intervals such that the resulting probe interval graph is a probe interval order. Falmagne and Ovchinnikov (Mediatic Graphs) discuss the concept of mediatic graphs that trace their study to "stochastic token theory" in mathematical psychology. They show that the sets of all interval orders and semiorders on a finite set each can be represented as mediatic graphs. Poljak and Roberts (An Application of Stahl's Conjecture About the k-tuple Chromatic Numbers of Knesser Graphs) analyze the chromatic number in graph coloring problems and apply known results about Stahl's Conjecture to answer two open questions about the relation between the n-tuple chromatic number of a graph and n times the size of the largest clique in the graph. Hwang and Dou (Optimal Reservation-Scheme Routing for Two-Rate Wide-sense Nonblocking Three-Stage Clos Networks) study interconnection networks that are widely used in data communications and parallel computing. In particular, they are interested in using these networks for different media to communicate. By using reservation-scheme routing, they show that such networks can require much less hardware. Sahi (The Harris Inequality for Partially Ordered Algebras) deals with inequalities concerning increasing functions on a distributive lattice. Partially ordered algebras are associative algebras over the reals with a nonempty subset closed under addition, multiplication and multiplication by positive real numbers. In special cases, the results relate to the Harris inequality that arises in percolation on random graphs and to the more general FKG inequality, both topics on which Fishburn has made important contributions.

Three papers tie in to a variety of issues at the intersection among combinatorics, probability, number theory, and linear programming. Lagarias, Rains and Vanderbei (The Kruskal Count) analyze a well-known (at least among mathematicians) card trick that relies on the high likelihood that two processes with different starts (one chosen by the subject, one by the magician) will converge before the deck runs out, enabling the latter to appear clairvoyant. The trick is modeled by a Markov chain; two different value distributions (geometric and uniform) are studied, the second for the first time; and then the results are compared to MC simulations of a real deck. Applegate, LeBrun, and Sloane (Descending Dungeons and Iterated Base-Changing) study the special sequences where each term arises from interpreting the previous term in a different base. These iterated base changes (dungeons) are distinguished from iterated exponentiation (or towers). They prove a theorem about the asymptotic value of the nth term in such a sequence. Shepp (Updating Hardy, Littlwood and Polya with Linear Programming) discusses ideas dating back to the famous 1934 book, *Inequalities*, by the authors named in his title. He studies inequalities that can be proven using linear programming or convexity arguments.

No tribute to Peter Fishburn would be complete without saying something about him as a person. The three of us have collaborated with him over many years on a variety of topics. Peter is not only conscientious and responsible to a fault, but he is also a delight to work with, always doing more than his fair share quickly and efficiently. We marvel at his ability to come up with new ideas, develop extensions of old ones, and demonstrate linkages–all cheerfully, with no fuss and bother. We have great admiration for this brilliant scholar, and we take enormous pleasure in having had the opportunity to work with him on so many exciting projects and to interact with him as a colleague and a friend.

Piscataway, NJ
November 2008

Steven J. Brams
William V. Gehrlein
Fred S. Roberts

References

Brams SJ, Fishburn PC (1983) Approval voting. Birkhäuser Publishers, Boston.
Fishburn PC (1964) Decision and value theory. Wiley, New York.
Fishburn PC (1970) Utility theory for decision making. Wiley, New York.
Fishburn PC (1973a) Mathematics of decision theory. Mouton Publishing, The Hague.
Fishburn PC (1973b) The theory of social choice. Princeton University Press, Princeton, NJ.
Fishburn PC (1982) The foundations of expected utility. D. Reidel Publishing Company, Dordrecht.
Fishburn PC (1985) Interval orders and interval graphs: a study of partially ordered sets. Wiley, New York.
Fishburn PC (1987) Interprofile conditions and impossibility. Harwood Academic Publishers, New York.
Fishburn PC (1988) Nonlinear preferences and utility theory. Johns Hopkins University Press, Baltimore, MD.

Contents

Part II Social Choice, Voting, and Social Welfare

Part III Posets, Graphs, Combinatorics, and Related Applied and Mathematical Topics

Graphs

Various Applied Mathematics Topics

Contributors

David Applegate AT&T Shannon Labs, 180 Park Ave., Florham Park, NJ 07932-0971, USA, david@research.att.com

David E. Bell 171 Morgan Hall, Harvard Business School, Boston, MA 02163, USA, dbell@hbs.edu

James N. Bodurtha Jr McDonough School of Business, Georgetown University, Washington, DC 20057, USA, bodurthj@georgetown.edu

Denis Bouyssou CNRS–LAMSADE, Université Paris Dauphine, Place du Maréchal de Lattre de Tassigny, F-75775 Paris Cedex 16, France, bouyssou@lamsade.dauphine.fr

Steven J. Brams Department of Politics, New York University, New York, NY 10012, USA, steven.brams@nyu.edu

David E. Brown Department of Mathematics and Statistics, Utah State University, Logan, UT 84322, USA, david.e.brown@usu.edu

Donald E. Campbell Department of Economics and The Program in Public Policy, The College of William and Mary, Williamsburg, VA 23187-8795, USA, decamp@wm.edu

Wenqing Dou Department of Mathematics, Shanghai Second Polytechnic University, Shanghai 201209, People's Republic of China, wenqingdou@yahoo.com.cn

James S. Dyer The McCombs School of Business, The University of Texas at Austin, Austin, TX 78712, USA, j.dyer@mail.utexas.edu

Jean-Claude Falmagne Department of Cognitive Sciences, University of California, Irvine, CA 92697, USA, jcf@uci.edu

Marc R. Feix Deceased

William V. Gehrlein Department of Business Administration, University of Delaware, Newark, DE 19716, USA, wvg@udel.edu

Brian Hopkins Department of Mathematics, Saint Peter's College, Jersey City, NJ 07306, USA, bhopkins@spc.edu

Frank K. Hwang Department of Applied Mathematics, National Chiaotung University, Hsinchu, Taiwan 30050, Republic of China, fkhwang@gmail.com

Garth Isaak Department of Mathematics, Lehigh University, Bethlehem, PA 18015, USA, gisaak@lehigh.edu

Jianmin Jia Faculty of Business Administration, The Chinese University of Hong Kong, Shatin, N.T. Hong Kong, jia@baf.msmail.cuhk.edu.hk

Michael A. Jones Mathematical Reviews, 416 South Fourth Street, Ann Arbor, MI 48104, USA, maj@ams.org

Ralph L. Keeney Fuqua School of Business, Duke University, Durham, NC 27708, USA, KeeneyR@aol.com

Jerry S. Kelly Department of Economics, Syracuse University, Syracuse, NY 13244-1090, USA, jskelly@maxwell.syr.edu

Jeffrey C. Lagarias Department of Mathematics, University of Michigan, Ann Arbor, MI 48109-1043, USA, lagarias@umich.edu

Larry J. Langley Department of Mathematics, University of the Pacific, Stockton, CA 95211, USA, llangley@pacific.edu

Marc LeBrun Fixpoint Inc., 448 Ignacio Blvd. #239, Novato, CA 94949, USA, mlb@well.com

Dominique Lepelley CERESUR, Department of Economics, Université de La Réunion, 97715 Saint-Denis cedex 9, France, dominique.lepelley@univ-reunion.fr

R. Duncan Luce Institute for Mathematical Behavioral Sciences, University of California, Irvine, CA 92697-5100, USA, rdluce@uci.edu

Thierry Marchant Ghent University, Dunantlaan 1, 9000 Ghent, Belgium, thierry.marchant@ugent.be

Anthony J. Marley Department of Psychology, University of Victoria, PO Box 3050, STN CSC, Victoria, BC V8W 3P5, Canada, ajmarley@uvic.ca

Vincent Merlin CREM, Faculté de Sciences Économiques et de Gestion, Université de Caen, 19 rue Claude Bloch, 14032 Caen cedex, France, vincent.merlin@unicaen.fr

Bernard Monjardet CES, Université Paris I Panthéon Sorbonne, Maison des Sciences Économiques, 106–112 bd de l'Hopital, 75647 Paris Cédex 13, France, Bernard.Monjardet@univ-paris1.fr

Yutaka Nakamura University of Tsukuba, Graduate School of Systems and Information Engineering, Division of Social Systems and Management 1-1-1 Tennnoudai, Tsukuba, Ibaraki 305-8573, Japan, nakamura@sk.tsukuba.ac.jp

Che Tat Ng Department of Pure Mathematics, University of Waterloo, Waterloo, ON N2L 3G1, Canada, ctng@uwaterloo.ca

Sergei Ovchinnikov Mathematics Department, San Francisco State University, San Francisco, CA 94132, USA, sergei@sfsu.edu

Marc Pirlot Faculté Polytechnique de Mons, 9 rue de Houdain, 7000 Mons, Belgium, marc.pirlot@fpms.ac.be

Svata Poljak Deceased

Eric Rains Department of Mathematics, California Institute of Technology, Pasadena, CA 91125, USA, rains@caltech.edu

Fred S. Roberts Department of Mathematics and DIMACS, Rutgers University, Piscataway, NJ 08854, USA, froberts@dimacs.rutgers.edu

Jean-Louis Rouet Institut des Sciences de la Terre d'Orléans, Université d'Orléans, 1A rue de la Férollerie, 45071 Orléans Cedex 2, France, jean-louis.rouet@univ-orleans.fr

Donald G. Saari Institute for Mathematical Behavioral Sciences, University of California, Irvine, CA 92697-5100, USA, dsaari@uci.edu

Siddhartha Sahi Department of Mathematics, Rutgers University, New Brunswick, NJ 08903, USA, sahi@math.rutgers.edu

M. Remzi Sanver Department of Economics, Istanbul Bilgi University, 80310 Kustepe, Istanbul, Turkey, sanver@bilgi.edu.tr

Qi Shen Market Risk Management, Wachovia Bank Corporation, 301 South College Street, Charlotte, NC 28288, USA, qi.shen@wachovia.com

Larry Shepp Rutgers University, Statistics Department, Piscataway, NJ 08854, USA, shepp@stat.rutgers.edu

Alan Shuchat Department of Mathematics, Wellesley College, Wellesley, MA 02481, USA, ashuchat@wellesley.edu

Randy Shull Department of Computer Science, Wellesley College, Wellesley, MA 02481, USA, rshull@wellesley.edu

Arkadii Slinko Department of Mathematics, University of Auckland, Auckland, New Zealand, a.slinko@auckland.ac.nz

N.J.A. Sloane AT&T Shannon Labs, 180 Park Ave., Florham Park, NJ 07932-0971, USA, njas@research.att.com

Ann N. Trenk Department of Mathematics, Wellesley College, Wellesley, MA 02481, USA, atrenk@wellesley.edu

Robert J. Vanderbei Department of Operations Research and Financial Engineering, Princeton University, Princeton, NJ 08544, USA, rvdb@princeton.edu

William S. Zwicker Department of Mathematics, Union College, Schenectady, NY 12308, USA, zwickerw@union.edu

Part I
Utility, Preference, Individual Choice, Subjective Probability, and Measurement

Utility Theory

Entropy-Related Measures of the Utility of Gambling

R. Duncan Luce, Anthony J. Marley, and Che Tat Ng

1 Background of Work Reported

1.1 Roles of Peter Fishburn on this topic

The first author has known Peter for a very long time, dating back some 45 years to when we met at a colloquium he gave at the University of Pennsylvania. After that our paths crossed fairly often. For example, in the early 1970s, he spent a year at the Institute for Advanced Study where Luce spent three years until the attempt to establish a program in scientific social science was abandoned for a more literary approach favored by the humanists and, surprisingly, the mathematicians then at the Institute. The second author has learnt a tremendous amount about both substantive and technical issues from Peter's work, beginning with Peter's book "Utility Theory for Decision Making" (Fishburn, 1970), which he reviewed for *Contemporary Psychology* (see Marley, 1972).

Peter's volume on interval orders (Fishburn, 1985) was a marvelous development of various ideas related to the algebra of imperfect discrimination that elaborated the first author's initial work on semiorders (Luce, 1956).

Beginning in 1988, Peter made a major contribution in his integrative book "Nonlinear Preference and Utility Theory." And in the first half of the 1990s, Fishburn and Luce collaborated on three efforts to understand better the rank-dependent generalizations of expected utility that had attracted considerable notice in the 1980s (Fishburn & Luce, 1995; Luce & Fishburn, 1991, 1995). It was here that we first came up with the so-called p-additive form for the utility of joint receipts. All of that played a major role in Luce's (2000) attempt to pull together many of the results about utility, both experimental and theoretical, of the period starting in 1979.

R.D. Luce (✉)
Institute for Mathematical Behavioral Sciences, University of California, Irvine, CA, USA, 92697-5100
e-mail: rdluce@uci.edu

S.J. Brams et al. (eds.), *The Mathematics of Preference, Choice and Order: Essays in Honor of Peter C. Fishburn*, Studies in Choice and Welfare,

And joint receipts play a key role in our attempt to incorporate a concept of the utility of gambling into the situation, which is described in this paper.

To our knowledge, Peter directly addressed the issue of the utility of gambling just once (Fishburn, 1980), where he presented the first, but very restricted, formal model of it (Sect. 1.2).

So, in sum, we have learned much from Peter and are still tilling grounds that he was, in very many ways, influential in developing in modern utility theory. This chapter pays tribute to Peter not by commenting directly on his contributions, but by summarizing some generalizations found in several articles cited below.

1.2 Utility of Gambling

The founders of "modern" utility theory, Ramsey (1931), von Neumann and Morgenstern (1947), and, less explicitly, Savage (1954, pp. 13–17) all noted that their theories could not accommodate the existence of utility of gambling (UofG) per se. For example Ramsey (1931, p. 172) contended that the method of establishing beliefs in terms of bets is "...inexact...partly because the person may have a special eagerness or reluctance to bet, because he either enjoys or dislikes excitement...The difficulty is like that of separating two different cooperating forces." This 1931 essay was actually dated 1926. Over two decades later von Neumann and Morgenstern (1947, p. 28) remarked: "Since [our axioms] secure that the necessary construction can be carried out, concepts like a 'specific utility of gambling' cannot be formulated free of contradiction on this level." Adjoined is the footnote: "This may seem to be a paradoxical assertion. But anybody who has seriously tried to axiomatize that elusive concept, will probably concur with it."

Furthermore, the sharp partition in these theories of valueless events and valued consequences is often not the case in reality. Insurance on an airplane trip represents such a separation, but not all of the events that might occur are valueless in their own right – for instance, a crash of your flight.

Most theoretical work has ruled out UofG by incorporating in some fashion a version of idempotence, namely, that attaching the same consequence x to each chance event arising from a chance "experiment" is perceived as indifferent to receiving x with certainty. Savage (1954) called such gambles "constant acts." That indifference means that no utility or disutility accrues either to the events themselves or to the execution of the experiment, as such.

Ignoring both the value of events and the utility associated with uncertainty and/or risk is a major idealization that has only rarely been questioned or addressed. Some formal models of UofG appearing in the utility literature focused on the risky cases[1], and typically involved modifications of the expected utility representation. Conlisk (1993) summarizes them from an economic perspective and Luce and Marley (2000) from a more psychological one, but with important economic

---

[1] Those for which each of the possible consequences of the gamble occurs with a specified probability.

influences. Fishburn (1980) gave the first formal model in which he appended a UofG term to the expected utility of a risky gamble in such a way that this term affects the choice between a pure consequence (sure-thing) and a risky gamble, but does not influence the choice between two risky gambles. He also axiomatized several possible forms for the UofG term, including the case where it is constant for all gambles. Diecidue, Schmidt and Wakker (2004) generalized Fishburn's formulation, but, in the main, they continued to assume that preferences between risky gambles agree with expected utility. Bleichrodt and Schmidt (2002) present a related model, with preferences between risky gambles again agreeing with expected utility but with different utility functions depending upon whether or not one of the alternatives is a pure consequence. Luce and Marley (2000) considered uncertain[2] gambles, with a UofG term that depends on the events, but not the consequences. In that model the UofG term can affect the choice between two gambles when they are based on different events. They also motivate, but do not axiomatize, several possible forms for the UofG term for binary gambles. Le Menestrel (2001) and Pope (1995 and earlier papers) offer process models for the utility of gambling. So far as we know, no one before our work has dealt explicitly with valued uncertain events, often because the underlying structure has been one of risk.

Meginniss (1976) seems to have been the first author to arrive at, in the context of risk, a sensible theory incorporating UofG. Until quite recently, his result appears to have been unknown, ignored, and/or forgotten by utility theorists[3], and its ability to account for many anomalies has not been widely recognized. His result is that the overall utility of a risky gamble is given by a linear weighted utility term plus an (information-theoretic) entropy (Shannon, 1948) term dependent only on the probabilities. His clever proof of the result rested on quite special, unaxiomatized, representational assumptions. Unaware of Meginniss' article, Yang and Qiu (2005) proposed a closely related nonaxiomatized representation involving Shannon's entropy, explored some of its properties, and applied it to some of the well known anomalies. We summarize similar explanations of several such anomalies in Sect. 5.

Ng, Luce, and Marley (2008a) generalized Meginniss' approach in several ways, fundamentally following his general ideas, whereas Luce, Ng, Marley and Aczél (2008 a,b) and Ng, Luce, and Marley (2008b) take an axiomatic approach. Specifically, Luce et al. (2008a), summarized in Sect. 2, treat uncertain gambles and Luce et al. (2008 b), summarized in Sect. 3, extend those results to risky gambles. Ng et al. (2008b), summarized in Sect. 4, further extend the results to obtain representations of the UofG term that involve a weighted value function over events, plus an entropy term involving the same weights. The resulting representations include the "rational" expected utility (EU) and subjective expected utility (SEU) representations as very special cases, with no UofG term.

Section 5 applies a special case of these representations to several sets of data. And Sect. 6 summarizes the results reported in this paper and states four major open problems.

---

[2] Those where the events have no readily agreed upon probabilities.

[3] It was brought to our attention in 2004 by our collaborator János Aczél.

1.3 Formulation of Gambles and Utility Representations

We begin with the general concept of uncertain gambles, then extend the results to risky gambles and gambles involving valued events. Because we anticipate that most of our readers are already familiar with standard notations in this domain and need no more than reminders, we are not fully formal – that can be found in Luce et al. (2008 a,b).

The set of pure consequences – no risk or uncertainty – is denoted X. Included in X is a singular element e, called *no change from the status quo*, whose special properties are given below. The set of pure consequences is assumed to be closed under the binary[4] commutative and associative operation of joint receipt, $\oplus$. We postulate a (preference) ordering, $\succsim$, over $\langle X, \oplus \rangle$ that is assumed to be a weak order that is strictly increasing in each argument of $\oplus$. As usual, strict preference is denoted by $\succ$ and indifference by $\sim$. The latter is an equivalence relation. We assume that e is an identity of $\oplus$: for all $x \in X$, $x \oplus e \sim e \oplus x \sim x$. Moreover, X is assumed to satisfy the structural restriction of solvability, namely, for each x, y, there exists z such that $x \sim y \oplus z$. We define $x \ominus y := z$.

Assume that the axioms of the theory of extensive measurement are satisfied (Krantz et al., 1971, Chap. 3) leading to a mapping $U : X \to \mathbb{R}$ such that:

$$x \succsim y \Leftrightarrow U(x) \geq U(y), \tag{1}$$

$$U(x \oplus y) = U(x) + U(y). \tag{2}$$

It follows immediately that $U(e) = 0$ and that $U(x \ominus y) = U(x) - U(y)$.

Let Ω denote a state space of the chance outcomes from some chance "experiment." Let $(C_1, \ldots, C_i, \ldots, C_n)$ denote a typical nontrivial, finite partition of Ω, i.e., $C_i \cap C_j = \varnothing$ if $i \neq j$, $C_i \neq \varnothing$, $\cup_{i=1}^{n} C_i = \Omega$. Unlike Savage (1954) and many subsequent treatments, we do not assume a single universal state space; rather we produce a more versatile model in which Ω is a variable, as is typical of both concrete examples of gambles, e.g., alternate modes of travel from A to B, and equally well of the experimental realizations of gambles in various experiments, e.g., spin of a wheel, withdrawal of a colored ball from a randomized urn, etc. One can, and airlines do it all the time, subtract and/or add alternatives to an existing set of flight alternatives. The versatility is essential to our approach using gamble decompositions.

An *uncertain alternative*, often called a *gamble* but with a far broader scope than ordinary usage, is defined inductively: A first-order one is a mapping $g_{[n]}$ from such a finite partition into X, a second-order one is a mapping to the union of X and first-order ones, which are not of first order, etc. We use only these two levels. The structure $\langle X, \oplus, \succsim \rangle$ can be extended to include all gambles and their joint receipts, $\mathcal{G}$, and we assume that the extended preference order continues to be a weak order, still called $\succsim$. And the additive representation over $\oplus$ also extends in the obvious way. With no loss of generality, we choose the indices so that the consequences of the gamble are ordered, i.e., $x_1 \succsim x_2 \succsim \cdots \succsim x_n$, and we assume that gambles are comonotonic in the sense of ordinary monotonicity so long as the ordering of

[4] Inductively, one constructs an algebraic version of commodity bundles of any size.

consequence is unchanged (Wakker, 1990). We may write a gamble explicitly in either of two equivalent ways:

$$g_{[n]} = \begin{pmatrix} C_1, \ C_2, \ \ldots, \ C_i, \ \ldots, \ C_n \\ x_1, \ x_2, \ \ldots, \ x_i, \ \ldots, \ x_n \end{pmatrix} \tag{3}$$

$$= (x_1, C_1; x_2, C_2; \ldots; x_i, C_i; \ldots; x_n, C_n). \tag{4}$$

We use which ever notation seems more useful at the occasion. Each consequence-event pair (x_i, C_i) is called a *branch* of the gamble. Thus, a gamble is a collection of n disjoint branches.

Although gambles are stated in ranked form, we note that such rankings are only a matter of convenience in stating both some axioms (e.g., comonotonicity) and some results (e.g., rank dependent representations). In fact, we assume that gambles differing only in a permutation of the branches are perceived as indifferent.

1.4 Assumptions about Kernel Equivalents and Elements of Chance

Following Luce and Marley (2000), any gamble for which every consequence is no change from the status quo, e, i.e., gambles of the form $(e, C_1; e, C_2; \ldots; e, C_n)$, is called an *element of chance.* This is simply the realization of a chance "experiment" with no assignment of consequences to the several events, meaning that the status quo is maintained, which we denote by e. A trivial example is watching a spin of a roulette wheel. For any gamble $g_{[n]} = (g_1, C_1; g_2, C_2; \ldots; g_n, C_n)$, where the g_i are first-order gambles, its *kernel equivalent,* denoted $KE(g_{[n]})$, is defined to be the pure consequence solution, which is assumed to exist, to the following indifference

$$g_{[n]} \sim KE(g_{[n]}) \oplus (e, C_1; e, C_2; \ldots; e, C_n). \tag{5}$$

Note that, because $KE(g_{[n]})$ is a pure consequence, the right hand expression involves only one realization of the experiment.

We see that (2) and (5) yield

$$U(g_{[n]}) = U(KE(g_{[n]})) + U(e, C_1; e, C_2; \ldots; e, C_n). \tag{6}$$

The utility of an element of chance is a possible measure of the UofG. Our goal is to discover something about its mathematical form. The first step in doing so is to weaken the classical assumptions about *idempotence*: The kernel equivalents are *idempotent* (KE-idempotent) if for any gamble, denoted $g_{[n]}(x)$, all of whose consequences are x,

$$KE(g_{[n]}(x)) \sim x. \tag{7}$$

The elements of chance are $e-$*idempotent* if

$$e \sim (e, C_1; e, C_2; \ldots; e, C_n). \tag{8}$$

Traditional theories of utility typically assume idempotence or prove it from other assumptions. We explicitly do *not* assume $e-$idempotence and it is not a consequence of our other assumptions.

Suppose that C_i, $i = 1,\ldots,n$, form a partition of a universal event Ω and that C_i' is the same partition but arising from an independent realization of the underlying experiment. We assume that

$$(e,C_1;e,C_2;\ldots;e,C_n) \sim (e,C_1';e,C_2';\ldots;e,C_n'). \tag{9}$$

This is obviously true if $e-$idempotence holds, but (9) does not imply $e-$idempotence.

Although Luce and Marley (2000) derived a number of properties about such a decomposition into KEs and elements of chance, they had no principled way of getting results about the utility of elements of chance, partly because they considered only binary gambles. We offer one remedy for that incompleteness.

1.5 Probabilities and Implicit Events

It is quite common to treat risky gambles in a fashion parallel to that for uncertain ones, but instead of providing the event structure, one simply replaces the state C_i by the probability p_i as, for example,

$$g_{[n]} = (x_1,p_1;\ldots;x_i,p_i;\ldots;x_n,p_n) \quad \left(\sum_{i=1}^{n} p_i = 1\right)$$

This is the form commonly invoked in most experiments and in most of the developments emanating from economics. Nevertheless, to provide a sound basis for such probabilities, there must be some implicit event structure – the risky gambles have to be realized in some fashion.

So we first summarize properties and results for event structures, and then specialize them to risky situations.

2 Key Properties

2.1 Separable Representations of Binary Gambles

For binary gambles, conjoint measurement assumptions are easily stated that lead to the following (multiplicative) *separable*, ordering preserving, (1), representation U^* over so-called *unitary binary gambles* in which one consequence is e:

$$U^*(KE(x,C;e,D)) = U^*(x)W_{C\cup D}(C), \tag{10}$$

where $W_{C\cup D}(C)$ is a subjective weighting of the event C, conditional on the event $C\cup D$. Because U of (1) and U^* each preserve the order $\succsim$, they are strictly monotonically related. Can one find a property linking the two underlying structures leading, respectively, to order preservation by U and additivity, (2), and to order preservation by U^* and separability, (10)? To that end we assume *simple joint-receipt decomposability*:

$$(f\oplus g,C;e,D)\oplus(e,C';e,D')\sim(f,C;e,D)\oplus(g,C';e,D'), \tag{11}$$

where the prime simply indicates an independent realization of the "experiment" underlying the partition (C,D). Then, the result is that there exists $\kappa>0$ such that $U=(U^*)^{\kappa}$ and so we have the (multiplicative) separable form

$$U(KE(x,C;e,D))=U(x)S_{C\cup D}(C) \tag{12}$$

where $S_{C\cup D}:=W^{\kappa}_{C\cup D}$. The weights S_{Ω} for an event Ω are also involved in the representation of UofGs, explicitly for uncertain gambles, implicitly for risky gambles.

Next come two steps: the first extending (12) to unrestricted binary gambles, and the second extending the representation of binary gambles to general ones.

2.2 Two Alternative Binary Decompositions: Segregation and Duplex Decomposition

Luce (1997, 2000) has studied two closely related, but distinct, forms for extending unitary gambles $(x,C;e,D)$ to full binary gambles. The first is *segregation*:

$$(x\oplus y,C;y,D)\sim(x,C';e,D')\oplus y. \tag{13}$$

Kahneman and Tversky (1979) invoked segregation during the preliminary editing phase of their prospect theory. Segregation with the earlier assumptions, where (10) holds for gains only, leads to: For $f\succsim g$

$$U[KE(f,C;g,D)]=U(f)S_{C\cup D}(C)+U(g)\left[1-S_{C\cup D}(C)\right]. \tag{14}$$

The alternative decomposition, *duplex decomposition,* which first appeared in Slovic (1967) and in Slovic and Lichtenstein (1968), is:

$$(x,C;y,D)\oplus(e,C';e,D')\sim(x,C;e,D)\oplus(e,C';y,D'). \tag{15}$$

This with the earlier assumptions, where (10) is for both gains and losses, and results leads to:

$$U[KE(f,C;g,D)]=U(f)S_{C\cup D}(C)+U(g)S_{C\cup D}(D). \tag{16}$$

Note that segregation is significantly more restrictive than duplex decomposition in that it associates $1 - S_{C\cup D}(C)$ to the (g,D) branch whereas duplex decomposition associates $S_{C\cup D}(D)$ with no tie to $S_{C\cup D}(C)$.

One empirical study, Cho, Luce and Truong (2002), suggests that some people, perhaps 75% of them, satisfy one of these properties although that study was conducted under the assumption of that $(e,C';e,D') \sim e$, which, of course, matters only for duplex decomposition.

2.3 Inductive Properties: Branching and Upper Gamble Decomposition

We invoke two inductive properties, neither of which has received experimental evaluation. They are both cases of the reduction of compound gambles in the context of events, not probabilities. Their mathematical role is to reduce the utility expressions for gambles of order $n > 2$ to the those for binary gambles, which were given in Sects. 2.1 and 2.2. In particular, they lead to equations characterizing the utility of gambling, UofG, terms. The first, called *upper gamble decomposition* (UGD), is:

$$g_{[n]} = \begin{pmatrix} C_1, & C_2, & \ldots, & C_i, & \ldots, & C_n \\ x_1, & x_2, & \ldots, & x_i, & \ldots, & x_n \end{pmatrix}$$

$$\sim \begin{pmatrix} C_1, & \Omega\backslash C_1 \\ x_1, & \begin{pmatrix} C_2, & \ldots, & C_i, & \ldots, & C_n \\ x_2, & \ldots, & x_i, & \ldots, & x_n \end{pmatrix} \end{pmatrix}. \tag{17}$$

One sees that if one is willing to consider compound gambles, it is highly rational in nature, the "bottom lines" being the same.

The second property, *branching*, is

$$\begin{pmatrix} C_1, & C_2, & \ldots, & C_i, & \ldots, & C_n \\ x_1, & x_2, & \ldots, & x_i, & \ldots, & x_n \end{pmatrix}$$

$$\sim \begin{pmatrix} C_1 \cup C_2, & C_3, & \ldots, & C_i, & \ldots, & C_n \\ \begin{pmatrix} C_1, & C_2 \\ x_1, & x_2 \end{pmatrix}, & x_3, & \ldots, & x_i, & \ldots, & x_n, \end{pmatrix}. \tag{18}$$

This, too, is highly rational.

Note that each property involves a binary gamble, the first with the partition $(C_1,\Omega\backslash C_1)$ and the second with (C_1,C_2). Thus, we are able to invoke either (14) or (16).

Under these two properties for $n = 3$, one is able to prove (Luce et al., in press, a) the *choice property*[5]: for events $C \subseteq D \subseteq E$,

$$S_E(C) = S_D(C)S_E(D). \tag{19}$$

One can construct a function μ from events to the real numbers such that for all $C \subseteq E, E \neq \varnothing$,

$$S_E(C) = \mu(C)/\mu(E). \tag{20}$$

2.4 Main General Result

Under these assumptions one is able to arrive at a number of representations depending on which decomposition is assumed and on whether or not S_Ω is finitely additive (FA).

A first, important, result is that, under segregation, the representation has to be p-additive in the sense that for an appropriate choice for the unit of μ, there exists a constant Δ such that

$$S_\Omega(C \cup D) = S_\Omega(C) + S_\Omega(D) + \Delta\mu(\Omega)S_\Omega(C)S_\Omega(D). \tag{21}$$

The weights are finitely additive iff $\Delta = 0$.

Then the resulting representations are summarized in Table 1, which is to be read as follows: It is the cell wise sum of two 2×2 matrices corresponding, respectively, to the utility of kernel equivalents and to the utility of gambling terms. The matrix rows are whether or not S_Ω is finitely additive. The columns are by whether segregation or duplex decomposition is assumed. The cell entries are the representations listed below the table itself.

Table 1 Summary of representations for uncertain gambles*

Codes: DD = Duplex Decomposition, FA = Finitely Additive, KE = Kernel Equivalent, Seg = Segregation, UofG = Utility of Gambling

		U(KE)			UofG	
		Seg (13)	DD (15)		Seg	DD
S_Ω	FA	SEU	SEU	+	H	H
	Not FA	RDU	LWU		$0, A$	H

*Adapted from Table 1 of Luce, Ng, Marley, and Aczél (2008a), with kind permission of Springer Science+Business Media.

[5] With finitely additive weights, which we do not yet have, Luce (1959) called it the *choice axiom.* Here we use a more neutral term.

where

$$LWU\left(g_{[n]}\right) := \sum_{i=1}^{n} U(x_i) S_\Omega(C_i).$$

$$SEU(g_{[n]}) := LWU(g_{[n]}) \text{ with } \sum_{i=1}^{n} S_\Omega(C_i) = 1.$$

$$RDU(g_{[n]}) := \sum_{i=1}^{n} U(x_i)\left[S_\Omega(C(i)) - S_\Omega(C(i-1))\right]$$

$$= \sum_{i=1}^{n} U(x_i) S_\Omega(C_i)\left[1 + \Delta\mu(\Omega) S_\Omega(C(i-1))\right] \left(C(i) := \bigcup_{j=1}^{i} C_j\right),$$

and when Ω is maximal $H = A$, a constant; otherwise it is 0.

$$H(C_1,\ldots,C_n) := U(e,C_1;\ldots;e,C_n) = K(\Omega) - \sum_{i=1}^{n} K(C_i) S_\Omega(C_i).$$

These results are based on theorems reported in Davidson and Ng (1981), Ebanks (1982), and Ebanks, Kannappan and Ng (1988). The representation $SEU + H$ is known as generalized subjective expected utility (G-SEU) with H the utility of gambling. The function K that arises in the form of H, the last form listed, is not otherwise specified. The nonfinitely additive representation under segregation has *RDU* as its kernel equivalent and H is a constant A assigned to Ω that is 0 when Ω is not maximal. As we will see, the results under risk, given in Table 2, are far more specific.

3 Risky Elements of Chance and An Application

3.1 Risk and Implicit Events

Next, we discuss the more specific forms for the UofG that we have derived in the case of risky gambles (Luce et al., 2008 b), and later (Sect. 5) summarize the evaluation of one of those forms vis-a-vis available data. As already mentioned, the case of risk entails an explicit set of probabilities, p_i, and a risky gamble is a function assigning a consequence x_i to $p_i, i = 1,\ldots,n$. These cases are important because, first, they are the class of gambles most often postulated by economists, and second, more often than not, these cases are studied in laboratory experiments by both economists and psychologists. Usually in experiments, the events are implicit with no clear indication as to exactly how the probabilities are to be generated except to the extent that participants in the experiment are "educated" about how the probabilities might be realized through mechanisms such as spins of a color-coded pie chart or random draws from an urn of colored balls. In this sense, we might suggest that there is an "implicit" event space underlying the probability distributions. In fact, we now assume that, even when the probabilities are presented explicitly, the participant postulates an underlying implicit event space. Then, we add assumptions concerning the linkage between uncertain gambles over the event space and risky gambles over the probability space that allow us to use our previous results about the representation of uncertain gambles to obtain representations of risky gambles with specific entropy-based representations of the elements of chance.

3.2 Probabilities Realized by Implicit Events

Let $\mathbf{p}_n = (p_1, p_2, \ldots, p_n)$ be any nontrivial, complete probability distribution, i.e., $p_i > 0$ and $\sum_{i=1}^n p_i = 1$. We assume, as is standard in the foundations of probability theory, that in a particular decision making context of gambles with explicitly given probabilities, the decision maker postulates a fixed, implicit, underlying algebra of events that is associated with a maximal universal event Ω_0 and a probability measure Pr on that algebra such that there is an ordered partition $\mathbf{C}_n := (C_1, C_2, \ldots, C_n)$ of Ω_0, with $C_i \neq \varnothing$, in the algebra, and with[6] $\Pr(C_i|\,\Omega_0) = p_i$, $i = 1, \ldots, n$. This implicit algebra is assumed to be *fixed* for the decision making context, e.g., a state lottery, independent of any particular lotteries that the decision maker may confront. Of course, there may be another partition $\mathbf{D}_n = (D_1, \ldots, D_n)$ of Ω_0, with D_i in the algebra, such that $\Pr(D_i|\,\Omega_0) = p_i = \Pr(C_i|\,\Omega_0)$, $i = 1, \ldots, n$. Our assumptions, below, overcome this ambiguity.

The risky gamble is presented as $g_{[n]} = (x_1, p_1; \ldots; x_n, p_n)$. Let $\succsim$ denote the preference ordering over pure consequences and risky gambles, and assume that a preference ordering $\succsim_{\mathcal{G}}$ exists over the event-based gambles $\mathcal{G}$. We assume that $\succsim_{\mathcal{G}}$ agrees with $\succsim$ over the structure of pure consequences, risky gambles, and their joint receipt, so for simplicity we drop the subscript $\mathcal{G}$.

We make two observations about the assumption of the existence of an implicit algebra of events:

First, it is just that, an assumption. It is certainly conceivable that a decision maker may somehow deal with the probabilities without resorting at all to an underlying algebra of events, as for example in a binary gamble given as $(x, p; y, 1-p)$ where it is taken for granted that when carried out the decision maker gets exactly one of x and y.

Second, the assumption of an implicit algebra permits us to invoke the earlier assumptions about events and the corresponding results. As we shall see, this means that there are several quite different types of decision makers, which has important implications for the usual kind of data analysis that averages data over respondents instead of analyzing each respondent separately.

Now we need the preference ordering over event-based gambles to be compatible with the preference ordering over the conditional-probability-based risky gambles in the following sense where we write $C(n) = \cup_{i=1}^n C_i$, $D(m) = \cup_{i=1}^m D_i$:

---

[6] Usually $\Pr(C_i|\,\Omega_0)$ is abbreviated to $\Pr(C_i)$, but we think it best in this article to keep it explicit.

$$\begin{aligned}&(x_1,C_1;\ldots;x_n,C_n) \succsim (y_1,D_1;\ldots;y_m,D_m)\\ &\Leftrightarrow (x_1,\Pr(C_1|C(n));\ldots;x_n,\Pr(C_n|C(n)))\\ &\quad \succsim (y_1,\Pr(D_1|D(m));\ldots;y_m,\Pr(D_m|D(m))). \end{aligned} \tag{22}$$

Under the background conditions (but not including segregation or duplex decomposition) and with (22) and $p \mapsto U(e,p;e,1-p)$ continuous, Luce et al. (2008 b) show that there is a constant $\rho > 0$ such that

$$S_\Omega(C) = \Pr(C|\Omega)^\rho, \tag{23}$$

where S_Ω is the subjective weighting function in the representation of the uncertain gambles.

Under the above conditions and those leading to the results summarized in Table 1 for uncertain gambles, we obtain the representations for risky gambles that are summarized in Table 2, which is read in a fashion similar to Table 1.
where

$$EU(g_{[n]}) = \sum_{i=1}^{n} U(x_i)p_i,$$

and

$$I^{(\rho)}(p_1,\ldots,p_n) := \begin{cases} -\sum_{i=1}^n p_i \log_2 p_i, & \rho = 1 \\ \frac{1}{2^{1-\rho}-1}[\sum_{i=1}^n p_i^\rho - 1], & 0 < \rho \neq 1 \end{cases}.$$

The UofG term when $\rho = 1$ is a constant A times the well-known Shannon (1948) entropy. The sign of A determines whether UofG is positive or negative and the magnitude of A determines the importance of UofG relative to the expected utility term. The proof of these results rest upon the mathematical theory of information (entropy) discussed by Aczél and Daróczy (1975). The sum in Table 2 corresponding to $\rho = 1$, $EU + AI^{(1)}$, we call *entropy-modified expected utility* (EM-EU), and the sum corresponding to $\rho \neq 1$ under duplex decomposition, $\sum_{i=1}^n U(x_i)p_i^\rho + AI^{(\rho)}$, we call *linear power weighted utility* (LPWU), which, clearly, coincides with EM-EU when $\rho = 1$. As indicated in the table, the case where segregation holds and $\rho \neq 1$ cannot occur under our assumptions.

These results raise an interesting concern about the almost exclusive focus of many utility theorists on probabilities without any regard to the underlying event structure. Apparently, that focus can lead to overlooking cases with $\rho \neq 1$.

It is striking that we have not arrived at a risky version of RDU, such as cumulative prospect theory, plus a UofG term. This lack invites modifying the assumptions in some crucial way, in particular by replacing branching by some property, such as coalescing, satisfied by the kernel equivalent of such a form.

Although purely rational considerations favor segregation and so EM-EU over duplex decomposition, descriptively those considerations are not compelling and, as we shall see in Sect. 5, some data reject EM-EU. Other data (Cho, Luce, & Truong, 2002) strongly suggest that a substantial proportion of respondents are better described by duplex decomposition than segregation. In that case, individual

Table 2 Summary of representations for risky gambles*

Codes: DD = Duplex Decomposition, FA = Finitely Additive, KE = Kernel Equivalent, Seg = Segregation, UofG = Utility of Gambling

		U(KE)				UofG	
		Seg (13)	DD (15)			Seg	DD
$S_\Omega^{1/\rho}$	$\rho = 1$	EU	EU			$I^{(1)}$	$I^{(1)}$
is				$+$	$A\times$		
FA	$\rho \neq 1$	—	$\sum U(x_i)p_i^\rho$			—	$I^{(\rho)}$

*Adapted from Table 1 of Luce, Ng, Marley, and Aczél (2008 b), with kind permission of Springer Science+Business Media.

differences abound, depending on the value of ρ. Therefore, it only makes sense to look at data on an individual basis without averaging them. Despite that admonition, most of the available data are for sets of people, not individuals.

3.3 An Application: Short-Term Gambling

Let $b = b(g_{[n]})$ denote the maximum buying price of the gamble $g_{[n]} = (x_1, C_1; \ldots; x_n, C_n)$, where we have in mind a quick resolution of the uncertainty. Thus, we are not treating such long-term "gambles" as life insurance, long-term health disability, and long-term financial investments. Our theory is timeless and so no financial discounting is involved. The following definition of b is natural (Luce, 2000, and earlier references), where the subjective weights may or may not be finitely additive:

$$e \sim (x_1 \ominus b, C_1; \ldots; x_n \ominus b, C_n).$$

It is obvious that when one buys a gamble one acquires the gamble with each consequence reduced by the buying price.

In the following, to make clear that the utility and weighting functions belong to the buyer, who is the gambler, we use the subscript b.

In those cases where $S_{\Omega,b}$ is assumed to be finitely additive, as in this subsection, we know that $\sum S_{\Omega,b}(C_i) = 1$, and so this definition is equivalent to

$$\begin{aligned} U_b(b) &= \sum_{i=1}^{n} U_b(x_i) S_{\Omega,b}(C_i) + H_b(C_1, \ldots, C_n) \\ &= U_b(KE(g_{[n]})) + H_b(C_1, \ldots, C_n), \end{aligned} \tag{24}$$

which is equivalent to $b \sim g_{[n]}$.

The case of selling prices is a good deal more subtle and we do not take it up here.

Let us apply this to the issue of commercial gambling. Suppose that the seller is either a state (lottery) or a casino and the buyer, i.e., a gambler, is an individual.

Assume, as seems to be the case, that pricing money lotteries by both states and casinos is based on some factor times the expected rate of return, i.e., $s = (1+\alpha)EV$, $\alpha > 0$. Assuming the special case where the buyer's utility for money is the identity function, then (24) yields

$$\begin{aligned} b \geq s \Leftrightarrow U_b(b) &\geq U_b\left((1+\alpha)EV(g_{[n]})\right) = (1+\alpha)EV(g_{[n]}) \\ &\Leftrightarrow U_b(KE(g_{[n]})) + H_b(C_1,\ldots,C_n) \geq (1+\alpha)EV(g_{[n]}) \\ &\Leftrightarrow KE(g_{[n]}) + H_b(C_1,\ldots,C_n) \geq (1+\alpha)EV(g_{[n]}). \end{aligned}$$

Let us suppose that, except for enjoying gambling, the gambler is fully rational and identifies the kernel equivalent of the gamble with its expected value:

$$KE(g_{[n]}) = EV(g_{[n]}).$$

Then, s/he will gamble if and only if

$$H_b(C_1,\ldots,C_n) \geq \alpha EV(g_{[n]}),$$

namely, whenever the gambler's utility of gambling exceeds the profit to the seller. This suggests that the utility of gambling is a strong determinant of behavior, as, of course, has been widely recognized if not previously modeled so formally.

4 Utility of Gambling with Valued Uncertain Events

The problem to be addressed in this section is motivated by the obvious, but widely ignored, fact that in many important real-world situations not only do event partitions have consequences attached to the events, but some events themselves are inherently valued by the decision maker. An example is airplane travel in which some of the chance events, such as the trip being terminated in a crash, are themselves of (negative) value. Such a value is independent of any bet, – e.g., insurance on the flight – that is placed on the trip. Moreover, we know of no principled way that allows for the separate measurement of the inherent value of events. Nonetheless, by a novel conceptual device we are able to use the results of Table 1 to arrive at the more specific forms given below.

The conceptual device is an ordering $\succsim_X$, which has an additive representation over joint receipts, and a family $\mathcal{O}$ of order extensions of $\succsim_X$ to include gambles. Also, the model presumes, for each and every $\succsim \in \mathcal{O}$, the formulation of Sects. 1 and 2 and the results summarized in Table 1. A difference arises because the weights now depend on $\succsim$, i.e., we have $S_{\succsim,\Omega}$ rather than S_Ω. We make assumptions that are sufficient for there to be some $\sigma > 0$ such that, for each pair $(\mathbf{p}_n, \mathbf{C}_n)$, there is some $\succsim \in \mathcal{O}$ for which $S^{\sigma}_{\succsim,\Omega}(C_i) = p_i$. With these, and other assumptions, Ng, Luce, & Marley (2008b) show that, for each order $\succsim \in \mathcal{O}$ with additive $S^{\sigma}_{\succsim}$, we can define

$$H\begin{pmatrix} C_1, C_2, \ldots, C_n \\ p_1, p_2, \ldots, p_n \end{pmatrix} := H_{\succsim}(C_1,\ldots,C_n) := U_{\succsim}(e,C_1;\ldots;e,C_n).$$

Then, under the assumptions about the family of orders, this family of functions satisfies the conditions of what is known as inset entropy, introduced by Aczél and Daróczy and Aczél in 1975 and Kannappen in 1978. In particular, (18) with $x_i = e$, $i = 1,\ldots,n$, becomes

$$H\begin{pmatrix} C_1, C_2, \ldots, C_n \\ p_1, p_2, \ldots, p_n \end{pmatrix} = H\begin{pmatrix} C_1 \cup C_2, C_3, \ldots, C_n \\ p_1 + p_2, p_3, \ldots, p_n \end{pmatrix} + H\begin{pmatrix} C_1, & C_2 \\ \frac{p_1}{p_1+p_2} & \frac{p_2}{p_1+p_2} \end{pmatrix}, (p_1+p_2)^{1/\sigma}. \quad (25)$$

Using this, the utility of gambling becomes more specialized than in Table 1.

For both segregation and duplex decomposition with finitely additive S_Ω, the utility of gambling term becomes

$$\sum_{i=1}^{n} [V(\Omega) - V(C_i)]\, S_{\succsim,\Omega}(C_i) - A\sum_{i=1}^{n} S_{\succsim,\Omega}(C_i) \log_2 S_{\succsim,\Omega}(C_i), \quad (26)$$

where V maps events to numbers and A is a constant, both in the same units as U. We call the left term subjective expected value and, of course, the right term is the subjective Shannon entropy. As with risk, the sign of A determines whether the subjective entropy is seen as having utility or dis-utility, whereas the magnitude of A controls its importance relative to the two expectations.

For additive S_Ω^σ where $\sigma \neq 1$, the segregation case is impossible and the duplex decomposition one yields

$$V(\Omega) - \sum_{i=1}^{n} V(C_i) S_{\succsim,\Omega}(C_i) - A\left[1 - \sum_{i=1}^{n} S_{\succsim,\Omega}(C_i)\right]. \quad (27)$$

In this case, the term following A is called subjective entropy of degree $1/\sigma$ (Havrda and Charvát, 1967). The role of A is as before.

5 Data: Accommodated and Not Accommodated

In this section we focus mostly on the case of risk and illustrate the relation of EM-EU to existing data sets, although we do consider one case involving uncertainty (Sect. 5.2). Details for both the risky and the uncertain cases are presented in Luce et al. (2008 b). We focus on risk because, in the vast majority of experiments, the gambles are formulated as risky. Nonetheless, Luce et al. (2008 b) note that the concept of a purely risky gamble may be a fiction of the theorist and experimentalist in the sense that it need not really exist for a respondent. For instance, as discussed in Sect. 3.1, the experimenter often "educates" respondents about specific event spaces whereby the probabilities stated in the risky gambles might be realized. A respondent may have superstitions about the qualities, such as colors or numbers, used to identify such events and that may well affect behavior. Also, in naturalistic settings, people are confronted with valued events, such as a standard blood test, where the unpleasantness of the test is independent of the probability of the possible test results, and the above comments suggest that they may also impute values to events that an experimenter considers valueless. We are not aware of experimental studies of gambles, with human respondents, that explicitly involve valued events, nor have we thought through what impact imputing values to valueless events has for data analysis.

A number of "paradoxes" have been raised over the years, each of which casts doubt on the descriptive adequacy of progressively more general theories. The oldest and most famous, the St. Petersburg paradox, questioned the descriptive adequacy of expected value (EV); the Allais paradox questioned expected utility (EU); and the Ellsberg paradox questioned SEU. More recently Michael Birnbaum in collaboration with several others has explored a series of "independence" properties (for a summary and references, see Marley & Luce, 2005) that have cast considerable doubt on rank-dependent utility (RDU) — including, of course, cumulative prospect theory, SEU, and EU. The vast majority of these data are for the risky case, and Luce et al. (2008 b) show that EM-EU can handle many, but by no means all, of the empirical results. Here we summarize the results implied by EM-EU for the Allais paradox and the independence conditions, all situations of risk. For the Ellsberg paradox, which is based in part on uncertain events, we turn to the special case of G-SEU, given below as (29), where H is the subjective Shannon entropy. One can view this as a specialization of the finite additive cases of either Table 1 or of the representation (26) for which the value function V is a constant.

Two basic principles are useful in deriving the properties of EM-EU and in comparing them with those of EU and various data.[7] First, the properties of EM-EU agree with those of EU when either $A = 0$ or when the Shannon entropy terms $I^{(1)}$ in the various gambles under consideration are related in specific ways (some of which we illustrate below). Second, the properties of EM-EU are likely to differ from those of EU when $A \neq 0$ and the Shannon entropy terms $I^{(1)}$ in the various gambles under consideration are not equal and do not "cancel" in appropriate ways. We illustrate these principles with the Allais paradox, the Ellsberg paradox and one of Birnbaum's "independence" conditions.

As already mentioned, in the remainder of this section we develop most of the arguments for EM-EU, i.e., for

$$U(g_{[n]}) = EU\left(g_{[n]}\right) + AI^{(1)}(p_1, \ldots, p_n), \tag{28}$$

where $I^{(1)}$ is the Shannon (1948) entropy. This case arises under both segregation and duplex decomposition.

---

[7] Parallel principles apply to G-SEU, especially the special case that we apply to the Ellsberg paradox.

And, when gambles are based on uncertain events, – i.e., they are presented in terms of events C_i rather than probabilities p_i – we consider the following very special, but important, case of G-SEU:

$$U(g_{[n]}) = SEU\left(g_{[n]}\right) - A\sum_{i=1}^{n} S_{\Omega}(C_i)\log_2 S_{\Omega}(C_i), \tag{29}$$

where the UofG term is the Shannon (1948) entropy of the subjective probabilities.

5.1 The Allais Paradox

The classic example of the Allais paradox arises when an individual has the following pair of preferences (where M means million):

$$\begin{aligned}&\$1M \succ (\$5M, 0.10; \$1M, 0.89; \$0, 0.01),\\&(\$5M, 0.10; \$0, 0.90) \succ (\$1M, 0.11; \$0, 0.89),\end{aligned}$$

a pattern of choices that is shown easily to violate EU. However, note that each gamble is based on a different probability distribution, which means that the entropy terms do not in general "cancel" when $A \neq 0$. In fact, Luce et al. (2008 b) show that the above preference pattern is compatible with EM-EU for a sufficiently large negative A value. Such use of a negative A value makes sense as it corresponds to an aversion to gambling.

5.2 The Ellsberg Paradox

We now provide an explanation of the Ellsberg paradox in terms of the entropy-modified form of SEU given in (29).

The Ellsberg (1961) paradox in coalesced[8] form is of the following form with the choices between f vs. g and f' vs. g' where[9]

$$\begin{aligned}f &= (x, R; 0, G \cup Y) \equiv (x, p; 0, 1-p)\\g &= (x, G; 0, R \cup Y)\\f' &= (x, R \cup Y; 0, G)\\g' &= (x, G \cup Y; 0, R) \equiv (x, 1-p; 0, p)\end{aligned}$$

[8] If there are two (or more) branches (x, C), (x, D) in a gamble, with the common consequence x, then their coalesced form replaces the two by the single branch $(x, C \cup D)$. If the gambles are presented in uncoalesced form, then the following explanation of the paradox requires the additional assumption that the participants convert the gambles to their coalesced forms.

[9] The event notation R, G, Y arose from the interpretation of the chance experiment being a draw from an urn with red, green, and yellow balls.

with $x \succ e$. Note that the probability of G, and so of Y, is not specified beyond being bounded to the interval $(0, 1-p)$. In the classic example, where $\Pr(R) = p = 1/3$ and $\Pr(G \cup Y) = 1 - p = 2/3$, people typically pick f over g and g' over f'. It is checked easily that this pattern of choices is incompatible with SEU.

Paralleling the reasoning for the Allais paradox, note that the gambles f and g are based on different partitions of the events, as are f' and g'. This suggests that the entropy terms given by the entropy-modified form of SEU, (29), do not in general "cancel" when $A \neq 0$. In fact, Luce et al. (2008 b) show that the above preference pattern is compatible with (29) provided that, in (29), the Shannon entropy $I^{(1)}(S_\Omega(R), 1 - S_\Omega(R)) \neq I^{(1)}(S_\Omega(G), 1 - S_\Omega(G))$ and A is sufficiently large, either positively or negatively.

5.3 Independence Properties

Consider $n = 3$, (p_1, p_2, r) and (q_1, q_2, r) arbitrary nontrivial complete probability distributions, and consequences $x_1, y_1, x_2, y_2, z, z'$ with $y_1 \succ x_1 \succ x_2 \succ y_2 \succ e$ and $y_2 \succ z \succ e$, $y_2 \succ z' \succ e$. Then *branch independence of type*[10] $(3,3)^2$ states that:

$$f_{[3]} \sim (x_1, p_1; x_2, p_2; z, r) \succsim (y_1, q_1; y_2, q_2; z, r) \sim g_{[3]} \tag{30}$$

iff

$$f'_{[3]} \sim (x_1, p_1; x_2, p_2; z', r) \succsim (y_1, q_1; y_2, q_2; z', r) \sim g'_{[3]}. \tag{31}$$

Note that, under EM-EU, the above gambles are such that

$$\begin{aligned} EU(f_{[3]}) - EU(g_{[3]}) &= U(x_1)p_1 + U(x_2)p_2 - U(y_1)q_1 - U(y_2)q_2 \\ &= EU(f'_{[3]}) - EU(g'_{[3]}). \end{aligned} \tag{32}$$

Now, it is routine to show that, under EM-EU, (32) is sufficient for branch independence of type $(3,3)^2$ to hold, i.e., (30) iff (31). In fact, all cases of branch independence when $n = 3$ reduce to such a condition, and hence EM-EU predicts that they all hold, contrary to some data.

Applying similar arguments to other independence conditions, Luce et al. (2008b) show that EM-EU accommodates various, but by no means all, of the data obtained in tests of independence conditions not leading to simple cancellation of the UofG terms.

---

[10] The notation $(3,3)^2$ indicates that the consequence z (respectively, z') is the third consequence of the ranked gamble.

6 Conclusions

The major results, which are formally stated as theorems with proofs in our cited papers, are the four representations found in Table 1 for uncertain gambles and the three in Table 2 for risky ones. Those of Table 2 are, essentially, the same ones that Meginniss (1976) first discovered in his long ignored paper. The difference is that we have found an axiomatic basis for the results whereas he began by assuming a representation of the form $U(g_{[n]}) = \sum_{i=1}^{n} f(U(x_i), p_i)$, and that the common function f is differentiable. In the proof he invoked, with little comment, what amounts to GDU. His proof is far simpler and briefer than ours, but we feel it is less illuminating.

By using a conceptual construct of an (infinite) family $\mathcal{O}$ of order extensions of $\succsim_X$, plus other assumptions, we were able to develop, for each order extension, a representation of the UofG term as a subjectively weighted value of events plus a subjective entropy term involving the same weights.

Four major problems are worth mentioning that are unresolved here. First, the case where utility is p-additive rather than additive, i.e., $U(x \oplus y) = U(x) + U(y) + \delta U(x)U(y)$, is of considerable interest because the impact of the elements of chance is amplified by the utility of the kernel equivalents:

$$U(g_{[n]}) = U(KE(g_{[n]})) + U(e, C_1; \ldots; e, C_n)\left[1 + \delta U(KE(g_{[n]}))\right].$$

Ng, Luce, and Marley (2008c, submitted) obtains a very nice representation in the uncertain case for segregation but obtains essentially nothing interesting under duplex decomposition. Second, we need a fuller understanding of why RDU (including, of course, cumulative prospect theory), which has been so popular, admits only a very restricted UofG for uncertain gambles and simply does not arise for risky ones. To have a richer utility of gambling environment that permits rank dependent utility with utility of gambling must require some changes in the axioms.

Third, the conceptual device invoked in Sect. 4 cannot be empirically realized and tested because it applies to infinitely many orderings satisfying the same axioms and agreeing over $\langle X, \oplus \rangle$, whereas an individual generates just one. To make the conceptual device seem a bit more concrete, some people are comfortable in imagining an infinite family of individuals whose preference orders differ only due to differences in their assignment of probability distributions to event partitions. Others find it easier to think of a single individual whose extension is simply unknown to a theorist who must be prepared to model whatever extension happens to be true. The open problem is find some testable way to arrive at those results where the utility of a gamble was partitioned into the sum of three subjective terms: a linear weighted utility of consequences plus a linear weighted value of events per se plus an entropy term.

Fourth, although we have invoked the rank ordering induced by the consequences of a gamble, we have also assumed invariance under permutations and so that constraint actually imposed no real limitation. It was done merely as a convenience in stating certain assumptions and theorems. However, some of Birnbaum's data strongly suggest that whether an event underlies the best or the worst consequence actually matters greatly in how it is evaluated. Thus, a major open problem is to work out a theory for the inherently ordered case. One possibility is to try to arrive at weighted entropy,

$$\sum_{i=1}^{n} a_i S_\Omega(C_i) \log S_\Omega(C_i),$$

which has been mentioned in the literature. But this is certainly not the only possibility.

Acknowledgments The work of Luce and Marley was supported in part by National Science Foundation grants SES-0136431 and SES-0452756 to the University of California, Irvine, and by the Natural Sciences and Engineering Research Council (NSERC) of Canada Discovery Grant 8124 to the University of Victoria for Marley. That of Ng was supported in part by the NSERC of Canada Discovery Grant 8212. Any opinions, findings, and conclusions or recommendations expressed in this material are those of the authors and do not necessarily reflect the views of the National Science Foundation.

References

Aczél, J., & Daróczy, Z. (1975). *On Measures of Information and Their Characterizations*. New York, NY: Academic Press.

Aczél, J., & Daróczy, Z. (1978). A mixed theory of information. I. Symmetric, recursive and measurable entropies of randomized systems of events. *R.A.I.R.O. Informatique théorique (Theoretical Computer Science), 12*, 149–155.

Aczél, J., & Kannappan, P. (1978). A mixed theory of information. III. Inset entropies of degree β. *Information and Control, 39*, 315–322.

Bleichrodt, H., & Schmidt, U. (2002). A context-dependent model of the gambling effect. *Management Science, 48*, 802–812.

Cho, Y.-H., Luce, R.D., & Truong, L. (2002). Duplex decomposition and general segregation of lotteries of a gain and a loss: An empirical evaluation. *Organization Behavior and Human Decision Processes, 89*, 1176–1193.

Conlisk, J. (1993). The utility of gambling. *Journal of Risk and Uncertainty, 6*, 255–275.

Davidson, K.R., & Ng, C.T. (1981). Cocycles on cancellative semigroups. *Utilitas Mathematica, 20*, 27–34.

Diecidue, E., Schmidt, U., & Wakker, P. P. (2004). The utility of gambling reconsidered. *Journal of Risk and Uncertainty, 29*, 241–259.

Ebanks, B.R. (1982). The general symmetric solution of a functional equation arising in the mixed theory of information. *Comptes rendus mathématiques de l'Académie des sciences (Mathematical Reports of the Academy of Science), 4*, 195–200.

Ebanks, B.R., Kannappan, P.L., & Ng, C.T. (1988). Recursive inset entropies of multiplicative type on open domains. *Aequationes Mathematicae, 36*, 268–293.

Ellsberg, D. (1961). Risk, ambiguity and the Savage axioms. *Quarterly Journal of Economics, 75*, 643–669.

Fishburn, P.C. (1970). *Utility Theory for Decision Making*. New York, NY: Wiley.

Fishburn, P.C. (1980). A simple model of the utility of gambling. *Psychometrika, 45*, 435–338.

Fishburn, P.C. (1985). *Interval Orders and Interval Graphs*. New York, NY: Wiley.

Fishburn, P.C. (1988). *Nonlinear Preference and Utility Theory*. Baltimore, MD: Johns Hopkins.

Fishburn, P.C., & Luce, R.D. (1995). Joint receipt and Thaler's hedonic editing rule. *Mathematical Social Sciences, 29*, 33–76.

Havrda, J., & Charvát, F. (1967). Quantification method of classification processes. Concept of structural α-entropy. *Kybernetika, 3*, 30–35.
Kahneman, D., & Tversky, A. (1979). Prospect theory: An analysis of decision under risk. *Econometrica, 47*, 263–291.
Krantz, D.H., Luce, R.D., Suppes, P., & Tversky, A. (1971). *Foundations of Measurement, I.* New York, San Diego: Academic Press.
Le Menestrel, M. (2001). A process approach to the utility for gambling. *Theory and Decision, 50*, 249–262.
Luce, R.D. (1956). Semiorders and a theory of utility discrimination. *Econometrica, 24*, 178–191.
Luce, R.D. (1997). Associative joint receipts. *Mathematical Social Sciences, 34*, 51–74.
Luce, R.D. (2000). *Utility of Gains and Losses: Measurement-theoretical and Experimental Approaches*. Mahwah, NJ: Lawrence Erlbaum Associates, Errata: Luce web page at http//www.socsci.uci.edu.
Luce, R.D., & Fishburn, P.C. (1991). Rank- and sign-dependent linear utility models for finite first-order gambles. *Journal of Risk and Uncertainty, 4*, 29–59.
Luce, R.D., & Fishburn, P.C. (1995). A note on deriving rank-dependent utility using additive joint receipts. *Journal of Risk and Uncertainty, 11*, 5–16.
Luce, R.D., & Marley, A.A.J. (2000). On elements of chance. *Theory and Decision, 49*, 97–126.
Luce, R.D., Ng, C.T., Marley, A.A.J., & Aczél, J. (2008a) Utility of gambling I: Entropy-modified linear weighted utility. Economic Theory, *36*, 1–33.
Luce, R.D., Ng, C.T., Marley, A.A.J., & Aczél, J. (2008b) Utility of gambling II: Risk, Paradoxes, and Data. Economic Theory, *36*, 165–187.
Erratum to Luce, Ng, Marley, & Aczél (2008a, b): See http://www.imbs.uci.edu/personnel/luce/luce.html.
Marley, A.A.J. (1972). The logic of decisions: A review of P. C. Fishburn's *Utility Theory for Decision Making*. *Contemporary Psychology, 17*, 379–380.
Marley, A.A.J., & Luce, R.D.D (2005). Independence properties vis-à-vis several utility representations. *Theory and Decision, 58*, 77–143.
Meginniss, J.R. (1976). A new class of symmetric utility rules for gambles, subjective marginal probability functions, and a generalized Bayes' rule. *Proceedings of the American Statistical Association, Business and Economic Statistics Section*, 471–476.
Ng, C.T., Luce, R.D., & Marley, A.A.J. (2008a). Utility of gambling: extending the approach of Meginniss (1976). *Aequationes Mathematicae*.
Ng, C.T., Luce, R.D., & Marley, A.A.J. (2008b). Utility of gambling when events are valued: An application of inset entropy. *Theory and Decision*.
Ng, C.T., Luce, R.D., & Marley, A.A.J. (2008c, submitted). Utility of gambling under p-additive joint receipt.
Pope, R. (1995). Toward a more precise decision framework. *Theory and Decision,* 39, 241–265.
Ramsey, F.P. (1931). *The Foundations of Mathematics and Other Logical Essays.* New York, NY: Harcourt, Brace. In H. E. Kyburg, & H. E. Smokler (Eds.) *Studies in Subjective Probability*, Chap. VII. New York, NY: Wiley, 61–92 (1964).
Savage, L.J. (1954). *The foundations of probability*. New York, NY: Wiley.
Shannon, C.E. (1948). A mathematical theory of communication. *Bell System Technical Journal, 27*, 379–423, 623–656.
Slovic, P. (1967). The relative influence of probabilities and payoffs upon perceived risk of a gamble. *Psychonomic Science, 9*, 223–224.
Slovic, P., & Lichtenstein, S. (1968). The importance of variance preferences in gambling decisions. *Journal of Experimental Psychology, 78*, 646–654.
von Neumann, J., & Morgenstern, O. (1947). *Theory of Games and Economic Behavior*, 2nd Ed. Princeton, NJ: Princeton University.
Wakker, P.P. (1990). Characterizing optimism and pessimism directly through comonotonicity. *Journal of Economic Theory, 52*, 453–463.
Yang, J., & Qiu, W. (2005). A measure of risk and a decision-making model based on expected utility and entropy. *European Journal of Operations Research, 164*, 792–799.

Altruistic Utility Functions for Joint Decisions

David E. Bell and Ralph L. Keeney

1 Introduction

All of us make decisions that are not entirely self-centered; we voluntarily anticipate what we think to be the preferences of others and incorporate them into our decision making. We do this, not because of legal requirements or social norms, but because we are altruistic; we care intrinsically about the welfare of others. In this paper, we illustrate for these types of decisions how confusion may arise because the distinction between our personal (egotistical) preferences and our altruistic concerns is not carefully distinguished. We first define the distinction between personal and altruistic preferences, and then show how to use both of these kinds of preferences in prescriptive decision making methodologies.

We confine ourselves to the class of problems where two or more people must select a common course of action. The following story illustrates a simple example. Joan and Dan have decided to have dinner and must choose a restaurant. They quickly specify three possibilities: a Brazilian restaurant, a French restaurant, and a Thai restaurant. Joan is thoughtful and wishes to choose a restaurant that Dan will really like. Similarly, Dan wants to choose a restaurant that pleases Joan. So what happens? Joan, thinking about what might be Dan's preferences, decides that Dan would like the French restaurant, followed by the Brazilian restaurant, followed by the Thai restaurant. Dan, thinking about what Joan would like, also decides that the French restaurant would be best, followed by the Brazilian restaurant, and then the Thai restaurant. Joan speaks first and suggests the French restaurant. Dan, thinking that this is what Joan wants, agrees and off they go. During dinner discussion, Joan mentions that she would have preferred the Thai restaurant to the French restaurant. Somewhat surprised, Dan then says that he also would have preferred the Thai restaurant. They wonder how this state of affairs came about.

R.L. Keeney (✉)
Fuqua School of Business, Duke University, Durham, NC 27708, USA
e-mail: Keeney@duke.edu

S.J. Brams et al. (eds.), *The Mathematics of Preference, Choice and Order: Essays in Honor of Peter C. Fishburn*, Studies in Choice and Welfare,

Two compounding errors led to an inferior choice. First, each person guessed at the others preferences. Second, the stated preferences are mistakenly interpreted as those of the speaker. How could this have been avoided? Clearly both Dan and Joan could have written down their personal preference order for restaurants, assuming that they did not care about the other's preferences, and then compared notes. In our illustration, this would have led immediately to a mutually satisfactory decision. Our experience is that even decision analysts are rarely that explicit. What often happens instead is that through discussion, or generalized experience with the other person, each person informally updates their own preferences to take account of the other's likes and dislikes. There are many ways this informal approach can produce inadequate solutions.

There are many situations where a group of individuals must collectively choose among alternatives and where each individual wishes to please the others. Examples include parents making choices with their children, decisions by boards of directors, decisions by departments or groups within organizations, decisions by legislative or regulatory bodies, choices made by families, and decisions among friends. In many of these cases, parties to the decision will take account of the preferences of the others, not only for the expediency of arriving at a consensus, but often out of an altruistic interest in their happiness. An altruistic decision maker will be willing to forgo some direct personal gain to help others achieve their objectives.

The general problem of combining preferences of individuals into group preferences is not new. There is a large body of published work on this topic ((Arrow, 1951), (Harsanyi, 1955), (Diamond, 1967), (Sen, 1979), (Broome, 1984), and many others). Much of the work prior to 50 years ago is summarized in (Luce & Raiffa, 1957). Since that time, there has been work on risk sharing (e.g. Raiffa, 1968), group utility functions (e.g. Keeney & Raiffa, 1976), and utility functions where a seller incorporates the preferences of a buyer (Edgeworth, 1881), (Raiffa, 1982), (Keeney & Lilien, 1978), and (Keeney & Oliver, 2005). There has also been work on preference dependencies in multiattribute utility functions (Fishburn, 1965), (Bell, 1977), (Meyer, 1977), (Keeney, 1981). Several authors have discussed the adaptation of preferences in a group context (Zizzo, 2005), (Sobel, 2005), and (Cubitt & Sugden, 1998). Trautmann (2006) proposes a similar approach to ours, but his proposal is based on the descriptive criterion suggested by (Fehr & Schmidt, 1999), whereas ours is consistent with standard multiattribute approaches, and amenable to assessment as we discuss later.

We focus in this paper on one particular type of joint decision. One could think of this type as altruistic joint decisions, because each of the individuals has a fundamental preference for the other individuals being pleased. Section 2 defines an altruistic joint decision and discusses its relevance. As conceptual distinctions are extremely important in discussing problems with interpersonal dependence of preferences, Sect. 3 outlines the relevant concepts and terminology used to analyze altruistic joint decisions. In Sect. 4, we focus on altruistic joint decisions involving two individuals and illustrate the main results that collectively characterize a reasonable set of altruistic utility functions to use in analyzing joint decisions. Section 5 elaborates on the foundations for altruistic utility functions. Section 6 suggests how one might assess these utility functions, and Sect. 7 is a discussion of the insights from and uses of the concepts and results of the paper.

2 Altruistic Joint Decisions

The altruistic joint decisions that we investigate in this paper are characterized by six properties:

1. A group of individuals have a decision that they must make jointly,
2. The alternatives are exogenously given,
3. All individuals in the group bear the same consequences,
4. Each individual has utilities for the alternatives,
5. Each individual is altruistic about the others; they prefer them to be happy even at some cost to themselves and,
6. Each person is honest in revealing their preferences.

Property 3 rules out decisions that involve risk sharing or somehow dividing the consequences among the individuals. With regard to Property 4, the individuals may have utility functions over the consequences which can be used to derive utilities for the alternatives. Property 5 is the one that states the altruism assumption. Without it, we would have the more general situation sometimes referred to as the group decision problem. Property 6 eliminates the need to worry about strategic gaming; Property 2 is included to give the additional "safeguard" that individuals do not introduce irrelevant alternatives to skew the decision making procedures.

It is useful to analyze altruistic joint decisions for many reasons. First, as suggested above, they occur often in the real world. Second, the consequences are frequently important. Poor choices increase the likelihood of a disastrous vacation or a poor business outcome. Such consequences can contribute to dissolve what was previously a wonderful group of friends, a terrific marriage, or an exciting and productive business relationship. Third, *ad hoc* choices on altruistic decisions may contribute to poor choices and hence less desirable consequences. The reason this may occur is because there are sophisticated concepts necessary to take into account in altruistic joint decisions. Self-centered preferences for consequences can get confused or be confused with altruistic concerns for those same consequences. A little analysis can help define and distinguish these aspects.

3 Concepts and Terminology

We characterize an altruistic decision as follows: There are J alternatives a_j, $j = 1, \ldots, J$, one of which must be chosen by an altruistic group. The group has N individuals, referred to as $I_1, \ldots, I_N$. Each individual I_i has a personal utility function u_i

over the alternatives. This *egotistical utility function* only incorporates the value of the alternative directly to the individual and does not include any value to I_i due to his or her altruistic feelings for the happiness of others. Thus, for each alternative a_j, individual I_i assigns an egotistical utility $u_i(a_j)$.

Each individual also has what we refer to as an *altruistic utility function* U_i, $i = 1,\ldots,N$ which is the function that describes the preferences the person announces or acts upon, which takes into account both his or her personal concerns and concerns for the welfare of the others. For example, I_1's evaluation of alternative a_j might be expressed as $U_1(u_1(a_j), U_2(a_j), \ldots, U_N(a_j))$. An example of an individual altruistic utility function for individual I_1 is the additive form

$$U_1\left(u_1\left(a_j\right), U_2\left(a_j\right), \ldots, U_N\left(a_j\right)\right) = k_1 u_1\left(a_j\right) + \sum_{i=2}^{N} k_i U_i\left(a_j\right), \tag{1}$$

where u_1 and $U_i, i = 1,\ldots,N$ are scaled 0 to 1, $k_1 > 0$ (the person is not totally altruistic) and the scaling factors $k_2,\ldots,k_N$ are also non-negative to incorporate the altruism that individual I_1 feels for individuals $I_i, i = 2,\ldots,N$.

The *group altruistic utility function* U_G is a utility function that incorporates the preferences of each of the individuals in the group. In general the arguments in this utility function can be each individual's egotistical and/or altruistic utility function. A possible example is the additive utility function

$$U_G\left(a_j\right) = \sum_{i=1}^{N} K_i U_i\left(a_j\right), \tag{2}$$

where the scaling factors $K_i, i = 1,\ldots,N$ must be positive to incorporate altruism of each individual for the other individuals.

4 Main Results for Altruistic Decisions

In this section, we present our main analytical results. To focus on the conceptual ideas, all of the work in this section concerns a joint altruistic decision made by two individuals. We begin by stating our most important analytical results, though the assumptions we use for Result 1 are stronger than necessary. In Sect. 5, these assumptions are weakened. The ideas also extend to altruistic groups of more individuals as discussed in Sect. 7.

Result 1. An individual's altruistic utility function should have two attributes which are the egotistical utility functions of the two individuals. The resulting two-attribute function, should be multiplicative (or additive) in those attributes. Thus,

$$U_1\left(a_j\right) = k_1 u_1\left(a_j\right) + k_2 u_2\left(a_j\right) + k_3 u_1\left(a_j\right) u_2\left(a_j\right) \tag{3}$$

and

$$U_2\left(a_j\right) = k_4 u_1\left(a_j\right) + k_5 u_2\left(a_j\right) + k_6 u_1\left(a_j\right) u_2\left(a_j\right), \tag{4}$$

where all utility functions are scaled 0 to 1, all k_i scaling factors are positive, and $k_1+k_2+k_3=1$ and $k_4+k_5+k_6=1$. The scaling factors indicate the relative importance of the ranges of consequences possible on the corresponding utility function as discussed in the assessment Sect. 6.

Argument. As we will discuss in Sect. 5, it might be tempting to think that one person's altruistic utility function should be a function of the other person's altruistic function, but, as we shall see, this leads to problems. We believe that a fundamental property of altruism is that if individual I_1, say, is personally indifferent among the available alternatives then he or she would wish to select the alternative that maximizes the other individual's egotistical utility function. For example, if Dan personally regards all of the restaurant alternatives as equally preferable, surely he would wish to select the one that Joan most prefers. One might imagine that if Dan dislikes the available restaurants, then he might be jealous if Joan is delighted, but that does not meet our sense of altruism. Similarly if Joan is personally indifferent among the available restaurants, then surely Dan should feel comfortable selecting his own favorite, especially since he knows Joan is altruistic towards him (we assume all parties are altruistic). In the language of multiattribute utility, we have therefore concluded that individual I_1's altruistic utility function should have the two attributes and each should be utility independent of the other. Thus his altruistic utility function should have the form (3), and by symmetry, individual I_2's should have the form (4).

The factors k_1 and k_5 are positive as each individual certainly cares about their own direct consequences. Factors k_2 and k_4 are positive as both individuals are altruistic. We argue below that k_3 and k_6 should at least be non-negative and more likely are positive.

Suppose individual I_1 has a choice between two alternatives, one with $(u_1,u_2)=(x,y)$ and the other with $(u_1,u_2)=(x-b,y+c)$. Hence, I_1 must decide if for her the sacrifice of an amount of utility b is worth the improvement of an amount of utility c to individual I_2. Using her altruistic utility function (3), we see the answer is yes if $U_1(x-b,y+c)>U_1(x,y)$ which implies $k_1(x-b)+k_2(y+c)+k_3(x-b)(y+c)>k_1x+k_2y+k_3xy$, so

$$-k_1b+k_2c+k_3\,(cx-by-bc)>0. \tag{5}$$

Dividing (5) by bc yields

$$-k_1/c+k_2/b-k_3+k_3\,(x/b-y/c)>0. \tag{6}$$

If $k_3=0$, then this preference is independent of x and y. If $k_3>0$, then I_1 is relatively more altruistic when x is high or y is low. We believe this is more in line with how altruistic people will like to behave than when $k_3<0$. Thus, in general, it seems reasonable to have $k_3>0$, so $k_6>0$ also by the same argument. It is worth mentioning that all of our results hold for the cases when $k_3=0$ and/or $k_6=0$

though this is not required. It is quite possible that a person's level of altruism could vary depending on the actual disparity in egotistical utility each person derives from a consequence.

Result 2. The group altruistic utility function should be additive over the two arguments of the individual's altruistic utility functions, so

$$U_G(a_j) = C_1U_1(a_j) + C_2U_2(a_j), \tag{7}$$

where all utility functions are scaled 0 to 1 and $C_1 + C_2 = 1$.

Argument. The utility function U_G represents how the pair of individuals should evaluate joint decisions. It seems reasonable to suppose that if individual I_1 is indifferent among alternatives using U_1, then both individuals would be happy to let the joint decision be consistent with U_2. By symmetry the reverse would be true. Hence U_G should be multiplicative or additive in U_1 and U_2: $U_G(a_j) = C_1U_1(a_j) + C_2U_2(a_j) + C_3U_1(a_j)U_2(a_j)$.

Now we argue that C_3 should be zero. Consider two gambles involving lotteries over the alternatives. Suppose that individual I_1 has the same expected altruistic utility under either gamble. Suppose this is also true for individual I_2. Then both individuals are indifferent between the two gambles so it seems reasonable that U_G should reflect that indifference. As proven in (Harsanyi, 1955) and (Fishburn, 1984), this only occurs when $C_3 = 0$. If C_3 were greater than zero, for example, it would mean that the group might prefer an alternative with lower values of U_1 and U_2 in order to achieve more concordance between U_1 and U_2. But since U_1 and U_2 already, respectively, incorporate all of I_1's and I_2's altruistic concerns, any further sacrifice is counter-productive.

The conclusion that $C_3 = 0$, while not obvious, is consistent with the observation of (Keeney, 1981), namely that when the objectives are fundamental, complete, and do not overlap, an additive utility function is appropriate. The two individual altruistic utility functions are fundamental and a complete set in that they consider all objectives relevant to the decision (e.g. individual I_1's concerns are completely expressed by U_1) and do not overlap. Each individual altruistic utility function addresses both direct and altruistic preference concerns. It is also consistent with (Edgeworth, 1881) and (Harsanyi, 1955) who both argued that an altruistic solution could be determined by maximizing the sum of the affected individual's utilities.

Result 3. The group altruistic utility function is the multiplicative utility function with the egotistical utility functions of the individuals as the arguments, so

$$U_G(a_j) = K_1u_1(a_j) + K_2u_2(a_j) + K_3u_1(a_j)\,u_2(a_j), \tag{8}$$

where K_i, $i = 1,2,3$ are positive and $K_1 + K_2 + K_3 = 1$.

Argument. The argument in this case is a proof using Results 1 and 2. Substituting (3) and (4) into (7) and dropping the a_j's for clarity yields

$$\begin{aligned} U_G &= C_1 (k_1 u_1 + k_2 u_2 + k_3 u_1 u_2) + C_2 (k_4 u_1 + k_5 u_2 + k_6 u_1 u_2) \\ &= (C_1 k_1 + C_2 k_4) u_1 + (C_1 k_2 + C_2 k_5) u_2 + (C_1 k_3 + C_2 k_6) u_1 u_2. \end{aligned} \tag{9}$$

Equation (9) is (8) with $K_1 = C_1 k_1 + C_2 k_4$, and so on.

The group utility function (8) is not necessarily additive in the individual's personal utilities. This is because if the altruism of any member of the group (their willingness to give up utility to help someone else) depends on their own level of satisfaction, then the multiplicative term will be present in their individual altruistic utility function, and therefore in the group function also. The functional form (8) is mathematically identical to an analysis of the group decision problem (Keeney and Kirkwood, 1975) that posited a possible concern by the group for equity. In that development the multiplicative term reflects the desire by the group not to have disparate outcomes. It is possible for both phenomena to occur at the same time; someone could be altruistic but also concerned about equity.

5 Personal Utilities are Fundamental to Altruistic Decisions

In Result 1, we made a strong assumption that the arguments in an individual's altruistic utility function should be the egotistical utilities of the individuals. We did this so the important results in Sect. 4, and the logic supporting them would be clear. Here, from more basic reasoning, we provide support for having egotistical utilities as arguments in altruistic utility functions.

Result 4. The egotistical utility functions should be the arguments in the altruistic utility functions.

Argument. We asserted the truth of Result 4 in stating our Result 1. But why is that the case? It might seem reasonable to think that the altruistic utility function of individual I_1 might depend on hers and on I_2's altruistic utility functions. But that is circular. For example, if

$$U_1 (U_1, U_2) = h_1 U_1 + h_2 U_2 + h_3 U_1 U_2, \tag{10}$$

where we have deleted the a_j's for clarity, it is evident that $h_1 = 1$ and $h_2 = h_3 = 0$ is the only viable solution.

Another way to think about the appropriate attributes for U_1 is that it can be a function of u_1 and U_2 so individual I_1's altruistic utility function could be represented by

$$U_1 (u_1, U_2) = h_1 u_1 + h_2 U_2 + h_3 u_1 U_2, \tag{11}$$

and similarly for individual I_2,

$$U_2 (U_1, u_2) = h_4 U_1 + h_5 u_2 + h_6 U_1 u_2. \tag{12}$$

But (11) and (12) together lead to problems of double counting. One way to see this is by substituting (12) into (11) which yields

$$U_1(u_1, U_2) = h_1u_1 + h_2(h_4U_1 + h_5u_2 + h_6U_1u_2) + h_3u_1(h_4U_1 + h_5u_2 + h_6U_1u_2). \tag{13}$$

Substituting (11) into (13) results in squared terms of u_1 if either h_3 or h_6 is not zero, and squared terms are unreasonable. The problem stems from the fact that individual I_1, in trying to please individual I_2, who is trying to please individual I_1, ends up double counting his own interests. The intent of any utility function is to maximize its expected value, which is a simple calculation, not to maximize its square. Hence, h_3 and h_6 would necessarily have to be zero if (11) and (12) were reasonable.

Even if $h_3 = h_6 = 0$ in (11) and (12), there are still difficulties. Substituting (12) into (11) yields

$$U_1(u_1, U_2) = h_1u_1 + h_2(h_4U_1 + h_5u_2). \tag{14}$$

Solving (14) for U_1, we find

$$U_1 = (h_1u_1 - h_2h_5u_2)/(1 - h_2h_4). \tag{15}$$

As a numeric example suppose that $h_1 = h_4 = 0.2$, $h_2 = h_5 = 0.8$, and $h_3 = h_6 = 0$. That is, individual I_1 is very altruistic and assigns 80% of the weight to the preferences of individual I_2, whereas individual I_2 is less altruistic but does assign a 20% weight to individual I_1's preferences.

Substituting the values for the h's into (15) we find

$$U_1 = \frac{5}{21}u_1 + \frac{16}{21}u_2. \tag{16}$$

Similar calculations for I_2 yield

$$U_2 = \frac{1}{21}u_1 + \frac{20}{21}u_2. \tag{17}$$

Thus, although both individuals agree, in a sense, that 80% of the weight should be on the preferences of individual I_2, the calculations show that the double counting leads to a different outcome. It is possible that in selecting weights for (10), individual I_1 correctly anticipates the effect of the double counting, but we believe that for most individuals this would be challenging.

If we consider the group altruistic utility function (7) in this case, any choice of C_1 and C_2 necessarily leads to a weighting of individual I_1's personal utility of less than 20%. We conclude that altruistic utility functions should be based on individuals' egotistical utility functions rather than on other altruistic utility functions.

6 Assessment Issues

Based on the results in Sect. 4, the group altruistic utility function could be assessed based on either Result 2 or 3. The best way to make the necessary assessments is to use Result 2. This requires first assessing the two individuals' egotistical utility functions, then both individuals' altruistic utility functions (3) and (4), and then the scaling factors C_1 and C_2 in (7).

The individuals' egotistical utility functions should be assessed using standard procedures as outlined in Keeney and Raiffa (1976) and many other sources. There is nothing special about these utility functions as they are simply utility functions for an individual concerned with consequences directly to that individual.

Assessing the individuals' altruistic utility functions are also just individual assessments. Relative weights on the individual egotistical utility functions in (3) and (4) incorporate two separate issues. One of these is the well-known interpersonal comparison of utility problem (Luce & Raiffa, 1957) and the other is the altruistic value to each individual for pleasing each other. To make these assessments requires understanding the relative desirability of the impacts for each individual of going from the worst of their alternatives to the best of their alternatives. For instance, if the two individuals are selecting a restaurant together, the range for individual I_1 may be in qualitative terms from a poor restaurant that would be "acceptable food and pleasant atmosphere" to a best restaurant that would be "good food and pleasant atmosphere." For the second individual I_2, the range could go from "very unappealing food and objectionable atmosphere" to "excellent food and perfect atmosphere." In such a situation, each individual may decide to place greater weight on I_2's utilities as I_2 seems to have a much more significant difference in desirability of the worst and the best of restaurants.

To determine appropriate relative scaling factors (the k's) for the individuals' altruistic utility functions given by (3) and (4), each individual should consider the range of the desirability of the various consequences to each individual as well as how much weight she wants to place on pleasing the other individual. Consider the scaling factors k_1, k_2, and k_3 in (3). The best way to assess these factors is to compare specific alternatives in terms of their egotistical utilities to both individuals I_1 and I_2 and look for pairs of alternatives that the individual feels are equally desirable. Once two pairs of such joint consequences described by (u_1, u_2) are found indifferent, the individual's altruistic utility function (3) should equate the utilities of the pairs. This provides two equations with three unknowns, namely k_1, k_2, and k_3. The fact that $k_1 + k_2 + k_3 = 1$ is a third equation. These three equations can be solved to yield specific values for the three scaling factors. Note that the altruistic function just assessed is the function the individual would use if he or she were to make the group decision unilaterally. Put another way, it represents the preferences that this individual would use if the decision were left up to her.

Assessing C_1 and C_2 in the group altruistic utility function (7) is the only value judgment in the assessment process requiring agreement of the two individuals. The value judgments about C_1 and C_2 are basically assessments about the relative significance of each person to the group. With individuals who have altruistic feelings for each other, it seems reasonable to select $C_1 = C_2 = 0.5$. That is because all of the more conceptually difficult value judgments concerning altruism and strength of preferences are incorporated into each individual altruistic utility function. As a specific example, suppose individual I_1 selected $k_1 = 0.6$ and $k_2 = 0.4$ in her utility function (3), so $k_3 = 0$. This would mean that she thought her personal utility function counted 1.5 times as much as I_2's personal utility function. It would not then seem reasonable to underweight her altruistic preferences relative to those of individual I_2, with $C_1 < 0.5$, or to overweight them, with $C_2 > 0.5$. Obviously, similar reasoning holds for individual I_2.

7 Insights and Uses

The insights in this paper can be used informally or formally in making joint decisions. Indeed, we would expect that the more common use of the concepts would be in making decisions informally, but thoughtfully.

One basic finding is that the informal notion of agreement through discussion and iterated compromise, while intuitively attractive, is fraught with difficulty: even if the process converges, the compromise solution might not be the appropriate solution.

Though it may appear to be selfish, it is important for altruistic decision makers to focus initially on what they know best, their own personal (egotistical) utilities. These are the utilities the individual has for the direct consequences of an alternative. Each individual naturally knows much more about his or her own preferences than about the other individual's preferences. There is no reason for the guessing effort to occur in altruistic decisions. Each individual should honestly first express their own preferences for themselves. Once these are clearly laid out for both individuals to understand, then any appropriate weighting by each individual to account for the personal utilities and the altruistic concerns can more effectively occur.

An important insight from this work is that an altruistic utility function should be over the egotistical utility functions. In particular, a multiplicative utility model is a general model that can address these concerns for individuals and for joint decisions of two individuals. The altruistic values that each of the individuals have can be addressed in assessing the scaling factors in the multiplicative utility function.

So how would one use this theory on a simple decision like Joan and Dan's choice of a restaurant? First Joan and Dan should express their personal preferences for the restaurants to each other. If they agree on their first choice, choose it. If they disagree, eliminate any dominated alternatives. Then they should discuss their personal strengths of preference among the remaining contenders, and then jointly decide based on that information. Either the choice should be obvious or it should not matter as they are about equally desirable in the joint sense.

Results 2 and 3 together state that the group altruistic utility function is additive over the two individual's altruistic utility functions and also multiplicative over those two individual's egotistical utility functions. This demonstrates the significance of how framing a decision, in this case specifying the objectives explicitly included in the analysis of a decision, can and should influence the functional form of the appropriate utility function.

The insights discussed above generalize to joint decisions involving more than two individuals. Specifically, the multiplicative utility function is an appropriate formulation for a joint altruistic utility function and the arguments of that function should be the egotistical utility functions. The altruistic values of each of the individuals are addressed in assessing the scaling factors in that altruistic utility function.

References

Arrow, K. J. (1951). Social choice and individual values (2nd ed.). Wiley: New York.

Bell, D. E. (1977). A utility function for time streams having inter-period dependencies. *Operations Research, 25*, 448–458.

Broome, J. (1984). Uncertainty and fairness. *The Economic Journal, 94*, 624–632.

Cubitt, R. P., & Sugden, R. (1998). The selection of preferences through imitation. *Review of Economic Studies, 65*, 761–771.

Diamond, P. A. (1967). Cardinal welfare, individual ethics, and interpersonal comparison of utility: Comment. *Journal of Political Economy, 75*, 765–766.

Edgeworth, F. Y. (1881). *Mathematical physics: An essay on the application of Mathematics to the Moral Sciences*. London: Kegan Paul. (Reprinted in 1967 by M. Kelley, New York.)

Fehr, E., & Schmidt, K. M. (1999). A theory of fairness, competition, and cooperation. *Quarterly Journal of Economics, 114*, 817–868.

Fishburn, P. C. (1965). Markovian dependence in utility theory with whole product sets. *Operations Research, 13*, 238–257.

Fishburn, P. C. (1984). On harsanyi's utilitarian cardinal welfare theorem. *Theory and Decision, 17*, 21–28.

Harsanyi, J. C. (1955). Cardinal welfare, individualistic ethics, and interpersonal comparisons of utility. *Journal of Political Economy, 63*, 309–321.

Keeney, R. L. (1981). Analysis of preference dependencies among objectives. *Operations Research, 29*, 1105–1120.

Keeney, R. L., & Kirkwood, C. W. (1975). Group decision making using cardinal social welfare functions. *Management Science, 22*, 430–437.

Keeney, R. L., & Raiffa, H. (1976). *Decisions with multiple objectives*. New York: Wiley. (Reprinted in 1993, Cambridge University Press, New York).

Keeney, R. L., & Lilien, G. L. (1978). A utility model for product positioning. In S. Zionts (Ed.), *Multiple criteria problem solving* (pp. 321–334). Berlin: Springer.

Keeney, R. L., & Oliver, R. M. (2005). Designing win-win financial loan products for consumers and businesses. *Operational Research, 56*, 1030–1040.

Luce, R. D., & Raiffa, H. (1957). *Games and decisions*. New York: Wiley.

Meyer, R. F. (1977). State-dependent time preference. In D. E. Bell, R. L. Keeney, & H. Raiffa (Eds.), *Conflicting objectives in decisions* (pp. 232–243). London: Wiley.

Raiffa, H. (1968). *Decision analysis*. Reading, MA: Addison-Wesley.

Raiffa, H. (1982). *The art and science of negotiation*. Cambridge, MA: Harvard University Press.

Sen, A. (1979). Utilitarianism and welfarism. *The Journal of Philosophy, 76*, 463–489.

Sobel, J. (2005). Interdependent preferences and reciprocity. *Journal of Economic Literature, 43*, 392–436.

Trautmann, S. T. (2006). *Operationalizing individual fairness in Harsanyi's utilitarianism*. Unpublished manuscript, Erasmus University, Rotterdam.
Zizzo, D. J. (2005). Interdependent preferences. To appear In W Darity Jr (Ed.), *International encyclopedia of the social sciences* (2nd ed). New York: Macmillan.

SSB Preferences: Nonseparable Utilities or Nonseparable Beliefs

Yutaka Nakamura

1 Introduction

It is around 1980 that new era for decision making under risk/uncertainty began to uncover numerous alternative representations which generalize the traditional (subjective) expected utility maximization. The initial major contributors include Chew and MacCrimmon (1979), Chew (1983), Fishburn (1982), Kahneman and Tversky (1979), Machina (1982), Quiggin (1981), and Schmeidler (1988) (first appeared in 1981 as a discussion paper). One of Fishburn's works in this area is the discovery of an axiomatic structure of SSB (skew-symmetric bilinear) preferences in decision making under risk, and its numerical representation, dubbed SSB utility (see Fishburn, 1982). Since then, he published a series of papers which study SSB preferences and their numerical representations in various contexts in decision making under risk/uncertainty (see a survey, Fishburn, 1988b).

This paper further explores representational aspects of SSB preferences particularly in decision making under uncertainty and discusses their necessary and sufficient axiomatizations. Three representational forms will be examined. One of them is known as an SSA (skew-symmetric additive) representation first explored by Fishburn, 1984a. The other two are new in the literature, one of which seems to be a more natural application of SSB utility to decision making under uncertainty than SSA representation. A characteristic feature of the first two representations is nonseparability of utilities for decision outcomes. The last one is a generalization of subjective expected utility (SEU) which replaces subjective probabilities with nonseparable representation of comparative beliefs first discovered by Fishburn (1983a and b).

Y. Nakamura
University of Tsukuba, Graduate School of Systems and Information Engineering, Division of Social Systems and Management 1-1-1 Tennnoudai, Tsukuba, Ibaraki 305-8573, Japan
e-mail: nakamura@sk.tsukuba.ac.jp

S.J. Brams et al. (eds.), *The Mathematics of Preference, Choice and Order: Essays in Honor of Peter C. Fishburn*, Studies in Choice and Welfare,

There may be three formulations in the literature, discussed in the next section, to arrive at axiomatizations of preferences in decision making under uncertainty: pure-act formulation (Savage's approach), lottery-act formulation, and act-lottery formulation. Fishburn's axiomatizations of SSB preferences are based on the first two formulations. We adopt the third formulation to develop necessary and sufficient axiomatizations for the three nonseparable representations when the state space is finite.

The paper is organized as follows. The next section introduces nonseparable representations. In Sect. 3, axiomatic SSB preference structures and their numerical representations are presented. Section 4 studies necessary and sufficient axioms for two nonseparable utility representations. Then in Sect. 5, we explores necessary and sufficient axioms for SEU with nonseparable beliefs. Section 6 concludes the paper.

2 Nonseparable Representations

Let $\mathcal{A}$ be the set of all (pure) *acts* that are functions from the set S of states of the nature into the set X of outcomes. Each $a \in X$ will be identified with *constant act* $\boldsymbol{a}$ for which $\boldsymbol{a}(s) = a$ for all $s \in S$. Let $\succ$ be a binary strict preference relation on $\mathcal{A}$, read as 'is strictly preferred to'. The traditional SEU theories yield a utility function u on X and a probability measure π on an algebra $\mathcal{B}_S$ of subsets of S such that, for all $\boldsymbol{a},\boldsymbol{b} \in \mathcal{A}$,

$$\boldsymbol{a} \succ \boldsymbol{b} \Longleftrightarrow E(\boldsymbol{a},u,\pi) > E(\boldsymbol{b},u,\pi),$$

where $E(\boldsymbol{a},u,\pi)$ is expected utility of act $\boldsymbol{a}$ with respect to π. We may have three equivalent integral expressions of $E(\boldsymbol{a},u,\pi)$:

(i) $\boldsymbol{a} \succ \boldsymbol{b} \iff \int_S (u(\boldsymbol{a}(s)) - u(\boldsymbol{b}(s)))\, d\pi(s) > 0,$
(ii) $\boldsymbol{a} \succ \boldsymbol{b} \iff \int_S \int_S (u(\boldsymbol{a}(s)) - u(\boldsymbol{b}(t)))\, d\pi(s) d\pi(t) > 0,$
(iii) $\boldsymbol{a} \succ \boldsymbol{b} \iff \int_{-\infty}^{+\infty} (\pi(\{s : u(\boldsymbol{a}(s)) \geq \tau\}) - \pi(\{s : u(\boldsymbol{b}(s)) \geq \tau\}))\, d\tau > 0.$

The aim of this paper is to axiomatically characterize nonseparable generalizations of the SEU representation. The most general nonseparable representation yields a real valued bivariate function Ψ on $\mathcal{A} \times \mathcal{A}$ such that, for all $\boldsymbol{a},\boldsymbol{b} \in \mathcal{A}$,

$$\boldsymbol{a} \succ \boldsymbol{b} \Longleftrightarrow \Psi(\boldsymbol{a},\boldsymbol{b}) > 0.$$

We shall examine three specializations of Ψ. Two of them are concerned with the first and second integral expressions (i) and (ii), where each of the integrands in (i) and (ii), $u(\boldsymbol{a}(s)) - u(\boldsymbol{b}(s))$ and $u(\boldsymbol{a}(s)) - u(\boldsymbol{b}(t))$, are respectively replaced by nonseparable utility representations, $\psi(\boldsymbol{a}(s),\boldsymbol{b}(s))$ and $\psi(\boldsymbol{a}(s),\boldsymbol{b}(t))$, i.e.,

(I) $\Psi(\boldsymbol{a},\boldsymbol{b}) = \int_S \psi(\boldsymbol{a}(s),\boldsymbol{b}(s)) d\pi(s),$
(II) $\Psi(\boldsymbol{a},\boldsymbol{b}) = \int_S \int_S \psi(\boldsymbol{a}(s),\boldsymbol{b}(t)) d\pi(s) d\pi(t),$

where ψ is a skew-symmetric bivariate function on $X \times X$, i.e., for all $x,y \in X$, $\psi(x,y) + \psi(y,x) = 0$. Model (I) is known as an SSA representation. Bilinearity with respect to state probabilities is reflected in model (II), but not in model (I). Thus model (II) may be dubbed an SSB representation under uncertainty. Observe that model (I) satisfies the Savage's sure-thing principle, i.e., preferences for acts are independent of outcomes in some states as long as those outcomes are identical for acts under consideration. It is well known that this principle is often behaviorally violated. However, model (II) does not necessarily satisfy the principle.

The last one is concerned with the third integral expression (iii) in which the integrand $\pi(\{s : u(\boldsymbol{a}(s)) \geq \tau\}) - \pi(\{s : u(\boldsymbol{b}(s)) \geq \tau\})$ is replaced by the nonseparable belief representation $\rho(\{s : u(\boldsymbol{a}(s)) \geq \tau\}, \{s : u(\boldsymbol{b}(s)) \geq \tau\})$, i.e.,

(III) $\Psi(\boldsymbol{a},\boldsymbol{b}) = \int_{-\infty}^{+\infty} \rho(\{s : u(\boldsymbol{a}(s)) \geq \tau\}, \{s : u(\boldsymbol{b}(s)) \geq \tau\}) d\tau,$

where ρ is a bivariate function on $\mathcal{B}_S \times \mathcal{B}_S$ that satisfies:

(a) normalization: $\rho(S, \emptyset) = 1,$
(b) monotonicity: for all $A, B \in \mathcal{B}_S, A \supseteq B \Longrightarrow \rho(A,B) \geq 0,$
(c) skew-symmetry: for all $A, B \in \mathcal{B}_S, \rho(A,B) = -\rho(B,A),$
(d) conditional additivity: for all $A, B \in \mathcal{B}_S, A \cap B = \emptyset \Longrightarrow$

$$\rho(A \cup B, C) + \rho(\emptyset, C) = \rho(A,C) + \rho(B,C).$$

Note that when S is finite, conditional additivity implies that, for all $A, B \in 2^S$,

$$\rho(A,B) = \sum_{s \in A} \sum_{t \in B} \rho(\{s\},\{t\}) + (1 - |B|) \sum_{s \in A} \rho(\{s\},\emptyset) + (1 - |A|) \sum_{t \in B} \rho(\emptyset,\{t\}), \quad (1)$$

where $|A|$ denotes the number of elements of a set A (see Fishburn 1983b).

Let $\succ^*$ be a binary comparative belief judgement on $\mathcal{B}_S$, read as "is more probable than", which is defined by the preference relation $\succ$ for pure acts as follows: for all $A, B \in \mathcal{B}_S$, $A \succ^* B \iff \boldsymbol{a} \succ \boldsymbol{b}$ whenever

$$\boldsymbol{a}(s) = \begin{cases} a & \text{if } s \in A \\ b & \text{otherwise} \end{cases} \quad \text{and} \quad \boldsymbol{b}(s) = \begin{cases} a & \text{if } s \in B \\ b & \text{otherwise} \end{cases}$$

for some $a, b \in X$ with $a \succ b$. Then models (I) and (II) yield that, for all $A, B \in \mathcal{B}_S$, $A \succ^* B \iff \pi(A) > \pi(B)$. On the other hand, model (III) gives that, for all $A, B \in \mathcal{B}_S$,

$$A \succ^* B \iff \rho(A,B) > 0.$$

Thus ρ is a nonseparable representation of belief judgments for likelihoods of events. When ρ is separable, i.e., for all $A, B \in \mathcal{B}_S$, $\rho(A,B) = \pi(A) - \pi(B)$ for some probability measure π on $\mathcal{B}_S$, model (III) is reduced to SEU model.

There are many axiomatizations of SEU models (see a survey Fishburn, 1981). In Savage's (pure-act) formulation, preferences for pure acts in $\mathcal{A}$ are axiomatized to arrive at desired representations. On the other hand, in lottery-act and act-lottery formulations, we respectively enlarge X and $\mathcal{A}$ by randomization. A gamble on a set Y is a nonnegative real valued function f on Y for which $\{f(y) : y \in Y \text{ and } f(y) > 0\}$ is finite and $\sum_y f(y) = 1$. Each $f(y)$ is interpreted as (objectively known) probability number with which y obtains. Let $\boldsymbol{G}(Y)$ denote the set of all gambles on Y. In lottery-act formulation, we consider preferences for lottery-acts that map S into $\boldsymbol{G}(X)$. Thus the set of pure-acts is enlarged by changing 'internal' structure of pure-acts, that is, change from the set of pure-outcomes to the set of randomized outcomes. In act-lottery formulation, we consider preferences for gambles in $\boldsymbol{G}(\mathcal{A})$, i.e., the set of all randomized pure-acts. This means no alteration of internal structure of pure-acts. This point seems to be a conceptual advantage of act-lottery formulation.

Model (I) was axiomatized in various contexts (see Fishburn 1984a, 1988a; Fishburn & La Valle 1987a and b; Nakamura, 1998). Model (II) is new in the literature. When $|X| = 2$, Fishburn (1983a and b) provided two axiomatizations for the existence of ρ in model (III) under act-lottery formulation and pure-act formulation. Nakamura (1997) axiomatized a slightly more general representation of model (III) under pure-act formulation.

3 SSB Preferences

We adopt act-lottery formulation to develop necessary and sufficient conditions for the nonseparable representations (I), (II), and (III) when S is finite. Let $S = \{s_1,\ldots,s_n\}$. Then the set $\mathcal{A}$ of all acts can be identified with the n-Cartesian product of X, i.e., $X^n = X \times \cdots \times X$ (n times). Elements of X^n will be denoted by bold faced small letters, $\boldsymbol{x},\boldsymbol{y},\boldsymbol{z}$, and so forth. The i-component of $\boldsymbol{x}$ is written by x^i, so $\boldsymbol{x} = (x^1,\ldots,x^n)$, where each x^i is the consequence of act $\boldsymbol{x}$ when state s_i is true. Gambles in $\boldsymbol{G}(X^n)$ will be denoted by bold faced capital letters, $\boldsymbol{P},\boldsymbol{Q},\boldsymbol{R}$ and so forth. For $\boldsymbol{P} \in \boldsymbol{G}(X^n)$, let P^i denote the marginal probability distribution on i-component, i.e., for all $a \in X$,

$$P^i(a) = \sum \{\boldsymbol{P}(\boldsymbol{x}) : x^i = a \text{ and } \boldsymbol{x} \in X^n\}.$$

Each $\boldsymbol{x} \in X^n$ is identified with gamble $\boldsymbol{P}$ in $\boldsymbol{G}(X^n)$ for which $\boldsymbol{P}(\boldsymbol{x}) = 1$. For $\boldsymbol{P},\boldsymbol{Q} \in \boldsymbol{G}(X^n)$, the convex combination of $\boldsymbol{P}$ and $\boldsymbol{Q}$ with respect to probability number λ, denoted $\langle \boldsymbol{P},\lambda,\boldsymbol{Q}\rangle$, is a gamble that yields outcome $\boldsymbol{x}$ with probability $\lambda\boldsymbol{P}(\boldsymbol{x}) + (1-\lambda)\boldsymbol{Q}(\boldsymbol{x})$ for all $\boldsymbol{x} \in X^n$. The compound gamble of m gambles $\boldsymbol{P}_1,\ldots,\boldsymbol{P}_m$ with equal probabilities is denoted by $\langle \boldsymbol{P}_1,\ldots,\boldsymbol{P}_m\rangle$. In particular, $\langle \boldsymbol{P},\boldsymbol{Q}\rangle$ is tantamount to $\langle \boldsymbol{P},\frac{1}{2},\boldsymbol{Q}\rangle$.

We shall consider a binary preference relation $\succ$ on $\boldsymbol{G}(X^n)$. Two binary relations $\sim$ and $\succsim$ on $\boldsymbol{G}(X^n)$ are defined as usual, i.e., $\boldsymbol{P} \sim \boldsymbol{Q}$ if $\neg(\boldsymbol{P} \succ \boldsymbol{Q})$ and $\neg(\boldsymbol{Q} \succ \boldsymbol{P})$, and $\boldsymbol{P} \succsim \boldsymbol{Q}$ if $\neg(\boldsymbol{Q} \succ \boldsymbol{P})$. We say that a skew-symmetric function Φ on $X^n \times X^n$ *bilinearly represents* $\succ$ if, for all $\boldsymbol{P},\boldsymbol{Q} \in \boldsymbol{G}(X^n)$,

$$\boldsymbol{P} \succ \boldsymbol{Q} \iff \sum_{\boldsymbol{x}}\sum_{\boldsymbol{y}} \boldsymbol{P}(\boldsymbol{x})\boldsymbol{Q}(\boldsymbol{y})\Phi(\boldsymbol{x},\boldsymbol{y}) > 0,$$

where skew-symmetry means that $\Phi(\boldsymbol{x},\boldsymbol{y}) = -\Phi(\boldsymbol{y},\boldsymbol{x})$ for all $\boldsymbol{x},\boldsymbol{y} \in X^n$. We extend the domain of Φ to $\boldsymbol{G}(X^n) \times \boldsymbol{G}(X^n)$ by

$$\Phi(\boldsymbol{P},\boldsymbol{Q}) = \sum_x \sum_y \boldsymbol{P}(\boldsymbol{x})\boldsymbol{Q}(\boldsymbol{y})\Phi(\boldsymbol{x},\boldsymbol{y})$$

for all $\boldsymbol{P},\boldsymbol{Q} \in \boldsymbol{G}(X^n)$, so that Φ on $\boldsymbol{G}(X^n) \times \boldsymbol{G}(X^n)$ is skew-symmetric (i.e., $\Phi(\boldsymbol{P},\boldsymbol{Q}) = -\Phi(\boldsymbol{Q},\boldsymbol{P})$ for all $\boldsymbol{P},\boldsymbol{Q} \in G(X^n)$) and bilinear (i.e.,

$$\Phi(\langle \boldsymbol{P},\lambda,\boldsymbol{Q}\rangle,\boldsymbol{R}) = \lambda\Phi(\boldsymbol{P},\boldsymbol{R}) + (1-\lambda)\Phi(\boldsymbol{P},\boldsymbol{R}),$$
$$\Phi(\boldsymbol{R},\langle \boldsymbol{P},\lambda,\boldsymbol{Q}\rangle) = \lambda\Phi(\boldsymbol{R},\boldsymbol{P}) + (1-\lambda)\Phi(\boldsymbol{R},\boldsymbol{Q})$$

for all $\boldsymbol{P},\boldsymbol{Q},\boldsymbol{R} \in \boldsymbol{G}(X^n)$ and all $0 < \lambda < 1$). Φ on $\boldsymbol{G}(X^n) \times \boldsymbol{G}(X^n)$ thus defined is known as an SSB utility.

We say that $\succ$ on $\boldsymbol{G}(X^n)$ is an *SSB preference relation* if the following three axioms hold, which are understood as applying to all $\boldsymbol{P},\boldsymbol{Q},\boldsymbol{R} \in \boldsymbol{G}(X^n)$ and all $0 < \lambda < 1$,

Axiom A1 (Continuity). If $\boldsymbol{P} \succ \boldsymbol{Q}$ and $\boldsymbol{Q} \succ \boldsymbol{R}$, then $\boldsymbol{Q} \sim \langle \boldsymbol{P},\alpha,\boldsymbol{R}\rangle$ for some $0 < \alpha < 1$.

Axiom A2 (Convexity). If $\boldsymbol{P} \succ \boldsymbol{R}$ and $\boldsymbol{Q} \succeq \boldsymbol{R}$, then $\langle \boldsymbol{P},\lambda,\boldsymbol{Q}\rangle \succ \boldsymbol{R}$; if $\boldsymbol{R} \succ \boldsymbol{P}$ and $\boldsymbol{R} \succeq \boldsymbol{Q}$, then $\boldsymbol{R} \succ \langle \boldsymbol{P},\lambda,\boldsymbol{Q}\rangle$; if $\boldsymbol{P} \sim \boldsymbol{R}$ and $\boldsymbol{Q} \sim \boldsymbol{R}$, then $\langle \boldsymbol{P},\lambda,\boldsymbol{Q}\rangle \sim \boldsymbol{R}$.

Axiom A3 (Symmetry). If $\boldsymbol{P} \succ \boldsymbol{Q}$, $\boldsymbol{Q} \succ \boldsymbol{R}$, and $\boldsymbol{Q} \sim \langle \boldsymbol{P},\boldsymbol{R}\rangle$, then

$$\langle \boldsymbol{P},\boldsymbol{Q}\rangle \sim \langle \boldsymbol{P},\lambda,\boldsymbol{R}\rangle \iff \langle \boldsymbol{Q},\boldsymbol{R}\rangle \sim \langle \boldsymbol{R},\lambda,\boldsymbol{P}\rangle.$$

The representational implication of axioms A1–A3 is given by the following proposition (see Fishburn, 1982):

Proposition 1. *$\succ$ on $\boldsymbol{G}(X^n)$ is an SSB preference relation if and only if there is a skew-symmetric function Φ on $X^n \times X^n$ which bilinearly represents $\succ$. Furthermore, Φ is unique up to a multiplicative transformation by positive constants.*

Further generalizations of the proposition are found in (Fishburn & Nakamura 1991; Nakamura 1990, 2001).

We shall write $\boldsymbol{P} \approx \boldsymbol{Q}$ when $P^i = Q^i$ for $i = 1,\dots,n$. In what follows, we shall require the SSB preferences to satisfy that all gambles in $\boldsymbol{G}(X^n)$ that yield identical marginal probability distribution on each component are mutually indifferent. This condition, which is necessary for models (I)–(III) under act-lottery formulation, is stated in the following axiom, understood as applying to all $\boldsymbol{P},\boldsymbol{Q} \in \boldsymbol{G}(X^n)$.

Axiom A4 (Marginality–Equivalence). If $\boldsymbol{P} \approx \boldsymbol{Q}$, then $\boldsymbol{P} \sim \boldsymbol{Q}$.

In multiattributed decision problem, however, notice that this axiom is generally considered to be a restrictive assumption, where the i-th component x_i of act $\boldsymbol{x}$ is regarded as attribute i's level of decision alternative $\boldsymbol{x}$.

Although Φ in Proposition 1 is a multivariate function on X^{2n}, marginality-equivalence further decomposes Φ into sum of several bivariate functions on X^2 and univariate functions on X, dubbed here a *conditional additive decomposition*. To represent the decomposition, we need the following notations. Let $N = \{1,\dots,n\}$.

A partition of N is a collection of mutually disjoint subsets of N whose union equals N. When $\{A_1,\ldots,A_m\}$ is a partition of N, we shall allow some of $A_1,\ldots,A_m$ to be empty. When $a^1,\ldots,a^m \in X$ and $\{A_1,\ldots,A_m\}$ is a partition of N, let $a^1_{A_1}\cdots a^m_{A_m}$ denote act $\boldsymbol{x} \in X^n$ for which, for $i=1,\ldots,n$ and $k=1,\ldots,m$, $x^i = a^k$ if $i \in A_k$. Given $\boldsymbol{x} \in X^n$, $a \in X$, and $i \in N$, we shall let $\boldsymbol{x}_{\{i\}}a_{(i)}$ denote a vector $\boldsymbol{y} \in X^n$ for which, for $k=1,\ldots,n$,

$$y^k = \begin{cases} x^i & \text{if } k=i, \\ a & \text{otherwise,} \end{cases}$$

and let a_N denote a constant act $\boldsymbol{x} \in X^n$ for which $x^k = a$ for $k=1,\ldots,n$. Note that $\boldsymbol{x}_{\{i\}}a_{(i)}$ is the same as $x^i_{\{i\}}a_{\{i\}^c}$, where $\{i\}^c = N\setminus\{i\}$, the complement of N.

The conditionally additive decomposition of Φ in proposition 1 is given by the following proposition (see Fishburn, 1984b).

Proposition 2. *Let Φ be a skew-symmetric function on $X^n \times X^n$ which bilinearly represents $\succ$ on $\boldsymbol{G}(X^n)$. Axiom A4 holds if and only if, for all $\boldsymbol{x},\boldsymbol{y} \in X^n$ and all $a \in X$,*

$$\Phi(\boldsymbol{x},\boldsymbol{y}) = \sum_{i,j}\Phi(\boldsymbol{x}_{\{i\}}a_{(i)},\boldsymbol{y}_{\{j\}}a_{(j)}) - (n-1)\sum_i(\Phi(\boldsymbol{x}_{\{i\}}a_{(i)},a_N) - \Phi(\boldsymbol{y}_{\{i\}}a_{(i)},a_N)).$$

Fixing $a \in X$ in the proposition, we see that Φ is additively decomposed into bivariate functions $\Phi(\boldsymbol{x}_{\{i\}}a_{(i)},\boldsymbol{y}_{\{j\}}a_{(j)})$ on $X \times X$ and univariate functions $\Phi(\boldsymbol{x}_{\{i\}}a_{(i)},a_N)$ on X.

When $X=\{a,b\}$ and $a_N \succ b_N$, Proposition 2 gives that, for all $A,B \subseteq N$,

$$\begin{aligned}\Phi(a_Ab_{A^c},a_Bb_{B^c}) &= \sum_{i\in A}\sum_{j\in B}\Phi(a_{\{i\}}b_{(i)},a_{\{j\}}b_{(j)}) \\ &\quad + (1-|B|)\sum_{i\in A}\Phi(a_{\{i\}}b_{(i)},b_N) + (1-|A|)\sum_{j\in B}\Phi(b_N,a_{\{j\}}b_{(j)})\end{aligned} \tag{2}$$

This is equivalent to (1) by defining $\rho(A,B) = \Phi(a_Ab_{A^c},a_Bb_{B^c})$ for all $A,B \subseteq N$. Also, for all $A,B \subseteq N$,

$$A \succ^* B \iff \rho(A,B) > 0.$$

We see that ρ satisfies skew-symmetry (c) and conditional additivity (d). By the uniqueness of Φ, we can assume that $\rho(N,\emptyset)=1$, so normalization (a) is satisfied. If $a_{\{i\}}b_{(i)} \succsim b_N$ for $i=1,\ldots,n$, then ρ thus defined satisfies monotonicity (b).

Given a subset Y of X^n, we say that $\succ$ on $\boldsymbol{G}(Y)$ is *independent* if, for all $\boldsymbol{P},\boldsymbol{Q},\boldsymbol{R} \in \boldsymbol{G}(Y)$, $\langle\boldsymbol{P},\boldsymbol{R}\rangle \sim \langle\boldsymbol{Q},\boldsymbol{R}\rangle$ whenever $\boldsymbol{P} \sim \boldsymbol{Q}$. The representational implication of independent $\succ$ on $\boldsymbol{G}(Y)$ is given by the following proposition (see Fishburn, 1982 for the proof).

Proposition 3. *Let $Y \subseteq X^n$ and Φ be a skew-symmetric function on $X^n \times X^n$ which bilinearly represents $\succ$ on $\boldsymbol{G}(X^n)$. Then $\succ$ on $\boldsymbol{G}(Y)$ is independent if and only if, for all $\boldsymbol{x},\boldsymbol{y},\boldsymbol{z} \in Y$, $\Phi(\boldsymbol{x},\boldsymbol{y})+\Phi(\boldsymbol{y},\boldsymbol{z})+\Phi(\boldsymbol{z},\boldsymbol{x})=0$.*

Fixing $\boldsymbol{z} \in Y$ in the proposition, we obtain an additive decomposition of Φ on $Y \times Y$, i.e., for all $\boldsymbol{x},\boldsymbol{y} \in Y$, $\Phi(\boldsymbol{x},\boldsymbol{y}) = u(\boldsymbol{x}) - u(\boldsymbol{y})$, where $u(\cdot) = \Phi(\cdot,\boldsymbol{z})$.

4 Axioms for Nonseparable Utilities

Throughout the rest of the paper, let Φ be a skew-symmetric function on $X^n \times X^n$ which bilinearly represents marginality-equivalent $\succ$ on $\boldsymbol{G}(X^n)$. We shall fix $a^0 \in X$, so Proposition 2 yields that, for all $\boldsymbol{x},\boldsymbol{y} \in X^n$,

$$\Phi(\boldsymbol{x},\boldsymbol{y}) = \sum_{i,j} \Phi(\boldsymbol{x}_{\{i\}}a^0_{(i)},\boldsymbol{y}_{\{j\}}a^0_{(j)}) - (n-1)\sum_i \left(\Phi(\boldsymbol{x}_{\{i\}}a^0_{(i)},a^0_N) - \Phi(\boldsymbol{y}_{\{i\}}a^0_{(i)},a^0_N)\right). \tag{3}$$

Necessary and sufficient axioms for models (I) and (II) will be discussed in this section and those for model (III) will appear in the next section. A key axiom is domain-restricted independence which says that $\succ$ on $\boldsymbol{G}(Y)$ is independent for any subsets Y of X^n that are appropriately chosen for each model.

4.1 SSA Structures

We show necessary and sufficient axioms for model (I), which is stated in our framework as follows: there exist a skew-symmetric function ϕ on $X \times X$ and a probability vector $(\pi_1,\ldots,\pi_n) \in \mathbb{R}^n$ such that, for all $\boldsymbol{x},\boldsymbol{y} \in X^n$,

$$\Phi(\boldsymbol{x},\boldsymbol{y}) = \sum_i \pi_i \phi(x^i,y^i),$$

where $\pi_i \geq 0$ for $i = 1,\ldots,n$ and $\sum_i \pi_i = 1$.

The following axiom applies to all $a,b,c \in X$ and all distinct $i,j \in N$.

Axiom B1 (Domain-restricted Independence). *$\succ$ on $\boldsymbol{G}\left(\{a_{\{i\}}c_{(i)},b_{\{j\}}c_{(j)},c_N\}\right)$ is independent.*

The representational implication of axiom B1 is given as follows.

Theorem 1. *Axiom B1 holds if and only if there exist n skew-symmetric functions $\phi_1,\ldots,\phi_n$ on $X \times X$ such that, for all $\boldsymbol{x},\boldsymbol{y} \in X^n$, $\Phi(\boldsymbol{x},\boldsymbol{y}) = \sum_i \phi_i(x^i,y^i)$. Furthermore, ϕ_i's are unique up to a multiplicative transformations by common positive constants.*

Proof. The necessity of B1 easily follows. We show its sufficiency. Suppose axiom B1 holds. Since $\succ$ on $\boldsymbol{G}\left(\left\{\boldsymbol{x}_{\{i\}}a^0_{(i)},\boldsymbol{y}_{\{j\}}a^0_{(j)},a^0_N\right\}\right)$ is independent for all $\boldsymbol{x},\boldsymbol{y} \in X^n$, it follows from Proposition 3 that

$$\Phi(\boldsymbol{x}_{\{i\}}a^0_{(i)},\boldsymbol{y}_{\{j\}}a^0_{(j)}) = \Phi(\boldsymbol{x}_{\{i\}}a^0_{(i)},a^0_N) + \Phi(a^0_N,\boldsymbol{y}_{\{j\}}a^0_{(j)}).$$

We then substitute this additive decomposition for (3) and get

$$\begin{aligned}\Phi(\boldsymbol{x},\boldsymbol{y}) &= \sum_i \Phi(\boldsymbol{x}_{\{i\}}a^0_{(i)},\boldsymbol{y}_{\{i\}}a^0_{(i)}) + \sum_{i\neq j}\left(\Phi(\boldsymbol{x}_{\{i\}}a^0_{(i)},a^0_N)+\Phi(a^0_N,\boldsymbol{y}_{\{j\}}a^0_{(j)})\right)\\ &\quad -(n-1)\sum_i\left(\Phi(\boldsymbol{x}_{\{i\}}a^0_{(i)},a^0_N)-\Phi(\boldsymbol{y}_{\{i\}}a^0_{(i)},a^0_N)\right)\\ &= \sum_i \Phi(\boldsymbol{x}_{\{i\}}a^0_{(i)},\boldsymbol{y}_{\{i\}}a^0_{(i)}).\end{aligned}$$

Thus defining n skew-symmetric functions ϕ_i on $X \times X$ by $\phi_i(a,b) = \Phi(a_{\{i\}}a^0_{(i)}, b_{\{i\}}\ a^0_{(i)})$ for $i = 1,\ldots,n$, we obtain the desired representation. □

Given gambles $P_1,\ldots,P_n$ in $\boldsymbol{G}(X)$, let $(P_1,\ldots,P_n)$ denote gamble $\boldsymbol{Q}$ in $\boldsymbol{G}(X^n)$ for which $\boldsymbol{Q}(x) = P_1(x^1)\times\cdots\times P_n(x^n)$ for all $\boldsymbol{x}\in X^n$. Thus $Q^i = P_i$ for $i = 1,\ldots,n$. Each $a \in X$ is identified with gamble P in $\boldsymbol{G}(X)$ for which $P(a) = 1$. When $P^i = P$ and $P^j = a$ for $j \neq i$, we shall write $(P_{\{i\}}, a_{(i)})$ in place of $(a,\ldots,a,P,a,\ldots,a)$ whenever P is located at i-th position. We say that $i \in N$ is *null* if $(P_{\{i\}}, a_{(i)}) \sim (Q_{\{i\}}, a_{(i)})$ for all $P,Q \in \boldsymbol{G}(X)$ and all $a \in X$.

The following axiom, which applies to all $P,Q \in G(X)$ and all $a \in X$, says that preferences for marginal probability distributions are independent of the state in which those distributions are obtained whenever outcomes in other states are identical.

Axiom B2 (Interstate Consistency). *If $i, j \in N$ are not null, then $(P_{\{i\}}, a_{(i)}) \succ (Q_{\{i\}}, a_{(i)})$ iff $(P_{\{j\}}, a_{(j)}) \succ (Q_{\{j\}}, a_{(j)})$.*

Since the underlying outcome space X is the same under all states, this axiom seems to be plausible unless ex post evaluation of outcomes is state-dependent.

Model (I) is completely characterized by axioms B1 and B2 as follows.

Theorem 2. *Axiom B1 and B2 hold if and only if model (I) holds.*

Proof. Necessity of axioms B1 and B2 is trivial. We show their sufficiency. Suppose axioms B1 and B2 hold. Then by Theorem 1, $\Phi(\boldsymbol{x},\boldsymbol{y}) = \sum_i \phi_i(x^i,y^i)$. Assume that i, $j \in N$ are not null. Then

$$\Phi((P_{\{i\}},a_{(i)}),(Q_{\{i\}},a_{(i)})) = \sum_b\sum_c P(b)Q(c)\phi_i(b,c) > 0$$

if and only if

$$\Phi((P_{\{j\}},a_{(j)}),(Q_{\{j\}},a_{(j)})) = \sum_b\sum_c P(b)Q(c)\phi_j(b,c) > 0.$$

Thus by the uniqueness of SSB utilities on $\boldsymbol{G}(X)\times\boldsymbol{G}(X)$, $\phi_i = \alpha_{ij}\phi_j$ for a positive constant α_{ij}. This completes the proof. □

4.2 SSB Structures

We show necessary and sufficient axioms for model (II). The required decomposition of Φ in our framework is given as follows: there exist a skew-symmetric function ϕ on $X \times X$ and a probability vector $\pi \in \mathbb{R}^n$ such that, for all $\boldsymbol{x},\boldsymbol{y} \in X^n$,

$$\Phi(\boldsymbol{x},\boldsymbol{y}) = \sum_{i=1}^{n}\sum_{j=1}^{n} \pi_i \pi_j \phi(x^i, y^j).$$

The following axioms apply to all $a,b,c,d \in X$ and all $i,j \in N$.

Axiom C1 (Betweenness of Event-mixture). *If* $a_N \succ b_N$, *then* $a_N \succeq a_{\{i\}}b_{(i)} \succeq b_N$.

Axiom C2 (Consistent Comparative Beliefs). *If* $a_N \succ b_N$, $c_N \succ d_N$, *and* $a_{\{i\}}b_{(i)} \sim \langle a_N, \lambda, b_N \rangle$ *for some* $0 < \lambda < 1$, *then* $c_{\{i\}}d_{(i)} \sim \langle c_N, \lambda, d_N \rangle$.

Axiom C3 (Strong Domain-restricted Independence). *If* $a_{\{i\}}b_{(i)} \sim \langle a_N, \alpha, b_N \rangle$ *and* $c_{\{j\}}d_{(j)} \sim \langle c_N, \beta, d_N \rangle$ *for some* $\alpha, \beta \in (0,1)$, *then* $\langle a_{\{i\}}b_{(i)}, \langle c_N, \beta, d_N \rangle \rangle \sim \langle c_{\{j\}}d_{(j)}, \langle a_N, \alpha, b_N \rangle \rangle$.

Betweenness of event-mixture seems plausible, although it may be violated in some situations. Axiom C2 is crucial to derive subjective probabilities through preference judgments. However, it is argued that indifference judgments in the axiom might depend on selected pairs of outcomes.

Although axiom C3 does not look like domain-restricted independence condition, it does imply the independence on some restricted domains. To see this, we need the following lemma.

Lemma 1. *Suppose* $a_{\{i\}}b_{(i)} \sim \langle a_N, \alpha, b_N \rangle$ *and* $a_{\{j\}}b_{(j)} \sim \langle a_N, \beta, b_N \rangle$ *for some* $\alpha, \beta \in (0,1)$. *Then*

(1) $\Phi(a_{\{i\}}b_{(i)}, a_N) = (1-\alpha)\Phi(b_N, a_N)$.
(2) $\Phi(a_{\{i\}}b_{(i)}, b_N) = \alpha\Phi(a_N, b_N)$.
(3) $\Phi(a_{\{i\}}b_{(i)}, a_{\{j\}}b_{(j)}) = (\alpha - \beta)\Phi(a_N, b_N)$.

Proof. (1) Since $a_{\{i\}}a_{(i)} \sim \langle a_N, \lambda, a_N \rangle$, Axiom C3 implies $\langle a_{\{i\}}b_{(i)}, \langle a_N, \lambda, a_N \rangle \rangle \sim \langle a_{\{j\}}a_{(j)}, \langle a_N, \alpha, b_N \rangle \rangle$, so that

$$\begin{aligned}\Phi(a_{\{i\}}b_{(i)}, a_{\{j\}}a_{(j)}) &= \Phi(\langle a_N, \alpha, b_N \rangle, \langle a_N, \lambda, a_N \rangle) \\ &= (1-\alpha)\Phi(b_N, a_N).\end{aligned}$$

(2) and (3) similarly obtain. Q.E.D.

By Lemma 1(2) and Lemma 1(3), we have

$$\Phi\left(a_{\{i\}}b_{(i)}, a_{\{j\}}b_{(j)}\right) = \Phi\left(a_{\{i\}}b_{(i)}, b_N\right) + \Phi(b_N, a_{\{j\}}b_{(j)}).$$

Substituting this for (2), we obtain that, for all $A, B \subseteq N$,

$$\begin{aligned}
\Phi(a_A b_{A^c}, a_B b_{B^c}) &= \sum_{i\in A}\sum_{j\in B}\left(\Phi(a_{\{i\}}b_{(i)}, b_N) + \Phi(b_N, a_{\{j\}}b_{(j)})\right) \\
&\quad + (1-|B|)\sum_{i\in A}\Phi(a_{\{i\}}b_{(i)}, b_N) + (1-|A|)\sum_{j\in B}\Phi(b_N, a_{\{j\}}b_{(j)}) \\
&= \sum_{i\in A}\Phi(a_{\{i\}}b_{(i)}, b_N) + \sum_{j\in B}\Phi(b_N, a_{\{j\}}b_{(j)}) \\
&= \Phi(a_A b_{A^c}, b_N) + \Phi(b_N, a_B b_{B^c})
\end{aligned}$$

which implies that $\succ$ on $\boldsymbol{G}(\{a_A b_{A^c}, a_B b_{B^c}, b_N\})$ is independent.

The following theorem says that axioms C1–C3 completely characterize model (II).

Theorem 3. *Axioms C1–C3 hold if and only if model (II) holds.*

Proof. The necessity of axioms C1–C3 easily obtains. Thus we show their sufficiency. Let $a_N \succ b_N$. Then by axiom C1, $a_N \succsim a_{\{i\}}b_{(i)} \succsim b_N$. By Axioms A1 and A2, $a_{\{i\}}b_{(i)} \sim \langle a_N, \pi_i, b_N\rangle$ for a unique $0 \le \pi_i \le 1$. By axiom C2, $c_{\{i\}}d_{(i)} \sim \langle c_N, \pi_i, d_N\rangle$ whenever $c_N \succ d_N$. Since $\langle a_{\{1\}}b_{(1)}, \ldots, a_{\{n\}}b_{(n)}\rangle \approx \langle a_N, \frac{1}{n}, b_N\rangle$, axiom A4 implies $\langle a_{\{1\}}b_{(1)}, \ldots, a_{\{n\}}b_{(n)}\rangle \sim \langle a_N, \frac{1}{n}, b_N\rangle$. Thus,

$$\begin{aligned}
0 &= n\Phi\left(\langle a_{\{1\}}b_{(1)}, \ldots, a_{\{n\}}b_{(n)}\rangle, \langle a_N, \tfrac{1}{n}, b_N\rangle\right) \\
&= \sum_i \Phi\left(a_{\{i\}}b_{(i)}, \langle a_N, \tfrac{1}{n}, b_N\rangle\right) \\
&= \sum_i \left(\tfrac{1}{n}\Phi(a_{\{i\}}b_{(i)}, a_N) + \left(1-\tfrac{1}{n}\right)\Phi(a_{\{i\}}b_{(i)}, b_N)\right) \\
&= \sum_i \left(\tfrac{1}{n}(1-\pi_i)\Phi(b_N, a_N) + \left(1-\tfrac{1}{n}\right)\pi_i\Phi(a_N, b_N)\right) \quad \text{(by Lemma 1(1) and (2))} \\
&= \Phi(b_N, a_N)\sum_i\left(\pi_i - \frac{1}{n}\right)
\end{aligned}$$

so that $\sum_i \pi_i = 1$. Hence $\pi = (\pi_1, \ldots, \pi_n)$ is a probability vector.

Take any $\boldsymbol{x}, \boldsymbol{y} \in X^n$. Assume that $x_N^i \succsim a^0$ and $y_N^i \succsim a^0$ for $i = 1, \ldots, n$. Then by axiom C2, $\boldsymbol{x}_{\{i\}}a^0_{(i)} \sim \langle x^i, \pi_i, a^0\rangle$ and $\boldsymbol{y}_{\{i\}}a^0_{(i)} \sim \langle y^i, \pi_i, a^0\rangle$ for $i = 1, \ldots, n$. Thus by axiom C3, $\left\langle \boldsymbol{x}_{\{i\}}a^0_{(i)}, \left\langle y_N^j, \pi_j, a^0\right\rangle\right\rangle \sim \left\langle \boldsymbol{y}_{\{j\}}a^0_{(j)}, \langle x_N^i, \pi_i, a^0\rangle\right\rangle$ for all $i, j \in N$, which gives

$$\Phi(\boldsymbol{x}_{\{i\}}a^0_{(i)}, \boldsymbol{y}_{\{j\}}a^0_{(j)}) = \pi_i\pi_j\Phi(x_N^i, y_N^j) + \pi_i(1-\pi_j)\Phi(x_N^i, a_N^0) + \pi_j(1-\pi_i)\Phi(a_N^0, y_N^j).$$

Substituting this for (3), we obtain

$$\begin{aligned}\Phi(\boldsymbol{x},\boldsymbol{y}) &= \sum_{i,j}\Big(\pi_i\pi_j\Phi(x^i_N,y^j_N)+\pi_i(1-\pi_j)\Phi(x^i_N,a^0_N)+\pi_j(1-\pi_i)\Phi(a^0_N,y^j_N)\Big)\\ &\quad -(n-1)\sum_i\left(\pi_i\Phi(x^i_N,a^0_N)-\pi_i\Phi(y^i_N,a^0_N)\right)\\ &= \sum_{i,j}\pi_i\pi_j\Phi(x^i_N,y^j_N),\end{aligned}$$

which does not depend on choice of a^0. Letting $\phi(a,b)=\Phi(a_N,b_N)$ for all $a,b\in X$, we obtain the desired result. □

5 SEU with Nonseparable Beliefs

We show necessary and sufficient axioms for model (III), which is stated in the present framework as follows. Given any $\boldsymbol{x},\boldsymbol{y}\in X^n$, let $x^i,y^i\in\{a^1,\ldots,a^m\}$ for $i=1,\ldots,n$ and $a^1\succsim\cdots\succsim a^m$. For $k=1,\ldots,m$, let $A_k=\{i\in N:x^i=a^k\}$ and $B_k=\{i\in N:y^i=a^k\}$, so that $\{A_1,\ldots,A_m\}$ and $\{B_1,\ldots,B_m\}$ are partitions of N. Note that $\boldsymbol{x}$ and $\boldsymbol{y}$ are respectively represented by $a^1_{A_1}\cdots a^m_{A_m}$ and $a^1_{B_1}\cdots a^m_{B_m}$. Then model (III) yields a real valued function u on X and a bivariate set function ρ on $2^N\times 2^N$, satisfying (a)–(d), such that

$$\Phi(\boldsymbol{x},\boldsymbol{y})=\Phi\left(a^1_{A_1}\cdots a^m_{A_m},a^1_{B_1}\cdots a^m_{B_m}\right)=\sum_{i=1}^{m-1}\rho\left(\bigcup_{j=1}^{i}A_j,\bigcup_{j=1}^{i}B_j\right)\left(u(a^{i+1})-u(a^i)\right).$$

The following axiom applies to all $a,b,c\in X$ with $a_N\succsim b_N$ and $b_N\succsim c_N$ and all $A,B\subseteq N$.

Axiom D1 (Domain-restricted Independence). *$\succ$ on $\boldsymbol{G}(\{a_Ab_{A^c},b_N,b_Bc_{B^c}\})$ is independent.*

When $A=N$ and $B=\emptyset$, axiom D1 means that $\succ$ on $\boldsymbol{G}(\{a_N,b_N,c_N\})$ is independent, so that $\succ$ is a weak order (i.e., $\succ$ and $\sim$ are transitive) if $\succ$ is restricted to the set of all constant acts.

The important implication of axiom D1 is the following decomposition of Φ, whose proof will be deferred to the appendix.

Theorem 4. *Axiom D1 holds if and only if, for all positive integers m, all partitions $\{A_1,\ldots,A_m\}$ and $\{B_1,\ldots,B_m\}$, and all $a^1,\ldots,a^m\in X$,*

$$\begin{aligned}&\Phi(a^1_{A_1}\cdots a^m_{A_m},a^1_{B_1}\cdots a^m_{B_m})\\ &=\Phi(a^1_{A_1}a^2_{A_2\cup\cdots\cup A_m},a^1_{B_1}a^2_{B_2\cup\cdots\cup B_m})+\Phi(a^2_{A_1\cup A_2}a^3_{A_3}\cdots a^m_{A_m},a^2_{B_1\cup B_2}a^3_{B_3}\cdots a^m_{B_m})\end{aligned}$$

whenever $a^1\succsim\cdots\succsim a^m$.

By applying the decomposition of the above theorem consecutively, it immediately follows that

$$\begin{aligned}&\Phi(a^1_{A_1}\cdots a^m_{A_m}, a^1_{B_1}\cdots a^m_{B_m})\\&=\Phi(a^1_{A_1}a^2_{A_2\cup\cdots\cup A_m}, a^1_{B_1}a^2_{B_2\cup\cdots\cup B_m})+\Phi(a^2_{A_1\cup A_2}a^3_{A_3\cup\cdots\cup A_m}, a^2_{B_1\cup B_2}a^3_{B_3\cup\cdots\cup B_m})\\&+\cdots+\Phi(a^{m-1}_{A_1\cup\cdots\cup A_{m-1}}a^m_{A_m}, a^{m-1}_{B_1\cup\cdots\cup B_{m-1}}a^m_{B_m}).\end{aligned}\tag{4}$$

We need to further decompose each term in (4) as $\Phi(a_Ab_{A^c}, a_Bb_{B^c}) = \mu(a,b)\rho(A,B)$ for two bivariate functions, $\mu \geq 0$ on $X\times X$ and ρ on $2^N\times 2^N$, satisfying properties specified in the Theorem 5 below. This decomposition requires that preferences for gambles in $\boldsymbol{G}(\{a_Ab_{A^c} : A\subseteq N\})$ depend on subsets A of N but not on choice of $a,b\in X$ with $a_N\succ b_N$. This requirement is stated in the following axiom, understood as applying to all $a,b,c,d\in X$.

Axiom D2 (Consistent Comparative Probability). *If $a_N\succ b_N, c_N\succ d_N$, $\boldsymbol{P},\boldsymbol{Q}\in \boldsymbol{G}(\{a_Ab_{A^c} : A\subseteq N\})$, and $\boldsymbol{P}',\boldsymbol{Q}'\in \boldsymbol{G}(\{c_Ad_{A^c} : A\subseteq N\})$, then $\boldsymbol{P}\succ\boldsymbol{Q} \iff \boldsymbol{P}'\succ\boldsymbol{Q}'$ whenever $\boldsymbol{P}(a_Ab_{A^c}) = \boldsymbol{P}'(c_Ad_{A^c})$ and $\boldsymbol{Q}(a_Ab_{A^c}) = \boldsymbol{Q}'(c_Ad_{A^c})$ for all $A\subseteq N$.*

It may be argued that as in axiom C2, preference judgments in axiom D2 depend on selected pairs of outcomes in some situations. If this is the case, then likelihood judgments about events cannot be derived from preference judgments for randomized acts.

The implication of axioms D1 and D2 is given as follows.

Theorem 5. *Axioms D1 and D2 hold if and only if there exist a nonnegative bivariate function μ on $\{(a,b)\in X\times X : a_N\succeq b_N\}$ and a skew-symmetric, conditionally additive function ρ on $2^N\times 2^N$ such that, for all $a,b,c\in X$ with $a_N\succeq b_N\succeq c_N$, $\mu(a,c) = \mu(a,b)+\mu(b,c)$, and, for all positive integers m, all partitions $\{A_1,\ldots,A_m\}$ and $\{B_1,\ldots,B_m\}$, and all $a^1,\ldots,a^m\in X$,*

$$\Phi(a^1_{A_1}\cdots a^m_{A_m}, a^1_{B_1}\cdots a^m_{B_m}) = \sum_{k=1}^{m-1}\mu(a^k,a^{k+1})\rho\left(\bigcup_{i=1}^k A_i, \bigcup_{i=1}^k B_i\right).$$

Proof. Necessity of axioms D1 and D2 easily obtains. Thus we assume that axioms D1 and D2 hold. Assuming that the hypotheses of axiom D2 hold, we obtain that

$$\begin{aligned}&\sum_{A\subseteq N}\sum_{B\subseteq N}\boldsymbol{P}(a_Ab_{A^c})\boldsymbol{Q}(a_Bb_{B^c})\Phi(a_Ab_{A^c}, a_Bb_{B^c}) > 0\\&\qquad\iff \sum_{A\subseteq N}\sum_{B\subseteq N}\boldsymbol{P}(a_Ab_{A^c})\boldsymbol{Q}(a_Bb_{B^c})\Phi(c_Ad_{A^c}, c_Bd_{B^c}) > 0.\end{aligned}$$

By the uniqueness of SSB utility, we obtain that, for all $A,B\subseteq N$, there is a $\lambda>0$ such that $\Phi(a_Ab_{A^c}, a_Bb_{B^c}) = \lambda\Phi(c_Ad_{A^c}, c_Bd_{B^c})$.

Fix $a^0,b^0\in X$ with $a^0_N\succ b^0_N$. Define $\rho(A,B) = \Phi(a^0_Ab^0_{A^c}, a^0_Bb^0_{B^c})$ for all $A,B\subseteq N$. Then for all $a,b\in X$ with $a_N\succ b_N$, $\Phi(a_Ab_{A^c}, a_Bb_{B^c}) = \mu(a,b)\rho(A,B)$ for some $\mu(a,b)>0$. When $a_N\sim b_N$, let $\mu(a,b)=0$.

Skew-symmetry of ρ follows from skew-symmetry of Φ. Conditional additivity of ρ follows from (2). It follows from the decomposition of Theorem 4 that, for all $a,b,c \in X$ with $a_N \succsim b_N \succsim c_N$, $\mu(a,c) = \mu(a,b) + \mu(b,c)$. This completes the proof. □

Since model (III) requires that $\mu(a,b) = u(a) - u(b)$ for a real valued function u on X, we must have $\mu(a,b) = \mu(c,d)$ whenever $a_N \sim c_N$ and $b_N \sim d_N$. This is ensured by the following axiom, which applies to all $a,b,c,d \in X$ and all $A,B \subseteq N$.

Axiom D3 (Consistent Outcome Utility). *If $a_N \succ b_N$, $a_N \sim c_N$, and $b_N \sim d_N$, then $\langle a_A b_{A^c}, \lambda, c_B d_{B^c}\rangle \sim \langle c_A d_{A^c}, \lambda, a_B b_{B^c}\rangle$ for all $0 \leq \lambda \leq 1$.*

The last axiom, which applies to all $a,b \in X$ and all $A,B \subseteq N$, simply says that ρ on $2^N \times 2^N$ satisfies monotonicity (b).

Axiom D4 (Monotonicity). *If $a_N \succ b_N$ and $A \supseteq B$, then $a_A b_{A^c} \succsim a_B b_{B^c}$.*

The implication of axioms D3 and D4 is stated in the following theorem.

Theorem 6. *Axioms D1–D4 hold if and only if model (III) holds.*

Proof. Necessity of axioms D1–D4 easily obtains. Thus we assume that axioms D1–D4 hold. Let μ on $\{(a,b) \in X \times X : a_N \succsim b_N\}$ and ρ on $2^N \times 2^N$ be obtained in Theorem 5. If $a_N \succ b_N$ and $A \supseteq B$, then axiom D4 implies $\Phi(a_A b_{A^c}, a_B b_{B^c}) \geq 0$. By Theorem 5, $\Phi(a_A b_{A^c}, a_B b_{B^c}) = \mu(a,b)\rho(A,B)$, so that $\rho(A,B) \geq 0$. Hence ρ is monotonic. Since $a_N \succ b_N$, $\Phi(a_S b_\emptyset, a_\emptyset b_S) > 0$, so $\rho(S,\emptyset) > 0$. By the uniqueness of SSB utility, with no loss of generality, we can normalize $\rho(S,\emptyset) = 1$.

It remains to show that if $a_N \succ b_N$, $a_N \sim c_N$ and $b_N \sim d_N$, then $\mu(a,b) = \mu(c,d) > 0$. By axiom D3, $\Phi(a_A b_{A^c}, c_A d_{A^c}) = \Phi(a_B b_{B^c}, c_B d_{B^c}) = 0$ and, for all $0 < \lambda < 1$,

$$\begin{aligned}\Phi(\langle a_A b_{A^c}, \lambda, c_B d_{B^c}\rangle, \langle c_A d_{A^c}, \lambda, a_B b_{B^c}\rangle) = \lambda(1-\lambda)\,(&\Phi(a_A b_{A^c}, a_B b_{B^c}) \\ &+\Phi(c_B d_{B^c}, c_A d_{A^c})) = 0.\end{aligned}$$

Thus $\Phi(a_A b_{A^c}, a_B b_{B^c}) = \Phi(c_A d_{A^c}, c_B d_{B^c})$. By Theorem 5, $\mu(a,b)\rho(A,B) = \mu(c,d)\rho(A,B)$, so $\mu(a,b) = \mu(c,d)$. This completes the proof. □

6 Conclusions

We studied necessary and sufficient axiomatizations of three nonseparable representations in decision making under uncertainty when the state space is finite. The first two models (I) and (II) deal with nonseparability of outcome utilities but yield additive subjective probabilities. On the other hand, the last one (III) is concerned with nonseparability of subjective likelihood judgements but retains weakly ordered preferences for decision outcomes.

Our axiomatizations are based on act-lottery formulation, in which pure-acts are randomized. Thus internal structures of acts remain unchanged. Usual axiomatizations applying randomization in the literature adopt lottery-act formulation, which alters internal structures of acts by randomizing pure-outcomes. Although the former seems to have a conceptual advantage over the latter, marginality-equivalence condition imposed on SSB preferences is rather restrictive. At present, I have no idea how to escape from this restrictiveness. Thus it may be desirable to find axiomatizations in pure-act formulation, that is, without randomization, where only model (I) has such an axiomatization.

Another open problem is to explore the extensions to infinite S. There are a few axiomatizations for model (I), but no such an axiomatization for models (II) and (III) is discovered.

Appendix

This appendix proves Theorem 4. Since the necessity of axiom D1 easily follows, we show its sufficiency below. Assume that axiom D1 holds. We need the following decompositional implications of axiom D1.

Lemma 2. *Suppose that $a_N \succsim b_N \succsim c_N \succsim d_N$ and $k, \ell \in N$. Then*

(1) $\Phi(a_{\{k\}}c_{(k)}, b_{\{\ell\}}c_{(\ell)}) = \Phi(a_{\{k\}}b_{(k)}, b_N) + \Phi(b_{\{k\}}c_{(k)}, b_{\{\ell\}}c_{(\ell)})$.
(2) $\Phi(a_{\{k\}}c_{(k)}, c_N) = \Phi(a_{\{k\}}b_{(k)}, b_N) + \Phi(b_{\{k\}}c_{(k)}, c_N)$.
(3) $\Phi\left(a_{\{k\}}d_{(k)}, c_{\{\ell\}}d_{(\ell)}\right) = \Phi(a_{\{k\}}b_{(k)}, b_N) + \Phi\left(b_{\{k\}}d_{(k)}, c_{\{\ell\}}d_{(\ell)}\right)$.

Proof. First we show (1) and (2) which are combined into

$$\Phi(a_{\{k\}}c_{(k)}, b_I c_{(I)}) = \Phi(a_{\{k\}}b_{(k)}, b_N) + \Phi(b_{\{k\}}c_{(k)}, b_I c_{(I)}),$$

where I is either $\emptyset$ or $\{\ell\}$. Since $\langle a_{\{k\}}c_{(k)}, b_N, b_I c_{(I)}\rangle \approx \langle b_I c_{(I)}, a_{\{k\}}b_{(k)}, b_{\{k\}}c_{(k)}\rangle$, marginality-equivalence implies $\Phi\left(\langle a_{\{k\}}c_{(k)}, b_N, b_I c_{(I)}\rangle, \langle b_I c_{(I)}, a_{\{k\}}b_{(k)}, b_{\{k\}}c_{(k)}\rangle\right)$ $= 0$, which gives

$$\begin{aligned}\Phi(a_{\{k\}}c_{(k)}, b_I c_{(I)}) &+ \Phi(b_N, a_{\{k\}}b_{(k)}) + \Phi(b_I c_{(I)}, b_{\{k\}}c_{(k)}) \\ &= \Phi(a_{\{k\}}b_{(k)}, a_{\{k\}}c_{(k)}) + \Phi(b_{\{k\}}c_{(k)}, a_{\{k\}}c_{(k)}) + \Phi(b_I c_{(I)}, b_N) \\ &\quad + \Phi(b_{\{k\}}c_{(k)}, b_N) + \Phi(a_{\{k\}}b_{(k)}, b_I c_{(I)}).\end{aligned}$$

We show that the right-hand side vanishes. To show this, we need to have

$$\begin{aligned}\Phi(a_{\{k\}}b_{(k)}, a_{\{k\}}c_{(k)}) + \Phi(b_{\{k\}}c_{(k)}, a_{\{k\}}c_{(k)}) &= \Phi\left(b_N, a_{\{k\}}b_{(k)}\right) + \Phi(b_N, b_{\{k\}}c_{(k)}), \\ \Phi(a_{\{k\}}b_{(k)}, b_I c_{(I)}) &= \Phi(b_N, b_I c_{(I)}) + \Phi(a_{\{k\}}b_{(k)}, b_N).\end{aligned}$$

Since $\langle a_{\{k\}}b_{(k)}, b_{\{k\}}c_{(k)}\rangle \approx \langle a_{\{k\}}c_{(k)}, b_N\rangle$, marginality-equivalence implies the first. By axiom D1,

$$\Phi(a_{\{k\}}b_{(k)}, b_N) + \Phi(b_N, b_I c_{(I)}) + \Phi(b_I c_{(I)}, a_{\{k\}}b_{(k)}) = 0,$$

which is the second.

The claim (3) follows from (1) and (2) as follows:

$$\begin{aligned}\Phi\left(a_{\{k\}}d_{(k)},c_{\{\ell\}}d_{(\ell)}\right) &= \Phi(a_{\{k\}}c_{(k)},c_N)+\Phi\left(c_{\{k\}}d_{(k)},c_{\{\ell\}}d_{(\ell)}\right) \text{ (by (1))}\\ &= \Phi(a_{\{k\}}b_{(k)},b_N)+\Phi(b_{\{k\}}c_{(k)},c_N)+\Phi\left(c_{\{k\}}d_{(k)},c_{\{\ell\}}d_{(\ell)}\right) \text{ (by (2))}\\ &= \Phi(a_{\{k\}}b_{(k)},b_N)+\Phi\left(b_{\{k\}}d_{(k)},c_{\{\ell\}}d_{(\ell)}\right) \text{ (by (1))}\end{aligned}$$

This completes the proof. □

Sufficiency proof of Theorem 4. Assume that $a^1 \succeq \cdots \succeq a^m$ and $\{A_1,\dots,A_m\}$ and $\{B_1,\dots,B_m\}$ are partitions of N. By Proposition 2, we obtain

$$\begin{aligned}&\Phi(a^1_{A_1}\cdots a^m_{A_m},a^1_{B_1}\cdots a^m_{B_m})-\Phi(a^2_{A_1\cup A_2}a^3_{A_3}\cdots a^m_{A_m},a^2_{B_1\cup B_2}a^3_{B_3}\cdots a^m_{B_m})\\ &=\sum_{i=1}^m\sum_{j=1}^m\sum_{k\in A_i}\sum_{\ell\in B_j}\Phi\left(a^i_{\{k\}}a^m_{(k)},a^j_{\{\ell\}}a^m_{(\ell)}\right)-(n-1)\sum_{i=1}^m\sum_{k\in A_i}\Phi\left(a^i_{\{k\}}a^m_{(k)},a^m_N\right)\\ &\quad+(n-1)\sum_{j=1}^m\sum_{\ell\in B_j}\Phi\left(a^j_{\{\ell\}}a^m_{(\ell)},a^m_N\right)-\sum_{i=2}^m\sum_{j=2}^m\sum_{k\in A^*_i}\sum_{\ell\in B^*_j}\Phi\left(a^i_{\{k\}}a^m_{(k)},a^j_{\{\ell\}}a^m_{(\ell)}\right)\\ &\quad+(n-1)\sum_{i=2}^m\sum_{k\in A^*_i}\Phi\left(a^i_{\{k\}}a^m_{(k)},a^m_N\right)-(n-1)\sum_{j=2}^m\sum_{\ell\in B^*_j}\Phi\left(a^j_{\{\ell\}}a^m_{(\ell)},a^m_N\right)\\ &=\sum_{k\in A_1}\sum_{\ell\in B_1}\Phi\left(a^1_{\{k\}}a^m_{(k)},a^1_{\{\ell\}}a^m_{(\ell)}\right)+\sum_{j=2}^m\sum_{k\in A_1}\sum_{\ell\in B_j}\left(\Phi\left(a^1_{\{k\}}a^m_{(k)},a^j_{\{\ell\}}a^m_{(\ell)}\right)\right.\\ &\qquad\left.-\Phi\left(a^2_{\{k\}}a^m_{(k)},a^j_{\{\ell\}}a^m_{(\ell)}\right)\right)\\ &\quad+\sum_{i=2}^m\sum_{k\in A_i}\sum_{\ell\in B_1}\left(\Phi\left(a^i_{\{k\}}a^m_{(k)},a^1_{\{\ell\}}a^m_{(\ell)}\right)-\Phi\left(a^i_{\{k\}}a^m_{(k)},a^2_{\{\ell\}}a^m_{(\ell)}\right)\right)\\ &\quad-\sum_{k\in A_1}\sum_{\ell\in B_1}\Phi\left(a^2_{\{k\}}a^m_{(k)},a^2_{\{\ell\}}a^m_{(\ell)}\right)\\ &\quad-(n-1)\sum_{k\in A_1}\left(\Phi\left(a^1_{\{k\}}a^m_{(k)},a^m_N\right)-\Phi\left(a^2_{\{k\}}a^m_{(k)},a^m_N\right)\right)\\ &\quad+(n-1)\sum_{\ell\in B_1}\left(\Phi\left(a^1_{\{\ell\}}a^m_{(\ell)},a^m_N\right)-\Phi\left(a^2_{\{\ell\}}a^m_{(\ell)},a^m_N\right)\right),\end{aligned}$$

where $A^*_2=A_1\cup A_2, B^*_2=B_1\cup B_2, A^*_i=A_i$, and $B^*_i=B_i$ for $i=2,\dots,m$. We are to show that the last expression of the above equation, referred to as *LE* hereafter, exactly equals $\Phi(a^1_{A_1}a^2_{A_2\cup\cdots\cup A_m},a^1_{B_1}a^2_{B_2\cup\cdots\cup B_m})$.

By Lemma 2, for $j=2,\dots,m$,

$$\Phi\left(a^1_{\{k\}}a^m_{(k)},a^j_{\{\ell\}}a^m_{(\ell)}\right)-\Phi\left(a^2_{\{k\}}a^m_{(k)},a^j_{\{\ell\}}a^m_{(\ell)}\right)=\Phi\left(a^1_{\{k\}}a^2_{(k)},a^2_N\right).$$

We substitute this for *LE* and obtain

$$
\begin{aligned}
LE = &\sum_{k\in A_1}\sum_{\ell\in B_1}\Phi\left(a^1_{\{k\}}a^m_{(k)}, a^1_{\{\ell\}}a^m_{(\ell)}\right) - \sum_{k\in A_1}\sum_{\ell\in B_1}\Phi\left(a^2_{\{k\}}a^m_{(k)}, a^2_{\{\ell\}}a^m_{(\ell)}\right)\\
&+\sum_{j=2}^{m}\sum_{k\in A_1}\sum_{\ell\in B_j}\Phi\left(a^1_{\{k\}}a^2_{(k)}, a^2_N\right) + \sum_{i=2}^{m}\sum_{k\in A_i}\sum_{\ell\in B_1}\Phi\left(a^2_N, a^1_{\{\ell\}}a^2_{(\ell)}\right)\\
&-(n-1)\sum_{k\in A_1}\left(\Phi\left(a^1_{\{k\}}a^m_{(k)}, a^m_N\right) - \Phi\left(a^2_{\{k\}}a^m_{(k)}, a^m_N\right)\right)\\
&+(n-1)\sum_{\ell\in B_1}\left(\Phi\left(a^1_{\{\ell\}}a^m_{(\ell)}, a^m_N\right) - \Phi\left(a^2_{\{\ell\}}a^m_{(\ell)}, a^m_N\right)\right).
\end{aligned}
$$

Noting by skew-symmetry of Φ that

$$
\begin{aligned}
-\sum_{k\in A_1}\sum_{\ell\in B_1}\Phi\left(a^2_{\{k\}}a^m_{(k)}, a^2_{\{\ell\}}a^m_{(\ell)}\right) = &\sum_{k\in A_1}\sum_{\ell\in B_1^c}\Phi\left(a^2_{\{k\}}a^m_{(k)}, a^2_{\{\ell\}}a^m_{(\ell)}\right)\\
&+\sum_{k\in A_1^c}\sum_{\ell\in N}\Phi\left(a^2_{\{k\}}a^m_{(k)}, a^2_{\{\ell\}}a^m_{(\ell)}\right),
\end{aligned}
$$

LE is rearranged to give

$$
\begin{aligned}
LE = &\sum_{k\in A_1}\sum_{\ell\in B_1}\Phi\left(a^1_{\{k\}}a^m_{(k)}, a^1_{\{\ell\}}a^m_{(\ell)}\right) + \sum_{k\in A_1}\sum_{\ell\in B_1^c}\left(\Phi\left(a^1_{\{k\}}a^2_{(k)}, a^2_N\right)\right.\\
&\left.+\Phi\left(a^2_{\{k\}}a^m_{(k)}, a^2_{\{\ell\}}a^m_{(\ell)}\right)\right)\\
&+\sum_{k\in A_1^c}\sum_{\ell\in B_1^c}\Phi\left(a^2_{\{k\}}a^m_{(k)}, a^2_{\{\ell\}}a^m_{(\ell)}\right) + \sum_{k\in A_1^c}\sum_{\ell\in B_1}\left(\Phi\left(a^2_N, a^1_{\{\ell\}}a^2_{(\ell)}\right)\right.\\
&\left.+\Phi\left(a^2_{\{k\}}a^m_{(k)}, a^2_{\{\ell\}}a^m_{(\ell)}\right)\right)\\
&-(n-1)\left\{\sum_{k\in A_1}\Phi\left(a^1_{\{k\}}a^m_{(k)}, a^m_N\right) + \sum_{k\in A_1^c}\Phi\left(a^2_{\{k\}}a^m_{(k)}, a^m_N\right)\right\}\\
&+(n-1)\left\{\sum_{\ell\in B_1}\Phi\left(a^1_{\{\ell\}}a^m_{(\ell)}, a^m_N\right) + \sum_{\ell\in B_1^c}\Phi\left(a^2_{\{\ell\}}a^m_{(\ell)}, a^m_N\right)\right\},
\end{aligned}
$$

Hence this is easily modified to the exact expression of conditionally additive decomposition of $\Phi(a^1_{A_1}a^2_{A_2\cup\cdots\cup A_m}, a^1_{B_1}a^2_{B_2\cup\cdots\cup B_m})$, since, by Lemma 2,

$$
\Phi(a^1_{\{k\}}a^2_{(k)}, a^2_N) + \Phi(a^2_{\{k\}}a^m_{(k)}, a^2_{\{\ell\}}a^m_{(\ell)}) = \Phi(a^1_{\{k\}}a^m_{(k)}, a^2_{\{\ell\}}a^m_{(\ell)}).
$$

This completes the proof. □

References

Chew, S.H. & McCrimmon, K.R. (1979). Alpha-nu choice theory: a generalization of expected utility theory. Unpublished mimeograph.

Chew, S.H. (1983). A generalization of the quasilinear mean with applications to the measurement of income inequality and decision theory resolving the Allais paradox. *Econometrica* 51, 1065–1092.

Fishburn, P.C. (1981). Subjective expected utility: a review of normative theories. *Theory and Decision, 13*, 139–199.

Fishburn, P.C. (1982). Nontransitive measurable utility. *Journal of Mathematical Psychology, 26*, 31–67.

Fishburn, P.C. (1983a). Ellsberg revisited: a new look at comparative probability. *Annals of Statistics, 11,* 1047–1059.

Fishburn, P.C. (1983b). A generalization of comparative probability on finite sets. *Journal of Mathematical Psychology, 27,* 298–310.

Fishburn, P.C. (1984a). SSB utility theory and decision-making under uncertainty. *Mathematical Social Sciences, 8,* 63–94.

Fishburn, P.C. (1984b). Multiattribute nonlinear utility theory. *Management Science, 30,* 1301–1310.

Fishburn, P.C. (1988a). Nontransitive measurable utility for decision under uncertainty. *Journal of Mathematical Economics, 18,* 187–207.

Fishburn, P.C. (1988b). *Nonlinear preference and utility theory*. Baltimore: Johns Hopkins University Press.

Fishburn, P.C. & LaValle, I.H. (1987a). A nonlinear, nontransitive and additive probability model for decisions under uncertainty. *Annals of Statistics, 15,* 830–844.

Fishburn, P.C. & LaValle, I.H. (1987b). Transitivity is equivalent to independence for state-additive SSB utilities. *Journal of Economic Theory, 44,* 202–208.

Fishburn, P.C. & Nakamura, Y. (1991). Nontransitive measurable utility with constant threshold. *Journal of mathematical Psychology, 35,* 471–500.

Kahneman, D. & Tversky, A. (1979). Prospect theory; an analysis of decision under risk. *Econometrica, 47,* 263–291.

Machina, M.J. (1982). Expected utility analysis without the independence axiom. *Econometrica, 50,* 277–323.

Nakamura, Y. (1990). Bilinear utility and a threshold structure for nontransitive preferences. *Mathematics of Social Sciences, 19,* 1–21.

Nakamura, Y. (1997). *A generalization of subjective expected utility without additivity and transitivity*. IPPS discussion paper No. 719, University of Tsukuba.

Nakamura, Y. (1998). Skew-symmetric additive representations of preferences. *Journal of Mathematical Economics, 30,* 367–387.

Nakamura, Y, (2001). Totally convex preferences for gambles. *Mathematical Social Sciences, 42,* 295–305.

Quiggin, J. (1981). A theory of anticipated utility. *Journal of Economic Behavior and Organization, 3,* 323–343.

Schmeidler, D. (1988). Subjective probability and expected utility without additivity. *Econometrica, 57,* 571–587.

Decision Theory

Decision Making Based on Risk-Value Tradeoffs

Jianmin Jia and James S. Dyer

1 Introduction

This essay provides a review for measures of risk and risk-value models that we have developed for the past ten years. Risk-value models are a new class of decision making models based on the idea of risk-value tradeoffs. Intuitively, individuals may consider their choices over risky alternatives by trading off between risk and return, where return is typically measured as the mean (or expected return) and risk is measured by some indicator of dispersion or possible losses. This notion is prevalent in the literatures in finance, marketing and other areas.

Markowitz (1959, 1987, 1991) proposed variance as a measure of risk, and a mean-variance model for portfolio selection based on minimizing variance subject to a given level of mean return. But arguments have been made that mean-variance models are appropriate only if the investor's utility function is quadratic or the joint distribution of returns is normal. However, these conditions are rarely satisfied in practice.

Previous researchers usually consider expected utility theory as the foundation of mean-risk models and risk-return models (e.g., Fishburn, 1977; Meyer, 1987; Bell, 1988, 1995; Sarin & Weber, 1993). However, the expected utility theory has been called into question by empirical studies of risky choice (e.g., Allais, 1953, 1979; Kahneman and Tversky, 1979; Machina, 1987; Weber, 2001). This suggests that an alternative approach regarding the paradigm of risk-return tradeoffs would be of interest.

The notion of risk as a primitive concern has also been investigated extensively, and a number of perceived risk models have been proposed (e.g., Pollatsek & Tversky, 1970; Coombs & Lehner, 1981, 1984; Luce, 1980; Fishburn, 1982, 1984; Luce & Weber, 1986; Sarin, 1987, Lowenstein et al., 2001, Weber et al., 2004).

J.S. Dyer (✉)
The McCombs School of Business, The University of Texas at Austin, Austin, TX 78712-012, USA
e-mail: Jim.Dyer@mccombs.utexas.edu

S.J. Brams et al. (eds.), *The Mathematics of Preference, Choice and Order: Essays in Honor of Peter C. Fishburn*, Studies in Choice and Welfare,

These risk models have two major problems: first, the validity of most of these risk models as measures of perceived risk has not been supported by empirical studies (e.g., Coombs & Bowen, 1971; Coombs & Lehner, 1981, 1984; Weber, 1984; Keller, Sarin & Weber, 1986; Weber & Bottom, 1989; Weber 2001); second, it is not clear how to incorporate these risk measures into decision models because they were developed separately from preference measures. Thus, the usefulness of these risk measures is limited in efforts to model or to improve decision making.

In the main stream of decision research, the role of risk in determining preference is usually considered implicitly. For instance, in the expected utility model (von Neumann & Morgenstern, 1947), an individual's attitude toward the risk involved in choices among risky alternatives is defined by the shape of his or her utility function (Pratt, 1964); and in some non-expected utility models, risk (or "additional" risk) is also captured by some nonlinear functions over probabilities (e.g., see Kahneman and Tversky, 1979; Quiggin, 1982; Tversky & Kahneman, 1992, Wu & Gonzalez, 1996). Thus, these decision theories are not, at least explicitly, compatible with the choice behavior based on the intuitive idea of risk-return tradeoffs as often observed in practice. Therefore, they offer little guidance for this type of decision making.

In this essay, we review our risk-value studies and provide a framework that is compatible with choice behavior based on risk-value tradeoffs. In particular, our framework unifies two streams of research: one in developing preference models and the other in modeling risk judgments. This synthesis makes our risk-value models more descriptively powerful than other preference models and risk models that have been proposed separately.

The remainder of this paper is organized as follows. The next section provides a preference-dependent measure of risk with several useful examples. Section 3 develops the basic framework of our risk-value studies and related preference conditions. Section 4 presents three particular forms of risk-value models. Section 5 concludes our risk-value studies and discusses topics for future research.

2 The Standard Measure of Risk

In order to develop risk-value models, we first propose a preference-dependent measure of risk, called a standard measure of risk, that offers a new foundation for research regarding risk judgments and decision making by risk-value tradeoffs (Jia & Dyer, 1996). This general measure of risk is based on the converse expected utility of normalized lotteries with zero-expected values, so it is compatible with the measure of expected utility and provides the basis for linking risk with preference.

For lotteries with zero-expected values, we assume that the only choice attribute of relevance for them is risk. A riskier lottery would be less preferable and vice versa, by any risk averse decision maker. Therefore, the riskiness ordering of these lotteries should be simply the reverse of the preference ordering. We consider decomposing a lottery X (i.e., a random variable) into its mean $\bar{X}$ and its standard risk, $X' = X - \bar{X}$, and the standard measure of risk is defined as follows:

$$R(X') = -\mathrm{E}[u(X')] = -\mathrm{E}[u(X - \bar{X})], \tag{1}$$

where $u(\cdot)$ is a utility function (von Neumann & Morgenstern, 1947) and the symbol E represents expectation over the probability distribution of a lottery. The mean of the lottery serves as a status quo for measuring the standard risk.

One of the characteristics of our standard measure of risk is that it depends on an individual's utility function. When the form of the utility function is determined, then we can derive the associated standard measure of risk. More important, our standard measure of risk can offer a preference justification for some commonly used measures of risk so that the suitability of those risk measures can be evaluated.

If a utility function is quadratic, $u(x) = ax - bx^2$, where $a, b > 0$, then the standard measure of risk is characterized by variance, $R(X') = b\mathrm{E}[(X - \bar{X})^2]$. However, the quadratic utility function has a disturbing property; that is, it will be decreasing after a certain point and it exhibits increasing risk aversion. Since the quadratic utility function may not be an appropriate description of preference, it follows that variance may not be a good measure for risk (unless the distribution of a lottery is normal).

To obtain an increasing utility function based on the quadratic one, let us consider a third-order polynomial (or cubic) utility model, $u(x) = ax - bx^2 + c'x^3$, where $a, b, c' > 0$. When $b^2 < 3ac'$, the cubic utility model is increasing. This utility function is concave, and hence risk averse for low outcome levels (i.e., $x < b/(3c')$), and convex, and thus risk seeking for high outcome values (i.e., $x > b/(3c')$). Such a utility function may be used to model a preference structure consistent with the observation that a large number of individuals purchase both insurance (a moderate outcome-small probability event) and lottery tickets (a small chance of a large outcome) in the traditional expected utility framework (see Friedman & Savage, 1948). The associated standard measure of risk for this utility function can be obtained as follows:

$$R(X') = \mathrm{E}[(X - \bar{X})^2] - c\mathrm{E}[(X - \bar{X})^3], \tag{2}$$

where $c = c'/b > 0$. Model (2) provides a simple way to combine skewness with variance into a measure of risk. This measure of risk should be superior to variance alone since the utility function implied by (2) has a more intuitive appeal than the quadratic one implied by variance.

Markowitz (1952) noted that an individual with the utility function that is concave for low outcome levels and convex for high outcome values will tend to prefer positively skewed distributions (with large right tails) over negatively skewed ones (with large left tails). The standard measure of risk (2) clearly reflects this observation; i.e., a positive skewness will reduce risk and a negative skewness will increase risk.

If an individual's preference can be modeled by an exponential or the quadratic utility function, $u(x) = ax - bx^2$, where $a \geq 0$, and b, $c > 0$, then its corresponding standard measure of risk (with the normalization condition $R(0) = 1$) is:

$$R(X') = E[e^{-c(X-\bar{X})} - 1]. \tag{3}$$

Bell (1988) identified $E[e^{-c(X-\bar{X})}]$ as a measure of risk from the linear plus exponential utility model by arguing that the riskiness of a lottery should be independent of its expected value. Weber (1990) also modified Sarin's (1987) expected exponential risk model by requiring that the risk measure be location free.

If an individual is risk averse for gains but risk seeking for losses (Fishburn & Kochenberger, 1979; Kahneman and Tversky, 1979), then we can consider a piecewise power utility model as follows:

$$u(x) = \begin{cases} ex^{\theta_1}, & \text{when } x \geq 0 \\ -d|x|^{\theta_2}, & \text{when } x < 0 \end{cases} \tag{4}$$

where e, d, θ_1 and θ_2 are non negative constants. According to (1), the corresponding standard measure of risk is:

$$R(X') = d\mathrm{E}^-[|X-\bar{X}|^{\theta_2}] - e\mathrm{E}^+[|X-\bar{X}|^{\theta_1}], \tag{5}$$

where $\mathrm{E}^-[|X-\bar{X}|^{\theta_2}] = \int_{-\infty}^{\overline{X}} |x-\bar{X}|^{\theta_2} f(x)dx$,

$\mathrm{E}^+[|X-\bar{X}|^{\theta_1}] = \int_{\overline{X}}^{\infty} (x-\bar{X})^{\theta_1} f(x)dx$ and $f(x)$ is the probability density of a lottery.

The standard measure of risk (5) includes several commonly used measures of risk in the financial literature as special cases. When $d > e > 0$, $\theta_1 = \theta_2 = \theta > 0$ and the distribution of a lottery is symmetric, then we can have $R(X') = (d-e)\mathrm{E}|X-\bar{X}|^{\theta}$, which is associated with variance and absolute standard deviation if $\theta = 2$ and $\theta = 1$ respectively. This standard measure of risk is also related to the difference of d and e, which reflects the relative effect of loss and gain on risk. In general, if the distribution of a lottery is not symmetric, the standard measure of risk will not be connected with variance even if $\theta_1 = \theta_2 = 2$ but it is still related to the absolute standard deviation if $\theta_1 = \theta_2 = 1$ (Jia, Dyer & Butler, 2001).

Based on preference considerations, the absolute standard deviation should be a better choice than the variance as a measure of risk. In the financial literature, this point has been made by Konno & Yamazaki (1992). In statistics, the absolute standard deviation is also considered a more robust measure for dispersion than variance.

Another extreme case of (5) arises when $e = 0$ (i.e., the utility function is non increasing for gains); then the standard measure of risk $R(X') = d\mathrm{E}^-[|X-\overline{X}|^{\theta_2}]$, which is a lower partial moment risk model. When $\theta_2 = 2$, it becomes a semi-variance measure of risk (Markowitz, 1959); and when $\theta_2 = 0$, it reduces to the probability of loss.

In summary, some other proposed measures of risk are special cases of our standard measure of risk. The standard measure of risk is more normative in nature, as it is independent of the expected value of a lottery. To obtain more descriptive power and to capture perceptions of risk, we have also established measures of perceived risk that are based on a two-attribute structure: the mean of a lottery and its standard risk (Jia, Dyer & Butler, 1999).

3 Frameworks for Risk-Value Tradeoff

When we decompose a lottery into its mean and standard risk, then the evaluation of the lottery can be based on the tradeoff between mean and risk. We assume a risk-value preference function $f(\bar{X},R(X'))$, where f is increasing in $\bar{X}$ and decreasing in $R(X')$ if one is risk averse.

Consider an investor who wants to maximize his or her preference function f for an investment and also requires a certain level μ of expected return. Since f is decreasing in $R(X')$ and $\bar{X}=\mu$ is a constant, then maximizing $f(\bar{X},R(X'))$ is equivalent to minimizing $R(X')$; i.e., max $\{f(\bar{X},R(X'))|\bar{X}=\mu\} \Rightarrow \min\{R(X')|\bar{X}=\mu\}$. This conditional optimization model includes many financial optimization models as special cases by choosing different standard measures of risk; e.g., Markowitz's mean-variance model, the mean-absolute standard deviation model, and the mean-semivariance model. We can also propose some new optimization models based on our standard measures of risk (2) and (5).

In the conditional optimization problem, we do not need to assume an explicit form for the preference function f. The problem only depends on the standard measure of risk. However, we may argue that an investor should maximize his or her preference functions unconditionally in order to obtain the overall optimal portfolio. For an unconditional optimization decision, the investor's preference function must be specified. Here we consider two cases for the preference function f: (1) when it is consistent with the expected utility theory; and (2) when it is based on a two-attribute expected utility foundation.

Let **P** be a convex set of all simple probability or lotteries $\{X, Y, \ldots\}$ on a non-empty set **X** of outcomes, and Re be the set of real numbers (assuming $\mathbf{X} \in$ Re is finite). We define $\succ$ as a binary preference relation on **P**.

Definition 1. For two lotteries $X,Y \in \mathbf{P}$ with $\mathrm{E}(X)=\mathrm{E}(Y)$, if $w_0+X \succ w_0+Y$ for some $w_0 \in$ Re, then $w+X \succ w+Y$ for all $w \in$ Re.

This is called the risk independence condition. It means that for a pair of lotteries with a common mean, the preference order between the two lotteries will not be changed when the common mean changes; i.e., the preference order can be determined solely by the ranking of their standard risk.

Theorem 1. *Assume that the risk-value preference function f is consistent with the expected utility theory. Then f can be represented as the following standard risk-value form,*

$$f(\bar{X},R(X')) = u(\bar{X}) - \phi(\bar{X})[R(X') - R(0)], \tag{6}$$

if and only if the risk independence condition holds, where $\varphi(\bar{X}) > 0$ and $u(\cdot)$ is a von Neumann and Morgenstern utility function.

Model (6) shows that an expected utility model could have an alternative representation if the risk independence condition holds. If one is risk averse, then $u(\cdot)$ is a concave function and $R(X') - R(0)$ is always positive. $u(\bar{X})$ provides a measure of value for the mean, and $\varphi(\bar{X})$ is a tradeoff factor that may depend on the mean. If we further require the utility model to be continuously differentiable, then it must be either a quadratic, exponential, or linear plus exponential model (Jia & Dyer, 1996).

There are also some other alternative forms of risk-value models within the expected utility framework under different preference conditions (Sarin & Weber, 1993; Bell, 1995; Dyer & Jia, 1998). In addition, for non-negative lotteries such as those associated with the price of a stock, we propose a relative risk-value model that is compatible with the logarithmic (or linear plus logarithmic) and the power (or linear plus power) utility functions (Dyer & Jia, 1997).

However, the notion of risk-value tradeoffs within the expected utility framework is very limited. In particular, the risk measure and the value measure must be based on the same utility function. Intuitively, a decision maker may deviate from this "consistency" to have different measures for risk and value if his choice is based on risk-value tradeoffs.

In order to be more realistic and flexible in the framework of risk-value tradeoffs, we consider a two-attribute structure $(\bar{X}, X')$ for the evaluation of a risky alternative X. In this way we can explicitly base the evaluation of lotteries on two attributes, mean and risk, so that the mean-risk (or risk-value) tradeoffs are not necessarily consistent with the traditional expected utility framework.

We assume the existence of the von Neumann and Morgenstern expected utility axioms over the two-attribute structure $(\bar{X}, X')$ and require the risk-value model to be consistent with the two-attribute expected utility model, i.e., $f(\bar{X}, R(X')) = \mathrm{E}[U(\bar{X}, X')]$, where U is a two-attribute utility function. As a special case when the relationship between $\bar{X}$ and X' is a simple addition, the risk-value model reduces to a traditional expected utility model, i.e., $f(\bar{X}, R(X')) = \mathrm{E}[U(\bar{X}, X')] = \mathrm{E}[U(\bar{X} + X')] = \mathrm{E}[U(X)] = \mathrm{a}\,\mathrm{E}[u(X)] + \mathrm{b}$, where $\mathrm{a} > 0$ and b are constants.

To obtain some separable forms of the risk-value model, we need to have a risk independence condition for the two-attribute structure. Let P^0 be the set of normalized lotteries with zero-expected values, and $\succ$ a strict preference relation for the two-attribute structure.

Definition 2. For $X', Y' \in \mathrm{P}^0$, if there exists a $w_0 \in \mathrm{Re}$ for which $(w_0, X') \succ (w_0, Y')$, then $(w, X') \succ (w, Y')$ for all $w \in \mathrm{Re}$.

This two-attribute risk independence condition requires that if two lotteries have the same mean and one is preferred to the other, then transforming the lotteries by adding the same constant to all outcomes will not reverse the preference ordering. This condition is generally supported by our recent experimental studies (Butler, Dyer & Jia, 2005).

Theorem 2. *Assume that the risk-value preference function f is consistent with the two-attribute expected utility model. Then f can be represented as the following generalized risk-value form,*

$$f(\bar{X}, R(X')) = V(\bar{X}) - \phi(\bar{X})[R(X') - R(0)] \tag{7}$$

if and only if the two-attribute risk independence condition holds, where $\phi(\bar{X}) > 0$ and $R(X')$ is the standard measure of risk.

In contrast to the risk-value model (6), three functions $V(\bar{X}), R(X')$ and $\phi(\bar{X})$ in this generalized risk-value model (7) can be considered independently, which leads to a very flexible structure for risk-value tradeoffs. Thus we can choose different functions for the value measure $V(\bar{X})$ instead of the utility function. The expected utility measure is only used for the standard measure of risk. Even though expected utility theory has been challenged by some empirical studies for general lotteries, we believe that it should be appropriate for describing risky choice behavior within a special set of normalized probability distributions with the same expected values. In fact, the generalized risk-value model can capture a number of decision paradoxes that violate the traditional expected utility theory (Jia, 1995).

If the utility function u is strictly concave, then $R(X') - R(0) > 0$ and model (7) will reflect risk averse behavior. In addition, if $V(\bar{X})$ is increasing and twice continuously differentiable, $\phi(\bar{X})$ is once continuously differentiable and $\phi'(\bar{X})/\phi(\bar{X})$ is nonincreasing, then the generalized risk-value model (7) exhibits decreasing risk aversion if and only if $-V''(\bar{X})/V'(\bar{X}) < -\phi'(\bar{X})/\phi(\bar{X})$; and the generalized risk-value model (7) exhibits constant risk aversion if and only if $-V''(\bar{X})/V'(\bar{X}) = -\phi'(\bar{X})/\phi(\bar{X})$ is a constant. Thus, if a decision maker is decreasingly risk averse and has a linear value function, then we must choose a decreasing function for the tradeoff factor $\phi(\bar{X})$.

The basic form of the risk-value model may be further simplified if some stronger preference conditions are satisfied. When $\phi(\bar{X}) = k > 0$, model (7) becomes the following additive form:

$$f(\bar{X}, R(X')) = V(\bar{X}) - k[R(X') - R(0)]. \tag{8}$$

When $\phi(\bar{X}) = -V(\bar{X}) > 0$, then model (7) reduces to the following multiplicative form:

$$f(\bar{X}, R(X')) = V(\bar{X})R(X'), \tag{9}$$

where $R(0) = 1$ and $V(0) = 1$. In this multiplicative model, $R(X')$ serves as a value discount factor due to risk.

We also develop measures of perceived risk based on the converse interpretation of the axioms of risk-value models, and thus a negative linear transformation of the risk-value model (7) provides a measure of the perceived risk for an individual (Jia et al., 1999). Our risk-value framework offers a unified approach to both risk judgment and preference modeling.

4 Generalized Risk-Value Models

According to the generalized risk-value model (7), the standard measure of risk, the value function, and the tradeoff factor can be considered independently. Some examples of the standard measure of risk $R(X')$ are provided in Section 2. The value measure $V(\bar{X})$ should be chosen as an increasing function and may have the same functional form as a utility model. For appropriate risk averse behavior, the tradeoff factor $\phi(\bar{X})$ should be either a decreasing function or a positive constant; e.g., $\phi(\bar{X}) = ke^{-b\bar{X}}$, where $k > 0$ and $b \geq 0$. We consider three types of risk-value models, namely moments risk-value models, exponential risk-value models and generalized disappointment models as follows.

4.1 Moments Risk-Value Models

People often use mean and variance to make tradeoffs for financial decision making because of their operational advantages and because they provide a reasonable approximation for modeling decision problems (see Markowitz, 1959, 1987, 1991; Sharpe 1970, 1991). In the past, expected utility theory has been used as a foundation for mean-variance models. Now we can provide a better foundation, the risk-value theory, for developing moments models that include the mean-variance model as a special case.

As an example, the mean-variance model, $\bar{X} - k\mathrm{E}[(X-\bar{X})^2]$where $k > 0$, is a simple risk-value model with variance as the standard measure of risk and a constant tradeoff factor. Sharpe (1970, 1991) assumed this mean-variance model in his analysis for portfolio selection and the Capital Asset Pricing Model. However, under the expected utility framework, this mean-variance model is based on the assumptions that the investor has an exponential utility function and that returns are jointly normally distributed.

According to our risk-value theory, this mean-variance model is constantly risk averse. To obtain a decreasing risk averse mean-variance model, we can simply use a decreasing function for the tradeoff factor:

$$f(\bar{X},R(X')) = \bar{X} - ke^{-b\bar{X}}\mathrm{E}[(X-\bar{X})^2] \tag{10}$$

where b, $k > 0$.

For many decision problems, mean-variance models are an over simplification. Based on our risk-value framework, we can develop some richer moment models for risky decision making. First, let us consider the moment standard measure of risk (2) for the additive risk-value model (8):

$$f(\bar{X},R(X')) = \bar{X} - k\{\mathrm{E}[(X-\bar{X})^2] - c\mathrm{E}[(X-\bar{X})^3]\}, \tag{11}$$

where c, $k > 0$. The three moments model (11) can be either risk averse or risk seeking, depending on the distribution of a lottery. For symmetric bets or lotteries not highly skewed (e.g., an insurance policy) such that $\mathrm{E}[(X-\bar{X})^2] > c\mathrm{E}[(X-\bar{X})^3]$, model (11) will be risk averse. But for highly positive skewed lotteries (e.g., lottery tickets) such that the skewness overwhelms the variance, i.e., $\mathrm{E}[(X-\bar{X})^2] < c\mathrm{E}[(X-\bar{X})^3]$, then model (11) will exhibit risk seeking behavior.

Markowitz (1952) noticed that individuals of all wealth levels have the same tendency to purchase insurance and lottery tickets whether they are poor or rich. This observed behavior contradicts a common assumption of expected utility theory that preference ranking is defined over ultimate levels of wealth. For the three moments model (10), the change of wealth level just causes a parallel shift for the model, which will not affect the risk attitude and the choice behavior of this model. This is consistent with Markowitz's observation. In addition, the three moments model implies an nonlinear weight of probability that can be consistent with Kahneman and Tversky's (1979) prospect theory (Jia, 1995).

4.2 Exponential Risk-Value Models

If the standard measure of risk is based on exponential or linear plus exponential utility models, then the standard measure of risk is given by (3). To be compatible with the form of the standard measure of risk, we can also choose the same form of exponential functions, but with different parameters, for the value measure $V(\bar{X})$ and the tradeoff factor $\phi(\bar{X})$, which leads to the following model:

$$f(\bar{X},R(X')) = -he^{-a\bar{X}} - ke^{-b\bar{X}}\mathrm{E}[e^{-c(X-\bar{X})} - 1], \tag{12}$$

where a, b, c, h, and k are positive constants. When $a = b = c$ and $h = k$, this model reduces to an exponential utility model. Otherwise, these two models are different. When $b > a$, model (12) is decreasing risk averse even though the traditional exponential utility model exhibits constant risk aversion.

As a special case, when $a = b$ and $h = k$, model (12) reduces to the following simple multiplicative form:

$$f(\bar{X},R(X')) = ke^{-a\bar{X}}\mathrm{E}[e^{-c(X-\bar{X})}]. \tag{13}$$

This model is constantly risk averse, and therefore has the same risk attitude as an exponential utility model. It has more flexibility since there are two different parameters. This simple risk-value model can be used to explain some well known decision paradoxes (Jia, 1995).

Choosing a linear function or a linear plus exponential function for $V(\bar{X})$ leads to the following models:

$$f(\bar{X},R(X')) = \bar{X} - ke^{-b\bar{X}}\mathrm{E}[e^{-c(X-\bar{X})} - 1], \tag{14}$$

$$f(\bar{X},R(X')) = \bar{X} - he^{-a\bar{X}} - ke^{-b\bar{X}}\mathrm{E}[e^{-c(X-\bar{X})} - 1]. \tag{15}$$

Model (14) is decreasingly risk averse. Model (15) includes a linear plus exponential utility model as a special case when $a = b = c$ and $h = k$. It is decreasingly risk averse if $b \geq a$.

4.3 Generalized Disappointment Models

Bell (1985) proposed a disappointment model for decision making under uncertainty. Although Bell's development of the disappointment model has an intuitive appeal, his model is only applicable to lotteries with two outcomes.

Jia et al. (2001) use the risk-value framework to develop a generalized version of Bell's (1985) disappointment model. Consider the following piece-wise linear utility model:

$$u(x) = \begin{cases} ex & \text{when } x \geq 0 \\ dx & \text{when } x < 0 \end{cases} \tag{16}$$

where d, $e > 0$ are constant. Decision makers who are averse to downside risk or losses should have $d > e$, as illustrated in Fig. 1. The standard measure of risk for this utility model can be obtained as follows:

$$R(X') = d\mathrm{E}^-[|X - \bar{X}] - e\mathrm{E}^+[|X - \bar{X}|] = [(d-e)/2]\mathrm{E}[|X - \bar{X}|], \tag{17}$$

where $\mathrm{E}^-[|X - \bar{X}|] = \sum_{x_i < \bar{X}} p_i|x_i - \bar{X}|$ and $\mathrm{E}^+[|X - \bar{X}|] = \sum_{x_i > \bar{X}} p_i(x_i - \bar{X})$, and $\mathrm{E}[|X - \bar{X}|]$ is the absolute standard deviation. According to Bell's (1985) basic idea, $d\mathrm{E}^-[|X - \bar{X}|]$ should be a general measure of expected disappointment and $e\mathrm{E}^+[|X - \bar{X}|]$ a general measure of expected elation, and then overall psychological satisfaction is measured by - $R(X')$, which is the converse of the standard measure of risk (17).

If we assume a linear value measure and a constant tradeoff factor, then we can have the following risk-value model based on the measure of disappointment risk (17):

$$\begin{aligned} f(\bar{X}, R(X')) &= \bar{X} - \{d\mathrm{E}^-[|X - \bar{X}|] - e\mathrm{E}^+[|X - \bar{X}|]\} \\ &= \bar{X} - [(d-e)/2]\mathrm{E}[X - \bar{X}|]. \end{aligned} \tag{18}$$

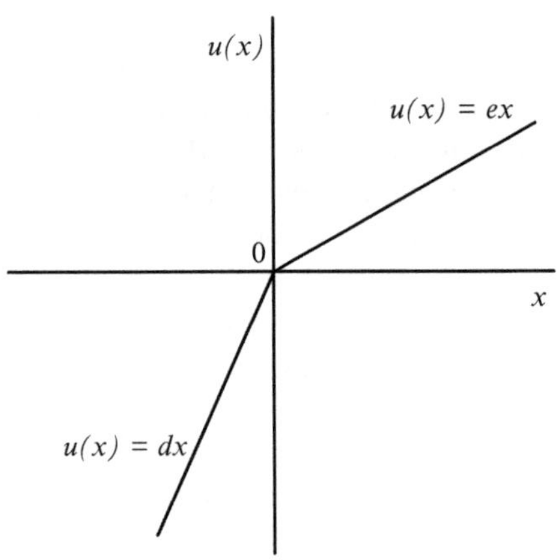

Fig. 1 A piece-wise linear utility function

For a two-outcome lottery, model (18) reduces to Bell's disappointment model. Thus, we call the risk-value model (18) a "generalized disappointment model." This model is a risk averse when $d > e$.

Using his two-outcome disappointment model, Bell (1985) gave an explanation for the common ratio effect. Our generalized disappointment model (18) can explain the Allais Paradox (Allais, 1953, 1979), which involves an alternative with three outcomes (Jia et al., 2001). Another concern for Bell's model and our model (18) is that they imply constant risk aversion. Thus, they are not appropriate for decreasing risk averse behavior. To obtain a disappointment model with decreasing risk aversion, we can use a decreasing function for the tradeoff factor:

$$f(\bar{X}, R(X')) = \bar{X} - ke^{-b\bar{X}}\mathrm{E}[|X - \bar{X}|]. \tag{19}$$

Bell's disappointment model and our model (18) imply that disappointment and elation are proportional to the difference between the expected value and an outcome. Then we should use some nonlinear functions for disappointment and elation such as the risk model (5), which leads to a more general form of disappointment model:

$$f(\bar{X}, R(X')) = \bar{X} - d\mathrm{E}^-[|X - \bar{X}|^{\theta_2}] - e\mathrm{E}^+[|X - \bar{X}|^{\theta_1}]. \tag{20}$$

When $\theta_1 = \theta_2 = 1$, this model reduces to model (18). When $e = 0$ and $\theta_2 = 2$, model (20) becomes a mean-semivariance model. This model also provides an interpretation for the decision weight in prospect theory based on the concept of disappointment (Jia et al., 2001).

Finally, our generalized disappointment models are different from Loomes and Sugden (1986) model, $\bar{X} + \mathrm{E}[D(X - \bar{X})]$, where $D(x - \bar{X}) = -D(\bar{X} - x)$, and D is continuously differentiable and convex for $x > \bar{X}$ (thus concave for $x < \bar{X}$). Even though this model is different from our generalized disappointment models (20), it is a special case of our risk-value model with a linear measure of value, a constant tradeoff factor, and a specific form of the standard measure of risk (i.e., $R(X') = -\mathrm{E}[D(X - \bar{X})]$, where $D(x - \bar{X}) = -D(\bar{X} - x)$). Loomes and Sugden (1986) used this model to provide an explanation for the choice behavior that violates Savage's (1954) sure-thing principle.

5 Conclusion

We have summarized our efforts to incorporate the intuitively appealing idea of risk-value tradeoffs into decision making under risk. The risk-value framework ties together two streams of research: one in developing preference models and the other in modeling risk judgments, and unifies a wide range of decision phenomena including both normative and descriptive aspects.

This development also refines and generalizes a substantial number of previously proposed decision theories and models, ranging from the mean-variance model in finance to disappointment models in decision science. It is also possible to create many new risk-value models. Specifically, we have discussed three classes of decision models based on this risk-value theory: moments risk-value models, exponential risk-value models and generalized disappointment risk-value models. These models are very flexible in modeling preferences. They also provide new resolutions for observed risky choice behavior and the decision paradoxes that violate the independence axiom of the expected utility theory.

The most important assumption in this study is the risk independence condition, which leads to a separable form of risk-value models. Although some other weaker condition could be used to derive a risk-value model that has more descriptive power, this reduces the elegance of the basic risk-value form, and increases operational difficulty. Butler et al. (2005) conducted an empirical of this key assumption, and found some support for it. This study also highlighted some additional patterns of choices indicating that the translation of lottery pairs from the positive domain to the negative domain often results in the reversal of preference and risk judgments. To capture this phenomenon, we have extended risk independence conditions to allow the tradeoff factor in the risk-value models to change sign, and therefore to infer risk aversion in the positive domain and risk seeking in the negative domain. These generalized risk-value models provide additional insights into the reflection effects in prospect theory (Kahneman and Tversky, 1979) and related empirical results (Fishburn & Kochenberger, 1979; Payne et al., 1980,1981).

Even though some other non-expected utility theories that have been proposed (e.g., Prospect Theory and rank dependent utility models) may produce the same predictions for the decision paradoxes as risk-value theory, it offers a new justification for them based on an appealing and realistic notion of risk-value tradeoffs. In particular, since the role of risk is merely considered implicitly in these decision theories and models, they are not compatible with the choice behavior that is based on risk and mean return tradeoffs as often faced in financial management and other applied fields. Therefore, these theories and models offer little guidance in practice for this type of decision making. We believe that the potential for contributions of these risk-value models in finance is very exciting. And also applications of our risk-value models in other fields such as economics, marketing, insurance and risk management should be promising.

Risk-value theory can be made compatible with traditional utility theory by restricting the choices of the components of model (6). However, the risk-value theory can be extended to model (7) by basing it on the two-attribute expected utility framework, which retains many appealing properties of the traditional expected utility theory. In particular, our risk-value models reduce to single-attribute expected utility models for lotteries that have the same expected values. Fishburn (1989) pointed out, "in view of the accumulated evidence for persistent and predictable violation of expected utility, new theories have been proposed to accommodate such violations without abandoning too much of the mathematical elegance of the traditional theories." Our risk-value theory is a further development toward achieving this goal.

References

Allais, M. (1953). Le comportement de l'homme rationnel devant le risque critique des postulats et axiomes de l'ecole americaine. *Econometrica*, 2, 503–546.
Allais, M. (1979). The foundations of a positive theory of choice involving risk and a criticism of the postulates and axioms of the American school. In M. Allais, O. Hagen (Eds.), *Expected utility hypotheses and the allais paradox* (pp. 27–145). Dordrecht: Holland: D Reidel.
Bell, D. E. (1985). Disappointment in decision making under uncertainty. *Operations Research, 33*, 1–27.
Bell, D. E. (1988). One-switch utility functions and a measure of risk. *Management Science, 34*, 1416–1424.
Bell, D. E. (1995). Risk return and utility. *Management Science, 41*, 23–30.
Butler, J. C., Dyer, J. S., and Jia, J. (2005). An empirical investigation of the assumptions of risk-value models. *Journal of Risk and Uncertainty, 30(2)*, pp. 133–156.
Coombs, C. H., & Bowen, J. N. (1971). A test of VE-theories of risk and the effect of the central limit theorem. *Acta Psychologica, 35*, 15–28.
Coombs, C. H., & Lehner, P. E. (1981). An evaluation of two alternative models for a theory of risk: Part 1. *Journal of Experimental Psychology Human Perception and Performance, 7*, 1110–1123.
Coombs, C. H., & Lehner, P. E. (1984). Conjoint design and analysis of the bilinear model: An application to judgments of risk. *Journal of Mathematical Psychology, 28*, 1–42.
Dyer, J. S., & Jia, J. (1997). Relative risk-value models. *European Journal of Operational Research, 103*, 170–185.
Dyer, J. S., & Jia, J. (1998). Preference conditions for utility models: A risk-value perspective. *Annals of Operations Research, 80*, 167–182.
Fishburn, P. C. (1977). Mean-risk analysis with risk associated with below-target returns. *American Economic Review, 67*, 116–126.
Fishburn, P. C. (1982). Foundations of risk measurement II Effects of gains on risk. *Journal of Mathematical Psychology, 25*, 226–242.
Fishburn, P. C. (1984). Foundations of risk measurement I Risk as probable loss. *Management Science, 30*, 296–406.
Fishburn, P. C. (1989). Generalizations of expected utility theories: A survey of recent proposals. *Annals of Operations Research, 19*, 3–28.
Fishburn, P. C., & Kochenberger, G. A. (1979). Two-piece von Neumann-Morgenstern utility functions. *Decision Sciences, 10*, 503–518.
Friedman, M., & Savage, L. P. (1948). The utility analysis of choices involving risk. *Journal of Political Economy, 56*, 279–304.
Jia, J. (1995). Measures of risk and risk-value theory. unpublished PhD Dissertation, University of Texas at Austin TX.
Jia, J., & Dyer, J. S. (1996). A standard measure of risk and risk-value models. *Management Science, 42*, 1961–1705.
Jia, J., Dyer, J. S., & Butler, J. C. (1999) Measures of perceived risk. *Management Science, 45*, 519–532.
Jia, J., Dyer, J. S., & Butler, J. C. (2001). Generalized disappointment models. *Journal of Risk and Uncertainty, 22*(1), 59–78.
Kahneman, D. H., & Tversky, A. (1979). Prospect theory: An analysis of decision under risk. *Econometrica, 47*, 263–290.
Keller, L. R., Sarin, R. K., & Weber, M. (1986). Empirical investigation of some properties of the perceived riskiness of gambles. *Organizational Behavior and Human Decision Process, 38*, 114–130.
Konno, H., & Yamazaki, H. (1992). Mean-absolute deviation portfolio optimization model and its applications to Tokyo stock market. *Management Science, 37*, 519–531.

Loomes, G., & Sugden, R. (1986). Disappointment and dynamic consistency in choice under uncertainty. *Review of Economic Studies, LIII*, 271–282.
Lowenstein, G., Weber, E. U., Hsee, C., & Welch, E. (2001). Risk as feelings. *Psychological Bulletin, 127*, 267–286.
Luce, R. D. (1980). Several possible measures of risk. *Theory and Decision, 12*, 217–228; Correction 1981:13 381.
Luce, R. D., & Weber, E. U. (1986). An axiomatic theory of conjoint expected risk. *Journal of Mathematical Psychology, 30*, 188–205.
Machina, M. J. (1987). Choice under uncertainty: Problems solved and unsolved. *Economic Perspectives, 1*(1), 121–154.
Markowitz, H. M. (1952). The utility of wealth. *Journal of Political Economy, 60*, 151–158.
Markowitz, H. M. (1959). *Portfolio selection*. New York: Wiley.
Markowitz, H. M. (1987). *Mean-variance analysis in portfolio choice and capital markets*. New York: Basil Blackwell.
Markowitz, H. M. (1991). Foundations of portfolio theory. *Journal of Finance, XLVI*, 469–477.
Meyer, J. (1987). Two-moment decision models and expected utility maximization. *American Economic Review, 77*, 421–430.
Payne, J. W., Laughhunn, D. J., & Crum, R. (1980). Translation of gambles and aspiration level effects in risky choice behavior. *Management Science, 26*, 1039–1060.
Payne, J. W., Laughhunn, D., & Crum, R. (1981). Further tests of aspiration level effects in risky choice behavior. *Management Science, 27*, 953–958.
Pollatsek, A., & Tversky, A. (1970). A theory of risk. *Journal of Mathematical Psychology, 7*, 540–553.
Pratt, J. W. (1964). Risk aversion in the small and in the large. *Econometrica, 32*, 122–136.
Quiggin, J. (1982). A theory of anticipated utility. *Journal of Economic Behavior and Organization, 3*, 323–343.
Sarin, R. K. (1987). Some extensions of Luce's measures of risk. *Theory and Decision, 22*, 25–141.
Sarin, R. K., & Weber, M. (1993). Risk-value models. *European Journal of Operational Research, 70*, 135–149
Savage, L. J. (1954). *The foundations of statistics*. New York: Wiley.
Sharpe, W. F. (1970). *Portfolio theory and capital markets*. New York: McGraw-Hill.
Sharpe, W. F. (1991). Capital asset prices with and without negative holdings. *Journal of Finance, XLVI*, 489–509.
Tversky, A., & Kahneman, D. H. (1992). Advances in prospect theory: Cumulative representation of uncertainty. *Journal of Risk and Uncertainty*, 5, 297–323.
von Neumann, J., & Morgenstern, O. (1947). *Theory of games and economic behavior*. Princeton, NJ: Princeton University Press.
Weber, E. U. (1984). Combine and conquer: a joint application of conjoint and functional approaches to the problem of risk measurement. *Journal of Experimental Psychology Human Perception and Performance, 10*, 179–194.
Weber, E. U. (2001). Decision and choice: Risk empirical studies. In N. Smelser, P. Baltes (Eds.), *International Encylopedia of the Social Sciences* (pp. 13347–13351). Oxford: Elsevier—.
Weber, E. U., Blais, A. -R., & Betz, N. (2004). Predicting risk sensitivity in humans and lower animals: Risk as variance or coefficient of variation. *Psychological Review, 111*, 430–445.
Weber, E. U., & Bottom, W. P. (1989). Axiomatic measures of perceived risk: Some tests and extensions. *Journal of Behavioral Decision Making, 2*, 113–131.
Weber, M. (1990). *Risikoentscheidungskalkule in der Finanzierungstheorie*. Stuttgart: Poeschel.
Wu. G., & Gonzalez, R. (1996). Curvature of the probability weighting function. *Management Science, 42*, 1676–1690.

Normally Distributed Admissible Choices are Optimal

James N. Bodurtha Jr and Qi Shen

1 Notation and Definitions

Generally accepted observable behavior has led to the following classes of continuously differentiable utility functions, $u(\bullet)$:

I. Nonsatiation axiom: $u' > 0$
II. Risk aversion: $u' > 0, u'' < 0$

Adopting the notation of (Bawa, 1975), let the uncertain prospects be characterized by random variables $x_i, i = 1, 2, \ldots, n+1$, with known continuous probability distribution functions defined over an open interval R^1 given by $(a,b), a < b$.

Let the following progressively restrictive set of utility functions, $u(\cdot)$, describe the decision maker's preferences. The utility functions are defined over the space R^1 of realizations of a random variable x:

$$U_1 = \{u(x)\,|u(x) \text{ is finite } u'(x) > 0, \text{ for all } x \in R\},$$
$$U_2 = \{u(x)\,|u(x) \in U_1, u''(x) < 0, \text{ for all } x \in R\}.$$

These definitions lead to the following well-known second-order stochastic dominance theorem and definition[1]:

[1] First Order Stochastic Dominance is developed assuming only non-satiation, (Quirk & Sapasnik, 1962) and (Fishburn, 1964). Assuming risk-aversion, several authors formulated second-order stochastic dominance, (Hadar & Russell, 1969, 1971), (Hanoch & Levy, 1969), and (Rothschild & Stiglitz, 1970, 1971). Third-Order Dominance (Whitmore, 1970), and decreasing absolute risk aversion (Vickson, 1975) treatments followed. Algorithms for the first three orders of stochastic dominancehave been specified, (Porter, Wart, & Ferguson, 1973), (Bawa, Lindenberg, & Rafsky, 1979) and (Aboudi & Thon, 1994). Levy provides a review of Stochastic Dominance (Levy, 1992). Convex Stochastic Dominance (CSD) identifies optimal choices among mutually exclusive alternatives (Fishburn, 1974, 1975), and the associated algorithm determines First-, Second- and Third-Order CSD (Bawa et al., 1985).

J.N. Bodurtha Jr (✉)
McDonough School of Business, Georgetown University, Washington, DC 20057, USA,
e-mail: bodurthj@georgetown.edu

S.J. Brams et al. (eds.), *The Mathematics of Preference, Choice and Order: Essays in Honor of Peter C. Fishburn*, Studies in Choice and Welfare,

Theorem 1. *Second-Order Stochastic Dominance (SSD). For any two cumulative distributions F_i and F_j, F_i is (strictly) preferred to F_j for all utility functions in U_2, if and only if*

$$\int_a^x F_i(t)\,dt \leq \int_a^x F_j(t)\,dt\,\forall x \in R \quad (\text{and } < \text{ for some} x \in R). \tag{1}$$

Definition 1. SSD Admissible Set - A subset C of choice set P, its members are not second-order stochastically dominated.

If a choice in P is not in subset C (not admissible), then all investors unanimously drop it from consideration. By dropping these choices, the SSD admissible set substantially reduces the full choice set. In the case of normally distributed choice alternatives, convex second-order dominance is an optimal choice rule.

Definition 2. Convex Second-Order Stochastic Dominance (CSSD) - A distribution function F_{n+1} is convex second-order stochastically dominated by $\{F_i, i = 1, 2, \ldots, n\}$, if $\forall u \in U_2$, there exists an $F_j \in \{F_1, F_2, \ldots, F_n\}$ such that

$$\int_a^b U(x)\,dF_j(x) \geq \int_a^b U(x)\,dF_{n+1}(x)$$

Correspondingly, we introduce the CSSD admissible set.

Definition 3. CSSD Admissible Set - A subset C of choice set P is CSSD admissible if $\forall u \in U_2$, choice a is not CSSD dominated by any other members of P. Since CSD admissibility is more restrictive than the usual SSD admissibility, the CSSD admissible set is generally smaller than the SSD admissible set.

Let $\lambda = (\lambda_1, \lambda_2 \ldots \lambda_n), \lambda \in \Lambda_n$ with $\lambda_i \geq 0, i = 1, 2, \ldots n$ and $\sum_{i=1}^n \lambda_i = 1$. We state the convex generalization of Theorem 1 (Fishburn, 1974).

Theorem 2. *Convex Second-Order Stochastic Dominance (CSSD). F_{n+1} is convex second-order stochastically dominated by $\{\mathrm{F_i, i = 1, 2, \ldots, n}\}$, iff $\lambda \in \Lambda_n$ such that*

$$\sum_{i=1}^n \lambda_i \int_a^x F_i(t)\,dt \leq \int_a^x F_{n+1}(t)\,dt \quad \forall x \in R$$
$$(\text{and } < \text{ for some } x \in R)$$

Conversely, if F_{n+1} is not convex second-order stochastically dominated by $\{\mathrm{F_i, i = 1, 2, \ldots, n}\}$ then it is optimal:

$$\forall \lambda \in \Lambda_n, \exists x \in R, \int_a^x F_{n+1}(t)\,dt < \textstyle\sum_{i=1}^n \lambda_i \int_a^x F_i(t)\,dt \Leftrightarrow$$
$$\exists u \in U_2, u(F_i) < u(F_{n+1}), \forall \{i = 1, 2, \ldots, n\}$$

Therefore, the CSSD admissible, C, is the optimal set.

We also define another important concept relevant to investment choice, the efficiency of a choice set.

Definition 4. Second-Order Efficient Set - A subset E of choice set P is second-order efficient if it contains the maximizers for all U_2.[2]

Obviously, investors with non-satiation and risk-aversion attributes should only evaluate the minimal second-order efficient choice set in order to make their investment decisions. We show that the minimal efficient choice set is the CSSD admissible set.

2 Optimal Choices Among Mutually Exclusive Alternatives

Convex Stochastic Dominance (CSD) identifies choice distribution mixtures that dominate other elements of the choice set (the dominated elements). Any choice dominated by a mixture of other alternatives will not be chosen (Fishburn, 1974). Conversely, any choice that is not so dominated is in the optimal set.

Our method of proof is straightforward. For normal distributions, the appropriate SD decision rule is second-order (SSD). Since normal distributions cross in most cases, first-order stochastic dominance (FSD) is precluded. Under Convex Second-Order Stochastic Dominance (CSSD), we show that the set of mixture distributions necessary to dominate any member of the admissible set is empty. Hence, the admissible set is optimal.

For mutually exclusive choices, the choice space may be written as the following:

$$P = \left\{ \sum_{i=1}^{n} \lambda_i F_i \,\middle|\, \lambda \in \Lambda_n, F_i \text{ is normal for } i = 1, 2, \ldots, n \right\},$$

In Appendix A, we prove two needed Lemmas.

The set of non-SSD dominated distributions (the admissible set) is no smaller than the set of non-CSSD dominated distributions (the optimal set). However, the following theorem shows that in the case of normal distributions, these two concepts coincide. In this case, the two choice sets are identical.

Proposition 1. *Given a set of normal distributions $\Phi = \{F_1, F_2, \ldots, F_n, F_{n+1}\}$, if Φ is a U_2 admissible set, then it is also the CSSD admissible set and optimal.*

Proof. Φ is an admissible set; therefore, distributions are mutually undominated. Since in the normal distribution case, SSD is equivalent to the mean-variance decision rule, we can order the distributions in Φ in such a way that

$$\sigma_1 < \sigma_2 < \cdots < \sigma_n, \text{ and } \mu_1 < \mu_2 < \cdots < \mu_n.$$

The mean and standard deviation of distribution F_{n+1} may be anywhere in the sequence of $F_1, F_2, \ldots F_n$.

[2] The equivalence between SD admissibility and efficiency for the portfolio allocation problem has been shown (Bawa & Goroff, 1983).

Case 1: $\sigma_{n+1} < \sigma_n = \max_{1\leq j\leq n}\{\sigma_j\}$. We divide the set Φ in two parts: $\Phi_1 = \{F_1,\ldots,F_k\}$, and $\Phi_2 = \{F_{k+1},\ldots,F_n\}$, such that $\mu_k < \mu_{n+1} < \mu_{k+1}$. and $\sigma_k < \sigma_{n+1} < \sigma_{k+1}$

We can take a degenerate distribution as a special case of the normal distribution, by defining its variance to be zero. We replace the set Φ_1 with another set $\hat{\Phi}_1$ such that

$$\mu\left(\hat{F}_i\right) - \mu\left(F_i\right), \quad \sigma\left(\hat{F}_i\right) = 0, \quad i = 1,2,\ldots k.$$

If F_{n+1} cannot be dominated by $\hat{\Phi}_1 \cup \Phi_2$, then F_{n+1} also can't be dominated by $\Phi_1 \cup \Phi_2$ (since each member of Φ_1 is dominated by the corresponding member in $\hat{\Phi}_1$). For members of set Φ_2, we choose a sufficiently small number, r, such that the Variance Dominance Rule can be applied to each element of Φ_2. For simplicity, we keep the notation of $F_i, i = 1,\ldots,k$, instead of $\hat{F}_i$.

From Lemma 1, for any given $\lambda_j > 0$, there exists an r_j such that

$$\int_{-\infty}^{r_j} F_{n+1}(t)\,dt < \lambda_j \int_{-\infty}^{r_j} F_j(t)\,dt, \quad j = k+1,\ldots n$$

Therefore, there exists a real number $\mathrm{r} \in \mathrm{R}$, $\mathrm{r} < \min\{\mu_\mathrm{i} : \mathrm{i} = 1,\ldots, \mathrm{k}, r_\mathrm{j} : \mathrm{j} = \mathrm{k}+1, \ldots, n\}$, for any given $\lambda \in \Lambda_n$,

$$\int_{-\infty}^{r} F_{n+1}(t)\,dt < \sum_{j=k+1}^{n} \lambda_j \int_{-\infty}^{r} F_j(t)\,dt =$$

$$\sum_{j=k+1}^{n} \lambda_j \int_{-\infty}^{r} F_j(t)\,dt + \sum_{j=1}^{k} \lambda_j \int_{-\infty}^{r} F_j(t)\,dt$$

$$\int_{-\infty}^{r} F_{n+1}(t)\,dt < \sum_{j=1}^{n} \lambda_j \int_{-\infty}^{r} F_j(t)\,dt$$

Here, we have used the fact that $\int_{-\infty}^{r} F_j(t)\,dt = 0$ for $j = 1,\ldots,k$, since $r < \mu_j$.

We have shown that F_{n+1} is not CSSD dominated by $\{F_1,\ldots F_n\}$.

Case 2: $\sigma_{n+1} > \max_{1\leq j\leq n}\{\sigma_j\} = \sigma_n$

In this case, from Lemma 1, there exists a sufficiently large number r_j, such that

$$\int_{r_j}^{+\infty} F_1(t)\,dt < \int_{r_j}^{+\infty} F_{n+1}(t)\,dt \quad j = 1,2,\ldots n.$$

Thus, $\int_{r}^{+\infty} F_{n+1}(t)\,dt = 1 - \int_{r}^{+\infty} F_{n+1}(t)\,dt$

$$< 1 - \sum_{j=1}^{n} \lambda_j \int_{r}^{+\infty} F_j(t)\,dt = \sum_{j=1}^{n} \lambda_j \int_{-\infty}^{r} F_j(t)\,dt$$

where $r > \max_{1\leq j\leq n}\{r_j\}$. In this case, we have shown that F_{n+1} can't be CSSD dominated by Φ. **Q.E.D.**

3 CSSD Portfolio Choices

For portfolio choices, a choice vector, π, dominates the associated mixed strategy, λ_π, for all strictly concave von Neumann-Morgenstern utility functions (Baron, 1977). We present a corollary to this result as Proposition 2.[3]

Our construct is, again, Fishburn's CSSD. Additionally, we need two more lemmas (3 and 4), which are also in Appendix A. Our CSSD efficient portfolio proposition follows:

Proposition 2. *The mean-variance efficient portfolio frontier choices are CSSD admissible.*

Proof. Given Lemma 4, any mixture of alternatives is dominated by an associated portfolio. Any portfolio not associated with the mean-variance efficient frontier is dominated by some element of the set of portfolios on the efficient frontier. Therefore, mean-variance efficient portfolio choices dominate mixtures of portfolio distributions, and all such portfolios are CSSD admissible.

Like mutually exclusive choice CSSD Proposition 1, Proposition 2 shows that the entire mean-variance efficient portfolio frontier is optimal.

4 Conclusion

For sets of investors with non-satiation and risk-aversion attributes, U_2, who face mutually exclusive normally distibuted investment returns, we have shown that the second-order stochastic dominance (SSD) admissible set is the optimal set (Bawa et al., 1985) and the strictly best set (Bawa & Goroff, 1982). By our CSSD methods (Fishburn, 1974), or from an analogous portfolio choice problem specification (Yitzhaki & Mayshar, 1997), we also know that efficient portfolio choices among normally-distributed alternatives are optimal. Therefore, we conclude that admissible sets of normally distributed choice elements are optimal.

In the absence of mean and variance parameter estimation risk, our results highlight Sharpe's classic mean-variance ratio as an optimal delegated financial management choice measure (Sharpe, 1966). In this context, a portfolio manager should

[3] This result has been proved in the context of Marginal Stochastic Dominance (Yitzhaki & Mayshar, 1997) for general discrete distributions and in an alternative context for normal distributions. As in our case, there results generalize to a broader class of exchangeable distribution functions. For the general discrete distribution cases, Post, like Yitzhaki–Mayshar, separates dominated and efficient portfolios (Post, 2003). For the dominated allocations, two works, (Kousmanen, 2004) and (Bodurtha, 2004), provide methods to identify efficient reallocations. The continuous distribution case has been treated as well.(Goroff & Whitt, 1980) For utility functions manifesting some risk-seeking preference, separation of dominated and efficient portfolio allocations have been analyzed, (Post & Levy, 2005).

identify inefficient or dominated choice set elements by this simple mean-variance rule and should not reduce the choice set further before presenting choices to investors.[4]

Appendix – Lemmas

For the mutually exclusive choice case, we now state and prove two lemmas.

Lemma 1. *Variance Dominance Rule*

Given two distributions F_1 andF_2 with finite variances σ_1^2 and σ_2^2, if we let $\sigma_1 < \sigma_2$, then there exist three numerals x^*, r_1, and r_2 (with $r_1 < r_2$), such that

I. the density functions $f_1(x)$ and $f_2(x)$ satisfy $f_1(x) < f_2(x)$, if $x < r_1$ or $x > r_2$
II. the distribution functions have the same value at x^* and satisfy: $F_1(x) < F_2(x) \quad \text{if } x < x^*$ or $F_1(x) > F_2(x) \quad \text{if } x > x^*$.

Proof. The proof has three steps.[5]

Step 1: There are exactly two intersection points for $f_1(x)$ and $f_2(x)$.
Therefore, the following equation must have exactly two real roots:

$$\sigma_1 e^{\frac{(x-\mu_1)^2}{2\sigma_1^2}} = \sigma_2 e^{\frac{(x-\mu_2)^2}{2\sigma_2^2}} \tag{A.1}$$

---

[4] Though we have noted that our results extend to some other continuous "location-scale" distributions [e.g. (Bawa, 1975)], It has been shown that the SSD admissible set and various "optimal" sets are not, in general, equal, (Peleg-Yaari 1975), (Peleg, 1975), (Bawa & Goroff, 1982), and (Dybvig & Ross, 1982). Further analysis of the respective "risk-aversely efficient" and "regular risk-aversely efficient" random variables, "strictly best choices," and "portfolio efficient sets" is needed. In the portfolio context and more generally for the mutually exclusive investment choices (Dybvig & Ross, 1982), the potential for non-convex choice sets raises particular difficulties in this analysis. Alternatively, the admissible set is dense in the optimal-strictly best set (Bawa & Goroff, 1982). Therefore, the delegated manager who provides decision makers with admissible choices is not grossly non-optimal.

[5] By replacing the mean and variance with the Generalized Location and Scale (ℓ,s) parameters, this proof will show that mean-scale admissible densities within the following classes cross twice: t distributions with the same degree of freedom, Cauchy distributions and log-normal distributions. In these cases, the densities are, like the normal, functions of a standardized random variable, $((x-\ell)/s)^2$. The differences between and ratios of any two admissible choices for these distributions satisfy Lemma 1 (the double crossing property defined in location and scale) and Lemma 2 (the distribution dominance condition.) Though no analytic density functions exist for Stable Distributions other than the normal and Cauchy, the densities associated with stable distributions with the same characteristic exponent and skewness parameter also cross-twice. While the uniform distribution is in the location scale family and admissible uniform distributions are optimal, the switching nature of the mean-scale admissible rule over the range of uniform random variables precludes our line of proof. (Bawa, 1975, 1979).

Taking a logarithm of both sides of this equation, and collecting terms, we have

$$0 = \left(\sigma_1^2 - \sigma_2^2\right)x^2 + 2x\left(\mu_1\sigma_2^2 - \mu_1\sigma_1^2\right) + \left(\sigma_1^2\mu_2^2 - \sigma_2^2\mu_1^2 - 2\sigma_1^2\sigma_2^2 \ln\frac{\sigma_1}{\sigma_2}\right)$$

We then define the determinant as:

$$\Delta = 4\left(\mu_1\sigma_2^2 - \mu_2\sigma_1^2\right)^2 - 4\left(\sigma_1^2 - \sigma_2^2\right)\left[\sigma_1^2\mu_2^2 - \sigma_2^2\mu_1^2 - 2\sigma_1^2\sigma_2^2 \ln\frac{\sigma_1}{\sigma_2}\right] \quad \text{(A.2)}$$

To show that this determinant is greater than zero, we show that the first term on the right-hand side of equation (A.2) is greater than a quantity that is, itself, greater than the second term on the right-hand side of equation (A.2).

Since $\ln\frac{\sigma_1}{\sigma_2} > 0$, we need only to show that, $\left(\mu_1\sigma_2^2 - \mu_2\sigma_1^2\right)^2 \geq \left(\sigma_1^2 - \sigma_2^2\right)\left[\sigma_1^2\mu_2^2 - \sigma_2^2\mu_1^2\right]$. This inequality is equivalent to $(\mu_2 - \mu_1)^2 \geq 0$, so we denote the two real roots as r_1 and r_2.

Step 2: To show (I), we reconsider equation (A.1). Let

$$h(x) \equiv \sigma_1 e^{(x-\mu_1)^2/2\sigma_1^2} - \sigma_2 e^{(x-\mu_2)^2/2\sigma_2^2}.$$

Following Step 1, it is straightforward to verify that

$$h'(x) < 0, x \in (-\infty, r_1) \text{ and } h'(x) > 0, x \in (r_2, +\infty)$$

Step 3: To show II, notice that $F_1(\infty) = F_2(\infty) = 1$.

Since $F_1(r_1) = \int_{-\infty}^{r_1} f_1(t)\,dt < \int_{-\infty}^{r_1} f_2(t)\,dt = F_2(r_1)$, and $\int_{r_2}^{\infty} f_1(t)\,dt < \int_{r_2}^{\infty} f_2(t)\,dt$, it must be that $\int_{r_1}^{r_2} f_1(t)\,dt > \int_{r_1}^{r_2} f_2(t)\,dt$.

Both $F_1(x)$ and $F_2(x)$ are increasing continuous functions on $(-\infty, \infty)$.[6] Therefore, there exists a unique $x^* \in (r_1, r_2)$, such that

$$F_1(x^*) = \int_{-\infty}^{x^*} f_1(t)\,dt = \int_{-\infty}^{x^*} f_2(t)\,dt = F_2(x^*), \text{ and}$$
$$F_1(x) < F_2(x)\, if\, x < x^* \ \textbf{Q.E.D.}$$

The first part of the Variance Dominance Rule states that the density function curve for the smaller variance distribution, F_1, always lies below the other one with larger variance, F_2, on the interval $(-\infty, r_1)$. However, a reversed relationship is true on an interval $(r_2, +\infty)$.

[6] In the log-normal case, $F_1(x)$ and $F_2(x)$ are increasing continuous functions on $(0, \infty)$. The log-normal density crossing points, r_1', r_2', are defined by location-scale and determined in the log space. While the distribution crossing point is a unique $x^{*'} \in \left(e^{r_1'}, e^{r_2'}\right)$, and distribution dominance follows in the return space.

Lemma 2. *Given two distribution functions, as in Lemma 1, the value of $F_1(x)$ is negligible compared to the value of $F_2(x)$ if x is sufficiently small.*[7]

Proof. By L'Hopital's Law, we show that

$$\lim_{x \to -\infty} \frac{F_2(x)}{F_1(x)} = \lim_{x \to -\infty} \frac{F_2'(x)}{F_1'(x)} = +\infty$$

Since $\frac{f_2(x)}{f_1(x)} = \frac{\sigma_1}{\sigma_2} e^{(x-\mu_1)^2/2\sigma_1^2 - (x-\mu_2)^2/2\sigma_2^2}$, we show that

$$\sigma_2^2 (x-\mu_1)^2 - \sigma_1^2 (x-\mu_2)^2 \to +\infty \tag{A.3}$$

This is self-evident since $\sigma_1 < \sigma_2$. Similarly, we show that

$$\lim_{x \to +\infty} \frac{f_2(x)}{f_1(x)} = +\infty \text{ Q.E.D.}$$

This Lemma is another interpretation of the Variance Dominance Rule, and states that the distribution curve of larger variance not only dominates the distribution curve with a smaller variance, but also that the magnitude of the latter one is actually negligible. In fact as $x \to -\infty$, $F_1(x)$ approaches 0 much faster than $F_2(x)$ does.

For the portfolio choice case, we now state and prove two additional lemmas.

Lemma 3. *The SSD integral, (1), is convex.*

Proof. The SSD integral is a twice continuously differentiable real-valued function on an open interval. Furthermore, its second derivative is the normal density and hence, non-negative throughout its domain. Convexity follows by Theorem 4.4 of Rockefellar, and essentially strict convexity follows by his Theorem 26.3 (Rockefellar, 1970). (The SSD integral gradient is the normal distribution and is positive over the real line.)

Lemma 4. *A portfolio of normally distributed choices SSD dominates the associated mixture of normally distributed choices.*

Proof. Given Lemma 3 [convexity of the SSD integral (1)], a convex combination (mixture) of these integrals is no less than the SSD integral defined over the linear combination (portfolio) of the associated random variables.

[7] These limits apply for mean-scale admissible t distributions with the same degree of freedom and stable distributions with the same characteristic exponent and skewness. For mean-scale admissible log-normal distributions, (A.3) is defined in location, ℓ, and scale parameters, s. The necessary SSD log-normal distribution mean condition is imposed with $\ell_2 + s_2^2/2 > \ell_1 + s_1^2/2$. As in the normal case for x, the terms that are quadratic in ln x are the difference in squared scale, which is positive. In this case, the limits are evaluated approaching zero from the right, and all other terms are linear in the natural logarithm of x.

With integration by parts, we have the following:

$$\int_{-\infty}^{x} F_i(t)\,dt=\sigma_i\left[\left(\frac{x-\mu_i}{\sigma_i}\right)\Phi\left(\frac{x-\mu_i}{\sigma_i}\right)+\phi\left(\frac{x-\mu_i}{\sigma_i}\right)\right], \text{ and } \Phi\left(\frac{x-\mu_i}{\sigma_i}\right) \text{ and } \phi\left(\frac{x-\mu_i}{\sigma_i}\right),$$

are the standard normal distribution and density, respectively.

For a portfolio to CSSD dominate a mixture requires

$$\begin{aligned}
&\sigma_p\left[\left(\frac{x-\mu_p}{\sigma_p}\right)\Phi\left(\frac{x-\mu_p}{\sigma_p}\right)+\phi\left(\frac{x-\mu_p}{\sigma_p}\right)\right]\\
&\quad\leq \alpha\sigma_1\left[\left(\frac{x-\mu_1}{\sigma_1}\right)\Phi\left(\frac{x-\mu_1}{\sigma_1}\right)+\phi\left(\frac{x-\mu_1}{\sigma_1}\right)\right]\\
&\qquad+(1-\alpha)\,\sigma_2\left[\left(\frac{x-\mu_2}{\sigma_2}\right)\Phi\left(\frac{x-\mu_2}{\sigma_2}\right)+\phi\left(\frac{x-\mu_2}{\sigma_2}\right)\right], \forall x\in .
\end{aligned}$$

Defining the portfolio weights to equal the mixture weights, we have

$$\begin{aligned}
&x_p=\alpha x_1+(1-\alpha)\,x_2, \mu_p=\alpha\mu_1+(1-\alpha)\,\mu_2, \text{ and}\\
&\sigma_p^2=\alpha^2\sigma_1^2+2\alpha(1-\alpha)\,\sigma_1\sigma_2\rho+(1-\alpha)^2\,\sigma_2^2\neq[\alpha\sigma_1+(1-\alpha)\,\sigma_2]^2,
\end{aligned}$$

However, setting the correlation equal to one implies that the portfolio standard deviation is a convex combination of the other two standard deviations, and that this standard deviation is an upper bound on the actual portfolio standard deviation:

$$\sigma_p\leq\sigma_{p|\rho=1}=\alpha\sigma_1+(1-\alpha)\,\sigma_2$$

Therefore,

$$\begin{aligned}
&\sigma_p\left[\left(\frac{x-\mu_p}{\sigma_p}\right)\Phi\left(\frac{x-\mu_p}{\sigma_p}\right)+\phi\left(\frac{x-\mu_p}{\sigma_p}\right)\right]\\
&\quad\leq\sigma_{p|\rho=1}\left[\left(\frac{x-\mu_p}{\sigma_{p|\rho=1}}\right)\Phi\left(\frac{x-\mu_p}{\sigma_{p|\rho=1}}\right)+\phi\left(\frac{x-\mu_p}{\sigma_{p|\rho=1}}\right)\right]\\
&\quad\leq \alpha\sigma_1\left[\left(\frac{x-\mu_1}{\sigma_1}\right)\Phi\left(\frac{x-\mu_1}{\sigma_1}\right)+\phi\left(\frac{x-\mu_1}{\sigma_1}\right)\right]\\
&\qquad+(1-\alpha)\,\sigma_2\left[\left(\frac{x-\mu_2}{\sigma_2}\right)\Phi\left(\frac{x-\mu_2}{\sigma_2}\right)+\phi\left(\frac{x-\mu_2}{\sigma_2}\right)\right],
\end{aligned}$$

$\forall x\in(-\infty,\infty)$ and $0<\alpha<1$. **Q.E.D.**

References

Aboudi, R., & Thon, D. (1994). Efficient algorithms for stochastic dominance tests based on financial data. *Management Science, 40*, 508–515

Afriat, S. (1967). The construction of a utility function from expenditure data. *International Economic Review, 8*, 67–77

Bawa, V. (1975). Optimal rules for ordering uncertain prospects. *Journal of Financial Economics, 2*, 95–121

Bawa, V. (1979). *Lecture notes: Generalized location and scale family of distributions*. NYU, unpublished

Bawa, V., Bodurtha, J., Jr., Rao, M. R., & Suri, H. (1985). On determination of stochastic dominance optimal sets. *Journal of. Finance, 40*, 417–431

Bawa, V., & Goroff, D. (1982). *Admissible efficient, and best choices under uncertainty*. University of Texas-Austin, Department of Finance working paper 81/82-2-8, 1982

Bawa, V., & Goroff, D. (1983). Stochastic dominance, efficiency and separation in financial markets. *Journal of Economic Theory, 30*, 410–414

Bawa, V., Lindenberg, E., & Rafsky, L. (1979). An efficient algorithm to determine stochastic dominance admissible sets. *Management Science, 25*(7), 609–622

Bodurtha, J., Jr., (2004) *Efficient portfolio reallocation for all risk-averse investors*. Georgetown University working paper

Dybvig, P., & Ross, S. (1982). Portfolio efficient sets. *Econometrica, 50*, 1525–1546

Fishburn, P. (1964). Decision and value theory. New York: Wiley

Fishburn, P. (1974). Convex stochastic dominance with continuous distribution functions. *Journal of Economic Theory, 7*, 143–158

Fishburn, P. (1975). Convex stochastic dominance. In W. Ziemba, & R. Vickson (Eds.), *Stochastic optimization models in finance*. New York: Academic.

Goroff, D., & Whitt, W. (1980). Approximating the admissible set in stochastic dominance. *Journal of Economic Theory, 23*, 218–235

Green, R., & Srivastiva, S. (1986). Expected utility maximization and demand behavior. *Journal of Economic Theory, 38*, 313–328

Green, R., & Srivastiva, S. (1985). Risk aversion and arbitrage. *Journal of Finance, 50*, 257–268

Hadar, J., & Russell, W. (1969). Rules for ordering uncertain prospects. *American Economic Review, 59*, 25–34

Hadar, J., & Russell, W. (1971). Stochastic dominance and diversification. *Journal of Economic Theory, 3*, 288–305

Hanoch, G., & Levy, H. (1969). The efficiency analysis of choices involving risk. *Review of Economic Studies, 36*, 335–346

Kuosmanen, T. (2004). Efficient diversification according to stochastic dominance criteria. *Management Science, 50*, 1390–1406

Meyer, J. (1979). Mean variance efficient sets and expected utility. *Journal of Finance, 34*, 1221–1229

Peleg, B. (1975). Efficient random variables. Journal of Mathematical *Economics, 2*, 243–252

Peleg, B., & Yaari, M. (1975). A price characterization of efficient random variables. *Econometric, 43*, 283–292

Porter, R. B., Wart, J., & Ferguson, D. (1973). Efficient algorithms for conducting stochastic dominance tests on large numbers of portfolios. *Journal of Financial and Quantitative Analysis, 8*, 71–81

Post, T. (2003). Empirical tests for stochastic dominance efficiency. *Journal of Finance, 58*, 1905–1931

Post, T. (2005). *Note on an alternative lpm-related dominating portfolio algorithm*. correspondence

Post, T., & Levy, H. (2005). Does risk-seeking drive asset prices. *Review of Financial Studies, 18*, 925–953

Quirk, J., & Sapasnik, R. (1962). Admissibility and measurable utility functions. *Review of Economic Studies, 29*, 140–146
Rothschild, M., & Stiglitz, J. (1970). Increasing risk I: A definition. *Journal of Economic Theory, 2*, 225–243
Rothschild, M., & Stiglitz, J. (1971). Increasing Risk II: Its economic consequences. *Journal of Economic Theory, 3*, 66–68
Sharpe, W. (1966). Mutual fund performance. *Journal of Business, 39*, 119–138
Varian, H. (1982). The non-parametric approach to demand analysis. *Econometrica, 50*, 945–973
Varian, H. (1983). Non-parametric tests of models of investor behavior. *Journal of Financial and Quantitative Analysis, 18*, 269–278
Vickson, R. (1975). Stochastic dominance for decreasing absolute risk aversion. *Journal of Financial and Quantitative Analysis, 10*, 799–811
Whitmore, G. (1970). Third order stochastic dominance. *American Economic Review, 50*, 457–459
Yitzhaki, S., & Mayshar, J. (1997). *Charaterizing efficient portfolios*. Hebrew University of Jerusalem working paper

A Conjoint Measurement Approach to the Discrete Sugeno Integral

Denis Bouyssou, Thierry Marchant, and Marc Pirlot

1 Introduction and Motivation

In the area of decision-making under uncertainty, the use of fuzzy integrals, most notably the Choquet integral and its variants, has attracted much attention in recent years. It is a powerful and elegant way to extend the traditional model of (subjective) expected utility (on this model, see Fishburn, 1970, 1982). Indeed, integrating with respect to a non-necessarily additive measure allows to weaken the independence hypotheses embodied in the additive representation of preferences underlying the expected utility model that have often been shown to be violated in experiments (see the pioneering experimental findings of Allais, 1953; Ellsberg, 1961). Models based on Choquet integrals have been axiomatized in a variety of ways (see Gilboa, 1987; Schmeidler, 1989; or Wakker, 1989, Chap. 6. For related works in the area of decision-making under risk, see Quiggin, 1982; and Yaari, 1987). Recent reviews of this research trend can be found in Chateauneuf and Cohen (2000), Schmidt (2004), Starmer (2000) and Sugden (2004).

More recently, still in the area of decision-making under uncertainty, Dubois, Prade, and Sabbadin (2000b) have suggested to replace the Choquet integral by a Sugeno integral (see Sugeno, 1974, 1977), the latter being a kind of "ordinal counterpart" of the former, and provided an axiomatic analysis of this model (special cases of the Sugeno integral are analyzed in Dubois, Prade, & Sabbadin 2001b. For a related analysis in the area of decision-making under risk, see Hougaard & Keiding, 1996). Dubois, Marichal, Prade, Roubens, and Sabbadin (2001a) offer a lucid survey of these developments.

Unsurprisingly, people working in the area of multiple criteria decision making (henceforth, MCDM) have considered following a similar path to build models weakening the independence hypotheses embodied in the additive value function

M. Pirlot (✉)
Faculté Polytechnique de Mons, 9 rue de Houdain, 7000 Mons, Belgium
e-mail: marc.pirlot@fpms.ac.be

S.J. Brams et al. (eds.), *The Mathematics of Preference, Choice and Order: Essays in Honor of Peter C. Fishburn*, Studies in Choice and Welfare,

model that underlies most of existing MCDM techniques. This offers an alternative to the decomposable and polynomial models studied in Krantz, Luce, Suppes, and Tversky (1971, Chap. 7). The work of Grabisch (1995, 1996) has widely popularized the use of Choquet and Sugeno integrals in MCDM. Since then, there has been many developments in this area. They are surveyed in Grabisch and Roubens (2000) and Grabisch and Labreuche (2004) (an alternative approach to weaken the independence hypotheses of the traditional model that does not use fuzzy integrals is suggested in Gonzales & Perny, 2005).

It is well known that decision-making under uncertainty and MCDM are related areas. When there is only a finite number of states of nature, acts may indeed be viewed as elements of a homogeneous Cartesian product in which the underlying set is the set of all consequences (this is the approach advocated and developped in Wakker, 1989, Chap. 4). In the area of MCDM, a Cartesian product structure is also used to model alternatives. However, in MCDM the product set is generally not homogeneous: alternatives are evaluated on several attributes that do not have to be expressed on the same scale.

The recent development of the use of fuzzy integrals in the area of MCDM should not obscure the fact that there is a major difficulty involved in the transposition of techniques coming from decision-making under uncertainty to the area of MCDM. In the former area, any two consequences can easily be compared: considering constant acts gives a straightforward way to transfer a preference relation on the set of acts to the set of consequences. The situation is vastly different in the area of MCDM. The fact that the underlying product set is not homogeneous invalidates the idea to consider "constant acts". Therefore, there is no obvious way to compare consequences on different attributes. Yet, such comparisons seem to be prerequisite for the application of models based on fuzzy integrals.

Traditional conjoint measurement models (see, e.g., Krantz et al., 1971, Chap. 6; or Wakker, 1989, Chap. 3) lead to compare *preference differences* between consequences. It is indeed easy to give a meaning to a statement like "the preference difference between consequences x_i and y_i on attribute i is equal to the preference difference between consequences x_j and y_j on attribute j" (e.g., because they exactly compensate the same preference difference expressed on a third attribute). These models do *not* lead to comparing in terms of preference consequences expressed on distinct attributes. Indeed, in the additive value function model a statement like "x_i is better than x_j" is easily seen to be meaningless (this is reflected in the fact that, in this model, the origin of the value function on each attribute may be changed independently on each attribute).

In order to bypass this difficulty, most studies involving fuzzy integrals in the area of MCDM postulate that the attributes are somehow "commensurate", while the precise content of this hypothesis is difficult to analyze and test (see, e.g., Dubois,

Grabisch, Modave, & Prade, 2000a). Less frequently, researchers have tried to build attributes so that this commensurability hypothesis is adequate. This is the path followed in Grabisch, Labreuche, and Vansnick (2003) who use the MACBETH technique (see Bana e Costa & Vansnick, 1994, 1997, 1999) to build such scales. Such an analysis requires the assessment of a neutral level on each attribute that is supposed to be "equally attractive". In practice, the assessment of such levels does not seem to be an easy task. On a more theoretical level, the precise properties of these commensurate neutral levels are not easy to devise.

A major breakthrough for the application of fuzzy integrals in MCDM has recently been done in Greco, Matarazzo, and Słowiński (2004) who give conditions characterizing binary relations on non-homogeneous product sets that can be represented using a discrete Sugeno integral, using this binary relation as the only primitive. This is an important result that paves the way to a measurement-theoretic analysis of fuzzy integrals in the area of MCDM (Greco et al., 2004, also relate the discrete Sugeno integral model to models based on decision rules that they have advocated in Greco, Matarazzo, & Slowinski, 1999, 2001). It allows to analyze the discrete Sugeno integral model without any commensurateness hypothesis, which is of direct interest to MCDM.

In the present paper, we will present a new model for the representation of preferences, inspired from the work of Bouyssou and Marchant (2007). This non-numerical model, called non-compensatory model, is slightly more general than the discrete Sugeno integral but, when the preference relation is a weak order that has a numerical representation, we will show that both models are equivalent. The analysis of this new model will thus help us to better understand the discrete Sugeno integral and, eventually, to answer some open questions. In particular, we will address the following issues:

- Besides the standard completeness, transitivity and order density conditions, Greco et al. (2004) used only one condition. We will show that it is possible to factorize this condition into two more elementary ones. This helps us to better understand the behavioural content of the conditions. It can also be useful for empirically testing the conditions. Finally, this will permit us to show that the discrete Sugeno integral model can be viewed as a particular case of a general decomposable representation, investigated in Bouyssou and Pirlot (2004) and Greco et al. (2004).
- The correspondence established between weak orders that are representable in the noncompensatory model and those representable by the discrete Sugeno integral model has an interesting byproduct. Starting from any (bounded) numerical representation of a weak order in the noncompensatory model, we provide formulae that allow to build a representation of the weak order by a Sugeno integral.
- Greco et al. (2004) used four conditions in their characterization of the discrete Sugeno integral. We will prove that they are independent.
- In the standard characterizations of the additive model for multi-attributed preferences (e.g., Wakker, 1989), no commensurateness hypothesis is made. Yet, it is well-known that the difference between two levels on attribute i can be compared to the difference between two levels on attribute j. So, in this model, differences

are commensurate and this can be derived from the axioms. This plays an important role in most elicitation techniques.

In their characterization, Greco et al. (2004) did not make any commensurateness hypothesis either. Yet, when we compute a discrete Sugeno integral, we compare levels on different attributes. So, just as with the additive model, it seems that commensurateness must be implied by the axioms and that this could be used in the elicitation. Unfortunately, we will show that the picture is more complex with the discrete Sugeno integral than with the additive model.

- Greco et al. (2004) have shown that, under some conditions, there exists utility functions (one per attribute) that can be used to represent the preferences by means of a discrete Sugeno integral. These utility functions are of course not unique; but to what extent? We will provide a partial answer to this question.

By the way, since the non-compensatory model and the discrete Sugeno integral are equivalent under some conditions, our proof of the characterization of the non-compensatory model can be used as a proof of the characterization of the discrete Sugeno integral. This can prove useful since no proof of it has been published so far.[1]

This paper is organized as follows. The result of Greco et al. (2004) is presented in Sect. 2. We there show how to factorize their main condition into two simpler conditions. Section 3 introduces and characterizes what we will call the noncompensatory model for weak orders. Section 4 analyzes the links between the noncompensatory model for weak orders and the discrete Sugeno integral model. Section 5 presents examples showing that the conditions used in the main result are independent. Section 6 discusses the uniqueness of the representation in the discrete Sugeno integral model and further investigates the commensurateness issue. Section 7 briefly concludes with the mention of some directions for future research.

2 The Discrete Sugeno Integral

2.1 Background on the Discrete Sugeno Integral

Let $\beta = (\beta_1, \beta_2, \dots, \beta_p) \in [0,1]^p$. Let $(\cdot)_\beta$ be a permutation on $P = \{1, 2, \dots, p\}$ such that $\beta_{(1)_\beta} \leq \beta_{(2)_\beta} \leq \dots \leq \beta_{(p)_\beta}$.

A capacity (see Choquet, 1953) on P is a function $v : 2^P \to [0,1]$ such that:

[1] It should be mentioned that a related result for the case of ordered categories is presented without proof in Słowiński, Greco, and Matarazzo (2002). This result is a particular case of the one presented in Greco et al. (2004) for weak orders with a finite number of distinct equivalence classes. A complete and quite simple proof for this particular case was proposed in Bouyssou and Marchant (2007), using comments made on an early version of the latter paper by Greco, Matarazzo, and Słowiński.

- $\nu(\varnothing) = 0$.
- $[A, B \in 2^P \text{ and } A \subseteq B] \Rightarrow \nu(A) \leq \nu(B)$.

The capacity ν is said to be normalized if, furthermore, $\nu(P) = 1$.

The discrete Sugeno integral of the vector $(\beta_1, \beta_2, \dots, \beta_p) \in [0,1]^p$ w.r.t. the normalized capacity ν is defined by

$$S_\nu[\beta] = \bigvee_{i=1}^{p} \left[\beta_{(i)_\beta} \wedge \nu(A_{(i)_\beta})\right],$$

where $A_{(i)_\beta}$ is the element of 2^P equal to $\{(i)_\beta, (i+1)_\beta, \dots, (p)_\beta\}$.

We refer the reader to Dubois, et al. (2001a) and Marichal (2000a, 2000b) for excellent surveys of the properties of the discrete Sugeno integral and its several possible equivalent definitions. Let us simply mention here that the reordering of the components of β in order to compute its Sugeno integral can be avoided noting that we may equivalently write

$$S_\nu[\beta] = \bigvee_{T \subseteq P} \left[\nu(T) \wedge \left(\bigwedge_{i \in T} \beta_i\right)\right]. \tag{1}$$

We will mainly use this presentation of the discrete Sugeno integral below.

2.2 The Model

Let $\succsim$ be a binary relation on a set $X = \prod_{i=1}^{n} X_i$ with $n \geq 2$. Elements of X will be interpreted as alternatives evaluated on a set $N = \{1, 2, \dots, n\}$ of attributes. The relations $\succ$ and $\sim$ are defined as usual. We denote by X_{-i} the set $\prod_{j \in N \setminus \{i\}} X_j$. We abbreviate $Not[x \succsim y]$ as $x \not\succsim y$.

We say that $\succsim$ has a representation in the *discrete Sugeno integral model* if there are a normalized capacity μ on N and functions $u_i : X_i \to [0,1]$ such that, for all $x, y \in X$,

$$x \succsim y \Leftrightarrow S_{\langle \mu, u \rangle}(x) \geq S_{\langle \mu, u \rangle}(y),$$

where $S_{\langle \mu, u \rangle}(x) = S_\mu[(u_1(x_1), u_2(x_2), \dots, u_n(x_n))]$.

2.3 Axioms and Result

A *weak order* is a complete and transitive binary relation. The set $Y \subseteq X$ is said to be dense in X for the weak order $\succsim$ if for all $x, y \in X$, $x \succ y$ implies $x \succsim z$ and $z \succsim y$, for some $z \in Y$. We say that the weak order $\succsim$ on X satisfies the *order-denseness condition* (condition *OD*) if there is a finite or countably infinite set $Y \subseteq X$ that is dense in X for $\succsim$. It is well-known (see Fishburn, 1970, p. 27; or Krantz et al., 1971, p. 40) that there is a real-valued function v on X such that, for all $x, y \in X$,

$$x \succsim y \Leftrightarrow v(x) \geq v(y),$$

if and only if $\succsim$ is a weak order on X satisfying the order-denseness condition.

Remark 1. Let $\succsim$ be a weak order on X. It is clear that $\sim$ is an equivalence and that the elements of $X/\sim$ are linearly ordered. We often abuse terminology and speak of equivalence classes of $\succsim$ to mean the elements of $X/\sim$. When $X/\sim$ is finite, we speak of the first equivalence class of $\succsim$ to mean the elements of $X/\sim$ that precede all others in the induced linear order.

The following condition was introduced in Greco et al. (2004). The relation $\succsim$ on X is said to be strongly 2-graded on attribute $i \in N$ (condition 2^*-graded$_i$) if, for all $x, y, z, w \in X$ and all $a_i \in X_i$,

$$\left.\begin{array}{c} x \succsim z \\ \text{and} \\ y \succsim w \\ \text{and} \\ z \succsim w \end{array}\right\} \Rightarrow \left\{\begin{array}{c} (a_i, x_{-i}) \succsim z \\ \text{or} \\ (x_i, y_{-i}) \succsim w, \end{array}\right.$$

where (a_i, x_{-i}) denotes the element of X obtained from $x \in X$ by replacing its ith coordinate by $a_i \in X_i$. The binary relation will be said to be *strongly 2-graded* (condition 2^*-graded) if it is strongly 2-graded on all attributes $i \in N$.

Although the above condition may look complex, it has a simple interpretation. Consider the particular case of condition 2^*-graded$_i$ in which $z = w$. Suppose that $(x_i, y_{-i}) \not\succsim w$. Since $(y_i, y_{-i}) \succsim w$ and $(x_i, y_{-i}) \not\succsim w$, this suggests that the level x_i is worse than y_i with respect to the alternative w. In this case, $(x_i, x_{-i}) \succsim w$ implies that $(a_i, x_{-i}) \succsim w$, for all $a_i \in X_i$. This means that, once we know that some level y_i is better than x_i w.r.t. to $w \in X$, there does not exist an element in X_i that could be worse than x_i, so that, if $(x_i, x_{-i}) \succsim w$, the same will be true replacing x_i by any element in X_i. This roughly implies that, for each $w \in X$, we can partition the elements of X_i into at most two categories of levels: the "satisfactory" ones and the "unsatisfactory" ones with respect to w. Condition 2^*-graded$_i$ implies these twofold partitions are not unrelated when considering distinct elements z and w in X.

Greco et al. (2004) state the following:

Theorem 1 (Greco et al. (2004, Theorem 3, p. 284)). *Let $\succsim$ be a binary relation on X. This relation has a representation in the discrete Sugeno integral model if and only if (iff) it is a weak order satisfying the order-denseness condition and being strongly 2-graded.*

The necessity of the conditions in this theorem is easy to establish. It is indeed clear that if $\succsim$ has a representation in the discrete Sugeno integral model, then it must be a weak order satisfying *OD*. It is not difficult to show that it must also satisfy 2^*-graded. Indeed, suppose that condition 2^*-graded$_i$ is violated, so that, for some $x, y, z, w \in X$ and some $a_i \in X_i$, we have $x \succsim z$, $y \succsim w$, $z \succsim w$, $(a_i, x_{-i}) \not\succsim z$ and $(x_i, y_{-i}) \not\succsim w$. Using $y \succsim w$ and $(x_i, y_{-i}) \not\succsim w$, we obtain $u_i(x_i) < S_{\langle\mu,u\rangle}(w)$. Because $z \succsim w$, we know that $S_{\langle\mu,u\rangle}(z) \geq S_{\langle\mu,u\rangle}(w)$, so that $S_{\langle\mu,u\rangle}(z) > u_i(x_i)$. Since $x \succsim z$

and $S_{\langle\mu,u\rangle}(z) > u_i(x_i)$, there is some $I \in 2^N$ such that $i \notin I$, $\mu(I) \geq S_{\langle\mu,u\rangle}(z)$ and $u_j(x_j) \geq S_{\langle\mu,u\rangle}(z)$, for all $j \in I$. This implies $S_{\langle\mu,u\rangle}((a_i,x_{-i})) \geq S_{\langle\mu,u\rangle}(z)$, so that $(a_i,x_{-i}) \succsim z$, a contradiction.

In Sect. 4, we give a proof of the sufficiency of the conditions, which links the discrete Sugeno integral model with the noncompensatory model studied in Sect. 3.

2.4 Factorization of 2^**-Graded*$_i$

We say that the relation $\succsim$ satisfies condition $AC1_i$ if, for all $x,y,z,w \in X$,

$$\left.\begin{array}{c} x \succsim y \\ \text{and} \\ z \succsim w \end{array}\right\} \Rightarrow \left\{\begin{array}{c} (z_i,x_{-i}) \succsim y, \\ \text{or} \\ (x_i,z_{-i}) \succsim w. \end{array}\right.$$

We say that $\succsim$ satisfies $AC1$ if it satisfies $AC1_i$ for all $i \in N$.

Condition $AC1$ was proposed and studied in Bouyssou and Pirlot (2004). It plays a central role in the characterization of binary relations (that may be incomplete or intransitive) admitting a decomposable representation of the type:

$$x \succsim y \Leftrightarrow G[u_1(x_1),\ldots,u_n(x_n),u_1(y_1),\ldots,u_n(y_n)] \geq 0,$$

with G being nondecreasing (resp. nonincreasing) in its first (resp. last) n arguments (see Bouyssou & Pirlot, 2004, Theorem 2). We refer to Bouyssou and Pirlot (2004) for a detailed interpretation of this condition. Let us simply mention here that condition $AC1_i$, independently of any transitivity or completeness properties of $\succsim$, allows to order the elements of X_i in such a way that this ordering is compatible with $\succsim$ (see Lemma 3 below).

We say that $\succsim$ is 2-graded on attribute $i \in N$ (condition 2-graded$_i$) if, for all $x,y,z,w \in X$ and all $a_i \in X_i$,

$$\left.\begin{array}{c} x \succsim z \\ \text{and} \\ (y_i,x_{-i}) \succsim z \\ \text{and} \\ y \succsim w \\ \text{and} \\ z \succsim w \end{array}\right\} \Rightarrow \left\{\begin{array}{c} (a_i,x_{-i}) \succsim z \\ \text{or} \\ (x_i,y_{-i}) \succsim w. \end{array}\right.$$

We say that $\succsim$ is *2-graded* (condition 2-graded) if it is 2-graded on all attributes $i \in N$. Condition 2-graded weakens condition 2^*-graded adjoining it the additional premise $(y_i,x_{-i}) \succsim z$. It has a similar interpretation. We have:

Lemma 1. *Let $\succsim$ be a weak order on the set X. Then $\succsim$ satisfies $AC1_i$ and 2-graded$_i$ iff it satisfies 2^*-graded$_i$.*

Proof. [$AC1_i$ & 2-graded$_i$ $\Rightarrow$ 2*-graded$_i$]. Suppose that $x \succsim z$, $y \succsim w$ $z \succsim w$. Using $AC1_i$, $x \succsim z$ and $y \succsim w$ implies either $(y_i, x_{-i}) \succsim z$ or $(x_i, y_{-i}) \succsim w$. In the latter case, one of the two conclusions of 2*-graded$_i$ holds. In the former case, we have $x \succsim z$, $(y_i, x_{-i}) \succsim z$, $y \succsim w$ and $z \succsim w$, so that 2-graded$_i$ implies either $(a_i, x_{-i}) \succsim z$, for all $a_i \in X_i$ or $(x_i, y_{-i}) \succsim w$, which is the desired conclusion.

[2*-graded$_i$ $\Rightarrow$ $AC1_i$ & 2-graded$_i$]. It is clear that 2*-graded$_i$ implies 2-graded$_i$ since 2-graded$_i$ is obtained from 2*-graded$_i$ by adding to it an additional premise. Suppose that $x \succsim y$ and $z \succsim w$. Since $\succsim$ is complete, we have either $y \succsim w$ or $w \succsim y$. If $y \succsim w$, we have $x \succsim y$, $z \succsim w$ and $y \succsim w$, so that 2*-graded$_i$ implies $(x_i, z_{-i}) \succsim w$ or $(a_i, x_{-i}) \succsim y$, for all $a_i \in X_i$. Taking $a_i = z_i$ shows that $AC1_i$ holds in this case. The proof is similar if it is supposed that $w \succsim y$. □

Why is this factorization interesting? First, it makes clear that the condition used by Greco et al. (2004) combines two distinct properties: (1) the elements of X_i can be ordered and (2) for each $w \in X$, we can partition the elements of X_i into at most two categories with respect to w. This helps us better understand the behavioural content of the conditions. It can also be useful for empirically testing the validity of the discrete Sugeno integral model. Indeed, if we run an experiment for testing whether a complex condition (like 2*-graded) is satisfied by subjects, it is likely that it will be rejected. This does not mean that the condition is completely wrong. It can happen that only part of it is wrong. Therefore, testing more elementary conditions can help identify what is wrong with a model. Finally, this factorization permit us to show that the discrete Sugeno integral model can be viewed as a particular case of a general decomposable representation, investigated and characterized in Bouyssou and Pirlot (2004) and Greco et al. (2004). Furthermore, thanks to the factorization, we know exactly what has to be imposed on the decomposable model in order to obtain the discrete Sugeno integral model.

3 The Noncompensatory Model for Weak Orders

This section presents and characterizes the noncompensatory model for weak orders. It will turn out to have intimate connections with the discrete Sugeno integral model.

The following non-numerical model is inspired from the work of Słowiński et al. (2002) and Bouyssou and Marchant (2007) who analyze ordered partitions of a Cartesian product using similar models. A similar model was first suggested in Fishburn (1978).

Definition 1. A weak order $\succsim$ on X has a representation in the *noncompensatory model* if for all $x \in X$, there are sets:

1. $A_i^x \subseteq X_i$, for all $i \in N$.
2. $F^x \subseteq 2^N$ such that

$$[I \in F^x \text{ and } I \subseteq J \in 2^N] \Rightarrow J \in F^x, \tag{2}$$

that are such that, for all $x, y \in X$,

$$x \succsim y \Rightarrow \begin{cases} A_i^x \subseteq A_i^y \\ \text{and} \\ F^x \subseteq F^y \end{cases} \tag{3}$$

and

$$x \succsim y \Leftrightarrow \{i \in N : x_i \in A_i^y\} \in F^y. \tag{4}$$

We often write $A(x,y)$ instead of $\{i \in N : x_i \in A_i^y\}$.

The noncompensatory model[2] can be interpreted as follows. For each $x \in X$ we isolate on each attribute a subset $A_i^x \subseteq X_i$ containing the levels on attribute i that are satisfactory for x. In order for an alternative to be at least as good as x, it must have evaluations that are satisfactory for x on a subset of attributes belonging to F^x. The subsets of attributes belonging to F^x are interpreted as subsets that are "sufficiently important" to warrant preference on x.

With this interpretation in mind, the constraint (3) means that if x is at least as good as y then every level that is satisfactory for x must be satisfactory for y. Furthermore, subsets of attributes that are "sufficiently important" to warrant preference on x must also be "sufficiently important" to warrant preference on y. Given the above interpretation of F^x, the constraint (2) simply says that any superset of a set that is "sufficiently important" to warrant preference on x must have the same property.

Suppose that $x \not\succsim y$ and that $x_i \in A_i^y$, for some $i \in N$. In the noncompensatory model, we have $(z_i, x_{-i}) \not\succsim y$, for all $z_i \in X_i$. It is therefore impossible, starting from x, to obtain an alternative that would be at least as good as y by modifying the evaluation of x on the ith attribute. In other terms, the fact that $A(x,y) \notin F^y$ cannot be compensated by improving the evaluation of x on an attribute in $A(x,y)$. Hence, our name for this model.

We first observe that a weak order having a representation in the noncompensatory model must satisfy $AC1$ and 2-graded.

Lemma 2. *If weak order $\succsim$ on X has a representation in the noncompensatory model, then it satisfies $AC1$ and 2-graded.*

Proof. [$AC1_i$]. Suppose that $x \succsim y$, $z \succsim w$, $(z_i, x_{-i}) \not\succsim y$ and $(x_i, z_{-i}) \not\succsim w$. It is easy to see that $x \succsim y$ and $(z_i, x_{-i}) \not\succsim y$ imply $x_i \in A_i^y$ and $z_i \notin A_i^y$. Similarly, $z \succsim w$ and $(x_i, z_{-i}) \not\succsim w$ imply $z_i \in A_i^w$ and $x_i \notin A_i^w$. Because $\succsim$ is complete, we have either $y \succsim w$ or $w \succsim y$. Hence, we have either $A_i^y \subseteq A_i^w$ or $A_i^w \subseteq A_i^y$, a contradiction.

[2-graded$_i$]. Suppose that 2-graded$_i$ is violated, so that, for some $x, y, z, w \in X$ and some $a_i \in X_i$, $(x_i, x_{-i}) \succsim z$, $(y_i, x_{-i}) \succsim z$, $(y_i, y_{-i}) \succsim w$, $z \succsim w$, $(a_i, x_{-i}) \not\succsim z$ and

[2] The noncompensatory model for weak orders must not be confused with "noncompensatory preferences" as introduced in Fishburn (1976). Noncompensatory preferences in the sense of Fishburn (1976) are preferences that result from an "ordinal aggregation" in the context of MCDM that is quite close from the type of aggregation studied in social choice theory in the vein of Arrow (1963) (for a recent analysis of such preferences, see Bouyssou and Pirlot (2005)). As first shown in Fishburn (1975), noncompensatory preferences that are weak orders are, except in degenerate cases, lexicographic.

$(x_i, y_{-i}) \not\succsim w$. Using the definition of the noncompensatory model, $(y_i, y_{-i}) \succsim w$ and $(x_i, y_{-i}) \not\succsim w$ imply $y_i \in A_i^w$ and $x_i \notin A_i^w$. Similarly, $(x_i, x_{-i}) \succsim z$ and $(a_i, x_{-i}) \not\succsim z$ imply $x_i \in A_i^z$ and $a_i \notin A_i^z$. Since $z \succsim w$, we have $A_i^z \subseteq A_i^w$, a contradiction. □

The main result of this section says that, for weak orders, the noncompensatory model is fully characterized by condition 2*-graded or, equivalently, by the conjunction of $AC1$ and 2-graded.

Proposition 1. *If a weak order on X satisfies AC*1 *and* 2*-graded then it has a representation in the noncompensatory model.*

Before proving Proposition 1, we will have to go through a few definitions and lemmas.

Consider an attribute $i \in N$. We define the *left marginal trace* on attribute $i \in N$ letting, for all $x_i, y_i \in X_i$, all $a_{-i} \in X_{-i}$ and all $z \in X$,

$$x_i \succsim_i^+ y_i \Leftrightarrow [(y_i, a_{-i}) \succsim z \Rightarrow (x_i, a_{-i}) \succsim z].$$

Similarly, given $a \in X$, we define the left marginal trace on attribute $i \in N$ with respect to $a \in X$, letting, for all $x_i, y_i \in X_i$ and all $z_{-i} \in X_{-i}$,

$$x_i \succsim_i^{+(a)} y_i \Leftrightarrow [(y_i, z_{-i}) \succsim a \Rightarrow (x_i, z_{-i}) \succsim a].$$

The symmetric and asymmetric parts of $\succsim_i^+$ (resp. $\succsim_i^{+(a)}$) are denoted $\sim_i^+$ and $\succ_i^+$ (resp. $\sim_i^{+(a)}$ and $\succ_i^{+(a)}$). It is clear that $\succsim_i^+$ and $\succsim_i^{+(a)}$ are always reflexive and transitive. They may be incomplete however.

We note a few useful obvious connections between $\succsim_i^{+(a)}$, $\succsim_i^+$ and $\succsim$ in the following lemma.

Lemma 3. *We have, for all $i \in N$, all $z, w \in X$ and all $x_i, y_i \in X_i$:*

1. $x_i \succsim_i^+ y_i \Leftrightarrow [x_i \succsim_i^{+(a)} y_i$, *for all* $a \in X]$.
2. $[z \succsim w, x_i \succsim_i^+ z_i] \Rightarrow (x_i, z_{-i}) \succsim w$.
3. *Furthermore, if* $\succsim$ *is reflexive then,* $[z_j \sim_j^+ w_j$, *for all* $j \in N] \Rightarrow z \sim w$.
4. *The relation* $\succsim_i^+$ *is complete iff* $AC1_i$ *holds.*

Proof. Parts 1 and 2 easily follow from the definitions. Part 3 follows from Part 2 and the fact that $w \succsim w$. It is obvious that negating the completeness of $\succsim_i^+$ is equivalent to negating $AC1_i$. □

Remark 2. When $\succsim$ is a weak order, condition $AC1_i$ is equivalent to supposing that, for all $x_i, y_i \in X_i$ and all $z_{-i}, w_{-i} \in X_{-i}$ $(x_i, z_{-i}) \succ (y_i, z_{-i}) \Rightarrow (x_i, w_{-i}) \succsim (y_i, w_{-i})$, i.e., that attribute i is weakly separable, using the terminology of Bouyssou and Pirlot (2004).

Indeed suppose that $\succsim$ satisfies $AC1_i$ and is such that attribute i is not weakly separable. Therefore there are $x_i, y_i \in X_i$ and $z_{-i}, w_{-i} \in X_{-i}$ such that $(x_i, z_{-i}) \succ (y_i, z_{-i})$ and $(y_i, w_{-i}) \succ (x_i, w_{-i})$. Since $\succsim$ is reflexive, we have $(x_i, z_{-i}) \succsim (x_i, z_{-i})$

and $(y_i, w_{-i}) \succsim (y_i, w_{-i})$. Using $AC1_i$, we have either $y_i \succsim_i^+ x_i$ or $x_i \succsim_i^+ y_i$, so that either $(y_i, z_{-i}) \succsim (x_i, z_{-i})$ or $(x_i, w_{-i}) \succsim (y_i, w_{-i})$, a contradiction.

Conversely, suppose that $\succsim$ is complete and transitive and that attribute i is weakly separable. Suppose that $AC1_i$ is violated so that, since $\succsim$ is complete, $(x_i, x_{-i}) \succsim y$, $(z_i, z_{-i}) \succsim w$, $y \succ (z_i, x_{-i})$ and $w \succ (x_i, z_{-i})$, for some $x, y, z, w \in X$. Since $\succsim$ is a weak order, we obtain $(x_i, x_{-i}) \succ (z_i, x_{-i})$ and $(z_i, z_{-i}) \succ (x_i, z_{-i})$, which violates the weak separability of attribute i.

We say that a weak order $\succsim$ is *weakly separable* if, for all $i \in N$, it is weakly separable for attribute i.

Hence, combining Lemma 1 with Theorem 1 shows that a relation has a representation in the discrete Sugeno integral model iff it is a weakly separable weak order satisfying *OD* and 2-graded.

Bouyssou and Pirlot (2004, Propositions 8 and B.3) have shown that, for weak orders satisfying *OD*, weak separability is a necessary and sufficient condition to obtain a general decomposable representation in which, for all $x, y \in X$,

$$x \succsim y \Leftrightarrow F[u_1(x_1), \ldots, u_n(x_n)] \geq F[u_1(y_1), \ldots, u_n(y_n)],$$

with F being nondecreasing in all its arguments (see also Greco et al., 2004, Theorem 1). Hence, condition 2-graded is exactly what must be added to go from this general decomposable representation to a representation in the discrete Sugeno integral model.

The following lemma makes precise the structure of the relations $\succsim_i^{+(a)}$ when $\succsim$ is a weak order satisfying $AC1_i$ and 2-graded$_i$.

Lemma 4. *Let $\succsim$ be a weak order on X satisfying $AC1_i$ and 2-graded$_i$. Then:*

1. *$\succsim_i^{+(a)}$ is complete for all $a \in X$.*
2. *$x_i \succ_i^{+(a)} y_i \Rightarrow [x_i \succsim_i^{+(b)} y_i$ for all $b \in X]$.*
3. *$\succsim_i^{+(a)}$ has at most two distinct equivalence classes, for all $a \in X$.*
4. *$[x_i \sim_i^{+(a)} z_i$ and $x_i \succ_i^{+(a)} y_i] \Rightarrow x_i \sim_i^{+(b)} z_i$, for all $b \in X$ such that $a \succsim b$.*
5. *If $a \succsim b$ and both $\succsim_i^{+(a)}$ and $\succsim_i^{+(b)}$ are nontrivial then the first equivalence class of $\succsim_i^{+(a)}$ is included in the first equivalence class of $\succsim_i^{+(b)}$.*

Proof. Parts 1 and 2 follow from Lemma 3 since $AC1_i$ implies that $\succsim_i^+$ is complete.

Part 3. Suppose that $\succsim_i^{+(a)}$ has at least three distinct equivalence classes. This implies that $(x_i, c_{-i}) \succsim a$, $(y_i, c_{-i}) \not\succsim a$, $(y_i, d_{-i}) \succsim a$ and $(z_i, d_{-i}) \not\succsim a$, for some $x_i, y_i, z_i \in X_i$, some $c_{-i}, d_{-i} \in X_{-i}$ and some $a \in X$. Using $AC1_i$, $(x_i, c_{-i}) \succsim a$, $(y_i, d_{-i}) \succsim a$ and $(y_i, c_{-i}) \not\succsim a$ imply $(x_i, d_{-i}) \succsim a$. Using 2-graded$_i$, $(y_i, d_{-i}) \succsim a$, $(x_i, d_{-i}) \succsim a$, $(x_i, c_{-i}) \succsim a$ and $a \succsim a$ imply $(y_i, c_{-i}) \succsim a$ or $(z_i, d_{-i}) \succsim a$, a contradiction.

Part 4. Suppose that $x_i \sim_i^{+(a)} z_i$, $x_i \succ_i^{+(a)} y_i$, $a \succsim b$ and $x_i \succ_i^{+(b)} z_i$ (the proof for the case $z_i \succ_i^{+(b)} x_i$ being similar). By construction, we have $(x_i, w_{-i}) \succsim b$, $(z_i, w_{-i}) \not\succsim b$, $(x_i, t_{-i}) \succsim a$ and $(y_i, t_{-i}) \not\succsim a$. Since $x_i \sim_i^{+(a)} z_i$, we must have $(z_i, t_{-i}) \succsim a$. Using $AC1_i$, $(x_i, w_{-i}) \succsim b$, $(z_i, t_{-i}) \succsim a$ and $(z_i, w_{-i}) \not\succsim b$ imply $(x_i, t_{-i}) \succsim a$. Using

2-graded$_i$, $(z_i,t_{-i}) \succsim a$, $(x_i,t_{-i}) \succsim a$, $(x_i,w_{-i}) \succsim b$ and $a \succsim b$ imply $(z_i,w_{-i}) \succsim b$ or $(y_i,t_{-i}) \succsim a$, a contradiction.

Part 5. Suppose that $a \succsim b$, $x_i \succ_i^{+(a)} y_i$ and $z_i \succ_i^{+(b)} x_i$. Using Part 2, we know that $z_i \succsim_i^{+(a)} x_i$. Because we know from Part 3 that $\succsim_i^{+(a)}$ has at most two equivalence classes, we must have $z_i \sim_i^{+(a)} x_i$. Using Part 4, $a \succsim b$, $z_i \sim_i^{+(a)} x_i$ and $x_i \succ_i^{+(a)} y_i$ imply $z_i \sim_i^{+(b)} x_i$, a contradiction. □

Let $\succsim$ be a weak order on X satisfying $AC1_i$ and 2-graded$_i$. Let $i \in N$. For all $a \in X$, we know that either $\succsim_i^{+(a)}$ is trivial or $\succsim_i^{+(a)}$ has two distinct equivalence classes. Define $B_i^a \subset X_i$ as the empty set in the first case and as the elements in the first equivalence class in the second case. Define C_i^a letting:

$$C_i^a = \bigcup_{\{x \in X : x \succsim a\}} B_i^x.$$

The following lemma studies the properties of the sets C_i^a.

Lemma 5. *Let $\succsim$ be a weak order on X satisfying AC1 and 2-graded. For all $x,y,z,w \in X$ and all $i \in N$:*

1. $z \succsim w \Rightarrow C_i^z \subseteq C_i^w$.
2. $\{j \in N : y_j \in C_j^z\} \subseteq \{j \in N : x_j \in C_j^z\} \Rightarrow [x_i \succsim_i^{+(z)} y_i$ for all $i \in N]$.
3. $C_i^x \subsetneq X_i$.

Proof. Part 1. We have $x_i \in C_i^z$ iff $x_i \in B_i^a$, for some $a \succsim z$. Because $z \succsim w$ and $\succsim$ is a weak order, we have $a \succsim z$. Hence, $x_i \in B_i^a$, for some $a \succsim w$, so that $x_i \in C_i^w$.

Part 2. If $\succsim_i^{+(z)}$ is trivial, we have by definition $x_i \sim_i^{+(z)} y_i$. If $\succsim_i^{+(z)}$ is not trivial, it follows from Part 5 of Lemma 4 that C_i^z is equal to the first equivalence class of $\succsim_i^{+(z)}$. If $y_i \in C_i^z$, we have $x_i \in C_i^z$, so that $x_i \sim_i^{+(z)} y_i$. If $y_i \notin C_i^z$, then we have $z_i \succsim_i^{+(z)} y_i$.

Part 3. By construction, B_i^y is strictly included in X_i. As the set C_i^x is obtained by taking the union of sets B_i^y, the conclusion follows. □

Lemma 6. *Let $\succsim$ be a weak order on X satisfying $AC1_i$ and 2-graded$_i$. Define, for all $x \in X$, the set $G^x \subseteq 2^N$ letting $I \in G^x$ whenever we have $\{i \in N : z_i \in C_i^x\} \subseteq I$, for some $z \in X$ such that $z \succsim x$. We have, for all $x,y \in X$:*

1. $x \succsim y \Leftrightarrow \{i \in N : x_i \in C_i^y\} \in G^y$.
2. $[I \in G^x$ and $I \subseteq J] \Rightarrow J \in G^x$.
3. $x \succsim y \Rightarrow G^x \subseteq G^y$.

Proof. Part 1. By construction, if $x \succsim y$ then $\{i \in N : x_i \in C_i^y\} \in G^y$. Let us show that the reverse implication is true. Suppose that $\{i \in N : x_i \in C_i^y\} \in G^y$. This implies that $\{i \in N : z_i \in C_i^y\} \subseteq \{i \in N : x_i \in C_i^y\}$, for some $z \in X$ such that $z \succsim y$. Using Part 2 of Lemma 5, $\{i \in N : z_i \in C_i^y\} \subseteq \{i \in N : x_i \in C_i^y\}$ implies $x_i \succsim_i^{+(y)} z_i$, for all $i \in N$. Hence, $z \succsim y$ implies $x \succsim y$.

Part 2 follows from the definition of the sets G^x.

Part 3. Suppose that $x \succsim y$ and let $I \in G^x$. Let us show that we must have $I \in G^y$. By construction, $I \in G^x$ implies that $\{i \in N : z_i \in C_i^x\} \subseteq I$, for some $z \in X$ such that $z \succsim x$. Consider the alternative $w \in X$ defined in the following way:

- If $z_i \in C_i^x$, let $w_i = z_i$. We have $w_i \in C_i^x$. Using Part 1 of Lemma 5, we know that this implies $w_i \in C_i^y$.
- If $z_i \notin C_i^x$. Using Part 3 of Lemma 5, we know that $C_i^y \subsetneq X_i$. We take w_i to be any element in $X_i \setminus C_i^y$. Because, we know that $C_i^x \subseteq C_i^y$, we have $w_i \notin C_i^x$.

By construction we have, for all $i \in N$, $z_i \in C_i^x \Leftrightarrow w_i \in C_i^x \Leftrightarrow w_i \in C_i^y$. Hence, we have $\{i \in N : z_i \in C_i^x\} = \{i \in N : w_i \in C_i^x\} = \{i \in N : w_i \in C_i^y\}$. The first equality implies $w \succsim x$. Using the fact that $\succsim$ is a weak order, we obtain $w \succsim y$. Hence, we have $\{i \in N : w_i \in C_i^y\} \subseteq I$ and $w \succsim y$. This implies $I \in G^y$. □

Defining $A_i^x = C_i^x$ and $F^x = G^x$, the sufficiency proof of Proposition 1 follows from combining Lemmas 5 and 6.

4 The Noncompensatory Model and the Discrete Sugeno Integral Model

The main result in this section says that if a weak order has a representation in the noncompensatory model and has a numerical representation, then it has a representation in the discrete Sugeno integral model. This will help to complete the proof of Theorem 1.

Proposition 2. *Let $\succsim$ be a weak order on X. Suppose that $\succsim$ can be represented in the noncompensatory model and that there is a real function v on X such that, for all $x, y \in X$,*

$$x \succsim y \Leftrightarrow v(x) \geq v(y). \tag{5}$$

Then $\succsim$ has a representation in the discrete Sugeno integral model.

Proof. Let $\succsim$ be a weak order representable in the noncompensatory model and such that there is a real-valued function v satisfying (5). We may assume w.l.o.g. that, for all $x \in X$, $v(x) \in [0,1]$. Furthermore, if there are minimal elements in X for $\succsim$, we may assume w.l.o.g. that v gives the value 0 to these elements. We consider now any such function v. For all $i \in N$, define u_i letting, for all $x_i \in X_i$,

$$u_i(x_i) = \begin{cases} \sup_{\{w \in X : x_i \in A_i^w\}} v(w) & \text{if } \exists w : x_i \in A_i^w, \\ 0 & \text{otherwise.} \end{cases} \tag{6}$$

Define μ on 2^N letting, for all $I \in 2^N$,

$$\mu(I) = \begin{cases} \sup_{\{w \in X : I \in F^w\}} v(w) & \text{if } \exists w : I \in F^w, \\ 0 & \text{otherwise.} \end{cases} \tag{7}$$

Since $I \in F^w$ and $J \supseteq I$ entails $J \in F^w$, we have that $\mu(J) \geq \mu(I)$. Hence, μ is a nondecreasing set function.

Let us show that $\mu(\varnothing) = 0$. If there is no $w \in X$ such that $\varnothing \in F^w$, then we have, by construction, $\mu(\varnothing) = 0$. Suppose that $X_\varnothing = \{w \in X : \varnothing \in F^w\} \neq \varnothing$. From the definition of the noncompensatory model, it follows that, for all $x \in X$ and all $w \in X_\varnothing$, we have $x \succsim w$. Hence, for all $w \in X_\varnothing$, w is minimal for $\succsim$. We therefore have $v(w) = 0$, for all $w \in X_\varnothing$ and, hence, $\mu(\varnothing) = 0$. This shows that μ defined by (7) is a capacity on 2^N. It is not necessarily normalized, i.e., we may not have that $\mu(N) = 1$.

Independently of the normalization of μ, we can compute, for all $x \in X$, $S_{\mu,\mathbf{u}}(x)$ letting:

$$S_{\langle\mu,u\rangle}(x) = \bigvee_{I \subseteq N} \left[\mu(I) \wedge \left(\bigwedge_{i \in I} u_i(x_i) \right) \right]. \tag{8}$$

It is clear that, for all $y \in X$, $S_{\langle\mu,u\rangle}(y) \in [0,1]$. Let us show that, for all $y \in X$, $S_{\langle\mu,u\rangle}(y) = v(y)$, which will complete the proof if μ happens to be normalized.

Let $x, y \in X$ be such that $x \succsim y$. This implies $A(x,y) = \{i \in N : x_i \in A_i^y\} \in F^y$. Hence, for all $i \in A(x,y)$, $y \in \{w \in X : x_i \in A_i^w\}$, so that $u_i(x_i) \geq v(y)$. Similarly, $y \in \{w \in X : A(x,y) \in F^w\}$, so that $\mu(A(x,y)) \geq v(y)$. Hence, for $I = A(x,y)$, we have

$$\mu(I) \wedge \left(\bigwedge_{i \in I} u_i(x_i) \right) \geq v(y).$$

In view of (8), this implies $S_{\langle\mu,u\rangle}(x) \geq v(y)$. Since $\succsim$ is reflexive, this shows that, for all $y \in X$, $S_{\langle\mu,u\rangle}(y) \geq v(y)$.

We now prove that, for all $y \in X$, $S_{\langle\mu,u\rangle}(y) \leq v(y)$. If y is maximal for $\succsim$ (i.e., $y \succsim x$, for all $x \in X$), we have $v(y) \geq v(x)$, for all $x \in X$. The definition of u_i and μ obviously implies that they cannot exceed the maximal value of v on X. Hence, in this case, we have $S_{\langle\mu,u\rangle}(y) \leq v(y)$.

Suppose henceforth that $y \in X$ is not maximal for $\succsim$, so that $x \succ y$, for some $x \in X$. This implies that $A(y,x) = \{i \in N : y_i \in A_i^x\} \notin F^x$. Define $A_y = \bigcup_{z \succ y} A(y,z)$. Because $A(y,z) \subseteq N$, N is a finite set, and $z' \succsim z$ implies $A(y,z') \subseteq A(y,z)$, there is an element $z_0 \in X$ with $z_0 \succ y$ that is such that $A(y,z_0) = A_y$ and $A(y,z) = A_y$, for all $z \in X$ such that $z_0 \succsim z \succ y$.

We claim the following:

Claim 1: for all $j \notin A_y$, $u_j(y_j) \leq v(y)$.
Claim 2: for all $I \subseteq A_y$, $\mu(I) \leq v(y)$.

Proof of Claim 1. Let $j \notin A_y$, so that $y_j \notin A_j^{z_0}$. If the set $\{w \in X : y_j \in A_j^w\}$ is empty, we have $u_j(y_j) = 0$ and the claim trivially holds. Otherwise, let $w \in X$ such that $y_j \in A_j^w$. If $w \succ z_0$, we have $A_j^w \subseteq A_j^{z_0}$, so that $y_j \in A_j^w$ implies $y_j \in A_j^{z_0}$, a contradiction. If $z_0 \succsim w \succ y$, we know that $A(y,w) = A(y,z_0)$. This is contradictory since $y_j \in A_j^w$ and $y_j \notin A_j^{z_0}$. Hence, when $j \notin A_y$, we must have $y \succsim w$, for all $w \in X$ such that $y_j \in A_j^w$. This implies that $u_j(y_j) = \sup_{\{w \in X : y_j \in A_j^w\}} v(w) \leq v(y)$, for all $j \notin A_y$.

Proof of Claim 2. Let $I \subseteq A_y$. If the set $\{w \in X : I \in F^w\}$ is empty, we have $\mu(I) = 0$ and the claim follows. Otherwise, let $w \in X$ such that $I \in F^w$. Suppose that $w \succ z_0$. This implies $F^w \subseteq F^{z_0}$, so that $I \in F^{z_0}$. Because $I \subseteq A_y$, we obtain $A_y \in F^{z_0}$. This is contradictory since $z_0 \succ y$ implies that $A_y = A(y, z_0) \notin F^{z_0}$. Suppose now that $z_0 \succsim w \succ y$. We have $A(y, w) = A_y \notin F^w$. But, since $I \in F^w$ and $I \subseteq A_y$, we obtain $A_y \in F^w$, a contradiction. Hence, for all $w \in X$ such that $I \in F^w$, we have $y \succsim w$. This implies $\mu(I) = \sup_{\{w \in X : I \in F^w\}} v(w) \leq v(y)$.

Using Claims 1 and 2, we establish that $S_{\langle \mu, u \rangle}(y) \leq v(y)$ for any $y \in X$ that is not maximal. Let $I \subseteq N$. We distinguish two cases in order to compute

$$\mu(I) \wedge \left(\bigwedge_{i \in I} u_i(x_i) \right).$$

1. If I is not included in A_y, we know that there is $j \in I$ such that $j \notin A_y$. Hence, using Claim 1, $u_j(y_j) \leq v(y)$ so that $\mu(I) \wedge (\bigwedge_{i \in I} u_i(y_i)) \leq v(y)$.
2. If I is included in A_y, using Claim 2, we have $\mu(I) \leq v(y)$. Hence, we know that $\mu(I) \wedge (\bigwedge_{i \in I} u_i(y_i)) \leq v(y)$.

Hence, for all $I \subseteq N$, we have $\mu(I) \wedge (\bigwedge_{i \in I} u_i(y_i)) \leq v(y)$, so that $S_{\langle \mu, u \rangle}(y) \leq v(y)$. This proves that, for all $y \in X$, $S_{\langle \mu, u \rangle}(y) = v(y)$.

It remains to show that we may always build a representation in the discrete Sugeno integral model using a *normalized* capacity, i.e., a capacity v such that $v(N) = 1$.

Using the above construction, the value of $\mu(N)$ is obtained using (7). We have $\mu(N) = \sup_{w \in X} v(w)$, since for all $w \in X$, $N \in F^w$. If the weak order $\succsim$ is not trivial, we have $\mu(N) > 0$. In order to obtain a representation leading to a normalized capacity, it suffices to apply the above construction to the function u obtained by dividing v by $\mu(N)$. If the weak order $\succsim$ is trivial, it is easy to see that it has a representation in the noncompensatory model such that, for all $x \in X$ and all $i \in N$, $A_i^x = X_i$ and $F^x = \{N\}$. Defining, for all $i \in N$ and all $x_i \in X_i$, $u_i(x_i) = 1$, $\mu(N) = 1$ and $\mu(A) = 0$, for all $A \subsetneq N$, leads to a representation of this trivial weak order in the discrete Sugeno integral model. □

The sufficiency proof of Theorem 1 follows from combining Lemma 1 with Propositions 1 and 2. This amounts to characterizing the discrete Sugeno integral model by the conjunction of any of the following three equivalent sets of conditions:

- Completeness, transitivity, *OD*, *AC*1 and 2-graded
- Completeness, transitivity, *OD*, weak separability and 2-graded
- Completeness, transitivity, *OD* and 2*-graded

The examples in the following section show no condition in the first set is redundant.

Remark 3. Consider a nontrivial weak order $\succsim$ on X that satisfies the hypotheses of Proposition 2. The proof of this proposition establishes that *any* function $v : X \to [0,1]$ satisfying (5) and giving a value 0 to the minimal elements in X for $\succsim$ (if any) can be used to define a representation in the Sugeno integral model. The functions u_i and the (non-necessarily normalized) capacity μ used in this representation can be defined on the basis of v using (6) and (7).

In other words, any (bounded) numerical representation v of a weak order representable in the noncompensatory model is essentially a Sugeno integral. By "essentially", we mean that a positive affine transformation may have to be applied first to the numerical representation v in order that the minimal elements in X (if any) receive the value 0 and that the supremum of v is 1. This transformation is only needed to ensure that $\mu(\varnothing) = 0$ and μ is a normalized capacity. Note that applying (6) and (7) to any bounded numerical representation of the preference would yield u_i's and μ such that formula (8) would restate the value of $v(x)$, even if μ does not satisfy $\mu(\varnothing) = 0$ or is not normalized.

Furthermore, as shown in this proof, (6) and (7) can be viewed as *inversion formulas* for the discrete Sugeno integral model in the following sense. If we know the value of $S_{\langle\mu,u\rangle}(x)$, for all $x \in X$, without knowing the functions μ and u_i, it is possible to use (6) and (7) to build functions u_j and a capacity μ that allow to reconstruct all these values using the discrete Sugeno integral formula (8).

5 Independence of Conditions

When strong 2-gradedness is factorized using $AC1$ and 2-gradedness, Theorem 1 uses five conditions: completeness, transitivity, $AC1$, 2-gradedness and order-denseness. The five examples below show that none of these conditions can be dispensed with.

Example 1. Let $X = \{x_1, y_1\} \times \{x_2, y_2\}$. Let $\succsim$ be identical to the weak order

$$(y_1, y_2) \succ [(x_1, y_2), (y_1, x_2)] \succ (x_1, x_2),$$

except that we have removed two arcs from $\succsim$, so as to have $(x_1, y_2) \not\succsim (y_1, x_2)$ and $(y_1, x_2) \not\succsim (x_1, y_2)$. It is clear that $\succsim$ is transitive but is not complete. Since X_1 and X_2 have only two elements, condition 2-graded trivially holds. It is not difficult to check that we have $y_1 \succ_1^+ x_1$ and $y_2 \succ_2^+ x_2$, so that $AC1$ holds.

Example 2. Let $X = \{x_1, y_1\} \times \{x_2, y_2\}$. Let $\succsim$ be identical to the trivial weak order except that we have removed one arc from $\succsim$, so as to have $(x_1, x_2) \not\succsim (y_1, y_2)$. It is not difficult to see that the resulting relation is complete but not transitive (it is a semi-order). Since X_1 and X_2 have only two elements, condition 2-graded trivially holds. It is not difficult to check that we have $y_1 \succ_1^+ x_1$ and $y_2 \succ_2^+ x_2$, so that $AC1$ holds.

Example 3. Let $X = \{x_1, y_1, z_1\} \times \{x_2, y_2\} \times \{x_3, y_3\}$. Let $\succsim$ be the weak order such that:

$$[(x_1, x_2, x_3), (y_1, x_2, x_3)]$$
$$\succ$$

$$[(x_1,x_2,y_3),(x_1,y_2,x_3),(y_1,x_2,y_3),(y_1,y_2,x_3),\\ (y_1,y_2,y_3),(z_1,x_2,x_3),(z_1,x_2,y_3),(z_1,y_2,x_3)]\\ \succ\\ [(z_1,y_2,y_3),(x_1,y_2,y_3)].$$

We have $y_1 \succ_1^+ x_1 \succ_1^+ z_1$, $x_2 \succ_2^+ y_2$ and $x_3 \succ_3^+ y_3$, which shows that $AC1$ holds.

Conditions 2-graded$_2$ and 2-graded$_3$ are trivially satisfied. Condition 2-graded$_1$ is violated since $(x_1,x_2,x_3) \succsim (y_1,x_2,x_3)$, $(y_1,x_2,x_3) \succsim (y_1,x_2,x_3)$, $(y_1,y_2,y_3) \succsim (x_1,x_2,y_3)$ and $(y_1,x_2,x_3) \succsim (x_1,x_2,y_3)$ but $(z_1,x_2,x_3) \not\succsim (y_1,x_2,x_3)$ and $(x_1,y_2,y_3) \not\succsim (x_1,x_2,y_3)$.

Example 4. Let $X = \{x_1,y_1\} \times \{x_2,y_2\} \times \{x_3,y_3\}$. Let $\succsim$ be the weak order such that:

$$[(x_1,x_2,x_3),(x_1,y_2,x_3),(y_1,y_2,x_3)]\\ \succ\\ [(y_1,y_2,y_3),(y_1,x_2,x_3)]\\ \succ\\ [(x_1,x_2,y_3),(x_1,y_2,y_3),(y_1,x_2,y_3)].$$

Condition 2-graded trivially holds. We have $y_2 \succ_2^+ x_2$ and $x_3 \succ_3^+ y_3$, so that conditions $AC1_2$ and $AC1_3$ hold. Since $(x_1,x_2,x_3) \succsim (y_1,y_2,x_3)$ and $(y_1,y_2,y_3) \succsim (y_1,x_2,x_3)$ but $(y_1,x_2,x_3) \not\succsim (y_1,y_2,x_3)$ and $(x_1,y_2,y_3) \not\succsim (y_1,x_2,x_3)$, condition $AC1_1$ is violated.

Remark 4. It is easy to check that the weak order in Example 4 satisfies the following condition

$$\left.\begin{array}{c} x \succsim y \\ \text{and} \\ z \succsim y \end{array}\right\} \Rightarrow \left\{\begin{array}{c} (z_i,x_{-i}) \succsim y, \\ \text{or} \\ (x_i,z_{-i}) \succsim y, \end{array}\right.$$

for all $x,y,z \in X$. This condition is a weakening of $AC1_i$ obtained by requiring that $y = w$ in the expression of $AC1_i$ (it is equivalent to requiring that all relations $\succsim_i^{+(a)}$ are complete). It is therefore not possible to weaken $AC1_i$ in this way.

Similarly, it is easy to check that the weak order in Example 3 satisfies the weakening of 2-graded$_i$ obtained by requiring that $z = w$ in the expression of 2-graded$_i$ (and, hence, removing the last redundant premise), i.e., for all $x,y,z \in X$ and all $a_i \in X_i$,

$$\left.\begin{array}{c} x \succsim z \\ \text{and} \\ (y_i,x_{-i}) \succsim z \\ \text{and} \\ y \succsim z \end{array}\right\} \Rightarrow \left\{\begin{array}{c} (a_i,x_{-i}) \succsim z \\ \text{or} \\ (x_i,y_{-i}) \succsim z, \end{array}\right.$$

Hence, condition 2-graded$_i$ cannot be weakened in this way.

Example 5. Let $X = 2^{\mathbb{R}} \times \{0,1\}$. We consider the weak order on X such that $(x_1,x_2) \succsim (y_1,y_2)$ if $[x_2 = 1]$ or $[x_2 = 0,\ y_2 = 0$ and $x_1 \geq^* y_1]$, where $\geq^*$ is any linear order on $2^{\mathbb{R}}$. It is easy to see that $\succsim$ is a weak order. It violates OD since the restriction of $\succsim$ to $2^{\mathbb{R}} \times \{0\}$ is isomorphic to $\geq^*$ on $2^{\mathbb{R}}$ and $\geq^*$ violates *OD*. The relation $\succsim$ has a representation in the noncompensatory model. Indeed, for all $x = (x_1, 1)$, take $A_1^x = \varnothing$, $A_2^x = \{1\}$ and $F^x = \{\{2\}, \{1,2\}\}$. For all $x = (x_1, 0)$, take $A_1^x = \{y_1 \in 2^{\mathbb{R}} : y_1 \geq^* x_1\}$, $A_2^x = \{1\}$ and $F^x = \{\{1\}, \{2\}, \{1,2\}\}$. It is easy to check that this defines a representation of the weak order $\succsim$ in the noncompensatory model. Using Lemma 2, this implies that $\succsim$ satisfies $AC1$ and 2-graded.

6 Uniqueness

This section briefly discusses the uniqueness of the representation in the noncompensatory model and the discrete Sugeno integral model. The "ordinal" character of these models makes them especially attractive to deal with finite sets of alternatives. We therefore restrict our attention to this case in what follows. When X is finite, combining Propositions 1 and 2 with Theorem 1, shows that a binary relation has a representation in the noncompensatory model iff it has a representation in the discrete Sugeno integral model.

6.1 Links Between Representations in the Noncompensatory Model and the Discrete Sugeno Integral Model

Let $\succsim$ be a non-degenerate weak order on a finite set X with $r > 1$ distinct equivalence classes. Suppose that $\succsim$ has a representation in the noncompensatory model using sets A_i^x and F^x. It is easy to deduce from this representation a representation of $\succsim$ in the discrete Sugeno integral model.

It follows from the definition of the noncompensatory model that, if x and y belong to the same equivalence class, we have $A_i^x = A_i^y$, for all $i \in N$, and $F^x = F^y$. Let $A_i^{(k)} = A_i^x$ and $F^{(k)} = F^x$, for some $x \in X$ belonging to the kth equivalence class of $\succsim$.

Take any numbers λ_k such that

$$\lambda_1 = 1 > \lambda_2 > \cdots > \lambda_{r-1} > \lambda_r = 0. \tag{9}$$

For all $i \in N$, define u_i letting, for all $x_i \in X_i$,

$$\begin{cases} u_i(x_i) = \lambda_1 & \text{if } x_i \in A_i^{(1)}, \\ u_i(x_i) = \lambda_2 & \text{if } x_i \in A_i^{(2)} \setminus A_i^{(1)}, \\ u_i(x_i) = \lambda_3 & \text{if } x_i \in A_i^{(3)} \setminus A_i^{(2)}, \\ \vdots & \\ u_i(x_i) = \lambda_{r-1} & \text{if } x_i \in A_i^{(r-1)} \setminus A_i^{(r-2)}, \\ u_i(x_i) = \lambda_r & \text{otherwise}, \end{cases} \tag{10}$$

and μ on 2^N letting, for all $A \in 2^N$,

$$\begin{cases} \mu(A) = \lambda_1 & \text{if } A \in F^{(1)}, \\ \mu(A) = \lambda_2 & \text{if } A \in F^{(2)} \setminus F^{(1)}, \\ \mu(A) = \lambda_3 & \text{if } A \in F^{(3)} \setminus F^{(2)}, \\ \vdots & \\ \mu(A) = \lambda_{r-1} & \text{if } A \in F^{(r-1)} \setminus F^{(r-2)}, \\ \mu(A) = \lambda_r & \text{otherwise.} \end{cases} \tag{11}$$

With such definitions, for all $x \in X$, the value $S_{\langle \mu,u \rangle}(x)$ belongs to $\{\lambda_1, \lambda_2, \ldots, \lambda_r\}$. It is easy to see that $x \in X$ belongs to the kth equivalence class of $\succsim$ iff $\{i \in N : x_i \in A_i^{(k)}\} \in F^{(k)}$ iff $S_{\langle \mu,u \rangle}(x) = \lambda_k$.

The above formulas therefore give a systematic way to build a representation in the discrete Sugeno integral model on the basis of a representation in the noncompensatory model.

Clearly, the real numbers λ_k may be chosen arbitrarily, provided that they satisfy (9). Given a particular choice of λ_k, the representation built above is "minimal" in the sense that it uses as few real numbers as possible in order to build the representation in the Sugeno integral model.

The minimal representation, given a particular choice of λ_k compatible with (9), envisaged above is not the only possible one. Given the numbers λ_k, we can, for instance, use them to define the values of μ through (11). When this is done, it is clear that for each distinct $x_i \in A_i^{(k)} \setminus A_i^{(k-1)}$ we can define $u_i(x_i)$ to take an arbitrary value in the interval $[\lambda_k, \lambda_{k-1})$. Other choices are clearly possible.

6.2 Uniqueness of Representations

It is easy to deduce from the results in Bouyssou and Marchant (2007) the uniqueness of the representation in the noncompensatory model. Consider the kth equivalence class of $\succsim$. We say that attribute $i \in N$ is influent for this equivalence class if there are $x_i, y_i \in X_i$ and $a_{-i} \in X_i$ such that (x_i, a_{-i}) belongs at least to the kth equivalence class of $\succsim$ and (y_i, a_{-i}) belongs to a strictly lower equivalence class. Using the results in Bouyssou and Marchant (2007), it is easy to show that, when each attribute $i \in N$ is influent for the kth equivalence class of $\succsim$, the sets $A_i^{(k)}$ and $F^{(k)}$ are uniquely determined. This condition is not necessary for such a uniqueness however. This is illustrated in the example below adapted from Bouyssou and Marchant (2007).

Example 6. Let $n = 3$, $X = \{x_1, y_1\} \times \{x_2, y_2\} \times \{x_3, y_3\}$. Let $\succsim$ be such that:

$$\begin{aligned} &(x_1,x_2,x_3) \succ (y_1,x_2,x_3) \succ [(x_1,x_2,y_3),(x_1,y_2,x_3)] \\ &\succ [(x_1,y_2,y_3),(y_1,x_2,y_3),(y_1,y_2,x_3),(y_1,y_2,y_3)]. \end{aligned}$$

It is easy to check that all attributes are influent for the first equivalence class of $\succsim$. We must have $A_1^{(1)} = \{x_1\}$, $A_2^{(1)} = \{x_2\}$, $A_3^{(1)} = \{x_3\}$ and $F^{(1)} = \{\{1,2,3\}\}$. Similarly, all attributes are influent for the third equivalence class. We must have $A_1^{(3)} = \{x_1\}, A_2^{(3)} = \{x_2\}, A_3^{(3)} = \{x_3\}$ and $F^{(3)} = \{\{1,2\},\{1,3\},\{2,3\},\{1,2,3\}\}$.

Attributes 2 and 3 are influent for the second equivalence class of $\succsim$ while attribute 1 is not. In order to satisfy the constraints of the noncompensatory model, we must take $A_1^{(2)} = \{x_1\}$, $A_2^{(2)} = \{x_2\}$, $A_3^{(2)} = \{x_3\}$ and $F^{(2)} = \{\{2,3\},\{1,2,3\}\}$. The conditions ensuring the uniqueness of the representation in the noncompensatory model are investigated in Bouyssou and Marchant (2007). Whenever this representation is not unique, we may use each of these representations as a basis for the analysis in Sect. 6.1.

In order to analyze the uniqueness of the representation in the discrete Sugeno integral model, two points should therefore be kept in mind. First, given a representation in the noncompensatory model, it is possible to deduce several distinct representations in the discrete Sugeno integral model. Second, the representation in the noncompensatory model may not be unique. Combining these two effects, it is clear that the uniqueness of the representation in the discrete Sugeno integral model is quite weak. Since its precise analysis does not seem to be informative, we do not develop this point.

6.3 Commensurateness

When we compute a Sugeno integral, we compare levels on different attributes. This seems to indicate that the axioms of the discrete Sugeno integral model imply the existence of a relation $\succ^c$ defined on $\bigcup_{i \in N} X_i$, with the following interpretation: $x_i \succ^c x_j$ iff x_i is better than x_j. Given a preference relation $\succsim$, there can exist several representations in the discrete Sugeno integral model and it can happen that $u_i(x_i) > u_j(x_j)$ in one representation while $u'_i(x_i) < u'_j(x_j)$ in another one. Hence, stating $x_i \succ^c x_j$ (or the converse) does not make sense for such a pair. So, let us define $\succ^c$ by $x_i \succ^c x_j$ iff $u_i(x_i) > u_j(x_j)$ in *all* representations. In the following proposition, we characterize this relation in terms of the primitive relation $\succsim$.

Proposition 3. *Let $\succsim$ be a weak order representable by means of a Sugeno integral. We have $z_i \succ^c z_j$ if and only if, for some $c, d \in X$, $w_i \in X_i$, $w_j \in X_j$, $a_{-i} \in X_{-i}$ and $b_{-j} \in X_{-j}$, we have*

$$\begin{cases} c \succsim d, \\ (z_i, a_{-i}) \succsim c, \quad (w_i, a_{-i}) \not\succsim c, \\ (w_j, b_{-j}) \succsim d, \quad (z_j, b_{-j}) \not\succsim d. \end{cases} \tag{12}$$

Proof. If (12) holds, then, in any representation, $u_i(z_i) \geq S_{\langle \mu, u \rangle}(c) \geq S_{\langle \mu, u \rangle}(d) > u_j(z_j)$. So, in any representation, $u_i(z_i) > u_j(z_j)$ and, therefore, $z_i \succ^c z_j$.

Suppose now $z_i \succ^c z_j$ and let $(u_i^*)_{i\in N}$ be one of the representations constructed by means of (9), (10) and (11). We therefore know that $u_i^*(z_i) > u_j^*(z_j)$. There is thus k and l with $k < l$ such that $u_j^*(z_j) = \lambda^l$ and $u_i^*(z_i) = \lambda^k$ (this follows from (10)). Hence, $z_i \in A_i^{(k)}$ and $z_j \notin A_j^{(k)}$. So, (12) holds for some $c = d$ belonging to the kth equivalence class of $\succsim$. □

From the definition of $\succ^c$, it is clear that this relation is transitive and asymmetric, i.e., $z_i \succ^c z_j$ implies $z_j \not\succ^c z_i$. We now show that it is also negatively transitive, i.e., $x_i \not\succ^c y_j$ and $y_j \not\succ^c z_l$ implies $x_i \not\succ^c z_l$. Hence, $\succ^c$ is the asymmetric part of a weak order on the set $\bigcup_{i\in N} X_i$. This is in line with the intuitive notion of commensurateness.

Proposition 4. *Let $\succsim$ be a weak order representable by means of a Sugeno integral. Then $\succ^c$ is negatively transitive.*

Proof. Let $(u_i^*)_{i\in N}$ be one of the representations constructed by means of (9), (10) and (11). Suppose $x_i \not\succ^c y_j$ and $y_j \not\succ^c z_l$. If $u_i^*(x_i) > u_j^*(y_j)$, then, as shown in the proof of Proposition 3, (12) holds and, by Proposition 3, $x_i \succ^c y_j$. This contradiction implies $u_i^*(x_i) \le u_j^*(y_j)$. The same reasoning yields $u_j^*(y_j) \le u_j^*(z_l)$. By transitivity, $u_j^*(x_i) \le u_j^*(z_l)$. Suppose now, contrary to negative transitivity, that $x_i \succ^c z_l$. This implies $u_i^*(x_i) > u_l^*(z_l)$, a contradiction. □

To conclude this section, note that the "derived commensurateness", i.e., the relation $\succ^c$, is not easy to interpret and analyze however. Indeed, the way the above relation combines with $\succsim$ remains complex. As shown in the example below, it is quite possible to have $(x_i, x_j, x_{-ij}) \succsim y$, $z_j \succ^c x_i$ and $z_i \succ^c x_j$, while $(z_i, z_j, x_{-ij}) \not\succsim y$. This calls for further analysis.

Example 7. Let $n = 4$ and $X_1 = X_2 = X_3 = X_4 = \{0, 0.01, 0.02, \ldots, 0.99, 1\}$. For all $i \in N$, let $u_i(x_i) = x_i$. Define a normalized capacity μ on N such that: $\mu(\varnothing) = 0$, $\mu(A) = 0.1$, for all $A \subseteq N$ such that $|A| = 1$, $\mu(\{1,2\}) = 0.1$, $\mu(\{1,3\}) = 0.2$, $\mu(\{1,4\}) = 0.301$, $\mu(\{2,3\}) = 0.31$, $\mu(\{2,4\}) = 0.2$, $\mu(\{3,4\}) = 0.3$, $\mu(\{1,2,3\}) = 0.55$, $\mu(\{1,2,4\}) = 0.39$, $\mu(\{1,3,4\}) = 1$, $\mu(\{2,3,4\}) = 0.31$, $\mu(N) = 1$. Define $\succsim$ on X as the relation obtained through the comparison of the values $S_{\langle\mu,u\rangle}(x) = S_\mu[x]$ using the utility functions and the capacity defined above.

We have

$$S_\mu[(0.2,0,0.5,0)] = 0.2 > S_\mu[(0.1,0,0.5,0)] = 0.1,$$

$$S_\mu[(0,0.2,0,0.5)] = 0.2 > S_\mu[(0,0.15,0,0.5)] = 0.15.$$

Since it is clear that $S_\mu[(0.2,0.2,0.2,0.2)] = 0.2$ we thus have

$$\begin{aligned}
(0.2,0,0.5,0) &\succsim (0.2,0.2,0.2,0.2) = c,\\
(0.1,0,0.5,0) &\not\succsim (0.2,0.2,0.2,0.2) = c,\\
(0,0.2,0,0.5) &\succsim (0.2,0.2,0.2,0.2) = d,\\
(0,0.15,0,0.5) &\not\succsim (0.2,0.2,0.2,0.2) = d,\\
c = (0.2,0.2,0.2,0.2) &\succsim (0.2,0.2,0.2,0.2) = d,
\end{aligned}$$

so that the level 0.2 on X_1 is better than the level 0.15 on X_2

Similarly, we have

$$S_\mu[(0,0.46,0.5,0)] = 0.31 > S_\mu[(0,0.3,0.5,0)] = 0.3,$$
$$S_\mu[(0.5,0,0,0.5)] = 0.301 > S_\mu[(0.3,0,0,0.5)] = 0.3.$$

Since we have $S_\mu[(0.31,0.31,0.31,0.31)] = 0.31$ and $S_\mu[(0.301,0.301,0.301,0.301)] = 0.301$, we obtain

$$\begin{aligned}
(0,0.46,0.5,0) &\succsim (0.31,0.31,0.31,0.31) = c',\\
(0,0.3,0.5,0) &\not\succsim (0.31,0.31,0.31,0.31) = c',\\
(0.5,0,0,0.5) &\succsim (0.301,0.301,0.301,0.301) = d',\\
(0.3,0,0,0.5) &\not\succsim (0.301,0.301,0.301,0.301) = d',\\
c' &\succsim d',
\end{aligned}$$

so that the level 0.46 on X_2 is better than the level 0.3 on X_1.

We have $S_\mu[(0.3,0.15,0.29,0.4)] = 0.3$. Since the level 0.2 on X_1 is better than the level 0.15 on X_2 and 0.46 on X_2 is better than the level 0.3 on X_1, we should obtain that $S_\mu[(0.2,0.46,0.29,0.4)] \geq 0.3$, whereas it is equal to 0.29.

7 Discussion

In this paper, we have analyzed the relations between the discrete Sugeno integral model and the noncompensatory model as well as proposed a factorization of the main condition used in Greco et al. (2004, Theorem 3). By the same token, we have presented a proof of Greco et al. (2004, Theorem 3). We have also discussed the uniqueness of the representation in the discrete Sugeno integral model and shown that the conditions used in Greco et al. (2004, Theorem 3) are independent. Besides, we have analyzed the commensurateness that is implied by the discrete Sugeno integral model and shown that it is more complex than what is usually thought in the literature. Many questions are nevertheless left open. Let us briefly mention here what seems to us the most important ones.

The result in Greco et al. (2004) is a first step in the systematic study of models using fuzzy integrals in MCDM. A first and major open problem is to derive a similar result for the discrete Choquet integral. This appears very difficult and we have no satisfactory answer at this time.

A second open problem is to use the above result as a building block to study particular cases of the discrete Sugeno integral. This was started in Greco et al. (2004) who showed how to characterize ordered weighted minimum and maximum. There are nevertheless many other particular cases of the discrete Sugeno integral that would be worth investigating.

A third problem is to investigate assessment protocols of the various parameters of the discrete Sugeno integral model using the above result and conditions. This will clearly require a deeper investigation of the commensurateness at work in our models.

Finally, it should be mentioned that we have mainly used here the noncompensatory model for weak orders as a tool for analyzing the discrete Sugeno integral model. The noncompensatory model that we introduced can be extended in many possible directions. This will be the subject of a subsequent paper.

Acknowledgments This work was presented at the 27th Linz Seminar on Fuzzy Sets (7–11 February 2006) and at the DIMACS / LAMSADE Workshop on "Voting Theory and Preference Modelling" (25–28 October 2006). We are grateful to the participants of these meetings for their helpful comments on our results.

References

Allais, M. (1953). Le comportement de l'homme rationnel devant le risque : critique des postulats et axiomes de l'école américaine. *Econometrica 21*, 503–546.

Arrow, K. J. (1963). *Social choice and individual values* (2nd ed.). Wiley, New York.

Bana e Costa, C. A. & Vansnick, J.-C. (1994). MACBETH: An interactive path towards the construction of cardinal value functions. *International Transactions in Operational Research, 1*, 489–500.

Bana e Costa, C. A. & Vansnick, J.-C. (1997). Applications of the MACBETH approach in the framework of an additive aggregation model. *Journal of Multi-Criteria Decision Analysis, 6*, 107–114.

Bana e Costa, C. A. & Vansnick, J.-C. (1999). The MACBETH approach: Basic ideas, software and an application. In N. Meskens & M. Roubens (Eds.), *Advances in decision analysis* (pp. 131–157). Dordrecht: Kluwer.

Bouyssou, D. & Marchant, Th. (2007). An axiomatic approach to noncompensatory sorting methods in MCDM, II: More than two categories. *European Journal of Operational Research 178*(1), 246–276.

Bouyssou, D. & M. Pirlot (2004). Preferences for multiattributed alternatives: Traces, dominance, and numerical representations. *Journal of Mathematical Psychology 48*(3), 167–185.

Bouyssou, D. & Pirlot, M. (2005). A characterization of concordance relations. *European Journal of Operational Research 167*(2), 427–443.

Chateauneuf, A. & Cohen, M. (2000). Choquet expected utility model: A new approach to individual behavior under uncertainty and to social welfare. In M. Grabisch, T. Murofushi, and M. Sugeno (Eds.), *Fuzzy Measures and Integrals: Theory and Applications*, (pp. 289–313). Heidelberg: Physica.

Choquet, G. 1953. Theory of capacities. *Annales de l'Institut Fourier, 5*, 131–295.

Dubois, D., Grabisch, M., Modave, F., & Prade, H. (2000a). Relating decision under uncertainty and multicriteria decision making models. *International Journal of Intelligent Systems, 15*, 967–979.

Dubois, D., Marichal, J.-L., Prade, H., Roubens, M., & Sabbadin, R. (2001a). The use of the discrete Sugeno integral in decision making: A survey. *International Journal of Uncertainty, Fuzziness and Knowledge-Based Systems, 9*, 539–561.

Dubois, D., Prade, H., & Sabbadin, R. (2000b). Qualitative decision theory with Sugeno integrals. In M. Grabisch, T. Murofushi, & M. Sugeno (Eds.), *Fuzzy measures and integrals: Theory and applications* (pp. 314–332). Heidelberg: Physica.

Dubois, D., Prade, H., & Sabbadin, R. (2000b). Qualitative decision theory with Sugeno integrals. In M. Grabisch, T. Murofushi, & M. Sugeno (Eds.), *Fuzzy Measures and Integrals: Theory and Applications*, (pp. 314–332). Heidelberg: Physica.

Dubois, D., Prade, H., & Sabbadin, R. (2001b). Decision-theoretic foundations of qualitative possibility theory. *European Journal of Operational Research 128*, 459–478.

Ellsberg, D. (1961). Risk, ambiguity and the Savage axioms. *Quarterly Journal of Economics 75*, 643–669.
Fishburn, P.C. (1970). *Utility theory for decision-making*. New-York: Wiley.
Fishburn, P.C. (1975). Axioms for lexicographic preferences. *Review of Economic Studies, 42*, 415–419.
Fishburn, P.C. (1976). Noncompensatory preferences. *Synthese, 33*, 393–403.
Fishburn, P.C. (1978). A survey of multiattribute/multicriteria evaluation theories. In S. Zionts (Ed.), *Multicriteria problem solving*, (pp. 181–224). Berlin: Springer Verlag.
Fishburn, P.C. (1982). *The Foundations of Expected Utility*. Dordrecht: D. Reidel.
Gilboa, I. (1987). Expected utility with purely subjective non-additive probabilities. *Journal of Mathematical Economics, 16*, 65–68.
Gonzales, Ch. & Perny, P. (2005). GAI networks for decision making under cetainty. In R. Brafman & U. Junker (Eds.), *Proceedings of the 19th International Joint Conference on Artificial Intelligence – Workshop on advances in preference handling*, (pp. 100–105).
Grabisch, M. (1995). Fuzzy integral in multicriteria decision making. *Fuzzy Sets and Systems, 69*(3), 279–298.
Grabisch, M. (1996). The application of fuzzy integrals to multicriteria decision making. *European Journal of Operational Research, 89*, 445–456.
Grabisch, M. & Labreuche, C. (2004). Fuzzy measures and integrals in MCDA. In J. Figueira, S. Greco, & M. Ehrgott (Eds.), *Multiple Criteria Decision Analysis*, (pp. 563–608). Kluwer Academic Publishers.
Grabisch, M., Labreuche, C., & Vansnick, J.-C. (2003). On the extension of pseudo-boolean functions for the aggregation of interacting criteria. *European Journal of Operational Research, 148*(1), 28–47.
Grabisch, M. & Roubens, M. (2000). Application of the Choquet integral in multicriteria decision making. In M. Grabisch, T. Murofushi, & M. Sugeno (Eds.), *Fuzzy Measures and Integrals: Theory and Applications*, (pp. 348–374). Heidelberg: Physica.
Greco, S., Matarazzo, B., & Słowiński, R. (1999). The use of rough sets and fuzzy sets in MCDM. In T. Gal, T. Hanne., & T. Stewart (Eds.), *Multicriteria decision making, Advances in MCDM models, algorithms, theory and applications*, Dordrecht, (pp. 14.1–14.59). Dordrecht: Kluwer.
Greco, S., Matarazzo, B., & Słowiński, R. (2001). Conjoint measurement and rough set approach for multicriteria sorting problems in presence of ordinal criteria. In A. Colorni, M. Paruccini, & B. Roy (Eds.), *A-MCD-A, Aide Multicritère à la Décision / Multiple Criteria Decision Aid*, (pp. 117–144). European Commission, Joint Research Centre.
Greco, S., Matarazzo, B., & Słowiński, R. (2004). Axiomatic characterization of a general utility function and its particular cases in terms of conjoint measurement and rough-set decision rules. *European Journal of Operational Research, 158*(2), 271–292.
Hougaard, J. L. & Keiding, H. (1996). Representation of preferences on fuzzy measures by a fuzzy integral. *Mathematical Social Sciences, 31*, 1–17.
Krantz, D. H., Luce, R. D., Suppes, P., & Tversky, A. (1971). *Foundations of measurement*, Vol. 1: Additive and polynomial representations. New-York: Academic.
Marichal, J.-L. (2000a). On Choquet and Sugeno integrals as aggregation functions. In M. Grabisch, T. Murofushi, and M. Sugeno (Eds.), *Fuzzy measures and integrals*, (pp. 247–272). Heidelberg: Physica.
Marichal, J.-L. (2000b). On Sugeno integrals as an aggregation function. *Fuzzy Sets and Systems, 114*, 347–343.
Quiggin, J. (1982). A theory of anticipated utility. *Journal of Economic Behaviour and Organization, 3*, 323–343.
Schmeidler, D. (1989). Subjective probability and expected utility without additivity. *Econometrica, 57*, 571–587. (first version: 1984)
Schmidt, U. (2004). Alternatives to expected utility: Some formal theories. In S. Barberà, P. J. Hammond, & C. Seidl (Eds.), *Handbook of Utility Theory*, (Vol. 2, pp. 757–837). Dordrecht: Kluwer.

Słowiński, R., Greco, S., & Matarazzo, B. (2002). Axiomatization of utility, outranking and decision-rule preference models for multiple-criteria classification problems under partial inconsistency with the dominance principle. *Control and Cybernetics, 31*(4), 1005–1035.
Starmer, C. (2000). Developments in non-expected utility theory: The hunt for a descriptive theory of choice under risk. *Journal of Economic Literature, XXXVIII*, 332–382.
Sugden, R. (2004). Alternatives to expected utility: Foundations. In S. Barberà, P. J. Hammond, & C. Seidl (Eds.), *Handbook of Utility Theory*, (Vol. 2, pp. 685–755). Dordrecht: Kluwer.
Sugeno, M. (1974). *Theory of fuzzy integrals and its applications*. Phd thesis, Tokyo Institute of Technology.
Sugeno, M. (1977). Fuzzy measures and fuzzy integrals: a survey. In M. M. Gupta, G. N. Saridis, & B. R. Gaines (Eds.), *Fuzzy automata and decision processes* (pp. 89–102). Amsterdam: North Holland.
Wakker, P. P. (1989). *Additive representations of preferences: A new foundation of decision analysis*. Dordrecht: Kluwer.
Yaari, M. E. (1987). The dual theory of choice under risk. *Econometrica, 55*, 95–115.

Measurement Theory

Additive Representability of Finite Measurement Structures

Arkadii Slinko

1 Introduction

The theory of additive conjoint measurement takes its roots in the papers by Debreu, (1960) and Luce and Tukey (1964). It is presented in books (Pfanzagl (1968); Fishburn (1970); Krantz, Luce, Suppes, & Tversky, 1971; Luce, Krantz, Suppes, & Tversky, 1998; Suppes, Krantz, Luce, & Tversky, 1988; Roberts, 1979; Narens, 1985) and excellent surveys, of which Fishburn's survey (1999) is the most recent. The goal of the present paper is twofold: we would like to describe some recent developments that took place after Fishburn's survey was published, and to attract attention to several questions posed by Fishburn that remain unanswered.

The main object of this theory is a Cartesian product of finitely many mutually disjoint sets A_i

$$A = A_1 \times A_2 \times \ldots \times A_n \tag{1}$$

equipped with an order $\preceq$. This product is usually interpreted as the set of alternatives under the consideration of a decision maker, or the set of outcomes that may result from her actions. We may also think that there are n criteria in place and each set A_i is identified with the set of levels of the ith criterion. The order represents the decision maker's preference on the set of alternatives.

A decision maker often faces some kind of optimization problem. A solution of this problem would be made feasible if it were possible to find an additive utility representation over criteria of the decision maker's preference order $\preceq$ on A. The central theme of the theory of additive conjoint measurement is finding conditions which imply the existence of such a representation. Another important question is about uniqueness of this representation. It appeared that, in many aspects, the most difficult case to study is the case of finite measurement structures, i.e. when A is

A. Slinko
Department of Mathematics, University of Auckland, Auckland, New Zealand
e-mail: a.slinko@auckland.ac.nz

S.J. Brams et al. (eds.), *The Mathematics of Preference, Choice and Order: Essays in Honor of Peter C. Fishburn*, Studies in Choice and Welfare,

finite. The main focus of this paper is on this case. In addition to that we restrict ourselves with $\preceq$ being a (strict) linear order, in which case the uniqueness question does not emerge.

Kraft, Pratt, and Seidenberg (1959) established (see also Scott (1964)) that additive utility representation of $\preceq$ is equivalent to a denumerable set of conditions, called cancellation conditions, which is not equivalent to any finite subset of them. However, for a finite Cartesian product of a particular size we need to check only finitely many cancellation conditions for $\preceq$ to establish its additive representability. Fishburn (see, e.g. his motivation of this in (1997, 1999)) considered that it is extremely important to know the exact number of cancellation conditions needed as a function of the size of the product, or at least a good lower and upper bounds for this number. He saw the absence of such bounds as a serious gap in understanding of additive representability of preferences on finite measurement structures. Fishburn made a significant contribution to this theory and formulated a large number of open problems, which have guided and undoubtedly will continue to guide investigators in this area. And although some recent progress has been made, only a few of the great many questions posed by Fishburn have been answered to date.

Let us briefly outline what will be covered in the subsequent sections. In Sect. 2 we introduce the main types of finite measurement structures considered in the literature to date. They are Cartesian product structure, power set structure, power multiset structure. Section 3 surveys the most general case, the Cartesian product structure. In this case no significant progress has been recently made, and we highlight a number of open questions.

Comparative probability orders, which represent one of the main cases of the power set structure, are surveyed in Sect. 4. This measurement structure emerges when $A_i = \{0,1\}$ for $i = 1,2,\ldots,n$, in which case any n-tuple of the Cartesian product can be identified with a subset of the set of atoms $[n] = \{1,2,\ldots,n\}$. Here we reformulate the cancellation conditions for comparative probability orders in terms of portfolios of desirable gambles. This framework allows for a better understanding of Fishburn's function $f(n)$, the main object of his investigations in Fishburn (1996, 1997). We show that $f(n)$ can be interpreted as a measure of rationality of a player required to correctly evaluate any portfolio of gambles with n states of the world. We report on the recent progress in estimation of $f(n)$ and the related function $g(n)$, which was introduced by Conder and Slinko (2004). The reason for introducing this new function is as follows. It is known that for comparative probability orders the absence of arbitrage does not imply additive representation and some cancellation conditions may still be violated. However the absence of arbitrage is a very important condition and $g(n)$ is a complete analogue of $f(n)$ in the situation of no arbitrage.

Fishburn showed by way of a sophisticated combinatorial construction that $f(n) \geq n-1$, which together with the bound $f(n) \leq n+1$ of Kraft–Pratt–Seidenberg (1959) gave quite a narrow range for this function. Fishburn conjectured that $f(n) = n-1$. Recently however Conder and Slinko (2004) showed that $f(7) \geq 7$ and Marshall (2005, 2007) showed that $f(p) \geq p$ for a large number of prime numbers $p \geq 131$. Conder showed that $f(n) \geq n$ for all $7 \leq n \leq 13$. Fishburn (1996,

1997) also paid attention to minimal violations of the cancellation conditions which he called *irreducible patterns*. Here we present a theorem of Auger (2005) which says that there are only finitely many of them.

In Sect. 5, devoted to power multiset structure, sets are generalised to multisets which allow multiple entry of identical elements. If $A_i = \{0, 1, \ldots, m_i\}$ and if the ith coordinate of an n-tuple from the Cartesian product is j, then we may think that the multiset associated with this tuple has j copies of atom i. We see great advantages in describing this measurement structure in multiset terms, because of the emerging analogies with comparative probability orders. Orders on submultisets of a multiset were first used in the computer science literature by Dershowitz (1979) to prove termination of rewrite systems. Sertel and Slinko (2002) showed some important applications of multisets in Economics and Political Science. In Economics ranking multisets can be used for ranking income streams and investment projects. In Political Science they can be used for ranking committees or parliaments.

Additive conjoint measurement on subsets of Cartesian products containing? rank-ordered? n-tuples was considered by Wakker in (1991, 1993). He established that, contrary rank-ordered to what has often been thought, additive conjoint measurement on subsets of Cartesian products has characteristics different from additive conjoint measurement on full Cartesian products.

Fishburn himself did not work with this preference structure but many of his ideas work in this case too. An analogue of de Finetti's axiom here is Independence of Equal Submultisets (IES) introduced in Sertel and Slinko (2002); Sertel and Slinko (2007). The analogues of functions $f(n)$ and $g(n)$ can be introduced and those analogues will have k as an additional parameter, i.e. we obtain functions $f(n,k)$ and $g(n,k)$. It is rather surprising that in this case better progress can be achieved in describing these functions than in the case of comparative probability orders (Conder, Marshall and Slinko, 2007). The function $g(n,k)$ is determined exactly: we have $g(n,k) = n-1$ for $(n,k) \neq (5,2)$ and $g(5,2) = 3$. We also have $n \geq f(n,k) \geq g(n,k)$ and we conjecture that $f(n,k) = g(n,k)$.

2 Types of Finite Measurement Structures

In this paper we assume that the Cartesian product (1) is finite. Let m_i denote the cardinality of A_i and in this case the cardinality of the Cartesian product will be $|A| = m_1 m_2 \ldots m_n$. We interpret $\preceq$ as a nonstrict preference relation on A, i.e. $\mathbf{a} \preceq \mathbf{b}$ means $\mathbf{a}$ is not preferred to $\mathbf{b}$. The corresponding strict preference relation $\prec$ and indifference $\sim$ are defined in the usual way.

Sometimes A_i $(i = 1, \ldots, n)$ are sets without any additional structure. This happens, when elements in each A_i belong to the same class but cannot be compared and measured in units of something, e.g. $A_1 = \{\text{apple}, \text{banana}\}$ and $A_2 = \{\text{pepsi}, \text{coca cola}\}$. Here the Cartesian product consists of pairs

$$A = \{(\text{apple}, \text{pepsi}), (\text{apple}, \text{coca cola}), (\text{banana}, \text{pepsi}), (\text{banana}, \text{coca cola})\}.$$

We say that we have a *Cartesian product structure*. The additive utility representation in this case will then take the following form.

Definition 1. A binary relation $\preceq$ on a Cartesian product structure (1) is said to be additively representable if there are n non-negative real-valued functions $u_i\colon A_i \to \mathbb{R}$ such that for all $\mathbf{a} = (a_1, \ldots, a_n)$ and $\mathbf{b} = (b_1, \ldots, b_n)$ in A

$$\mathbf{a} \preceq \mathbf{b} \Longleftrightarrow \sum_{i=1}^{n} u_i(a_i) \le \sum_{i=1}^{n} u_i(b_i). \tag{2}$$

An important case emerges when we have n types of goods which are divisible to a certain extent but not infinitely divisible (such as money, cars, houses, etc.). These goods can be measured only in whole units of some quantity which is further indivisible. If the total number of available units of the good of type i is m_i, then each A_i can be identified with the set $\{0, 1, \ldots, m_i\}$ which has the structure of the truncated monoid of nonnegative integers $\mathbb{N}_{m_i}$. A truncated monoid $\mathbb{N}_k = (\{0, 1, \ldots, k-1\}, \oplus)$ of positive integers is an algebraic system on the base set $\{0, 1, \ldots, k-1\}$, where the addition $\oplus$ is defined as

$$m \oplus n = \begin{cases} m+n & \text{if } m+n < k, \\ \text{undefined} & \text{if } m+n \ge k. \end{cases}$$

The representability of linear orders on such a Cartesian product must respect the structure on the A_i's, which means that for the ith utility function we must have

$$u_i(k) = k u_i(1)$$

and, in particular, $u_i(0) = 0$.

When $m_1 = \ldots = m_n = 2$, and each A_i has the structure of $\mathbb{N}_2$, this is the case of goods which are indivisible. A 1 in the ith position of an n-tuple $\mathbf{a} = (a_1, \ldots, a_n) \in A$ means that the ith good is present in this bundle. The Cartesian product A thus can be identified with the set of all indicator functions on $[n]$ or with the set of all subsets of $[n]$. Then the order $\preceq$ becomes an order on subsets of $[n]$. We call it the *power set structure*. We will deal only with linear, i.e. antisymmetric orders on subsets, since the general theory has not been developed yet. One obvious necessary condition for additive representability of the power set structure is the famous axiom introduced by de Finetti.

Definition 2. An order $\preceq$ on $2^{[n]}$ is said to satisfy the *de Finetti axiom* if for any $A, B \in 2^{[n]}$ and any $C \in 2^{[n]}$ such that $C \cap (A \cup B) = \emptyset$

$$A \preceq B \Longleftrightarrow A \cup C \preceq B \cup C. \tag{3}$$

If a linear order $\preceq$ satisfies de Finetti's axiom and $\emptyset \prec X$ for any non-empty subset $X \subseteq [n]$, then it is called a *comparative probability order*. Some significant progress has been recently achieved in understanding of additive representability of comparative probability orders. We report it in Sect. 4.

A *multiset* M on a base set X is a collection of elements of X, where multiple entries of the same element of X are possible (Stanley, 1997). In general, if $X = \{x_1, \ldots, x_k\}$ is a set, then a multiset on X is denoted as $M = \{x_1^{q_1}, x_2^{q_2}, \ldots, x_k^{q_k}\}$, where q_j is the number of occurrences of x_j in M, respectively. The number q_j is normally referred to as the *multiplicity* of x_j in M. As some q_j may be zero, not all elements of the base set may be present. The number of unique elements of X in M we call the *width* of M and the sum $\sum_{j=1}^{k} q_j$ we call the *cardinality* of M.

When the Cartesian product (1) is such that every A_i has a structure of $\mathbb{N}_{m_i}$, then A can be identified with all submultisets of the multiset $\{1^{m_i}, 2^{m_2}, \ldots, n^{m_n}\}$ on $[n]$. In the language of bundles of goods, we have n types of goods, denoted $1, 2, \ldots, n$, and exactly m_i copies of good i are available. We call it the *power multiset structure*. The power multiset model has numerous useful interpretations (see e.g. Sertel and Slinko (2007); Conder et al. (2007)). We report results on the power multiset structure in Sect. 5.

In some applications not all alternatives of the Cartesian product (1) are actually available for choice. In this case we have to consider orders on a subset of this Cartesian product. Section 6.5.5 of Krantz et al. (1971) points out the importance of additive conjoint measurement on subsets of Cartesian products. Interest in this topic has increased during the last decade because of new developments in the literature on decision making under risk/uncertainty where conditions like independence are often required to hold only within certain subsets. Sertel and Slinko (2002) showed that sometimes from the applications point of view it is necessary to restrict ourselves to the submultisets $\{1^{k_i}, 2^{k_2}, \ldots, n^{k_n}\}$ of $\{1^{m_i}, 2^{m_2}, \ldots, n^{m_n}\}$ of fixed cardinality k, i.e. those for which $\sum_{i=1}^{n} k_i = k$. The set of all submultisets of cardinality k we will denote as $\mathcal{P}_k([n])$. They gave several important examples of such applications (see also Conder et al. (2007)).

3 Cartesian Product Structure

When we deal with sequences of elements of the Cartesian product A, we will index them with superscripts, while leaving subscripts to numerate the coordinates of elements of A. For example, if $\mathbf{a}^1, \ldots, \mathbf{a}^s$ is the sequence of elements of A, then a_9^7 is the ninth coordinate of the seventh vector.

If $\preceq$ is a binary relation on the Cartesian product A and $\mathbf{a} \preceq \mathbf{b}$ is true, then, using the preference elicitation terminology Fishburn, Pekeč, and Reeds (2002), we will say that $\mathbf{a} \preceq \mathbf{b}$ is a **valid comparison** of the two tuples $\mathbf{a}$ and $\mathbf{b}$.

Definition 3. Let $\preceq$ be a relation on the Cartesian product A and

$$\mathbf{a}^1 \preceq \mathbf{b}^1, \quad \mathbf{a}^2 \preceq \mathbf{b}^2, \quad \ldots, \quad \mathbf{a}^q \preceq \mathbf{b}^q \tag{4}$$

be a sequence of valid comparisons of pairs of elements of A such that $\mathbf{a}^i \prec \mathbf{b}^i$ for at least one i. We say that this sequence has the *cancellation property* if, for each coordinate $i = 1, 2, \ldots, n$, the sequence $b_i^1, b_i^2, \ldots, b_i^q$ is a permutation of the sequence $a_i^1, a_i^2, \ldots, a_i^q$.

The number q of comparisons in the sequence (4) will be called its *cardinality* and the number of unique comparisons in (4) will be called the *width* of this sequence. Note that this is consistent with the multiset terminology. This is because, if we drop the order of elements in any sequence, it becomes a multiset.

Example 1. The following two sequences of comparisons

$$\begin{array}{cc} (1,2) \prec (3,4) & (1,2) \prec (2,3) \\ (3,4) \prec (1,2) & (2,3) \prec (3,4) \\ & (3,4) \prec (1,2) \end{array}$$

the first one, in the left column, of cardinality two, and the second, in the right column, of cardinality three, both have the cancellation property. If all the comparisons of the first sequence are valid for $\preceq$, then $\preceq$ is not antisymmetric, If all the comparisons of the second sequence are valid, then $\preceq$ is not transitive.

From the previous example we get a feeling that having a sequence of valid comparisons with the cancellation property is some kind of a pathology.

Definition 4. We say that a binary relation $\preceq$ on a Cartesian product (1) satisfies the *cancellation condition* C_k if every sequence of comparisons which satisfies the cancellation property has width greater than k. We say that a binary relation $\preceq$ satisfies the *cancellation condition* $C_k^\dagger$ if every sequence of comparisons which satisfies the cancellation property has cardinality greater than k.

The following example is taken from Fishburn (1999).

Example 2. Let $A = \{1,2,3\} \times \{a,b,c\}$. Then the linear order

$$1a \prec 1b \prec 2a \prec 2b \prec 3a \prec 1c \prec 2c \prec 3b \prec 3c$$

satisfies C_2 and $C_2^\dagger$ but fails both C_3 and $C_3^\dagger$ since the sequence of valid comparisons

$$1b \prec 2a, \quad 3a \prec 1c, \quad 2c \prec 3b$$

has the cancellation property.

As the width of a multiset is not greater than its cardinality, C_k always implies $C_k^\dagger$. Both C_k and $C_k^\dagger$ group together a large number of conditions but they do it differently. Both are introduced to help us better comprehend the great many cancellation conditions necessary for additive representability.

It is obvious that an additively representable binary relation does not have sequences of valid comparisons that satisfy the cancellation property and, hence, satisfies all cancellation conditions. The converse is also true Krantz et al. (1971). The basic rationality assumption for a preference relation on A is called *Independence of Equal Subalternatives*. It says that for four n-tuples $\mathbf{x}, \mathbf{y}, \mathbf{z}, \mathbf{w} \in A$

$$\mathbf{x} \preceq \mathbf{y} \Longleftrightarrow \mathbf{z} \preceq \mathbf{w}$$

whenever there exists a proper subset $S \subseteq [n]$ such that $x_i = z_i$ and $y_i = w_i$ for all $i \in S$, and $x_i = y_i$ and $z_i = w_i$ for all $i \notin S$. We take this terminology from Wakker (1989); Fishburn calls it the first order independence Fishburn (1997); in Krantz et al. (1971) this is called coordinate independence. Independence of Equal Subalternatives, being a consequence of C_2, is not generally sufficient for additive representability. However, as we shall see later, for a limited set of sizes it is true.

Given a relation $\preceq$ on A, we may associate the following two numbers with it. Let $f(\preceq)$ be the smallest k such that $\preceq$ violates the cancellation condition C_k and $f^\dagger(\preceq)$ be the smallest k such that $\preceq$ violates the cancellation condition $C_k^\dagger$. An obvious relation between these two functions is, of course, $f(\preceq) \leq f^\dagger(\preceq)$. However the minimal violation of $C_k^\dagger$ hypothetically may not have the smallest possible width. Knowing only $f(\preceq)$, we know only half of the story and knowing $f(\preceq)$ and $f^\dagger(\preceq)$ gives us the full picture.

Now we will introduce two functions that were of primary interest to Fishburn. We set

$$f(m_1, m_2, \ldots, m_n) = \max f(\preceq), \qquad f^\dagger(m_1, m_2, \ldots, m_n) = \max f^\dagger(\preceq), \tag{5}$$

where the maximum both times is taken over all binary relations on A. In other words, any relation $\preceq$ on A, which satisfies cancellation conditions C_k with $k \leq f(m_1, m_2, \ldots, m_n)$ is additively representable and $f(m_1, m_2, \ldots, m_n)$ is the smallest number with this property. The second function $f^\dagger(m_1, m_2, \ldots, m_n)$ can be similarly characterised. Fishburn concentrated his attention on the first function leaving the second for future research. In this section we will not consider the important case of $(m_1, m_2, \ldots, m_n) = (2, 2, \ldots, 2)$ since we will devote the whole next section to it.

Krantz et al. (1971) (see pp. 427–428), who made the initial contribution to this topic, proved that $f(2, m_2) = 2$ and that $f(3, 3) \geq 3$. Little else was known about these functions until Fishburn's papers (1997, 2001). One of the most significant results of Fishburn (1997) was the general upper bound for $f(m_1, m_2, \ldots, m_n)$.

Theorem 1 (Fishburn, 1997). $f(m_1, m_2, \ldots, m_n) \leq \sum_{i=1}^n m_i - (n-1)$.

As $f(2, m_2)$ is known, the case $n = 2$ with $\min(m_1, m_2) \geq 3$ naturally attracted much attention Fishburn (1996, 1997).

Theorem 2 (Fishburn, 1997, 2001).

1. $f(3,3) = 3$, $f(3,4) = f(4,4) = 4$.
2. $f(3, m_2) \geq m_2$ *for all even* $m_2 \geq 4$, *and* $f(3, m_2) \geq m_2 - 1$ *for all odd* $m_2 \geq 5$.
3. $f(m_1, m_2) \geq m_1 + m_2 - 10$.
4. $f(5, m_2) \geq m_2 + 1$ *for all odd* $m_2 \geq 5$.

We note that Theorem 1 gives us $f(3, m_2) \leq m_2 + 2$ so the bounds for $f(3, m_2)$ given by Theorem 2 are rather tight. Apart from obvious questions that these results prompt, Fishburn (1997, 2001) formulated the following interesting ones.

Problem 1. What can be said about $f^\dagger(m_1, m_2, \ldots, m_n)$?

Problem 2. We can narrow the class of relations and define the functions $f(m_1, m_2, \dots, m_n)$ and $f^{\dagger}(m_1, m_2, \dots, m_n)$ for strict linear orders. Will the values of these functions remain the same? Fishburn conjectured that they would (see Conjecture 1 in Fishburn (1997)).

An important paper by Fishburn and Roberts (1988) studied the uniqueness question, which we do not survey here due to lack of space.

4 Comparative Probability Orders

As we already noticed, in the case when $m_1 = \ldots = m_n = 2$, the Cartesian product A can be identified with the power set of n-element set $[n]$. Here we adopt de Finnetti's point of view (de Finnetti, 1931) and consider $[n]$ as the set of the states of the world in which case we can identify comparisons of subsets with gambles. This approach was further developed by Fine (1973), Walley and Fine (1979) and Walley (1991, 1999) who believed that there are considerable advantages of basing the theory of comparative probability on desirability of gambles. In our case orders on subsets and desirability of gambles provide two equivalent characterisations but there are some nuances. The shift from preference to desirability is subtle but important. The word "preference" has an optimality flavour while the word "desirability" is more in line with the concept of satisficing introduced by Simon (1982). The behavioral aspect that can be introduced to comparative probability through the introduction of gambles shed a new light on some old concepts of the theory. In particular, as will be demonstrated below, the functions introduced by Fishburn (1996, 1997) become measures of rationality of personal comparative probability.

4.1 Discrete Cones

Let $[n] = \{1, 2, \dots, n\}$ be the set of possible states of the world, one of which will materialise. We suppose that agents can somehow compare probabilities of events. This is their personal probability assessment and it is subjective. If an agent believes that B is more likely to occur than A, she should accept the gamble which pays 1 if the state $i \in A \setminus B$ materialises, -1 if the state $i \in B \setminus A$ materialises, and pays nothing in all other cases. On the other hand, if the agent considers this gamble desirable, she must believe that B is more likely to happen than A. Thus it is clear that comparative probability assessments of sets and desirability of gambles provide two equivalent languages to discuss orders on subsets. Below we will make this connection formal.

Let $T = \{-1, 0, 1\}$. Any vector of T^n represents a gamble. The gamble which pays $x_i \in T$ if the state i materialises will be denoted $\mathbf{x} = (x_1, \dots, x_n) \in T^n$. On appearance of a nonzero gamble $\mathbf{x} \in T^n$ a participating agent must be ready to accept either $\mathbf{x}$ or $-\mathbf{x}$. The zero gamble $\mathbf{0}$ is neutral (no loss, no profit). Let us agree that it is not desirable.

The following properties will be assumed as basic rationality assumptions that all agents possess:

C1. $\mathbf{e}_i = (0,\dots,1,\dots,0)$ is a desirable gamble for all $1 \le i \le n$;
C2. If $\mathbf{x}$ and $\mathbf{y}$ are two desirable gambles and if $\mathbf{x}+\mathbf{y} \in T^n$, then $\mathbf{x}+\mathbf{y}$ is a desirable gamble;
C3. For every nonzero gamble $\mathbf{x} \in T^n$, either $\mathbf{x}$ or $-\mathbf{x}$ (but not both) is desirable.

Definition 5. Any subset $\mathcal{C}$ of T^n which contains $\mathbf{0}$ and whose nonzero vectors satisfy C1 - C3 is called a *discrete cone*.

To summarise: the set of desirable gambles for an agent is the set of all nonzero vectors of a certain discrete cone.

For each subset $A \subseteq [n]$ we define the characteristic vector χ_A of this subset by setting $\chi_A(i) = 1$ if $i \in A$, and $\chi_A(i) = 0$ if $i \notin A$. For any pair of subsets $A, B \in 2^{[n]}$ we define a gamble:

$$\chi(A,B) = \chi_B - \chi_A \in T^n.$$

Given an agent whose set of desirable gambles is a discrete cone $\mathcal{C}$, the agent can compare events as follows:

$$A \preceq B \Longleftrightarrow \chi(A,B) \in \mathcal{C}. \tag{6}$$

Due to properties of $\mathcal{C}$, $\preceq$ is an order (reflexive, complete and transitive relation) on $2^{[n]}$. This order satisfies de Finetti's axiom (3) and hence is a comparative probability. This probability assessment is, of course, specific for this particular agent only.

The study of discrete cones as algebraic objects was initiated by Kumar[1] in his PhD thesis (Kumar, 1982). This approach was rediscovered by Fishburn (1996) who pioneered their combinatorial study. Further combinatorial properties of discrete cones were studied in Fine and Gill (1976); Fishburn (1997); Fishburn, Pekeč, and Reeds (2002); Maclagan (1999); Conder and Slinko (2004); Marshall (2005); Christian and Slinko (2005). In this section we concentrate on combinatorics of rationality assessment.

If $\mathbf{p} = (p_1,\dots,p_n)$ is a probability measure on $[n]$, where p_i is the probability of i, then we know the probability of every event A, by the rule $p(A) = \sum_{i\in A} p_i$. We may now define an order $\preceq_{\mathbf{p}}$ on $2^{[n]}$ by

$$A \preceq_{\mathbf{p}} B \Longleftrightarrow p(A) \le p(B).$$

Suppose the probabilities of all events are different. Then $\preceq_{\mathbf{p}}$ is a comparative probability order on $[n]$.

Definition 6. Any comparative probability order $\preceq$ on $[n]$ is called *additively representable* by a measure or simply *representable* if there exists a probability measure

[1] I am grateful to Terry Fine for this reference.

$\mathbf{p}$ on $[n]$ such that $\preceq = \preceq_{\mathbf{p}}$. A comparative probability order $\preceq$ on $[n]$ is said to be *almost representable* by a measure $\mathbf{p}$ if

$$A \preceq B \Longrightarrow p(A) \leq p(B).$$

In this case we will also say that $\preceq$ is *almost representable* without specifying the measure $\mathbf{p}$.

If an order $\preceq$ is almost representable but not representable, then at least for one pair of subsets A and B we must have $A \prec B$ and at the same time $p(A) = p(B)$.

4.2 Portfolios of Acceptable Gambles

Our way to measure rationality of an agent is to look at how consistent she was in accepting and rejecting various gambles. We need the following concept.

Definition 7. Let $\mathcal{C}$ be a discrete cone. A multiset

$$P = \{\mathbf{x}_1^{a_1}, \mathbf{x}_2^{a_2}, \ldots, \mathbf{x}_m^{a_m}\},$$

where $\mathbf{x}_i \in \mathcal{C}$ and $a_i \in \mathbb{N}$, is called a *portfolio of desirable gambles.*

Gambles are like risky securities. You may own different number of shares of the same company. Similarly, a portfolio can contain several identical gambles. If the personal comparative probability of an agent is representable by a measure, then all portfolios of desirable gambles are (in the long run) profitable.

Definition 8. The portfolio P is said to be *neutral* if

$$a_1\mathbf{x}_1 + a_2\mathbf{x}_2 + \cdots + a_m\mathbf{x}_m = \mathbf{0}. \tag{7}$$

The portfolio P is said to be a *sure loss* if

$$a_1\mathbf{x}_1 + a_2\mathbf{x}_2 + \cdots + a_m\mathbf{x}_m = \sum_{i=1}^{n} b_i\mathbf{e}_i \tag{8}$$

with $b_i < 0$ for all $i = 1, \ldots, n$.

If a sure-loss portfolio exists, an agent is said to provide an *arbitrage*. A fully rational agent cannot accept a neutral portfolio and, of course, cannot provide an arbitrage. Here is an example of a comparative probability order that has a neutral portfolio of desirable gambles.

Example 3. Let $n = 5$ and consider the following comparative probability order:

$$\emptyset \prec 1 \prec 2 \prec 3 \prec 12 \prec \underline{13 \prec 4} \prec \underline{14 \prec 23} \prec$$
$$5 \prec 123 \prec 24 \prec \underline{34 \prec 15} \prec 124 \prec \underline{25 \prec 134} \ldots$$

(further continuation is unique). The following four desirable gambles

$$\mathbf{x}_1 = (-1,0,-1,1,0), \quad \mathbf{x}_2 = (-1,1,1,-1,0),$$
$$\mathbf{x}_3 = (1,0,-1,-1,1), \quad \mathbf{x}_4 = (1,-1,1,1,-1)$$

(they correspond to the underlined comparisons) form a neutral portfolio since $\mathbf{x}_1 + \mathbf{x}_2 + \mathbf{x}_3 + \mathbf{x}_4 = \mathbf{0}$.

For an example of arbitrage we must have $|\Omega| \geq 6$. Such an example is given in Kraft et al. (1959). Conder and Slinko (2004) used a computer program to help them find that for $n = 6$ there are 5202 such comparative probability orders.

Definition 9. A comparative probability order satisfies *cancellation condition* C_k when no neutral portfolio (7) of desirable gambles of width k exist, and satisfies the *cancellation condition* $C_k^\dagger$ when no neutral portfolio (7) of desirable gambles of cardinality k exist.

The criterion of representability given by Kraft et al. (1959) can be reformulated as follows.

Theorem 3. *Suppose $\preceq$ is the agent's comparative probability order on $2^{[n]}$ and $\mathcal{C}$ be the corresponding discrete cone. Then*

1. *$\preceq$ is representable iff $\mathcal{C}$ has no neutral portfolios of desirable gambles;*
2. *$\preceq$ is almost representable iff there is no arbitrage.*

4.3 Fishburn's Functions as Measures of Rationality

Let $\preceq$ be the agent's comparative probability order. Let $f(\preceq)$ be the smallest width of a neutral portfolio of desirable gambles and $f^\dagger(\preceq)$ be the smallest cardinality of a neutral portfolio of desirable gambles, if such portfolios exist. Otherwise set $f(\preceq) = f^\dagger(\preceq) = \infty$.

The idea is to measure the agent's rationality by the minimum "size" of the portfolio that she cannot handle properly with accepting a neutral portfolio being the early sign of non-rationality. We have two measures for the size of a portfolio: its width and its cardinality. Each measure gives us a measure of an agent's rationality. They are $f(\preceq)$ and $f^\dagger(\preceq)$, respectively. The larger these functions are the more rational is the agent. Fishburn defined these functions in terms of cancellation conditions of two types Fishburn (1996). He and his coauthors used their combinatorial interpretations in terms of multilists Fishburn, Pekeč, and Reeds (2002). Conder and Slinko (2004) used their algebraic reformulation of cancellation conditions in terms of linear dependencies of vectors of discrete cones. However in both cases the real meaning of cancellation conditions is hard to grasp due to the intricacies of those definitions. Portfolios clarify the real meaning of cancellation conditions.

Let $\mathcal{L}_n$ be the set of all comparative probability orders on $2^{[n]}$, and let $\mathcal{R}_n$ be the set of all almost representable comparative probability orders on $2^{[n]}$. Define

$$f(n) = \max_{\preceq \in \mathcal{L}_n} f(\preceq), \qquad f^{\dagger}(n) = \max_{\preceq \in \mathcal{L}_n} f^{\dagger}(\preceq),$$

These two functions were introduced and studied by Fishburn (1996, 1997). Also we define

$$g(n) = \max_{\preceq \in \mathcal{R}_n} f(\preceq), \qquad g^{\dagger}(n) = \max_{\preceq \in \mathcal{R}_n} f^{\dagger}(\preceq).$$

These functions were introduced by Conder and Slinko (2004). They are defined similarly to Fishburn's functions, but only for comparative probability orders which do not admit arbitrage. By temporarily setting all orders with arbitrage aside, Conder and Slinko showed that it is possible to achieve some progress and to answer some questions of Fishburn about $f(\preceq)$ and $f^{\dagger}(\preceq)$. The relationships between $f(\preceq)$ and $f^{\dagger}(\preceq)$ and their no arbitrage analogues $g(\preceq)$ and $g^{\dagger}(\preceq)$ are not completely clear. All we can state is that $g(n) \le f(n)$ and $g^{\dagger}(n) \le f^{\dagger}(n)$.

Some initial values for these functions are known Kraft et al. (1959); Fishburn (1996); Fishburn (1997); Conder and Slinko (2004):

$$f(n) = f^{\dagger}(n) = \infty, \qquad (n \le 4),$$

$$g(5) = g^{\dagger}(5) = f(5) = f^{\dagger}(5) = 4,$$
$$g(6) = g^{\dagger}(6) = f(6) = f^{\dagger}(6) = 5.$$

It is also known that $g(n) \le n$ Conder and Slinko (2004) and we will see later that $g(7) = 7$. The following bounds are known for $f(n)$, where the upper bound was established by Kraft et al. (1959) and the lower by Fishburn (1996, 1997).

Theorem 4 (Kraft et al., 1959, Fishburn, 1997). $n-1 \le f(n) \le n+1$.

The upper bound here is a rather trivial fact, the lower bound was obtained by a non-trivial construction. Fishburn (1996, 1997) conjectured that $f(n) = n-1$. However, since $f(n) \ge g(n)$, the first part of the following theorem refutes Fishburn's conjecture.

Theorem 5 (Conder & Slinko, 2004). $g(7) = 7$ *and* $g^{\dagger}(7) \ge 8$.

This result is based on the following construction theorem.

Theorem 6 (Conder & Slinko, 2004). *Let* $X = \{\mathbf{x}_1, \ldots, \mathbf{x}_m\} \in T^n$ *(*$m \ge 4$*), such that* $\sum_{i=1}^m a_i \mathbf{x}_i = \mathbf{0}$ *for some positive integers* a_i*, and either*

no proper subsystem $X' \subset X$ *is linearly dependent with positive coefficients*

or

the sum $\sum_{i=1}^m a_i$ *is minimal possible.*

Suppose further that the $m \times n$ *matrix* A *having the vectors* $\mathbf{x}_1, \ldots, \mathbf{x}_m$ *as its rows has the property that* $A\mathbf{b} = \mathbf{0}$ *for some positive integer-valued vector* $\mathbf{b} = (b_1, \ldots, b_n)$ *with* $b_1 > b_2 > \ldots > b_n > 0$*, and that*

$$\mathbf{b}^{\perp} \cap T^n = \{\pm\mathbf{x}_1, \ldots, \pm\mathbf{x}_m\}.$$

Let $\mathbf{p} = (b_1 + \ldots + b_n)^{-1}\mathbf{b}$ *and* $C = \{\mathbf{x} \in T^n \mid (\mathbf{x}, \mathbf{p}) \geq 0\}$. *Then the discrete cone*

$$C' = C \setminus \{-\mathbf{x}_1, \ldots, -\mathbf{x}_m\}$$

corresponds to an almost representable comparative probability order $\preceq$ *which almost agrees with* $\mathbf{p}$, *with either*

$$f(\preceq) = m \qquad or \qquad f^{\dagger}(\preceq) = \sum_{i=1}^{m} a_i,$$

respectively.

To prove the second part of Theorem 5 one may take the following 7×7 matrix:

$$A = \begin{bmatrix} 1 & -1 & -1 & 1 & 0 & 1 & -1 \\ 1 & 0 & -1 & -1 & 1 & -1 & -1 \\ 1 & 0 & -1 & -1 & -1 & 0 & 1 \\ -1 & 1 & -1 & 1 & 1 & 0 & 1 \\ 0 & -1 & 1 & 1 & 0 & -1 & 1 \\ 0 & -1 & 1 & -1 & 1 & 1 & 1 \\ -1 & 1 & 1 & 0 & -1 & 0 & -1 \end{bmatrix},$$

and let $\mathbf{x}_1, \ldots, \mathbf{x}_7$ denote its rows. It is easy to check that $\operatorname{rank}(A) = 6$,

$$\mathbf{x}_1 + \mathbf{x}_2 + \mathbf{x}_3 + \mathbf{x}_4 + \mathbf{x}_5 + \mathbf{x}_6 + 2\mathbf{x}_7 = \mathbf{0},$$

$$\mathbf{p}^{\perp} = \operatorname{Span}\{\mathbf{x}_1, \ldots, \mathbf{x}_7\} \cap T^7 = \{\pm\mathbf{x}_1, \ldots, \pm\mathbf{x}_7\},$$

for the probability measure

$$\mathbf{p} = \frac{1}{148}(48, 40, 27, 16, 12, 10, 7).$$

4.4 Extremal Cones and Comparative Probability Orders. Marshall's Theorem

In the previous section we saw that discrete cones and comparative probability orders with the property $g(n) = n$ do exist. Since this is the maximal possible value of $g(n)$, Marshall (2005) calls such objects *extremal*. He constructed a great many other extremal comparative probability orders by using some clever algebra and number theory. Before formulating Marshall's theorem we remind the reader that, given a prime p, an integer a is called a *quadratic residue* if there exists a b such that $a = b^2 \pmod{p}$; otherwise it is called a *quadratic non-residue*. The Legendre symbol $\left(\frac{a}{p}\right)$ is 0 if a is a multiple of p, 1 if a is a quadratic residue mod p, and -1 if a is a quadratic non-residue.

Theorem 7 (Marshall, 2005). *Let p be a prime greater than 131. If*

$$\left(1+\sqrt{\left(\frac{-1}{p}\right)p}\right)^p - 1 = a+b\sqrt{\left(\frac{-1}{p}\right)p},$$

where $gcd(a,b) = p$, then there exists an almost representable discrete cone in T^p with $g(p) = p$.

The odd primes satisfying the above equation he calls *optimus* primes. The first few non-optimus primes are

$$3, 23, 31, 137, 191, 239, 277, 359, \ldots.$$

Calculations that he and McCall conducted showed that 1,725 of the 1,842 primes between 132 and 16,000 are optimus primes.

Problem 3. Is the number of optimus primes infinite?

The idea of Marshall's construction is as follows. He uses the construction of Theorem 6 (changing rows into columns) and constructs the matrix needed there by altering the vector of Legendre quadratic residue symbols in the first two coordinates as follows:

$$\mathbf{q} = \left(1, \left(\frac{1}{p}\right) - 1, \left(\frac{2}{p}\right), \ldots, \left(\frac{p-1}{p}\right)\right)^T.$$

Then he forms a circulant matrix

$$Q = \left[\mathbf{q}, S\mathbf{q}, S^2\mathbf{q}, \ldots, S^{p-1}\mathbf{q}\right]$$

from $\mathbf{q}$, where S is the standard matrix of the circular shift operator. Finally he forms $A = Q - E_{11} + E_{1p}$ which is Marshall's matrix for prime p.

Theorem 8 (Conder, 2005). $g(n) = n$ *for* $7 \leq n \leq 13$.

This result was proved with the help of the MAGMA system Bosma and Cannon (1997) and announced in Marshall (2005). In the course of achieving it, Conder found that Marshall's matrices work not just for primes p satisfying the conditions given in Theorem 7, but also for some others, including all primes p in the range $5 \leq p \leq 23$.

A number of questions remain open. The most important ones are:

Problem 4.

1. Is $f(7) = 7$ or is $f(7) = 8$?
2. What is $g(14)$?
3. Is $g(n) = n$ for $n \geq 7$?
4. Is it true that $f(n) = g(n)$?
5. Does Marshall's construction work for all primes $p \geq 5$?

4.5 Patterns of Minimal Neutral Portfolios

Definition 10. Let $\preceq$ be a comparative probability order on $2^{[n]}$ and $\mathcal{C}$ be the corresponding discrete cone. Let

$$P = \{\mathbf{x}_1^{a_1}, \mathbf{x}_2^{a_2}, \dots, \mathbf{x}_n^{a_n}\}, \tag{9}$$

be a neutral portfolio of desirable gambles satisfying

1. $\mathrm{width}(P) = m$ is minimal possible for a neutral portfolio,
2. for neutral portfolios of desirable gambles of width m the cardinality $\mathrm{card}(P) = \sum_{i=1}^{m} a_i$ is minimal.

In this case we say that $(a_1, \dots, a_m)$ is an *irreducible pattern*. The set of all irreducible patterns of width m in $2^{[n]}$ is denoted as $\mathcal{A}_{m,n}$. Let us denote

$$\mathcal{A}_m = \bigcup_{n=4}^{\infty} \mathcal{A}_{m,n}.$$

Theorem 9 (Fishburn, 1996).

$$\mathcal{A}_4 = \{(1,1,1,1)\},$$
$$\mathcal{A}_5 = \{(1,1,1,1,1), (1,1,1,1,2)\}.$$

Moreover, $\mathcal{A}_{5,5} = \emptyset$, *and* $\mathcal{A}_5 = \mathcal{A}_{5,9}$.

Theorem 10 (Conder–Slinko, 2004).

$$\mathcal{A}_{5,6} = \{(1,1,1,1,1)\},$$
$$\mathcal{A}_{7,7} \supseteq \{(1,1,1,1,1,1,1), (1,1,1,1,1,1,2)\}.$$

This means that we don't know $\mathcal{A}_{5,7}$ and $\mathcal{A}_{5,8}$. We don't know $\mathcal{A}_6$ either. An unpublished recent result in this direction is the following theorem by Auger (2005), for which we provide here a short proof.

Theorem 11 (Auger, 2005). *For any positive integer m there are only finitely many irreducible patterns of length m.*

Proof. Let us consider the set of all vectors of $\mathbb{R}^m$ with non-negative integer coordinates. Let us denote it Z_m. All irreducible patterns from $\mathcal{A}_m$ belong to Z_m. For an arbitrary $\mathbf{a} = (a_1, \dots, a_m) \in Z_m$ we denote $h(\mathbf{a}) = \sum_{i=1}^{m} a_i$. We also define a set

$$R(\mathbf{a}) = \left\{ \{I,J\} \mid I,J \subseteq [m], \quad I \cap J = \emptyset, \quad \sum_{i \in I} a_i = \sum_{j \in J} a_j \right\}.$$

The set $R(\mathbf{a})$ has a cardinality smaller than the cardinality of the set of all pairs of subsets $\{I,J\}$ with $I \cap J = \emptyset$, which is $(3^m - 1)/2$. Hence $R(\mathbf{a})$ is finite. So it is sufficient to prove that there are only finitely many irreducible patterns $\mathbf{a}$ with the same $R(\mathbf{a})$.

Suppose now that we have two irreducible patterns $\mathbf{a}$ and $\mathbf{b}$ with $R(\mathbf{a}) = R(\mathbf{b})$. Let $\{\mathbf{x}_1, \ldots, \mathbf{x}_m\} \subseteq T^n$ such that $\sum_{i=1}^m a_i \mathbf{x}_i = \mathbf{0}$. Then each of the n coordinates of this vector equation will give us an element of $R(\mathbf{a})$ (they will not be necessarily distinct). Hence if another vector $\mathbf{b} = (b_1, \ldots, b_m) \in Z_m$ will satisfy $R(\mathbf{a}) = R(\mathbf{b})$, then $\sum_{i=1}^m a_i \mathbf{x}_i = \mathbf{0}$ will always imply $\sum_{i=1}^m b_i \mathbf{x}_i = \mathbf{0}$ and vice versa. Thus, if $\mathbf{a} \in Z_m$ and $\mathbf{b} \in Z_m$ are both irreducible patterns of $\mathcal{A}_m$, then we must have $h(\mathbf{a}) = h(\mathbf{b})$. Since there are only finitely many vectors $\mathbf{c}$ in Z_m with the given $h(\mathbf{c})$, we see that the set of irreducible patterns $\mathbf{a}$ with fixed $R(\mathbf{a})$ is finite and hence $\mathcal{A}_m$ is finite. □

Problem 5. Let $\mathcal{C}$ be a discrete cone and $P = \{\mathbf{x}_1^{a_1}, \mathbf{x}_2^{a_2}, \ldots, \mathbf{x}_m^{a_m}\}$ be the neutral portfolio of desirable gambles with the smallest height $\sum_{i=1}^m a_i$. Is $(a_1, \ldots, a_m)$ an irreducible pattern? Or, in other words, will the width of P also be smallest?

Axioms for unique additive representation of a comparative probability order (which in this case cannot be strict) were given by Fishburn and Roberts (1989).

5 Orders on Submultisets of a Multiset

In this section we will consider multisets on the base set $[n]$. Every such multiset $M = \{1^{m_i}, 2^{m_2}, \ldots, n^{m_n}\}$ is uniquely determined by its *multiplicity function* $\mu \colon [n] \to \mathbb{N}$ such that $\mu(i) = m_i$. We say that $M_1 = ([n], \mu_1)$ is a submultiset of $M_2 = ([n], \mu_2)$, if $\mu_1(i) \le \mu_2(i)$ for all $i \in [n]$, and we denote this by $M_1 \subseteq M_2$. We remind the reader that the set of all submultisets of cardinality k will be denoted as $\mathcal{P}_k([n])$.

5.1 Independence of Equal Submultisets and Additive Representability

Definition 11. An order $\preceq$ on $\mathcal{P}_k[n]$ is said to be *(additively) representable* if there exist nonnegative real numbers $u_1, \ldots, u_m$ (utilities) such that for all $M_1 = ([n], \mu_1)$ and $M_2 = ([n], \mu_2)$ belonging to $\mathcal{P}_k[n]$,

$$M_1 \preceq M_2 \iff \sum_{i=1}^n \mu_1(i) u_i \le \sum_{i=1}^n \mu_2(i) u_i. \tag{10}$$

The following basic rationality condition adopted for this situation was suggested by Sertel and Slinko (2002, 2007), who called it *consistency*. Here we follow Conder et al. (2007) who give a slightly different (but equivalent) definition of this concept, which makes it a close relative to the concept of the Independence of Equal Subalternatives and de Finetti's axiom.

Definition 12. An order $\preceq$ on $\mathcal{P}_k([n])$ is said to satisfy the *Independence of Equal Submultisets* condition (IES) if, for all $1 \le j \le k-1$, for every two multisets U,V of cardinality j and for every two multisets W_1,W_2 of cardinality $k-j$,

$$U \cup W_1 \preceq V \cup W_1 \Longleftrightarrow U \cup W_2 \preceq V \cup W_2. \tag{11}$$

Certainly every additively representable order must satisfy IES. The converse as we will see later is not true. However, it appeared that IES alone implies additive representability on $\mathcal{P}_k([3])$ for all k. The following theorem was proved first in Sertel and Slinko (2002) and later appeared in Sertel and Slinko (2007). We remind the reader of the definition of one of the main number-theoretic functions ϕ, which is *Euler's totient function*. For any positive integer n, $\phi(n)$ is the number of positive integers which are smaller than n and relatively prime to n. Also, the famous sequence of Farey fractions $\mathbf{F}_k$ is the increasing sequence of all fractions in lowest possible terms between 0 and 1, whose denominators do not exceed k. For example, the sequence of Farey fractions $\mathbf{F}_6$ will be:

$$\frac{0}{1}, \frac{1}{6}, \frac{1}{5}, \frac{1}{4}, \frac{1}{3}, \frac{2}{5}, \frac{1}{2}, \frac{3}{5}, \frac{2}{3}, \frac{3}{4}, \frac{4}{5}, \frac{5}{6}, \frac{1}{1}.$$

The standard reference for Farey fractions is Hardy and Wright (1960).

Theorem 12 (Sertel and Slinko, 2002). *Any order $\preceq$ on $\mathcal{P}_k([3])$ satisfying IES is additively representable. There are $2\Phi(k)-1$ of them, where $\Phi(k)=\sum_{h=1}^{k}\phi(h)$ and $\phi(h)$ is the Euler totient function, with exactly $\Phi(k)$ orders being strict (antisymmetric). Moreover, if utilities of 1 and 3 are normalized so that $u_1=1$, $u_3=0$, then the ith strict order occurs when u_2 belongs to the ith interval between consecutive Farey fractions in the kth sequence of Farey fractions $\mathbf{F}_k$.*

Here we will choose a combinatorial way to introduce cancellation conditions similar to Scott's approach Scott (1964).

Definition 13. Let $\preceq$ be an order on $\mathcal{P}_k[n]$ and let

$$A_1 \preceq B_1, \quad A_2 \preceq B_2, \quad \ldots, \quad A_q \preceq B_q \tag{12}$$

be a sequence of valid set comparisons such that $A_i \prec B_i$ for at least one $i = 1,2,\ldots,q$. We say that this sequence satisfies the *cancellation property* if the following two multiset unions coincide

$$A_1 \cup \ldots \cup A_q = B_1 \cup \ldots \cup B_q. \tag{13}$$

Definition 14. We say that an order $\preceq$ on $\mathcal{P}_k[n]$ satisfies the kth *cancellation condition C_k* if no sequence of comparisons (12) of width $\le k$ satisfy the cancellation property and we say that it satisfies the kth *cancellation condition $C_k^{\dagger}$* if no sequence of comparisons (12) of cardinality $\le k$ satisfy the cancellation property.

As in (Kraft et al., 1959, Theorem 2) it is easy to show that for an order $\preceq$ on $\mathcal{P}_k([n])$ to be additively representable, it is necessary and sufficient that all cancellation conditions $C_2, C_3, \ldots, C_\ell, \ldots$ are satisfied or alternatively all cancellation conditions $C_2^\dagger, C_3^\dagger, \ldots, C_\ell^\dagger, \ldots$ are satisfied.

Example 4 (Sertel and Slinko (2002)). The following linear order on $\mathcal{P}_2[4]$

$$1^2 \succ 12 \succ \underline{13 \succ 2^2} \succ \underline{23 \succ 14} \succ \underline{24 \succ 3^2} \succ 34 \succ 4^2$$

satisfies IES but is not representable. It does not satisfy the condition C_3, since it contains the following comparisons:

$$\{1,3\} \succ \{2^2\}, \quad \{2,3\} \succ \{1,4\}, \quad \{2,4\} \succ \{3^2\}. \tag{14}$$

Indeed, the union of the multisets on the right and the union of the multisets on the left are both equal to the multiset $\{1, 2^2, 3^2, 4\}$. Thus C_3 is violated with $a_1 = a_2 = a_3 = 1$, and hence $C_3^\dagger$ is also violated.

Definition 15. An order $\preceq$ on $\mathcal{P}_k[n]$ is said to be *almost (additively) representable* if there exist nonnegative real numbers $u_1, \ldots, u_m$, not all of which are equal, such that for all $M_1 = ([n], \mu_1)$ and $M_2 = ([n], \mu_2)$ belonging to $\mathcal{P}_k[n]$,

$$M_1 \preceq M_2 \implies \sum_{i=1}^n \mu_1(i) u_i \leq \sum_{i=1}^n \mu_2(i) u_i. \tag{15}$$

If the only way to get $u_1, \ldots, u_n$ which satisfy (15) is to set $u_1 = u_2 = \ldots = u_n$, then the order fails to be almost representable. Papers Sertel and Slinko (2002); Sertel and Slinko (2007) present such an order belonging to $\mathcal{P}_3[4]$.

Let $\mathcal{L}_{n,k}$ be the set of all orders on $\mathcal{P}_k[n]$ satisfying the IES and $\mathcal{R}_{n,k}$ be the set of all almost representable comparative probability orders on $\mathcal{P}_k[n]$ satisfying the IES. As in the case of comparative probability orders we define

$$f(n,k) = \max_{\preceq \in \mathcal{L}_{n,k}} f(\preceq), \qquad f^\dagger(n,k) = \max_{\preceq \in \mathcal{L}_{n,k}} f^\dagger(\preceq).$$

Also we define

$$g(n,k) = \max_{\preceq \in \mathcal{R}_{n,k}} f(\preceq), \qquad g^\dagger(n,k) = \max_{\preceq \in \mathcal{R}_{n,k}} f^\dagger(\preceq).$$

These functions have the same meaning as in the comparative probability orders case. Conder et al. (2007) fully characterized the function $g(n,k)$ as follows:

Theorem 13. *For all $n > 3$ and $k \geq 2$,*

$$g(n,k) = \begin{cases} n-2 & \textit{if } (n,k) = (5,2), \\ n-1 & \textit{otherwise}. \end{cases}$$

This result leaves very little room for the function $f(n,k)$, i.e. $n-1 \leq f(n,k) \leq n$ whenever $(n,k) \neq (5,2)$. Computer-assisted calculations show that $g(n,k) = f(n,k)$

for small values of n and k (namely, for $(n,k) = (4,2), (4,3), (5,2), (5,3), (6,2)$ and $(7,2)$), and so Conder, Marshall, and Slinko conjecture that this is true in general.

Problem 6. Is it true that $f(n,k) = g(n,k)$ for all $n \geq 4$ and $k \geq 1$?

Problem 7. What can be said about the relationship between $g^{\dagger}(n,k)$ and $f^{\dagger}(n,k)$?

Orders on the infinite set $\mathcal{P}[n]$ of all multisets on $[n]$ satisfying the analogue of the de Finetti axiom (3), where the union is understood as the multiset union and the condition $C \cap (A \cup B) = \emptyset$ is not assumed, were considered by Danilov (1987) and Martin (1989). Both independently prove that all orders on $\mathcal{P}[n]$ satisfying this axiom are additively representable. For the set $\mathcal{P}_{\leq k}[n]$ of all multisets on $[n]$ of cardinality $\leq k$, Danilov gives an example of nonrepresentable orders on $\mathcal{P}_{\leq k}[5]$ satisfying the modified de Finetti axiom.

Apart from the aforementioned paper by Danilov, the representability of orders on $\mathcal{P}_{\leq k}[n]$ has largely escaped the attention of researchers. However some interesting things have been observed. For example, it can be easily checked that the linear order on $\mathcal{P}_{\leq 2}[3]$

$$1^2 \succ 12 \succ 2^2 \succ 13 \succ 1 \succ 23 \succ 3^2 \succ 2 \succ 3 \succ \emptyset$$

is not representable. Hence the analogue of Theorem 12 is not true.

Problem 8. Develop an additive representation theory for orders on $\mathcal{P}_{\leq k}[n]$.

Acknowledgements I thank participants of The 5th International Simposium on Imprecise Probability: Theories and Applications (ISIPTA 07), Prague, Czech Republic, 16–19 July 2007, and several seminar audiences for comments and suggestions. Financial support from the Economic Design Network (ARC) and the New Zealand Institute of Mathematics and its Applications (NZIMA) is gratefully acknowledged.

References

Auger, M. (2005). Finite linear qualitative probability. *Report on The Summer Scholarship Project*. Department of Mathematics, The University of Auckland.

Bosma, W., Cannon, J., & Playoust, C. (1997). The MAGMA algebra system I: The user language. *Journal of Symbolic Computation, 24*, 235–265.

Christian, R., & Slinko, A. (2005). Answers to two questions of Fishburn on subset comparisons in comparative probability orderings. In *Proceedings of the fourth international symposium on imprecise probabilities and their applications (ISIPTA 05)* (pp. 117–124). Pittsburg, PA.

Christian, R., Conder, M., & Slinko, A. (2007). Flippable pairs and subset comparisons in comparative probability orderings.*Order, 24*(3), 193–213.

Conder, M., Marshall, S., & Slinko, A. (2007). Orders on multisets and discrete cones. *Order, 24*(4), 277–296.

Conder, M., & Slinko, A. (2004). A counterexample to Fishburn's conjecture on finite linear qualitative probability. *Journal of Mathematical Psychology, 48*, 425–431.

Danilov, V. I. (1987). Aggregation of dichotomic preferences. *Mathematical Social Sciences, 13*, 49–58.

Debreu, G. (1960). Topological methods in cardinal utility theory. In K.J. Arrow, S. Karlin, & P. Suppes (Eds.), *Mathematical Methods in the Social Sciences* (pp. 16–26). Stanford: Stanford University Press.

Dershowitz, N. (1979). Proving termination with multiset orderings. *Communications of the ACM, 22*(8), 465–476.

Fine, T., & Gill, J., (1976). The enumeration of comparative probability relations. *Annals of Probability, 4*, 667–673.

Fine, T. L. (1973). *Theories of probability: An examination of foundations*. New York: Academic.

de Finnetti, B. (1931). Sul significato soggetivo della probabilità. *Fundamenta Mathematicae, 17*, 298–329.

Fishburn, P. C. (1970). *Utility theory for decision making*. New York: Wlley.

Fishburn, P. C. (1996). Finite linear qualitative probability. *Journal of Mathematical Psychology, 40*, 64–77.

Fishburn, P. C. (1997). Failure of cancellation conditions for additive linear orders. *Journal of Combinatorial Designs, 5*, 353–365.

Fishburn, P. C., Pekeč A., & Reeds, J. A. (2002). Subset comparisons for additive linear orders. *Mathematics of Operations Research, 27*, 227–243.

Fishburn, P. C. (1999). Preference structures and their numerical representations. *Theoretical Computer Science, 217*, 359–383.

Fishburn, P. C. (1997). Cancellation conditions for multiattribute preferences on finite sets. In M. H. Karwan, J. Spronk, & J. Wallenius (Eds.), *Essays in decision making* (pp. 157–167). Berlin: Springer.

Fishburn, P. C. (2001). Cancellation conditions for finite two-dimensional additive measurement. *Journal of Mathematical Psychology, 45*, 2–26.

Fishburn, P. C., & Roberts, F. S. (1988). Unique finite conjoint measurement. *Mathematical Social Sciences* 16, 107–143.

Fishburn, P.C., & Roberts, F. S. (1989). Axioms for unique subjective probability on finite sets. *Journal of Mathematical Psychology, 33*, 117–130.

Hardy, G. H., & Wright, E. M. (1960). *An introduction to the theory of numbers.* Oxford: Oxford University Press.

Krantz, D. H., Luce, R. D., Suppes, P., & Tversky, A. (1971). *Foundations of measurement* (Vol. 1). New York: Academic.

Kraft, C. H., Pratt, J. W., & Seidenberg, A. (1959). Intuitive probability on finite sets. *Annals of Mathematical Statistics, 30*, 408–419.

Kumar, A. (1982). *Lower probability on infinite spaces and instability of stationary sequences.* PhD Thesis. Cornell University.

Luce, R. D., Krantz, D. H., Suppes, P., & Tversky, A. (1988). *Foundations of measurement* (Vol. 3). New York: Academic.

Luce, R. D., & Tukey, J. W. (1964). Simultaneous conjoint measurement: A new type of fundamental measurement, *Journal of Mathematical Psychology, 1*, 1–27.

Maclagan D. (1999). Boolean term orders and the root system B_n. *Order, 15*, 279–295.

Marshall, S. (2005). On the existence of extremal cones and comparative probability orderings. In *Proceedings of the fourth international symposium on imprecise probabilities and their applications (ISIPTA 05)* (pp. 246–255). Pittsburg, PA.

Marshall, S. (2007). On the existence of extremal cones and comparative probability orderings. *Journal of Mathematical Psychology, 51*(5), 319–324.

Martin, U. (1989). A geometric approach to multiset ordering. *Theoretical Computer Science, 67*, 37–54.

Narens, L. (1985). *Abstract measurement theory*. Cambridge, MA: MIT.

Pfanzagl, J. (1968). *Theory of measurement*. New York: New York.

Roberts, F. S. (1979). *Measurement theory with applications to decision making, utility, and the social sciences*. Reading, MA: Addison-Wesley.

Scott, D. (1964). Measurement structures and inequalities, *Journal of Mathematical Psychology, 1*, 233–247.

Sertel, M., & Slinko, A. (2002). *Ranking committees, words or multisets. Nota di Laboro 50.2002.* Center of Operation Research and Economics. The Fundazione ENI Enrico Mattei, Milan.

Sertel, M., & Slinko, A. (2007). Ranking committees, income streams or multisets. *Economic Theory, 30*(2), 265–287.

Stanley, R. P. (1997). *Enumerative combinatorics* (Vol. 1). Cambridge: Cambridge University Press.

Suppes, P., Krantz, D. H., Luce, R. D., & Tversky, A. (1988). *Foundations of measurement* (Vol. 2). New York: Academic.

Simon, H. A. (1982). *Models of bounded rationality* (Vol. 1). In Economic analysis and public policy (pp. 235–441). Cambridge, MA : MIT.

Wakker, P. P. (1989). *Additive representation of preferences*. Dordrecht: Kluwer.

Wakker, P. P. (1991). Additive representations on rank-ordered sets. I. The algebraic approach. *Journal of Mathematical Psychology, 35*, 260–266.

Wakker, P. P. (1993). Additive representations on rank-ordered sets. II. The topological approach. *Journal of Mathematical Economics, 22*, 1–26.

Walley, P., & Fine, T. (1979). Varieties of modal (classificatory) and comparative probability. *Synthese, 41*(3), 321–374.

Walley, P. (1991). *Statistical reasoning with imprecise probabilities*. London: Chapman and Hall.

Walley, P. (1999). Towards a unified theory of imprecise probability. In *First international symposium on imprecise probabilities and their applications*. Ghent, Belgium.

Part II
Social Choice, Voting, and Social Welfare

Condorcet Domains and Probabilities

Acyclic Domains of Linear Orders: A Survey

Bernard Monjardet

1 Notations and Preliminaries

$A = \{1,2\ldots i,j,k\ldots n\}$ is a finite set of n elements that I will generally call *alternatives* (but which could also be called issues, decisions, outcomes, candidates, objects, etc.). The elements of A will be also denoted by letters like x,y,z etc. A subset of cardinality p of A will be called a p-set.

A^2 (respectively, A^3) denotes the set of all ordered pairs (x,y) (respectively, ordered triples (x,y,z) written for convenience as xyz) of A. When the elements of A are denoted by the n first integers, $P^2(n)$ denotes the set of the $n(n-1)/2$ ordered pairs $(i<j)$.

A binary relation on A is a subset R of A^2 and we write xRy or $(x,y) \in R$ when x is in the relation R with y. For ℓ integer ≥ 2, a *cycle of length* ℓ of R, called also a ℓ*-cycle*, is a subset $\{x_1,x_2,\ldots .x_\ell\}$ of A such that $x_1Rx_2\ldots\ldots x_\ell Rx_1$. For $B \subseteq A$, the restriction of a relation R to B is denoted by $R_{/B}$.

A *strict linear order* on A is an irreflexive, transitive and complete ($x \neq y$ implies xRy or yRx) binary relation on A. Henceforth, we will omit the qualifier strict and sometimes, when there is no ambiguity, the qualifier linear. Linear orders on A are in a one-to-one correspondence with *permutations* of A. So if L is a linear order on A one can write it as a permutation $x_1\ldots x_kx_{k+1}\ldots x_n$. Then one says that x_k has *rank* k and is *covered* by x_{k+1} and that x_k and x_{k+1} are *consecutive* in L. I denote by τ_k the transposition which exchange x_k and x_{k+1} in L: $\tau_k(L) = x_1\ldots x_{k+1}x_k\ldots x_n$.

The set of all linear orders on A is denoted by $\mathbb{L}$ or $\mathbb{L}_n$ if $|A| = n$. $\mathbb{D}$ denotes any subset of $\mathbb{L}$.

In all of this paper the preferences of what I will call a *voter* (but what could also be called agent, person, individual, criterion, etc.) on a set A of alternatives is

B. Monjardet
CES, Université Paris I Panthéon Sorbonne, Maison des Sciences Économiques, 106–112 bd de l'Hopital 75647 Paris Cédex 13, France, and CAMS, EHESS
e-mail: Bernard.Monjardet@univ-paris1.fr

S.J. Brams et al. (eds.), *The Mathematics of Preference, Choice and Order: Essays in Honor of Peter C. Fishburn*, Studies in Choice and Welfare,

represented by a linear order $L = x_1x_2...x_n$ where x_1 is assumed to be the last preferred alternative, x_2 the next-to-last, etc. So, yLx or $(y,x) \in L$ means that alternative x is preferred to alternative y in the linear order L.

Remark 1. One could consider that the notation yLx should mean that y is preferred to x. But we are working in this paper with posets and, unfortunately, this choice would be not in accordance with the usual convention of poset theory. Indeed in this theory the symbol used for a (strict) order is generally $<$ what means that yLx is interpreted as $y < x$, and so as y is less than x. The reader must keep in mind a consequence of our choice: in a linear order of preference $L = x_1x_2...x_n$, the worst alternative x_1 (respectively, the best alternative x_n) has rank 1 (respectively, n).

The problem of getting a collective preference from various voters' preferences was tackled by Borda and Condorcet at the end of 18th century. Condorcet criticized Borda's rank method and proposed the use of the majority rule on the pairs of alternatives. Before we recall the definition of this rule, I introduce some notations. I consider v voters, which express their preferences on the alternatives by linear orders taken in a set $\mathtt{D}$ of linear orders ($\mathtt{D} \subseteq \mathtt{L}$). The state of their preferences is given by a *v-profile* $\pi = (L_1,.L_q,.,L_v)$ where L_q is the linear order of $\mathtt{D}$ representing the preference of voter q. $\mathtt{D}^v$ denotes the set of all these v-profiles. For a subset B of alternatives, $\pi_{/B} = (L_{1/B},.L_{q/B},.,L_{v/B})$ denotes the profile of voters' preferences restricted to B.

For a v-profile $\pi = (L_1,.L_q,.,L_v)$ and two alternatives x and y, one denotes by $v_\pi(y,x)$ the number of voters preferring x to y in this profile.

In his "*Essai sur l'application de l'analyse à la probabilité des décisions rendues à la pluralité des voix*" (1785) Condorcet recommended the rule now called Condorcet's majority rule.[1] This rule associates with a profile π the collective preference defined as the *strict (simple) majority relation*[2] $R_{SMAJ}(\pi)$:

$$yR_{SMAJ}(\pi)x \text{ if } v_\pi(y,x) > v/2$$

i.e., alternative x is collectively preferred to alternative y if it is preferred by a (strict) majority of voters. It is clear that this majority relation is asymmetric i.e., has no 2-cycles. But Condorcet discovered that majority relations can have cycles of length $\ell \geq 3$: $x_1R_{SMAJ}x_2......x_\ell R_{SMAJ}x_1$. This fact that was rediscovered for instance by Dodgson, Black and Arrow has been called the "Condorcet effect" by Guilbaud (1952) and is also known as the "voting paradox".[3] I prefer the first appellation, which emphasizes the fact that this occurrence of cycles is not really a paradox (see Guilbaud, 1952 or Monjardet, 2006).

---

[1] Condorcet uses other terms like "plurality".

[2] The *(simple) majority relation* is the relation defined by $yR_{MAJ}(\pi)x$ if $v_\pi(y,x) \geq v/2$. Observe that since π is a profile of linear orders one has for $x \neq y (y,x) \in R_{MAJ}(\pi)$ if and only if $(x,y) \notin R_{SMAJ}(\pi)$.

[3] Condorcet speaks of the "contradictory case". Dodgson and Black speak of "cyclical majorities" and I do not know who used the term paradox the first time (it appears in Arrow's 1951 book).

The simplest cases of the Condorcet effect occur when $A = \{i, j, k\}$ and $v = 3$, with the profiles (*ijk*, *jki*, *kij*) and (*jik*, *ikj*, *kji*)) since then majority relations are the 3-cycles $iR_{\text{SMAJ}}jR_{\text{SMAJ}}kR_{\text{SMAJ}}i$ and $jR_{\text{SMAJ}}iR_{\text{SMAJ}}kR_{\text{SMAJ}}j$. I say that such profiles are 3-*cyclic profiles*. More generally, for an integer $\ell \geq 3$, I say that a profile like $\pi = (x_1x_2x_3\ldots x_\ell; x_2x_3\ldots x_\ell x_1;\ldots; x_\ell x_1x_2\ldots x_{\ell-1})$ is a ℓ-*cyclic profile*. The strict majority relation associated with such a profile is a ℓ-cycle. Observe that arbitrary profiles can contain the same linear order several times, but that ℓ-cyclic profiles are subsets of $\mathtt{L}$.

A subset $\mathtt{D}$ of the set $\mathtt{L}$ of all linear orders on A is an *acyclic domain* (of linear orders) if for every integer v and every profile $\pi = (L_1, L_2, ., L_v) \in \mathrm{D}^v$, $R_{\text{SMAJ}}(\pi)$ has no cycles.[4]

Several classical characterizations of acyclic domains are given in the theorem below. I need some definitions. For ℓ integer greater than 2, I say that a set $\mathtt{D}$ of linear orders contains a ℓ-cyclic profile if there exists a subset $B = \{x_1, x_2, \ldots x_\ell\}$ of A and a subset $\{L_1, \ldots L_q, \ldots, L_\ell\}$ of ℓ linear orders in $\mathtt{D}$ such that the profile $\pi_{/B} = (L_{1/B}, .L_{q/B} \ldots L_{\ell/B})$ is a ℓ-cyclic profile. When a set of three alternatives is linearly ordered as $i < j < k$, then[5] i has rank 1, j has rank 2 and k has rank 3. I say that a set $\mathtt{D}$ of linear orders is *value-restricted* if for every subset $\{i, j, k\}$ of A, there exists an alternative which either never has rank 1 or never has rank 2 or never has rank 3 in the set $\mathtt{D}_{/\{i,j,k\}}$. Finally in condition (7) of the theorem I use the majority relation defined in footnote 2.

Theorem 1. *Let* $\mathtt{D}$ *be a subset of the set* $\mathtt{L}$ *of all linear orders on a set A. The following conditions are equivalent:*

1. $\mathtt{D}$ *is acyclic (i.e., for every integer v and every profile* $\pi \in D^v$, $R_{SMAJ}(\pi)$ *has no cycles),*
2. *For every integer v and every profile* $\pi \in D^v$, $R_{SMAJ}(\pi)$ *is a (strict) partial order,*
3. *For every odd integer v and every profile* $\pi \in D^v$, $R_{SMAJ}(\pi)$ *is a linear order,*
4. *For every integer* $\ell \geq 3$, $\mathtt{D}$ *does not contain* ℓ*-cyclic profiles,*
5. $\mathtt{D}$ *does not contain 3-cyclic profiles,*
6. $\mathtt{D}$ *is value-restricted,*
7. *For every integer v, every profile* $\pi \in D^v$ *and every* $B \subseteq A$, $\{a \in B$: *for every* $b \in B \backslash \{a\}$, $bR_{MAJ}(\pi)a\} \neq \emptyset\}$.

Condition (2) means that when voters' preferences belong to an acyclic domain, the collective preference that is given by majority rule is transitive (and asymmetric) which in particular implies that it can be extended into a linear order. For a given

[4] Acyclic domains have been also called consistent profiles (Ward, 1965), valued-restricted domains (Kim & Roush, 1980), transitive simple majority domains or consistent sets (Abello & Johnson, 1984), "états d'opinion fortement condorcéens" (Chameni-Nembua, 1989), acyclic sets (Fishburn, 1992,1997), majority-consistent sets (Craven, 1996) or Condorcet domains (Monjardet, 2006).

[5] See the Remark on the ranks of linearly ordered alternatives in the previous page.

profile I say that an alternative is a *Condorcet winner* if it is preferred to all other alternatives in the majority relation (see footnote 2) associated with this profile. Condition (7) means that for every profile and every subset of candidates there exists at least a Condorcet winner. Condition (5) means that D is acyclic if and only if for every subset $C = \{L_1, L_2, L_3\}$ of three different linear orders of D and every subset $\{i, j, k\}$ of three different alternatives, C is not a 3-cyclic profile on $\{i, j, k\}$. It was introduced by Ward (1965) which proved the equivalence of conditions (1), (4) and (5) of the above theorem. He called it the condition of *Latin-square-lessness* since a 3-cyclic profile forms a Latin square when it is disposed in a 3×3 array. Condition (6) of value-restriction was introduced by Sen (1966).[6]

In what follows I will use Fishburn's formulation of the condition of value restriction. One assumes that the n alternatives of A are ranked in an arbitrary linear order, which in fact will be the natural order $1 < 2 < \ldots i < j < k < \ldots . n$. There are two 3-cyclic profiles on a 3-element set $\{i, j, k\}$, namely $\{ijk,\ jki,\ kij\}$ and $\{jik,\ ikj,\ kji\}$. In each of these 3-cyclic profiles each element h of $\{i, j, k\}$ appears at rank 1, 2 and 3 in one of the three linear orders of the profile. In order to avoid a 3-cyclic profile on $\{i, j, k\}$, it suffices to assume that one of the linear orders in $\{ijk,\ jki,\ kij\}$ and one in $\{jik,\ ikj,\ kji\}$ never occurs. There are $3 \times 3 = 9$ different ways to do that. But each of these ways comes back to assume that an element h of $\{i, j, k\}$ never appears at rank 1, 2 or 3 in a linear order on $\{i, j, k\}$. For instance, to exclude ijk and *jik* comes back to assume that k never has rank 3 in the restrictions to $\{i, j, k\}$ of the linear orders of D. I will write this condition $kN_{\{i,j,k\}}3$. More generally for h in $\{i, j, k\}$ and r in $\{1, 2, 3\}$, the *Never Condition* $hN_{\{i,j,k\}}r$ means that h never has rank r in the restrictions to $\{i, j, k\}$ of the linear orders of D. With these definitions a set of linear orders is an acyclic domain if and only if for every ordered triple $i < j < k$ there exists $h \in \{i, j, k\}$ and $r \in \{1, 2, 3\}$ such that $hN_{\{i,j,k\}}r$. Since $1 < 2 \ldots < n$ contains n(n-1)(n-2)/6 ordered triples and that for each ordered triple $i < j < k$, one can choose one of the nine possible Never Condition $hN_{\{i,j,k\}}r$, one sees that there are many ways to get acyclic domains.[7] I will say that an acyclic domain satisfies the Never Condition hNr if for every ordered triple $i < j < k$, the same Never Condition $hN_{\{i,j,k\}}r$ is satisfied. For instance D satisfies $jN1$ if for every ordered triple $i < j < k$, j never has rank 1 (i.e., is never last) in the restrictions to $\{i, j, k\}$ of the orders of D. I will say that an acyclic domain satisfies the Never Condition $ijkNr$ if for every ordered triple $i < j < k$, one has either iNr or jNr or kNr (one of the three alternatives never has rank r).

An obvious but useful observation is that the Never Conditions are "hereditary". Firstly if a set D of linear orders satisfies a set of Never Conditions any subset of D satisfies the same set of Never Conditions. Secondly if a set D of linear orders

[6] In fact Sen's value-restriction condition is more general since it bears on the case where voters' preferences are represented by weak orders (transitive and complete binary relations). But Sen has immediately pointed out that when voters' preferences are represented by linear orders his condition is equivalent to Ward's Latin-square-lessness condition. In this case Ward's result and Arrow's theorem are "dual" (see Monjardet, 1978).

[7] But the set of Never Conditions chosen must be satisfied by at least a linear order. For instance, Raynaud (1981) has shown that for $n \geq 5$ there does not exist a linear order satisfying $jN2$ for every ordered triple $i < j < k$ (and that this condition is satisfied by only four orders for $n = 4$).

defined on A satisfies a set of Never Conditions then for every $B \subseteq A$, $\mathtt{D}_{/B}$ (the set of linear orders restrictions to B of the linear orders of $\mathtt{D}$) satisfies the same set of Never Conditions. It is also interesting to mention the following fact on these conditions. Let us denote by L^d the *dual* linear order of the linear order $L : xL^dy$ if and only if yLx, and for $\mathtt{D} \subseteq \mathtt{L}$, call $\mathtt{D}^d = \{L^d, L \in \mathtt{D}\}$ the *dual domain* of $\mathtt{D}$. Then a domain satisfies the Never Condition *hNr* if and only if its dual satisfies the Never Condition *hN*(4-*r*).

Now the interesting problem is: how large can domains of linear orders where Condorcet's majority rule works well be? Or more concisely, how large can acyclic domains be? Observe that the problem becomes a purely combinatorial problem: to construct large sets of linear orders satisfying the above restriction conditions. I introduce some definitions and notations. An acyclic domain $\mathtt{D}$ is *maximal* if for any linear order L not in $\mathtt{D}$, $\mathtt{D} \cup \{L\}$ is no longer an acyclic domain. Moreover a (maximal) acyclic domain contained in $\mathtt{L}_n$ is *maximum* if it has the maximum size, denoted by ***f(n)***, among all acyclic domains in $\mathtt{L}_n$. An acyclic domain $\mathtt{D} \subset \mathtt{L}_n$ is *connected* if there always exists a *path*[8] of $\mathtt{L}_n$ included in $\mathtt{D}$ between any two linear orders in $\mathtt{D}$; such a connected domain is of *diameter d* if the maximum length of a shortest path between two linear orders of $\mathtt{D}$ is d. One can observe that the diameter of $\mathtt{L}_n$ is n(n-1)/2. I denote by ***g(n)*** the maximum size of a connected acyclic domain of diameter n(n-1)/2 contained in $\mathtt{L}_n$. It has been shown that $g(n) = f(n)$ for $n \leq 6$, but it seems to be less than $f(n)$ for $n \geq 16$.

The problem of determining $f(n)$ or $g(n)$ for all n is daunting. Up to now these numbers are known only for $n \leq 6$ (where they are equal). Then one has to search good lower or upper bounds for them instead. Lower bounds are obtained by producing (maximal) acyclic domains. The first maximal connected acyclic domain obtained by Black contains only 2^{n-1} linear orders (compare to the $n!$ possible linear orders). For a long time the other maximal acyclic domains found were also connected and contained no more orders. I will present some of them in Sect. 2. This perhaps raised up the conjecture $f(n) = 2^{n-1}$; but this was unfortunate since it can be disproved for $n = 4$ (see footnote 13 and Fig. 4). Breakthroughs came first in the eighties with Abello and Chameni-Nembua's works which I will present in Sects. 3 and 4. They use the order on the "permutoèdre" and do not explicitly use Never Conditions. For instance for $n = 6$, maximal connected acyclic domains with 44 or 45 linear orders were obtained (instead of $32 = 2^5$). A clever use of the Never Conditions by Fishburn and Craven allowed them to find larger maximal connected acyclic domains for $n > 6$ (all of diameter $n(n-1)/2$). They will be presented in Sect. 5 along with Fishburn's construction that allows still larger, but not connected, maximal acyclic domains. Finally in Sect. 6, I will state Galambos and Reiner's work which allows to get a unified version of almost all the known results on maximal connected acyclic domains of diameter $n(n-1)/2$. In the conclusion I will point out two conjectures. The Appendix contains a Table giving numerical results on lower or upper bounds for $f(n)$ and $g(n)$.

---

[8] A path in $\mathtt{L}_n$ is a sequence of different linear orders $L_1 \ldots L_k L_{k+1} \ldots L_s$ such that for $k = 1, 2 \ldots s$-1, L_k and L_{k+1} differ only by a transposition (of two consecutive elements). In fact it is a path in the "permutoèdre graph" defined in Sect. 3.

2 The Beginnings: Small Maximal Acyclic Domains

As already noted the first maximal (connected) acyclic domain was produced by Black (1948, 1958, 1988) who called it the domain of the single peaked preferences. Assume that the set of alternatives is linearly ordered as $1 < 2 < \ldots < p \ldots < n$ by a "reference" order. Let L be a linear order of preference on A for which p is the preferred alternative. L is said *single-peaked* (with respect to the reference order $<$) if $i < j < p$ implies $iLjLp$ and $p < q < r$ implies $rLqLp$. This condition means that given that p is the preferred alternative of the voter, he prefers alternative x to alternative y if x is "closer" to p than y in the reference order (for instance such a condition can be satisfied for political preferences, when the political parties can be ranked from extreme left to extreme right). Now it is not difficult to see that a linear order L is single-peaked (w.r.t. $<$) if and only if for every ordered triple $i < j < k, jLi$ implies kLj and jLk implies iLj, which is true if and only if L satisfies the condition *jN1*, i.e., for every ordered triple $i < j < k$, $jN_{\{i,j,k\}}1$ (in other words, the middle alternative of the triple is never the least preferred). Then the domain of single-peaked (w.r.t. $<$) linear orders is the domain of all linear orders satisfying $jN1$. It is also easy to see that for n alternatives its size is 2^{n-1} (see for instance Kreweras (1962) who used the fact, already observed by Ward that no more than two alternatives can have rank 1 in these single-peaked linear orders). The set of the eight single-peaked linear orders on $\{1,2,3,4\}$ w.r.t. the linear order 1234 ($= 1<2<3<4$) is $\{1234, 1243, 1423, 1432, 4123, 4132, 4312, 4321\}$. The permutoèdre L_4 is represented at Fig. 4 and on this figure a black square is attached to each of these eight orders.

Black's single-peakedness condition is a subcase of *Arrow-Black's single-peakedness condition*[9] (1951), which is the condition $ijkN1$ i.e., for every 3-subset $\{i,j,k\}$, there exists h in $\{i,j,k\}$ such that $hN_{\{i,j,k\}}1$. An acyclic domain satisfying Arrow-Black's single-peakedness condition does not necessarily satisfy Black's single-peakedness condition. But such an acyclic domain contains also at most 2^{n-1} linear orders. This results immediately from the point already mentioned that a Never Condition is hereditary and from another easy observation: the set of elements ranked 1 in the linear orders belonging to a domain satisfying Arrow-Black's single-peakedness condition has size at most 2.

Some other interesting domains satisfying Arrow-Black's condition have been investigated. For instance let be L and L' denote two linear orders which rank the alternatives of A according two different criteria. A decision maker can rank the alternatives from the last by using alternatively the two criteria: he gives rank 1 to

---

[9] The terminology of these conditions depends on authors. For instance what I call Black's single-peakedness condition (respectively, Arrow-Black's single-peakedness condition) has been called unimodality condition by Romero 1978 (respectively, pseudo-unimodality condition by Romero and single-peakedness on the triples by Kelly 1978). In fact, as it was observed by Inada (1964), Arrow-Black's single-peakedness appears only implicitly in the proof of Theorem 4 in Arrow's book. This condition appears also in Dumett & Farquharson (1961). What is somewhat confusing is that the term single–peakedness condition is sometimes used without making it clear as to which of the two contexts above the term is being used.

an alternative ranked 1 by one of the criteria (i.e., to the worst alternative according to this criteria); then he deletes this alternative from the two linear orders and he uses the same procedure on the restrictions obtained to determine his next to last alternative, and so on. Romero (1978) said that a set of linear orders obtained by this procedure satisfies the *quasi-unimodality condition* and he proved that this set satisfies Arrow-Black's single-peakedness condition. When the two linear orders L and L' are dual (xLy if and only if $yL'x$) one gets again the set of all single-peaked linear orders (w.r.t. L).

It is obvious that the dual of an acyclic domain is also an acyclic domain. For instance the dual of Black's (respectively, Arrow-Black's) single–peaked linear orders, i.e., the set of linear orders satisfying $jN3$ (respectively, $ijkN3$) was called by Vickrey (1960) the domain of *single-troughed* (respectively, by Inada (1964) the domain of *single-caved*) linear orders. One can find a systematic study of the domains of linear orders satisfying one of the Never Conditions in Arrow and Raynaud's book 1986 (see also Kohler, 1978; Romero, 1978; Raynaud, 1981; Blin (1973)).

Another type of acyclic domains was discovered by Blin (1973) under the name of *multidimensional consistency*: the chains of the "permutoèdre lattice". It will be described in the following section but one can already mention that the size of such a domain is at most $n(n-1)/2+1$ and so less than 2^{n-1} (for $n>3$).

3 Abello's Work

I begin with Abello's contributions that are contained in his doctoral dissertation (1985) and several papers (1981, 1984 with Johnson, 1985, 1987, 1988, 1991, 2004). In all these papers Abello works with S_n the set of all permutations on a set of cardinality n. I will describe some of his results but I will continue to rather speak of linear orders belonging to $\mathbb{L}_n$. These results use the partial order known as the *weak Bruhat order* (on S_n).[10] Let L be an arbitrary linear order of $\mathbb{L}_n$; it will be convenient to take $L = 1 < 2 < \ldots .n$. For $L' \in \mathbb{L}_n$, one sets $\mathrm{Inv}L' = \{\{i,j\} \subseteq A$ such that iLj and $jL'i\}$ (i.e., the set of pairs $\{i,j\}$ on which L and L' "disagree"). For L', $L'' \in \mathbb{L}_n$, one sets $L'' \leq L'$ if $\mathrm{Inv}L' \subseteq \mathrm{Inv}L''$. It has been shown by Guilbaud and Rosenstiehl (1963) that the poset $(\mathbb{L}_n, \leq)$ denoted henceforth simply by $\mathbb{L}_n$ is a lattice[11] called the "*permutoèdre*" *lattice* in French tradition (see for instance Barbut

[10] S_n the symmetric group of all permutations on $\{1,2\ldots,n\}$ is an example of a finite Coxeter group. All Coxeter groups can be partially ordered by the so-called weak Bruhat order (and also by the strong Bruhat order).

[11] That is two linear orders have a least upper bound and a greatest lower bound in this partial order. Some authors attribute this result to Yanagimoto & Okamoto (1969). One can admit than a paper published in French will be less known that a paper written in English. But Guilbaud and Rosenstiehl's paper which precedes Yanagimoto and Okamoto's paper has been quoted in many English-written papers; moreover its proof that S_n is a lattice is reproduced in Principles of combinatorics (Berge, 1971) and above all Yanagimoto and Okamoto's paper does not contain a real proof of their assertion (read it !). One can add that properties of the permutoèdre lattice are studied in Barbut & Monjardet (1970), Le Conte de Poly-Barbut (1990), Duquenne and Cherfouh (1994),

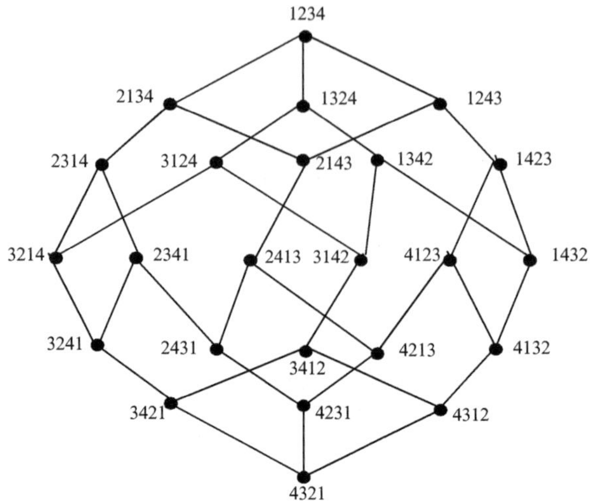

Fig. 1 The permutoèdre lattice L_4

& Monjardet, 1970). Its maximum element is $1 < 2 < \ldots n$ denoted by ϖ, and its minimum element is the dual linear order $n < \ldots 2 < 1$ denoted by α.

The lattice L_4 is represented on Fig. 1 by a (Hasse) diagram giving its covering relation. The undirected covering relation of this lattice is the *adjacency relation* between linear orders where a linear order is adjacent to another one if they differ on a unique pair of elements. The set of all linear orders endowed with this adjacency relation is called the *permutoèdre graph.*

Come back to acyclic domains. The first easy observation is that the set of ordered triples *ijk* contained in the linear orders of an acyclic domain D of L_n has size at most $4n(n-1)(n-2)/6$ (if not D contains a 3-cyclic profile). So when one adds to an acyclic domain D all the linear orders, which do not increase the set of ordered triples already present in D one gets a maximal acyclic domain. More generally the map, which adds to an arbitrary set of linear orders all the linear orders that do not increase the set of ordered triples, is a closure operator on the subsets of L_n.[12]

The second –also easy but significant– observation is that a maximal chain of L_n is an acyclic domain (a fact already observed by Blin (1973) as noted above) which contains exactly $4n(n-1)(n-2)/6$ ordered triples. So by applying the above closure operator to a maximal chain one obtains a maximal acyclic set. Now Abello has proved several significant results and in particular the following ones:

1. A maximal acyclic domain D obtained by the closure operator applied to a maximal chain of L_n is a connected subset of L_n of diameter $n(n-1)/2$ and an upper semimodular sublattice of the permutoèdre lattice;

Markowsky (1994) and Caspard (2000) and that more generally Bjorner (1984) proved that all finite Coxeter groups partially ordered by the weak Bruhat order are lattices.

[12] This closure operator appears already in Kim and Roush's 1980 book (see Definition 5.12).

2. For any maximal connected acyclic domain of $\mathtt{L}_n$ of diameter $n(n-1)/2$, there exists a maximal acyclic domain with the same size obtained by the closure of a maximal chain;
3. Let us say that two maximal chains of the permutoèdre lattice $\mathtt{L}_n$ are equivalent if they have the same closure (and so are two maximal chains of the associated lattice). One goes from one of these chains to the other by "quadrangular transformations" of linear orders: let $L = x_1 \ldots x_k x_{k+1} \ldots x_i x_{i+1} \ldots x_n$ be a linear order such that $x_k x_{k+1}$ and $x_i x_{i+1}$ are four different alternatives; then L is transformed into $L' = x_1 \ldots x_{k+1} x_k \ldots x_{i+1} x_i \ldots x_n$ $(= \tau_i \tau_k(L) = \tau_k \tau_i(L))$.

Property 2 means that to search maximal connected acyclic domains of diameter $n(n-1)/2$ with large size, it suffices to consider those obtained by the closure of a maximal chain. Abello gives an algorithm to get the maximal connected acyclic domain obtained from a maximal chain $L_0 \prec L_1 \ldots \prec L_{n(n-1)/2}$ of $\mathtt{L}_n$. The algorithm constructs a sequence $\mathtt{D}_0 = \{L_0\}, \mathtt{D}_1, \ldots \mathtt{D}_{n(n-1)/2}$ of acyclic domains. One goes from $\mathtt{D}_s$ to $\mathtt{D}_{s+1}$ by adding to $\mathtt{D}_s$ the linear order L_{k+1} and the set of linear orders obtained by applying to all the linear orders of a subset $\mathtt{E}_s$ of $\mathtt{D}_s$ the transposition τ_i (of x_i and x_{i+1}) used to obtain L_{k+1} from L_k; a linear order M is in E_s if there exists in $\mathtt{D}_s \cup \{L_{k+1}\}$ a maximal chain from M to L_{k+1}, for which none of the transpositions along this chain act on x_i or x_{i+1}.

A similar algorithm can be used with other acyclic domains to get maximal connected acyclic domains. With this algorithm Abello and Johnson (1984) show that $f(n) \geq 3(2^{n-2}) - 4$ (for $n \geq 4$). Except for $n = 4$, where one gets a lower bound of 8 and where a maximal acyclic domain of size 9 has been already found,[13] the acyclic domains so found were the first of size greater than 2^{n-1}. One will see in the following sections that there exist maximal connected acyclic domains with a much greater size.

4 Chameni-Nembua's Work

Chameni-Nembua's work on acyclic domains is contained in his 1970 "thèse de 3ème cycle" and in a paper that appeared the same year. I was his thesis' director and his work has answered some questions that I had asked him to investigate. The origin of these questions comes back to Guilbaud's paper in 1952. In this paper one finds an analysis of Black's domain showing that the set of single-peaked linear orders has a distributive lattice structure and that the majority relation of a profile taken in this domain is the *median* of the elements of the profile in this lattice.[14] In particular one finds (page 32 of the English translation) a figure showing the distributive lattice

[13] An acyclic domain of size 9 in $\mathtt{L}_4$ is given in Kim and Roush's book (1980) or in Raynaud (1982). Such an acyclic domain is represented Fig. 4 as AS(4) (see Sect. 5).

[14] The fact that in this case majority relation is both a metric and an algebraic median is a special case of median's theory in distributive lattices (or more generally in median semilattices). One will find elements of this theory and references in Barthélemy & Monjardet (1981), Monjardet (2006a) and in Day and McMorris 2005 book.

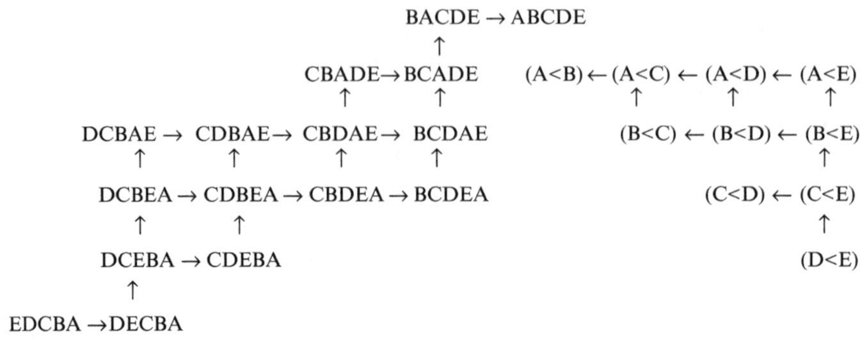

2a **2b**

Fig. 2 The distributive lattice of the 16 single-peaked linear orders on a 5-set and the associated poset of the ordered pairs

of the sixteen single-peaked linear orders on a set of five alternatives. This figure is reproduced below at Fig. 2a. One can observe that this lattice is a *covering sublattice* of the permutoèdre lattice L_5 that means that the covering relation in this sublattice is the same as the covering relation in L_5. Indeed, a single-peaked linear order is covered by another single-peaked linear order if and only if they differ on a unique pair of elements.

Several other acyclic domains that are covering distributive sublattices of the permutoèdre lattice were given in Frey (1971) and in Frey and Barbut's 1971 book. For instance, the so-called "fuseaux bipolaires d'insertions" which are in fact the sets of all linear orders containing a partial order formed by the (cardinal) sum of two unrelated chains. Figure 3 here reproduces the figure on page 121 of Frey and Barbut's book that shows the case where the two unrelated chains are $1 < 2 < 3$ and $4 < 5 < 6$ (I have replaced letters by integers); one obtains a (not maximal) covering distributive sublattice of L_6. Other examples given in this book are the so-called "faisceaux d'indifférence" which are the set of linear orders which differ from a given linear order L only on consecutive elements of L[15] and the set of "co-blackiens" (= single-troughed) linear orders.

So I asked Chameni-Nembua to answer the following question: is any covering distributive sublattice of the permutoèdre lattice an acyclic domain? His answer was positive, based on the properties of meet and join in this lattice and the fact that a distributive lattice must not contain some sublattices (see any book on lattice theory and Monjardet, 1971 for the case of L_n). Moreover, he showed that maximal covering distributive sublattices are maximal acyclic domains which contain the minimum and the maximum elements of L_n (i.e., $n < \ldots 2 < 1$ and $1 < 2 < \ldots .n$)

[15] Like the "fuseaux bipolaires", the "faisceaux d'indifférence" are also the set of linear extensions of some posets P of width 2 (where the width is the maximum number of incomparable elements of P). More generally the set of linear extensions of any poset of width 2 is a covering distributive sublattice of L_n (Chameni-Nembua, 1989).

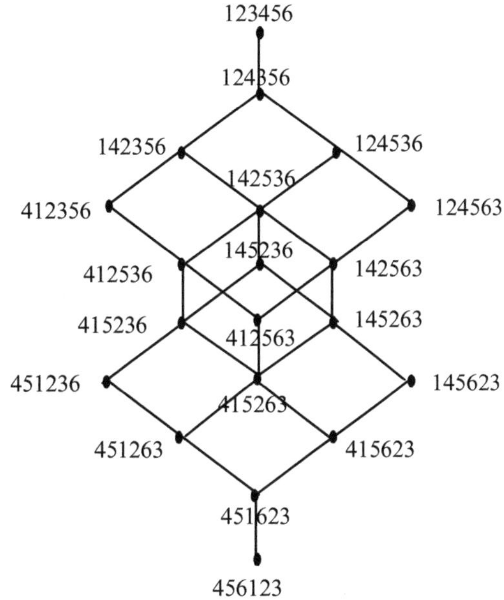

Fig. 3 The distributive lattice of the linear extensions of a poset sum of two chains

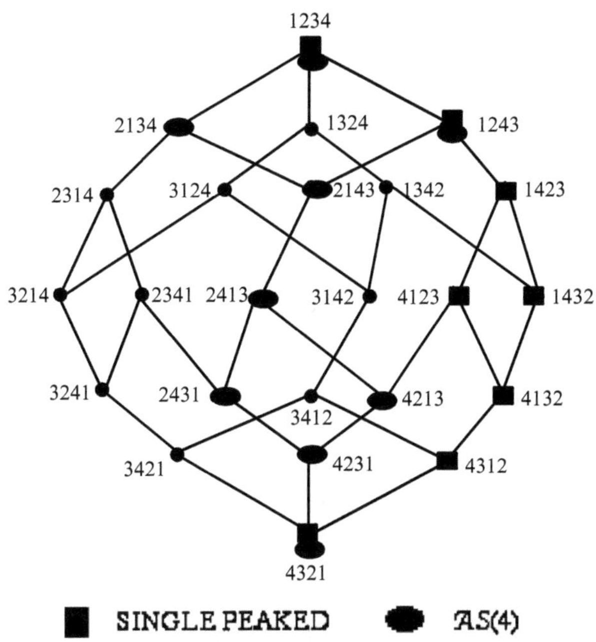

Fig. 4 Two distributive lattices acyclic domains on a 4-set

and so a maximal chain of $\mathtt{L}_n$. These results led us to search such large maximal covering distributive sublattices of $\mathtt{L}_n$. For $n = 4$, one founds the sublattice AS(4), of size 9 represented on Fig. 4 (the linear orders with a black ellipsoid).

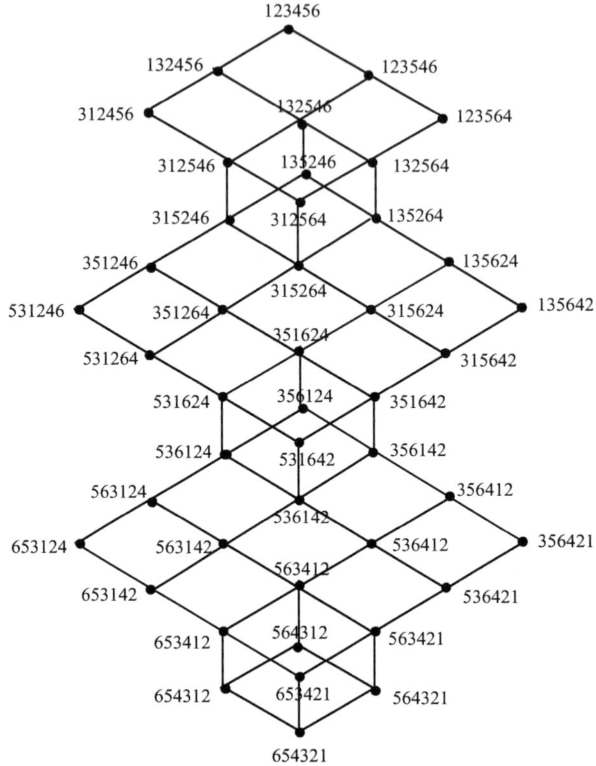

Fig. 5 AS(6), a distributive lattice acyclic domain of size 45 on a 6-set

For $n = 5$, we found such a sublattice of size 20, and for $n = 6$, I found a sublattice of size 45 which is the last Figure in Chameni-Nembua's paper and which is reproduced here at Fig. 5 (with integers instead of letters for the elements of A). This last sublattice showed that it was possible to surpass the best Abello and Johnson's lower bound known at this date ($f(6) \geq 45 > 44 = 3(2^6–2) - 4$). I was pretty sure that there was a general construction to get such large acyclic domains but since I didn't find it, I sent these examples to Peter who was already working on the topic and (obviously) found the construction described in the next section.[16]

[16] I should be ashamed to have not having found this construction since as it will seen in section 5 it was sufficient to look the triples, and in fact it was also found by Dridi (1994 private letter). But Fishburn achieved a much more difficult task: to compute the size of the corresponding acyclic domains for n up to 25 (Dridi computed this size up to $n = 8$ with the exact values for $n = 7$ but he found 220 instead of 222 for $n = 8$). By the way, it is worthwhile to mention here Fishburn's practice, which should be more wide-spread in our scientific world. In his works on acyclic domains, he always quoted the example that I sent to him. He always did the same in other circumstances and/or for some other authors when I indicated to him a result that preceded one of his works.

5 Fishburn's and Craven's Works

It seems that Peter's interest for acyclic domains was motivated by Craven's conjecture that was reported in Kelly's 1991 paper. In his 1992 book Craven conjectures that $f(n) = 2^{n-1}$ and he gives an example of an acyclic domain of size 8 for $n = 4$ (but see[17]). Kelly exhibits for $n \geq 4$ a maximal acyclic domain of size 2^{n-1} generalizing Craven's example (in fact, this domain is a maximal Arrow-Black's single-peaked domain). In his 1992 note Fishburn mentions that the above conjecture is false for $n \geq 4$ (see footnote 13) and that in fact $f(n)/2^{n-1} \to \infty$. This is proved by using an iterative construction of acyclic domains where the first one is the domain of size 9 on a 4-element set and one goes from an acyclic domain of size p on a n-set to an acyclic domain of size $2p^2$ on a $2n$-set. Fishburn's paper also contains a replacement construction which for $n = 2m$ and $m \geq 4$ that gives a much better lower bound than Abello and Johnson's lower bound: $f(16) \geq 59{,}049 > 3(2^{14} - 2) - 4 = 49{,}148$. In fact when Peter wrote his Notes on Craven's conjecture he didn't remember that Abello had worked on the topic. He remembered only after he read Kim, Roush and Intriligator's 1992 *Overview of Mathematical Social Sciences* where the problem to find $f(n)$ was mentioned. Therefore, when (in January 1993) I sent him Chameni-Nembua's paper with my example of Fig. 5 they were welcomed. A week later he sent me a seven page memo containing the first elements of what became his 1996 and 1997's papers (for which I was referee or editor) and the personal details mentioned above. These papers contain many significant results.

Firstly, Peter defines the alternating scheme which is the construction allowing a generalization of my example. Let $1 < 2 \ldots < p \ldots < n$ be a linear order on A. An acyclic domain $\mathtt{D}$ of $\mathtt{L}_n$ satisfies the *alternating scheme*, if for all $i < j < k$ either (1) $jN1$ if j is even and $jN3$ if j is odd, or (2) $jN3$ if j is even and $jN1$ if j is odd (observe that these two domains are dual). So to define such a domain, denoted by $\mathtt{AS}(n)$, one combines the Never Conditions used for the single-peaked and single-troughed domains. The size of $\mathtt{AS}(n)$ is computed by recursion up to $n = 25$. Concerning these sizes, Peter writes that he was unable to find a closed formula for them. Such a formula has been since obtained by Galambos and Reiner (2008 see next section). The number of linear orders satisfying the alternating scheme is:

$$2\mathrm{n} - 3(\mathrm{n}+3) - \mathrm{C}(\mathrm{n}-2, \mathrm{n}/2-1)(\mathrm{n}-3/2), \text{ for even n} > 2$$
$$2\mathrm{n} - 3(\mathrm{n}+3) - \mathrm{C}(\mathrm{n}-1, (\mathrm{n}-1)/2)(\mathrm{n}-1)/2), \text{ for odd n} > 1$$

where $\mathrm{C}(p,q) = p!/(p-q)!q!$ is the binomial coefficient.

Secondly, Fishburn proves that $f(4) = 9$, $f(5) = 20$ and that for $n \leq 5$, an acyclic domain is maximum if and only if satisfies the alternating scheme. He conjectured the same for $n = 6$ and 7 the first conjecture having been proved in his 2002 paper (a difficult task!).

[17] This is another example of the bad circulation of some scientific results, since this conjecture had already been made by Johnson (1978) and disproved at least since 1980 (see footnote 13 and Fig. 4).

Thirdly, it is shown that at least for $n \geq 16$, the alternating scheme is not optimal since the *replacement scheme* is better. This scheme uses two acyclic domains $\mathtt{D}$ defined on $\{0,1,2\ldots m\}$ and $\mathtt{D}'$ defined on $\{m+1,\ldots m+p\}$. For every order in $\mathtt{D}$ one replaces 0 by each of the orders in $\mathtt{D}'$. It is easy to check that the domain of linear orders obtained on $\{0,1,2\ldots m,m+1,\ldots m+p\}$ is acyclic. Hence one gets $f(m+p) \geq f(p)f(m+1)$ and in particular $f(16) \geq 108,336 > 105,884$ the size of the acyclic domain given by the alternating scheme. Another result allows to show that $f(n) > (2.17)^n$ for all large n and that $|\mathtt{AS}(n)|/f(n) \to 0$ as $n \to \infty$ i.e., that the lower bound given by the alternating scheme becomes more and more inaccurate.

Finally, the paper attacks the "major challenge" to find good upper bounds for $f(n)$. The only upper bound already known $2[(n-1)!]$ had been given in Arrow and Raynaud's book, but for instance it gives $f(9) \leq 103.698$ whereas a clever Fishburn's Lemma allows us to obtain $f(9) \leq 22.680$. The paper raises two conjectures. The first one is $f(n+m) \leq f(n+1)f(m+1)$ for all n, $m \geq 1$ and in Fishburn's 2002 paper it is shown that it would imply $f(n) < (2.591)^{n-2}$. The second conjecture is $f(n) \leq c^n$ for some constant c and this was proved later by Raz (2000).

I come back now to Craven's works. In his 1994 note he gives a *partition scheme* which generalizes a construction given in Fishburn's 1992 note and which in a particular case is equivalent to Fishburn's replacement scheme. So he obtains the same formula $f(m+p) \geq f(p)f(m+1)$ allowing him to improve some lower bounds of Fishburn's note. In his 1996 paper, after reproving the fact that there are 2^{n-1} single-peaked linear orders on a n-set (see Sect. 2), he studies the acyclic domains that are generated by Fishburn's alternating scheme. In particular he makes the linear orders that are generated by this scheme more precise and he gives some recurrence relations allowing him to obtain the sizes of the corresponding acyclic domains up to $n = 15$.

6 Galambos and Reiner's Work

In this section I consider the problem of computing $g(n)$ or rather good lower bounds to this number, i.e., to provide large connected acyclic domains. We have seen that Abello had constructed such domains by applying a closure operator to some maximal chains of the permutoèdre lattice. I gave an example showing that it was possible to find larger such domains that are covering distributive sublattices of the permutoèdre lattice (shown to be acyclic domains Chameni-Nembua). Generalizing this example by means of his alternating scheme using the two Never Conditions jN3 and jN1, Fishburn obtained the up to now best lower bound known for $g(n)$. I present now the link between these various results, as it is established in recent Galambos and Reiner's 2008 work (and anticipated in Guilbaud's, 1952 paper; see Remark later).

Abello constructs maximal connected acyclic domains which are (upper) semimodular sublattices of the permutoèdre lattice by using the fact that the maximal chains of these lattices have an invariant, namely the set of the ordered triples of elements appearing in the orders of the chain. Galambos and Reiner show that these lattices are the same as Chameni-Nembua's lattices, i.e., that they are (maximal) covering distributive sublattices of the permutoèdre lattice and that their maximal chains have another invariant, namely a poset defined on $P^2(n)$ (the set of $n(n-1)/2$ ordered pairs $(i<j)$). The fact that Abello's maximal connected acyclic domains are distributive lattices is significant since it allows to use the well-known Birkhoff's duality between posets and distributive lattices.

We need some notions of lattice theory. A join-irreducible element of a lattice is an element covering a unique element and an ideal (respectively, a filter) of a poset $(X,<)$ is a subset I of X such that $x \in I$ and $y < x$ implies $y \in I$ (respectively, a subset F of X such that $x \in F$ and $x < y$ implies $y \in F$). Now by Birkhoff's duality between posets and distributive lattices, a distributive lattice D is isomorphic to the set ordered by inclusion of all the ideals of the poset J_D of its join-irreducible elements (or to the set ordered by $\supseteq$ of all the filters of J_D). It is well-known that in this duality the maximal chains of a distributive lattice are in a one-to-one correspondence with the linear extensions of the poset J_D (i.e., with the linear orders containing the partial order between the join-irreducible elements); indeed when x_k is covered by x_{k+1} in a maximal chain of a distributive lattice then there exists a unique join-irreducible element j_k such that $x_{k+1} = x_k \vee j_k$; so the covering relation $x_k \prec x_{k+1}$, can be labeled by j_k and the linear order $j_1 j_2 \ldots . j_{|J_D|}$ obtained on J_D is a linear extension of the poset J_D.

What are the join-irreducible elements of a covering distributive sublattice of the permutoèdre lattice? I consider a covering distributive sublattice $\mathtt{D}$ containing a maximal chain of $\mathtt{L}_n$ (then containing the maximum element $\varpi = 1 < 2 < \ldots n$ and the minimum element $\alpha = n < \ldots 2 < 1$ of the permutoèdre lattice). A linear order L is a join-irreducible element of $\mathtt{D}$ if it covers a unique other element L' of $\mathtt{D}$. Then one has $L = x_1 \ldots x_k x_{k+1} \ldots x_n = \tau_k(L' = x_1 \ldots x_{k+1} x_k \ldots x_n)$ with $x_k < x_{k+1}$ (in the order $1 < 2 < \ldots . n$). Yet, since on a maximal chain between α and ϖ any of the $n(n-1)/2$ ordered pairs $j > i$ of α has to be transposed exactly once to get ϖ, the transposition of the elements x_k and x_{k+1} appears for the first time in any maximal chain between α and $x_1 \ldots x_k x_{k+1} \ldots x_n$. So we can identify the join-irreducible $L = x_1 \ldots x_k x_{k+1} \ldots x_n$ with the ordered pair (x_k, x_{k+1}), and finally the poset of join-irreducible elements of $\mathtt{D}$ is isomorphic to a poset $P_{\mathtt{D}} = [P^2(n), <_{\mathtt{D}}]$ defined on the set $P^2(n)$ of all the ordered pairs $i < j$. Now, any linear order L in $\mathtt{D}$ corresponds to an ideal of $P_{\mathtt{D}}$: L is obtained from $\alpha = n < \ldots 2 < 1$ by applying all the transpositions of the ordered pairs belonging to this ideal. And any maximal chain of $\mathtt{D}$ corresponds to a linear order on $P^2(n)$, which is a linear extension of the poset $P_{\mathtt{D}}$.

Using more general results on Bruhat orders (Ziegler,1993) Galambos and Reiner characterize the linear orders on $P^2(n)$ which are *admissible* i.e., which correspond to the sequence of transpositions of a maximal chain $\mathtt{C}$ of $\mathtt{L}_n$: a linear order λ on $P^2(n)$ is admissible if and only if it contain only triples (of ordered pairs) ordered in the lexicographic order or in its dual, i.e., triples of the form $ij < ik < jk$ or $jk < ik < ij$ (with $i < j < k$). Moreover, these two sets of ordered triples are the same for the linear orders corresponding to any maximal chain of the distributive lattice $\mathtt{D}$ closure of the chain $\mathtt{C}$. For instance, a maximal chain of the domain of single peaked-linear orders of $\mathtt{L}_4$ is $4321 \prec 4312 \prec 4132 \prec 1432 \prec 1423 \prec 1243 \prec 1234$, the associated linear order on $P^2(4)$ is $12 \prec 13 \prec 14 \prec 23 \prec 24 \prec 34$ and the set of ordered triples corresponding to any of the maximal chains in this domain is $\{(12,13,23), (12,14,24), (13,14,34), (23,24,34)\}$ (so it does not contain triples dually lexicographically ordered). The domain $\mathtt{AS}(4)$ contains the maximal chain $4321 \prec 4231 \prec 4213 \prec 2413 \prec 2143 \prec 2134 \prec 1234$; the associated linear order on $P^2(4)$ is $23 \prec 13 \prec 24 \prec 14 \prec 34 \prec 12$; the associated set of lexicographically (respectively, dually lexicographically) ordered triples is $\{(13,14,34), (23,24,34)\}$ (respectively, $\{(23,13,12), (24,14,12)\}$).

When one takes an arbitrary maximal chain $C = \alpha \prec L_1 \prec L_2 \ldots \prec \varpi$ of $\mathtt{L}_n$ it is a maximal chain in a maximal covering distributive sublattice $\mathtt{D}$ of the permutoèdre lattice. In order to determine $\mathtt{D}$ it suffices to determine the poset $P_{\mathtt{D}}$ associated to this maximal chain. Galambos and Reiner constructs $P_{\mathtt{D}}$ by using a notion of "arrangement of pseudolines" allowing to represent $P_{\mathtt{D}}$ and its ideals and so to recover the linear orders in $\mathtt{D}$. Another algorithm to get $P_{\mathtt{D}}$ is proposed in Monjardet (2006b).

When $P_{\mathtt{D}}$ is known, computing the size of $\mathtt{D}$ comes back to computing the numbers of ideals of this poset, a difficult task in general, since this computation is known to be $\#\mathcal{P}$-complete (Provan and Ball 1983). In the case when $\mathtt{D}$ is given by the alternating scheme, the corresponding poset has a very regular structure (its covering relation is given in Monjardet 2006b). Galambos and Reiner describe it by means of a certain arrangement of pseudolines and show that computing the ideals of this poset comes back computing some lattice paths. By cleverly using path enumeration techniques they get the formula for $|\mathtt{AS}(n)|$ given in the previous section.

Another significant Galambos and Reiner's result is the characterization of the maximal covering distributive sublattices $\mathtt{D}$ of $\mathtt{L}_n$ by a set of Never Conditions. Let $\mathtt{C}$ be a maximal chain of $\mathtt{D}$ and λ be the corresponding linear order admissible on $P^2(n)$, i.e., the linear order corresponding to the sequence of transpositions of this maximal chain. It has been noted above that the restrictions of λ to any subset $\{(ij),(ik),(jk)\}$ of three ordered pairs are ordered either lexicographically $(ij < ik < jk)$ or dually lexicographically $(jk < ik < ij)$. Let us denote by $\mathrm{LEX}_3\lambda$ respectively, $\mathrm{ALEX}_3\lambda$) the set of triples $i < j < k$ for which the set $\{(ij),(ik),(jk)\}$ is lexicographically ordered (respectively, dually lexicographically ordered) in λ. As also already noted, $\mathrm{LEX}_3\lambda$ and $\mathrm{ALEX}_3\lambda$ are the same for any other maximal chain of $\mathtt{D}$. Then, $\mathtt{D}$ is the set of all linear orders satisfying the following Never Conditions:

$$jN1, \forall i < j < k \text{ with } ijk \in \mathrm{LEX}_3\lambda$$
$$jN3, \forall i < j < k \text{ with } ijk \in \mathrm{ALEX}_3\lambda$$

For instance, for any linear order λ associated to a maximal chain of $|\mathtt{AS}(4)$, $\mathrm{LEX}_3\lambda = \{134, 234\}$ and $\mathrm{ALEX}_3\lambda = \{123, 124\}$ and one gets again the Never Conditions $3N1$ and $2N3$ of formula (2) in Sect. 5.

6.1 Remark

As noted before, Guilbaud's paper contains an anticipation of a Galambos and Reiner's result in a particular case. Indeed Guilbaud not only pointed out the distributive lattice structure of the domain of single-peaked linear orders but he also gave an explanation for it. He writes (page 29, English translation): "These observations focus attention on a sort of hierarchy of judgments[18]; one judgment dominates several others... This subordination is easy to designate in the form of an ordered network" (he adds in note: "This is a partially ordered structure, called a lattice"[19]). He represents this partial order by a triangular tableau for the domain of single-peaked linear orders on a 6-element set (this tableau is reproduced here Fig. 2b) and he adds below it: "Note that the affirmation of any one of these judgments implies the affirmation of all the "consequents"; that is, the affirmation of those located either in the same row and to the left, or in the same column and thus of all the judgments located to the left and above". He concludes that single-peaked orders corresponds to frontiers separating judgments + (i.e., $x > y$) and judgments – (i.e., $x < y$) in the triangular tableau. In other terms he shows that single-peaked orders correspond to filters in the partial order defined between the ordered pairs.

7 Conclusion

The search for large acyclic domains appears as a fascinating quest all the more that I have not said all. For instance, maximal chains of the permutoèdre lattice are in one-to-one correspondence with other significant combinatorial objects the standard Young tableaux and the balanced tableaux (see Edelman & Greene, 1987; Abello 2004) and this allows other interpretations of the problems that have been raised.[20]

There are also interesting algorithmic problems to answer the question of recognizing acyclic domains. Some answers have been given, especially for Black's single peaked domains, by Romero 1978, see also Arrow & Raynaud, 1986), Bartholdi and Trick (1986) and Doignon and Falmagne (1994).

---

[18] In Guilbaud's paper a (simple) judgment is an ordered pair of alternatives expressing a preference between them; for example, $x > y$ (see page 24ff of the translation).

[19] Indeed in the case of the covering distributive sublattice corresponding to single-peaked orders, it is not difficult to prove that the associated poset on $P^2(n)$ is the lattice where $(i,j) \vee (k,l) = (\max(i,k), \max(j,l))$ and $(i,j) \wedge (k,l) = (\min(i,k), \min(j,l))$. See also Monjardet (2006b).

[20] A *balanced tableau* is a staircase tableau T of $n(n-1)/2$ cases – corresponding to the ordered pairs $(i < j)$ – containing the integers from 1 to $n(n-1)/2$ and satisfying for every $i < j < k$, $t(i,k)$ between $t(i,j)$ and $t(k,j)$. Such a tableau codes a maximal chain of L_n by coding the linear order λ on $P^2(n)$ associated to this chain: the integer in the case corresponding to (i,j) is the rank of (i,j) in λ. Conversely a balanced tableau induces the maximal chain obtained by effecting the sequence of transpositions of the ordered pairs in the order of the cases of the tableau. The much more sophisticated bijection between maximal chains of L_n and standard Young tableaux allows to Edelman and Greene to give a formula for computing the number of these chains.

I end this paper by noting a final result and two conjectures. Instead of searching for the maximal covering distributive sublattices of the permutoèdre lattice which have a maximum size, one can ask what are those that have a minimum size. Since such a sublattice is the closure of a maximal chain, one gets the answer if there exist maximal chains that are closed. It's actually the case as it is shown in Monjardet 2006b). This paper contains also some results on the distributive lattices given by Fishburn's alternating scheme and by Black's single-peakedness condition.[21]

Conjecture 1 (Fishburn 1996, 1997)

$$f(n+m) \leq f(n+1) \text{ for all } n, m \geq 1$$

The proof of this conjecture would imply $(2.17)^n < f(n) < (2.591)^{n-2}$ for all large n since Fishburn (1997, 2002) proved the lower bound and the implication for the upper bound. Then, if true, it would give a much better upper bound that the bound 4^{n-1} conjectured by Abello (1991). In the same paper Abello conjectures $g(n) \leq 3^{n-1}$ for which the conjectured upper bound $(2.591)^{n-2}$ would still be much better.

Let $|\mathrm{AS}(n)|$ be the size of the acyclic domain given by the alternating scheme.

Conjecture 2 (Galambos & Reiner, 2008)

$$g(n) = |\mathrm{AS}(n)|$$

This conjecture is true for $n \leq 6$ since in this case $f(n) = |\mathrm{AS}(n)|$ and Galambos and Reiner checked it for $n = 7$.

Acknowledgements I warmly thank the anonymous referee and William Gehrlein as editor for their suggestions and corrections on the first version of the paper.

References

Abello, J. M. (1981). Toward a maximum consistent set. *Technical Report TRCS11–81*, Computer Science Department, University of California.

Abello, J. M. (1985a). *A study of an independence system arising in group choice via the weak Bruhat order*. Ph.D. thesis, University of California, San Diego, CA.

Abello, J. M. (1985b). Intrinsic limitations of the majority rule, an algorithmic approach. *SIAM Journal of Algebraic and Discrete Methods*, 6, 133–144.

Abello, J. M. (1987). Algorithms for consistent sets. *Congressus Numerantium*, 53, 23–38.

Abello, J. M. (1988). An extremal problem on S_n. Advanced Research Institute in Discrete Applied Mathematics (ARIDAM) Workshop.

Abello, J. M. (1991). The weak Bruhat order of S_n, consistent sets, and Catalan numbers. *SIAM Journal of Algebraic and Discrete Methods*, 4(1), 11–16.

Abello, J. M. (2004). The majority rule and combinatorial geometry (via the Symmetric Group). *Annales du LAMSADE*, 3, 1–13.

[21] But in this paper the conjecture on the size of the covering distributive sublattices of L_n is false as soon as $n = 5$ (contrary to what is written): indeed, there does not exist such sublattices of L_5 with size 13 or 18.

Abello, J. M., & Johnson, C. R. (1984). How large are transitive simple majority domains? *SIAM Journal of Algebraic and Discrete Methods*, 5(4), 603–618.

Arrow, K. J. (1951,1970). *Social choice and individual values*. New York: Wiley.

Arrow, K. J., & Raynaud, H. (1986). *Social choice and multicriterion decision-making*. Cambridge: MIT.

Barthélemy, J. P., & Monjardet, B. (1981). The Median procedure in cluster analysis and social choice theory. *Mathematical Social Science*, *1*, 235–268.

Barbut, M., & Monjardet, B. (1970). *Ordre et Classification, Algèbre et Combinatoire, tomes I et II*. Paris: Hachette.

Bartholdi, J. J., III., & Trick, M. A. (1986). Stable matching with preferences derived fom a psychological model. *Operations Research Letters*, 5(4), 165–169.

Berge, C. (1971). *Principles of combinatorics*. New York: Academic.

Björner, A. (1984). Orderings of Coxeter groups. In C. Greene, (Ed.), *Combinatorics and algebra. Contemporary Mathematics 34* (pp. 175–195), Providence R.I.: American Mathematical Society.

Black, D. (1948). On the rationale of group decision-making. *Journal of Political Economy*, *56*, 23–34.

Black, D. (1958). *The theory of committees and elections*. Cambridege: Cambridge University Press.

Black, D. (1998). The theory of committees & Committee decisions with complementary valuation by Duncan Black and R.A. Newing, Revised 2nd ed. In I. S. McLean, A. McMillan, & B. L. Monroe (Eds.) Boston: Kluwer.

Blin, J. -M. (1973). The general concept of multidimensional consistency: Some algebraic aspects of the aggregation problem. In *Conference proceedings of the south carolina seminar on multiple criteria decision making*. University of South Carolina Press.

Caspard, N. (2000). The lattice of permutations is bounded. *International Journal of Algebra and Computation*, 10(4), 481–489.

Chameni-Nembua, C. (1989a). *Permutoèdre et choix social*. Third cycle thesis, Université de Paris V.

Chameni-Nembua, C. (1989b). Régle majoritaire et distributivité dans le permutoèdre. *Mathématiques Informatique et Sciences humaines*, *108*, 5–22.

Condorcet, M. J. A. M. (1785). Essai sur l'application de l'analyse à la probabilité des décisions rendues à la pluralité des voix. Paris.

Craven, J. (1992a). *Social choice: a framework for collective decisions and individual judgements*. Cambridge: Cambridge University Press.

Craven, J. (1992b). Further notes on Craven's conjecture. *Social Choice and Welfare*, 11(3), 283–285.

Craven, J. (1996). Majority consistent preference orderings. *Social Choice and Welfare*, 13(3), 259-267.

Day, W. H. E., & McMorris, F. R. (2005). *Axiomatic consensus theory in group choice and biomathematics*. Philadelphia: SIAM.

Dumett, M., & Farquharson, R. (1961). Stability in voting. *Econometrica*, *29*, 33–41.

Duquenne, V., & Cherfouh, A. (1994). On permutation lattices. *Mathematical Social Sciences*, *27*(1), 73–89.

Doignon, J. P., & Falmagne, J. C. (1994). A polynomial algorithm for unidimensional unfolding representations. *Journal of Algorithms*, *16*, 218–233.

Edelman, P., & Greene, C. (1987). Balanced tableaux. *Advances in Mathematics*, *63*(1), 42–99.

Fishburn, P. C. (1992). Notes on Craven's conjecture. *Social Choice and Welfare*, *9*(3), 259–262.

Fishburn, P. C. (1996). Decision theory and discrete mathematics. *Discrete Applied Mathematics*, *68*(3), 209–221.

Fishburn, P. C. (1997). Acyclic sets of linear orders. *Social Choice and Welfare*, *14*(1), 113–124.

Fishburn, P. C. (2002). Acyclic sets of linear orders: *A progress report. Social Choice and Welfare*, *19*(2), 431–447.

Frey, L. (1971). *Parties distributives du treillis des permutations, In Ordres totaux finis Mathématiques et Sciences de l'Homme XII* (pp. 115–126). Paris: Gauthier-Villars.

Frey, L., & Barbut, M. (1971). *Technique ordinales en analyse des données, Algèbre et combinatoire*. Paris: Hachette.

Galambos, A., & Reiner, V. (2008). Acyclic sets of linear orders via the Bruhat order, *Social Choice and Welfare, 30*(2), 245–264.

Guilbaud, G. T. H. (1952). Les théories de l'ntérêt général et le problème logique de l'agrégation. Economie Appliquée 5(4) reprinted in *Eléments de la théorie des jeux*, Dunod, Paris, 1968. English translation: Theories of the general interest and the logical problem of aggregation. Electronic Journal for History of Probability and Statistics, 4, 2008.

Guilbaud, G. T. h., & Rosenstiehl, P. (1963). Analyse algébrique d'un scrutin. *Mathématiques et Sciences humaines, 4*, 9–33.

Inada, K. (1964). A note on the simple majority decision rule. *Econometrica, 32*(4), 525–531.

Johnson, C. R. (1978). *Remarks on mathematical social choice*. Working paper 78–25, Department of Economics, University of Maryland, College Park.

Kelly, J. S. (1991). Craven's conjecture. *Social Choice and Welfare, 8*(3), 269–274.

Kim, K. H., & Roush, F. W. (1980). *Introduction to mathematical consensus theory*. New York: Marcel Dekker.

Kim, K. H., Roush, F. W., & Intriligator, M. D. (1992). Overview of mathematical social sciences. *American Mathematical Monthly, 99*, 838–844.

Köhler, G. (1978). *Choix multicritère et analyse de données ordinales*. Third cycle thesis, Université scientifique et médicale de Grenoble.

Le Conte de Poly-Barbut, C. (1990) .Le diagramme du treillis permutoèdre est intersection des diagrammes de deux produits directs d'ordres totaux. *Mathématiques, Informatique et Sciences humaines, 112*, 49–53.

Markowsky, G. (1994). Permutation lattices revisited. *Mathematical Social Sciences, 27*(1), 59–72.

Monjardet, B. (1971). *Treillis d'ordres, In Ordres totaux finis, Mathématiques et Sciences de l'Homme XII* (pp. 29–45). Paris: Gauthier-Villars.

Monjardet, B. (1978). An axiomatic theory of tournament aggregation. *Mathematical and Operation Research, 3*(4), 334–351.

Monjardet, B. (2006a). Social choice theory and the "Centre de Mathématique Sociale". Some historical notes. *Social choice and Welfare, 25*(2–3), 433–456.

Monjardet, B. (2006b). Condorcet domains and distributive lattices. *Annales du LAMSADE, 6*, 285–302.

Provan, J. S., & Ball, M. O. (1983). The complexity of counting cuts and of computing the probability that a graph is connected. *SIAM Journal on Computing, 12*, 777–788.

Raynaud, H. (1981a). Paradoxical results from Inada's conditions for majority rule. *Technical Report 331*. Institute for mathematical studies in the social sciences, Standford University, Standford, CA.

Raynaud, H. (1981b). Conditions for transitivity of majority rule with algorithmic interpretations. *Technical Report 347*. Institute for mathematical studies in the social sciences, Standford University, Standford, CA.

Raynaud, H. (1981c). How restrictive actually are the value restriction conditions. *Technical Report 348*. Institute for mathematical studies in the social sciences, Standford University, Standford, CA.

Raynaud, H. (1982). The individual freedom allowed by the value restriction conditions. *Technical Report 360*. Institute for mathematical studies in the social sciences, Standford University, Standford, CA.

Raz, R. (2000). VC-dimension of sets of permutations. *Combinatoria, 20*, 1–15.

Romero, D. (1978). *Variation sur l'effet Condorcet*. Third cycle thesis, Université scientifique et médicale de Grenoble.

Sen, A. K. (1966). A possibility theorem on majority decisions. *Econometrica, 34*(2), 491–499.

Vickrey, W. (1960). Utility, strategy and social decision rules. *Quaterly Journal of Economics, 74*, 507–535.

Ward, B. (1965). Majority voting and the alternative forms of public enterprise. In J. Margolis, (Ed.), *The public economy of urban communities*. Baltimore, MD: John Hopkins University Press.
Yanagimoto, T., & Okamoto, M. (1969). Partial orderings of permutations and monotonicity of a rank correlation statistic. *Annals Institute of Statistics*, *21*, 489–506.
Ziegler, G. M. (1993). Higher Bruhat orders and cyclic hyperplanes arrangements. *Topology*, *32*(2), 259–279.

Appendix

Table 1 Table Exact values and bounds for **g(n)** (maximum size of a connected acyclic domain of maximum diameter) and **f(n)** (maximum size of an acyclic domain)

	A	B	C	D	E	F	G	H
n	2^{n-1}	$2^{n-1}+2^{n-3}-1$	$3.2^{n-2}-4$	$\mathtt{AS}(n)$	***g(n)***	$C(n)$	$RS(n)$	***f(n)***
3	4	4	2	4	4	5	4	4
4	8	9	8	9	9	14	8	9
5	16	19	20	20	20	42	16	20
6	32	39	44	45	45	132	36	45
7	64	79	92	100	100	429	81	?
8	128	159	188	222	?	1,430	180	?
9	256	319	380	488	?	4,862	400	?
10	512	639	764	1,069	?	16,796	900	?
11	1,024	1,279	1,532	2,324	?	58,786	2,025	?
12	2,048	2,559	3,068	5,034	?	208,012	4,500	?
13	4,096	5,119	6,140	10,840	?	742,900	10,000	?
14	8,192	10,239	12,284	23,266	?	2,674,440	22,200	?
15	16,384	20,479	24,572	49,704	?	9,694,845	49,284	?
16	32,768	40,959	49,148	105,884	?	35,357,670	108,336	?
17	65,536	81,919	98,300	224,720	?		238,144	?
18	131,072	163,840	196,604	475,773	?		521,672	?
19	262,144	826,680	393,216	1,004,212	?		1,142,761	?
20	524,288	671,359	805,628	2,115,186	?		2,484,356	?

Exact Values

E: $n \leq 4$ folklore, $n = 5, 6$ Fishburn 1997, 2002, $n = 7$ Galambos and Reiner
H: $n \leq 4$ folklore, $n = 5, 6$ Fishburn 1997, 2002

Lower Bounds

A: Craven's conjecture, 1992 (!)
B: Kim and Roush, 1980
C: Abello and Johnson 1984 (N.B. $3.2^{n-2} - 4 = 2^{n-1} + 2^{n-2} - 4$)

D: Fishburn 1997 (Alternating scheme, $n = 6$ BM 1989)
G: Fishburn 1997 (Replacement scheme $\mathrm{f}(n+m) \geq \mathrm{f}(n).\mathrm{f}(m+1)$)

For all large n, $(2.17)^n < f(n)$ (Fishburn 1997)

Upper Bounds

F: $g(n) < C(n) =$ Catalan number$2n!/n!(n+1)!$ (Abello 1991)
For all n, $f(n) < c^n$ for some $n > 0$ (Raz 2000)

Condorcet Domains: A Geometric Perspective

Donald G. Saari

1 Introduction

One of the several topics in which Fishburn (1997, 2002) has made basic contributions involves finding maximal Condorcet Domains. In this current paper, I introduce a geometric approach that identifies all such domains and, at least for four and five alternatives, captures Fishburn's clever alternating scheme (described below), which has advanced our understanding of the area.

To explain "Condorcet Domains" and why they are of interest, start with the fact that when making decisions by comparing pairs of alternatives with majority votes, the hope is to have decisive outcomes where one candidate always is victorious when compared with any other candidate. Such a candidate is called the *Condorcet winner.* The attractiveness of this notion, where someone beats everyone else in head-to-head comparisons, is why the Condorcet winner remains a central concept in voting theory. For a comprehensive, modern description of the Condorcet solution concept, see Gehrlein's recent book (2006).

But Condorcet also proved that pairwise rankings can lead to cycles, where a Condorcet winner cannot exist. His three voter example Condorcet (1785),[1] now called the *Condorcet triplet,* has the preferences

$$A_1 \succ A_2 \succ A_3, \quad A_2 \succ A_3 \succ A_1, \quad A_3 \succ A_1 \succ A_2 \tag{1}$$

("$A_1 \succ A_2 \succ A_3$" means that the voter prefers A_1 to A_2 and A_3, and A_2 to A_3). The majority vote generates the cycle where A_1 beats A_2, A_2 beats A_3, and A_3 beats A_1 each with a 2:1 tally. The trouble with cycles is that they frustrate society's ability to

[1] Condorcet's example in his Éssai Condorcet (1785) is not as concise; it involves about sixty voters. But, I expect that somewhere in his writings, Condorcet explicitly stated this triplet.

D.G. Saari
Institute for Mathematical Behavioral Sciences, University of California, Irvine, CA 92697-5100, USA
e-mail: dsaari@uci.edu

S.J. Brams et al. (eds.), *The Mathematics of Preference, Choice and Order: Essays in Honor of Peter C. Fishburn*, Studies in Choice and Welfare,

make a decision; e.g., who should be selected with this example? Not A_1 because a majority prefers A_3. Not A_3 because a majority prefers A_2. Not A_2 because a majority prefers A_1.

The difficulties associated with this behavior are much worse because the ways in which cyclic behavior can be manifested extend beyond frustrating the majority vote decision process to cause fundamental theoretical concerns. As we now know (Saari (2001a)), for instance, aspects of cyclic outcomes are totally responsible for Arrow's seminal theorem Arrow (1951), which purportedly shows that no non-dictatorial voting rule can satisfy seeming innocuous conditions, and Sen's Paretian Liberal Theorem Sen (1970), which identifies what is called a fundamental conflict between individual and societal decisions. (For different interpretations of Arrow's and Sen's theorems, see Saari (2001a); also see Saari and Petron (2006) and Li and Saari (2004).)

A standard way to handle these difficulties is to restrict the preferences that voters are permitted to have. (See Gaertner (2001) for other restricted domain conditions in choice theory.) This leads to the Condorcet Domain problem: it is to identify sets of preference rankings whereby, no matter how many voters have each of the specified rankings, the outcome never admits a cycle. A goal is to find or characterize all such domains – all sets of these preference rankings – and find the ones with a maximum number of rankings.

This Condorcet Domain challenge has captured the attention of influential contributers to this area; for a brief history with references see Fishburn (1997), Monjardet (2006b), Monjardet's survey (2006), and Monjardet (2008) that appears in this volume. Indeed, it was Monjardet's interesting presentation Monjardet (2006a,b) at an October 2006 DIMACS/LAMSADE conference in Paris that awakened my interest in this issue and led to this paper. As Monjardet explained, Fishburn's paper (1997) contains some of the deepest conclusions about this issue. Fishburn credits his discovery of the "alternating scheme" to clever examples that Monjardet created.

Fishburn's and Monjardet's approaches are essentially combinatoric. So, after introducing the basic problem, I will introduce a geometric approach to describe Fishburn's alternating scheme. My expectation is that the symmetries, which become apparent by use of geometry, will lead to other mathematical tools that can be used to analyze this and other pressing questions. Then, after showing how my geometric approach fits into a broader research theme, I generalize the Condorcet Domain problem by replacing sets of "individual rankings" with sets of "specific configurations of individual rankings." Namely, instead of finding specific rankings that avoid cycles, the new goal is to find configuration of rankings whereby if any number of groups of voters adopt any of these configurations, cycles never occur. Although this generalized problem appears to be far more complicated, the complete solution is in Sect. 4. The original problem, however, remains wide open.

2 Early Solutions

The *Condorcet Domain* problem is to identify subsets of preference rankings so that, no matter how many voters are assigned to each ranking, the pairwise majority vote outcomes never admit a cycle. As the number of voters with each ranking is not restricted, each Condorcet Domain defines a subspace of profiles with which not only majority pairwise voting, but several other voting issues avoid the difficulties of pairwise comparisons; this includes Arrow's Theorem (1951) as well as some decision problems from engineering (Saari and Sieberg (2004)). With these advantages, it is natural to find Condorcet Domains that have the maximal number of rankings; after all, such a domain defines a maximal dimensional profile subspace with these desired properties.

An early Condorcet Domain solution, which continues to be widely used, is Black (1948) single peaked condition. While his condition is slightly more general than described next, a flavor of it can be obtained by placing each alternative at a distinct point on a line. Next, place an individual's "ideal point" anywhere on the line; this individual's preference ranking is defined by the distance from his ideal point to each alternative where "closer is better." It is not difficult to show how and why this ordering of the voters' preference rankings always results in orderly pairwise outcomes. (See, for instance, Black (1948), Saari (2001a) among many other references.)

To see what happens with three candidates, notice that the alternative in the middle never is bottom-ranked by any voter. For special cases, if all ideal points are on one side of the alternatives, some candidate never is top-ranked; if the voters are split into polarized left-right regimes, some candidate never is middle-ranked. Black's condition probably motivated the Condorcet Domain solution advanced by Ward (1965) and later generalized by Sen (1966). Namely, with three candidates at least one of the following conditions is satisfied:

1. There is some candidate who never is bottom-ranked.
2. There is some candidate who never is middle-ranked.
3. There is some candidate who never is top-ranked.

As I indicate next with a geometric representation, when any of these conditions are satisfied, a majority vote pairwise cycle cannot occur.

2.1 Geometry of Triangles

My preferred way Saari (2001b) to represent three-candidate profiles is with an equilateral triangle, where the name of each candidate is assigned to a distinct vertex as illustrated in Fig. 1. The ranking assigned to a point in the triangle is determined by its distance to each vertex, where closer is better. Thus the vertical line represents all $A_1 \sim A_2$ tied rankings; the remaining two indifference lines connect a vertex with the midpoint on the opposite edge. What results is a partitioning of the triangle

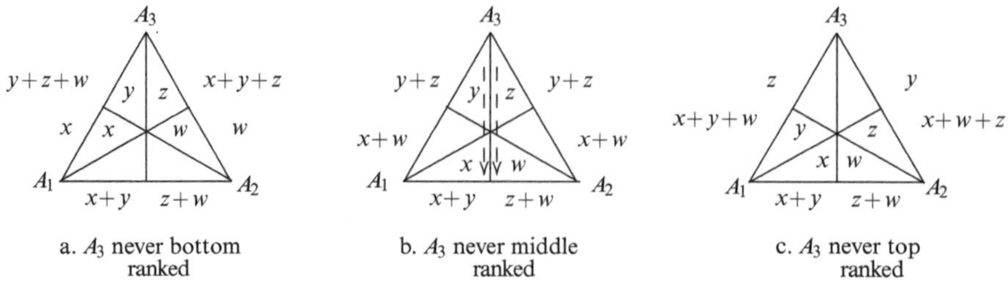

Fig. 1 Ranking triangles

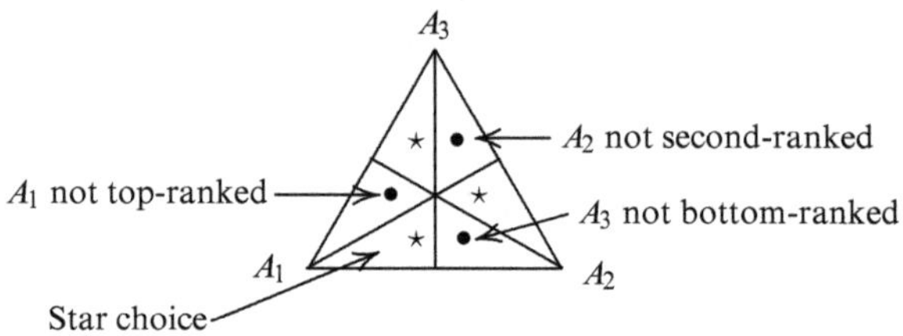

Fig. 2 Condorcet triplets

into what I call "ranking regions;" the open triangles represent strict rankings. For instance, any point in the small Fig. 1a triangle with an x is closest to A_1, next closest to A_3, and farthest from A_2, so it has the $A_1 \succ A_3 \succ A_2$ ranking.

A way to represent a profile is to place the number of voters with each ranking in the associated ranking region. The z in Fig. 1a, for instance, means that z voters have the $A_3 \succ A_2 \succ A_1$ ranking. A candidate never is bottom-ranked when there are no entries in the two regions farthest from the candidate's vertex; e.g., the Fig. 1a profile never has A_3 bottom-ranked. Figure 1b, c represent the remaining two Ward conditions with respect to A_3.

An indifference line associated with a particular pair divides the six rankings into two regions with the two possible pairwise rankings; e.g., the vertical line in Fig. 1a separates the three rankings on the left with $A_1 \succ A_2$ from the three rankings on the right with $A_2 \succ A_1$. Thus a quick way to tally majority votes is to project the numbers from the triangle to the appropriate edge and then add them; the sums are listed next to each edge. This projection and summing process is indicated by the dashed arrows in Fig. 1b, which represents all profiles where A_3 is never middle-ranked. Notice that the A_1, A_2 tallies are, respectively, $x+y$ and $z+w$. As the A_1, A_3 and A_2, A_3 tallies agree, A_3 must be either the Condorcet winner or loser; in either case, it follows that cycles cannot occur when some candidate never is middle-ranked. A similar argument holds for the other figures; e.g., in Fig. 1a, if A_1 beats A_2, then $x+y > z+w$, so A_3 beats A_2: as A_2 is the Condorcet loser, cycles cannot occur.

The complementary relationship between Ward conditions and the Eq. (1) Condorcet triplet is illustrated with Fig. 2. There are two possible Condorcet triplets; the Eq. (1) choice is illustrated with stars in the appropriate ranking regions, the other with bullets. To avoid the cycles caused by Condorcet triplets, it is worth examining what happens if one ranking with a star and one with a bullet are prohibited. By symmetry, it does not matter which star choice is selected, so choose the one indicated in Fig. 2. Next, select one of the three bullets; as the figure indicates, each choice corresponds to satisfying some Ward condition.

Ward's conditions, then, should be viewed as being the sharpest possible restrictions that avoid a Condorcet triplet. Namely, a way to restate Ward's conditions is that they identify *any* set of rankings from which a Condorcet triplet cannot be created. To complete the complementary connection between Ward and Condorcet, each Condorcet triplet consists of the smallest number of rankings that violate all of Ward's conditions. In summary, a three-candidate Condorcet Domain is any set of rankings from which a Condorcet triplet cannot be created; i.e., it is any set that satisfies one of Ward's conditions. As each candidate defines three different Condorcet Domains, nine different four-dimensional subspaces in the six-dimensional profile space are spared the problems of cyclic behavior.

2.2 More Candidates

What happens with more candidates? With four candidates, for instance, can pairwise cycles be avoided whenever some candidate never is bottom-ranked? As illustrated with the Eq. (2) example, where A_3 never is bottom-listed, the answer is no.

$$A_1 \succ A_2 \succ A_3 \succ A_4, \quad A_2 \succ A_3 \succ A_4 \succ A_1, \quad A_3 \succ A_4 \succ A_1 \succ A_2. \qquad (2)$$

Here, A_4 beats A_1, A_1 beats A_2, A_2 beats A_3 (each by $2:1$), and A_3 beats A_4 (unanimously) to form a cycle. Notice how this profile defines the $A_1 \succ A_2, A_2 \succ A_3, A_3 \succ A_1$ cycle with the familiar $2:1$ tallies. Indeed, by focussing attention on the *relative position* of these three candidates, we find that they create a Condorcet triplet, which means that all of Ward's conditions are violated. This insight explains Sen's condition Sen (1966) that a necessary and sufficient requirement for a set of rankings to define a Condorcet Domain is that, when restricting the rankings to any triplet, one of Ward's conditions holds. So for $\{A_1,\dots,A_n\}$, a set of rankings is a Condorcet Domain if and only if when restricted to each triplet, at least one candidate never is top-ranked, or middle-ranked, or bottom-ranked; i.e, these relative rankings can never be used to create a Condorcet triplet.

By knowing what creates Condorcet Domains, the next step is to find examples and maximal Condorcet Domains. This is where Fishburn (1997) alternating scheme and "never" conditions play a dominant role. To explain his condition with an example, consider the five candidates $\{A,B,C,D,E\}$. Select a ranking; say $E \succ A \succ D \succ C \succ B$. Assign temporary A_j names according to the ranking's order; e.g., E is called A_1, A is called A_2, ..., B is called A_5.

Fishburn's alternating scheme is as follows:

List each triplet in the order of their temporary names; e.g., list $\{A_3, A_1, A_4\}$ in the order of their subscripts as $\{A_1, A_3, A_4\}$. If the subscript for the middle alternative is odd, as it is here (it is 3), use the never-top ranked rule for this alternative with the triplet. If it is even, use the never bottom-ranked rule with the alternative. Apply this rule to all triplets. Alternatively, the rule used with all triplets could be that if the subscript for the middle alternative is odd, then use the never-bottom ranked rule; if it is even, use the never-top ranked rule.

The value of this algorithm comes from Fishburn's result stated next; proofs are in his papers:

Theorem 1 *(Fishburn (1997, 2002)) For $n = 4,5,6$ alternatives, a Condorcet Domain has the maximal number of rankings if and only if the set satisfies the alternating scheme. For $n \geq 16$, the alternating scheme does not define the maximal Condorcet Domain.*

What a delightful result! Beyond contributing to a long studied question, his theorem creates a mystery that begs to be investigated. Why does it work? What underlying mathematical structures permit this condition? Is there an intuitive way to appreciate his alternating condition? What is magical about the $n = 16$ cutoff? What happens between 7 and 15? As my objective is to develop insight and intuition, I describe the Ward–Sen and Fishburn conditions in a geometric framework.

3 Geometry

To find all four-candidate Condorcet Domains by using elementary geometry, replace the equilateral triangle with the Fig. 3a equilateral tetrahedron. Again, a ranking is assigned to a point based on its distances to the vertices. To create a two-dimensional representation of the tetrahedron, select a vertex (A_4 in Fig. 3), cut the three tetrahedron edges from the vertex to its base, and open the flaps to create the Fig. 3b object. Each of the 24 small triangles, or ranking regions, represents a

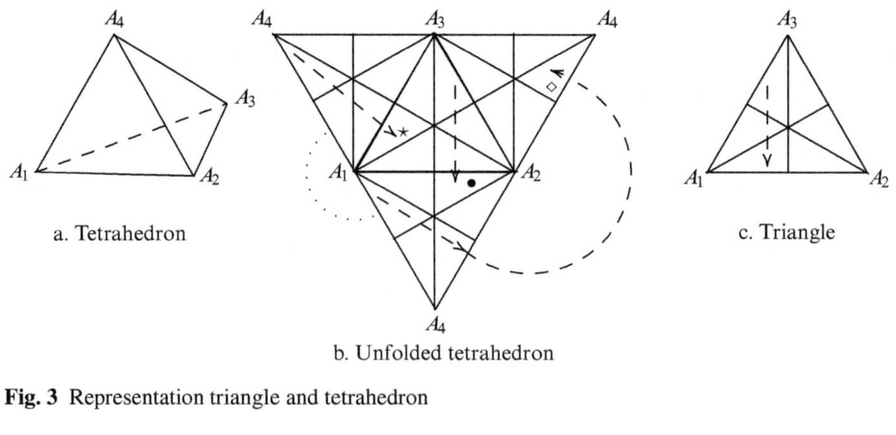

Fig. 3 Representation triangle and tetrahedron

particular ranking; e.g., using distances to vertices, the region with the bullet has the $A_2 \succ A_1 \succ A_4 \succ A_3$ ranking, while the one with the diamond has $A_4 \succ A_2 \succ A_3 \succ A_1$. The four large equilateral triangles are the four original tetrahedron faces; alternatively, they represent where one alternative is removed. For instance, the central equilateral triangle with vertices A_1, A_2, A_3 can be used to represent rankings when A_4 is dropped.

To motivate what is done next, recall that to construct a Condorcet Domain we need to find all rankings where after dropping A_4, the remaining triplet satisfies the "A_3 is never middle ranked" or some other Ward–Sen condition. Namely, we need to *avoid* all rankings whereby dropping A_4 leads to either $A_1 \succ A_3 \succ A_2$ or $A_2 \succ A_3 \succ A_1$. More generally, we need to find a way to identify all rankings that have a specified relative ranking after dropping a particular candidate.

To do this by using geometry, start with the three alternative setting of Fig. 3c. Similar to tallying elections, ignoring A_3 has the effect of projecting the rankings to the A_1-A_2 edge; e.g., the dashed arrow in the triangle represents projecting all three rankings with the $A_1 \succ A_2$ relative ranking to the $A_1 \succ A_2$ portion of the bottom edge. (So, to find all rankings with $A_1 \succ A_2$, just follow that dotted line backwards.) A similar projection occurs with Fig. 3a when an alternative is dropped, but we need help to see the projections. Assistance is provided by Fig. 3b.

Figure 3b easily handles projections when A_4 is ignored and a $\{A_1, A_2, A_3\}$ ranking results. For instance, the starred region has the ranking $A_1 \succ A_3 \succ A_2 \succ A_4$, with the $A_1 \succ A_3 \succ A_2$ relative ranking when ignoring A_4. The four rankings with this $A_1 \succ A_3 \succ A_2$ relative ranking are in the ranking regions with the dashed arrow pointing to the star; i.e., ignoring A_4 effectively projects these four rankings into the starred region. Indeed, "above" (i.e., directly away from the center point of the central triangle) each ranking region in the central equilateral triangle are the four four-candidate rankings with the same relative ranking of the triplet.

Now consider a ranking that is not in the central triangle; e.g., treating the region with a bullet as a triplet, the ranking is $A_2 \succ A_1 \succ A_4$. As A_3 is the missing candidate, one way to handle to geometry is to return to the tetrahedron and open it from the A_3 vertex. Doing so would create four attached equilateral triangles with the A_1, A_2, A_4 triangle in the center; each adjacent triangle has the vertex A_3. But this approach is not satisfactory for our needs as we want to compute the rankings to be removed for all triplets with one diagram. So, an equivalent way to create the same figure that is formed by slicing the tetrahedron open from vertex A_3 is to rotate (the circular dotted line) the A_1, A_3, A_4 triangle about vertex A_1 so that the two A_1-A_4 edges meet, and rotate the A_2, A_3, A_4 triangle about A_2 so that the two A_2-A_4 edges meet. By doing so, it is clear that the ranking regions with the dashed arrow pointing to the bullet are projected to this region. (Here, we did not need to rotate the faces.)

As a final example, the three-candidate ranking for the region with a diamond is $A_4 \succ A_2 \succ A_3$ where A_1 is the ignored alternative. To find all rankings with this relative ranking, rotate the A_1, A_4, A_2 triangle about the A_2 vertex, find the projection, and then rotate back again to show that the desired ranking regions are those with the dashed arrow combined with the circular arrow.

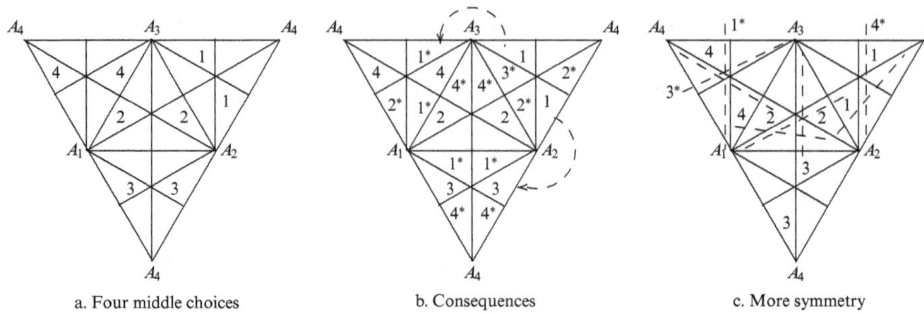

Fig. 4 No middle ranked alternatives

3.1 Excluding Rankings

To illustrate how to use this geometry, the "never-middle ranked" condition is imposed in Fig. 4a for each triplet. With the A_2, A_3, A_4 triplet, for instance, the 1's indicate those rankings where, when restricted to this triplet, A_4 is middle-ranked; thus, these two rankings are to be excluded and the other four are admitted. Similarly the regions with 4's indicate where A_1 is middle-ranked when restricting admissible rankings to the A_1, A_3, A_4 triplet, so the other four rankings satisfy Ward's condition where A_1 never is middle-ranked.

All rankings that satisfy these conditions, i.e., all ranking regions that project to any of the marked Fig. 4a regions, are marked in Fig. 4b. The top "1" in Fig. 4a, for instance, excludes the three regions indicated by the top circular arrow; one Fig. 4b region already is excluded as it has a 4, the other two regions, marked with 1*, are excluded because they are projected to a 1. Similarly, the lower circular Fig. 4b arrow identifies the three regions that project to the other 1; one region already is excluded with its 3, and the two with 1* are excluded by being projected to this 1.

Doing this for all four numbers leaves only four ranking regions without a label; these rankings, $\{A_4 \succ A_3 \succ A_1 \succ A_2, A_4 \succ A_3 \succ A_2 \succ A_1, A_1 \succ A_2 \succ A_3 \succ A_4, A_2 \succ A_1 \succ A_3 \succ A_4\}$ enjoy obvious symmetry relationships made apparent with the figure. (For instance, notice that each ranking is accompanied by its reversal.) They define a "complete Condorcet Domain" in that by adding any other ranking to the set, the new set no longer is a Condorcet Domain.

In general, for each of the four triangles, select a Ward–Sen condition for some alternative. Then, cross off all regions identified by the selected Ward–Sen choices, and all regions that project onto one of these regions. In this manner, all possible four-alternative complete Condorcet Domains can be found. As this approach shows, in profile space (which can be represented by the 24 dimensional Euclidean space $\mathbb{R}^{24}$) the Condorcet Domain is orthogonal to the space of regions that are eliminated by the Ward–Sen conditions.

The geometric challenge, which has the flavor of a Sudoku or crossword puzzle, is to determine which combinations of Ward–Sen structures leave the largest number of blank spaces after the projected regions are crossed off. Thus, finding a Condorcet

Domain with a maximal number of rankings requires finding combinations of Ward–Sen conditions with the minimal number of crossed off regions. Clearly, for this to occur, we need to select conditions so that some regions are eliminated by multiple conditions. For instance the regions with a 3 on the right in the A_1,A_2,A_4 triangle is excluded twice; first by being the indicated middle ranking for that triangle and then by being projected to a 1. The goal, then, is to determine which combinations of the Ward–Sen conditions minimize, and which maximize, multiple counting of ranking regions. The answer must involve the geometric structure and its associated symmetries.

An example using symmetry is depicted in Fig. 4c where the four not-middle choices, given by the numbers 1 to 4, are selected in a band. Notice, some numbers are in regions that are projected to other numbers. The projection regions are depicted by dashed lines leading out of regions with a number; three dashed lines are labeled with the donor number n^*. With this choice, five marked ranking regions are used three times, six twice, and only five once. This arrangement leaves eight blank regions that define a complete Condorcet Domain: the first part has A_3 bottom-ranked,

$$\{A_1 \succ A_2 \succ A_4 \succ A_3, A_1 \succ A_4 \succ A_2 \succ A_3, A_4 \succ A_2 \succ A_1 \succ A_3, A_2 \succ A_4 \succ A_1 \succ A_3\}$$

and the second part has A_3 is top-ranked

$$\{A_3 \succ A_1 \succ A_4 \succ A_2, A_3 \succ A_1 \succ A_2 \succ A_4, A_3 \succ A_2 \succ A_4 \succ A_1, A_3 \succ A_4 \succ A_2 \succ A_1\}.$$

Also notice, accompanying each ranking in this Condorcet Domain is its reversal.

3.2 Calculus of Ward–Sen Conditions

One of my contributions for this Condorcet Domain problem is to indicate how to create a calculus to determine which ranking regions should be eliminated. To do so, the Ward–Sen conditions are related to the geometry of a tetrahedron. Using the bottom face of Fig. 3a, with vertices $\{A_1,A_2,A_3\}$, which is the central face of Fig. 5a,

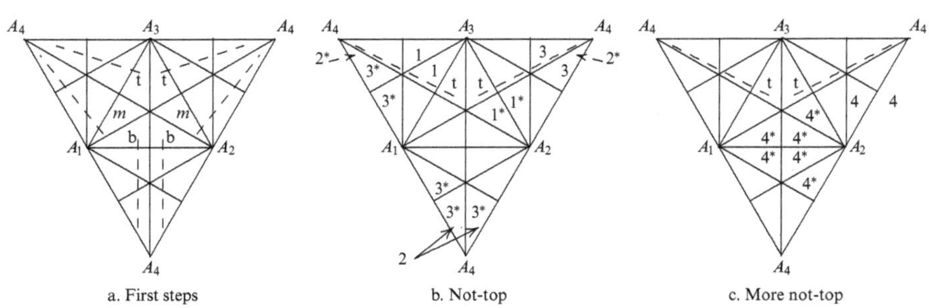

Fig. 5 Calculus

the never-bottom condition defines an edge; e.g., the b's in Fig. 5a are on the A_1-A_2 edge. Applying this condition to define a Condorcet Domain, it follows from the dashed lines moving out of the "b" regions that it eliminates all rankings in the other face that shares this edge; in Fig. 5a, it is the triangle $\{A_1, A_2, A_4\}$. Thus, a never-bottom condition defines one of the face's edges; it eliminates the two specified never-bottom rankings and *all rankings* in the face sharing the same edge.

A never-top condition defines two regions sharing a vertex; in Fig. 5a, the regions are denoted by t's, and the vertex is A_3. As illustrated by the dashed lines moving out of the two "t" regions, this condition eliminates all six rankings that share the same vertex and two more along the *"designated edge"* that connects the specified vertex with the vertex not in this face; here it is the A_4 vertex. Notice that this A_3-A_4 edge is depicted on two of the faces; the reason is that this is one of the edges along which the tetrahedron was cut open. What is not so obvious is that this edge connects A_3 from the A_1, A_2, A_3 face to A_4 from the A_1, A_2, A_4 face. After all, the same A_4 is on three faces; to see that this is so, just fold up the faces into a tetrahedron.

As indicated by the m's in Fig. 5a, the never-middle condition defines a face base and two adjacent faces; the excluded regions are the two selected rankings and three each in the adjacent faces. These eliminated rankings come in pairs; a ranking and its reversal. Also notice how four of the Fig. 5a rankings are below the $A_1 \sim A_3$ line, the other four are below the $A_2 \sim A_3$ line.

The next step is to identify what rankings disappear by combining these conditions; the ideas can be illustrated by using the same condition with two faces α and β; the remaining two faces (equilateral triangles) are called γ and δ. The easiest case is the never-bottom condition, which emphasizes selected edges.[2] (See Figs. 3a, 5a.) All possible combinations follow:

- If the never-bottom condition used with the α and β faces has the α identified edge bordering face γ and the β identified edge bordering face δ, then there is no overlap of eliminated regions. Thus 16 regions are eliminated; they are all of the γ, δ ranking regions and the four initiating regions. To illustrate with Fig. 5a, let the α face be given by the vertices A_1, A_2, A_3, and the bordering γ face be A_1, A_2, A_4. Then the *b*'s in the α face eliminate all γ face rankings. Let the β face be given by A_2, A_3, A_4 where the two bottom ranked rankings are on the A_3-A_4 edge. These two choices eliminate all rankings in the δ face defined by vertices A_4, A_1, A_3. In total, all rankings from the γ and δ faces, 12 of them, are eliminated along with the four selected rankings for a total of 16.
- If the α edge is on the β face, but the β edge is on face γ, then 14 regions are eliminated – the β face condition eliminates all γ rankings, the α face condition eliminates all β rankings including the two that drop all of the γ rankings, and the two initiating regions in the α face. Again, illustrating with Fig. 5a with the same α face but where the β face now is A_1, A_2, A_4, the *b*'s in Fig. 5a satisfy the first condition; all β face rankings are dropped. Now let the γ face be defined by A_2, A_3, A_4. To satisfy the specified conditions, the two bottom ranked rankings

[2] The approach becomes clear and fairly easy with some experience. Therefore I *strongly recommend* that the reader creates versions of the Fig. 5 triangles and carries out the described calculus.

from β must be on the A_2-A_4 edge. These choices eliminate all β and γ rankings (12 of them in total) and the two b rankings in the first face for a total of 14 rankings.

- If both conditions define the same edge, which connects the α and β faces, then both faces, or 12 regions are eliminated. To illustrate, let α and β be as in the last illustration. Let the b be as in Fig. 5a, and let the two choices for β be directly across the A_1-A_2 edge. These choices eliminate all α and β rankings; 12 of them.
- Finally if the α and β edges both border on face γ, only 10 regions are eliminated; each γ region is eliminated by both conditions; the remaining four are the initiating choices in α and β. To illustrate, let α be as above, β the A_2,A_3,A_4 face, and γ the A_1,A_2,A_4 face. Let the b's be as in Fig. 5a, so they eliminate all of the γ rankings. Chose the bottom ranked rankings in β along the A_2-A_4 edge; they, too, eliminate all γ rankings. So, the eliminated rankings are the six in face γ and the four selected ones for a total of ten.

Using the "never-top" conditions with the α, β faces characterizes all combinations of vertices that identify the never-top candidate and the interaction of designated edges.

- If both conditions use the same vertex, there will be overlap in the regions that are eliminated. Here, there are only 10 dropped regions – both "never-top" choices eliminate the six regions around the shared vertex, and each condition eliminates two more regions along the designated edges. This is illustrated in Fig. 5b where face α is given by A_1,A_2,A_3; the two t's eliminate three rankings along the dashed line in the β face of A_1,A_3,A_4 and three rankings along the dashed line in the γ face of A_2,A_3,A_4. The rankings selected in the β face are indicated in Fig. 5b with the 1's. In the γ face, this choice eliminates three rankings, but two of them were already eliminated by t. Similarly, the other 1 eliminates three rankings in the α face, but two of them are t's. Thus this choice eliminates only two additional regions; they are given by the 1^*'s. A total of 10 regions are eliminated.
- If the conditions use two different vertices that share the same designated edge, some overlap occurs meaning that 12 regions are eliminated. In Fig. 5b, the α face is defined by A_1,A_2,A_3 and selected rankings are given by the t's. Thus the designated edge connects vertices A_3 and A_4. To find the other vertex, as the designated edge is to be the same, the face cannot include vertex A_3. Thus this β face must be defined by A_1,A_2,A_4. Moreover, to have the same designated edge, A_4 is the selected vertex, thus the selected regions must be given by the 2's in this face. One 2 eliminates three regions in γ defined by A_1,A_3,A_4, but two of these regions have a dashed line meaning they already were eliminated by the t's. The same behavior occurs in δ defined by A_2,A_3,A_4. Thus the 2^*'s show the two regions not already eliminated by the t's, leading to a total of 12 dropped regions.
- If the designated vertices differ and the designated edges meet only in a single point, then the smaller overlap causes 14 eliminated regions. To illustrate why and what this means, using the same α face and t's, the designated edge connects

A_3 with A_4. What we need is to select the β face and its identified never-top rankings in a manner so that the designated line is the A_j-A_4 edge, where A_j is either A_1 or A_2. Suppose it is A_1. Now there is a choice; do we have A_1 or A_4 as the "never-top" ranked candidate? If it is A_1, the face is A_1,A_2,A_3, which is the α face where the never top condition already is specified. Thus, the choice must be A_4 where, as in Fig. 5b, the β face must be given by A_2,A_3,A_4 and the selected never-top rankings must be given by the 3's. The regions newly eliminated are given by the 3^*'s in Fig. 5b. In total, 14 regions are eliminated.

- The remaining condition is for the two vertices differ and the two designated edges not to meet. To see what this means, start with the same α face and the t's. This choice defines the designated line connecting vertices A_3 and A_4. Thus, the other designated line must connect A_1 and A_2; one of these vertices identifies the "never-top ranked" candidate for a particular triplet. If it is A_2, then the choice of the designated line means that the triplet cannot contain A_1; the β face is A_2,A_3,A_4. In Fig. 5c, this situation is given by the 4's. As the eliminated regions do not meet, this last situation drops 16 regions.

The analysis for the never-middle condition is similarly easy. Using the never-middle with faces α and β where both have the same edge as a base, the number of eliminated regions is 12. If the ranking regions for two never-middle choices are adjacent, so they share a portion of an edge of the tetrahedron, the number of excluded regions also is 12. Otherwise, the number of excluded regions is 14. No combination eliminates 16 rankings. Incidentally, for any $n \geq 3$, for each ranking not eliminated by applying the condition to a triplet, its reversal also is not eliminated; i.e., any Condorcet Domain defined strictly with never-middle conditions has an even number of rankings.

Similar straightforward computations hold for other combinations; e.g., when combining a never-bottom with a never-top, emphasize how the never-bottom edge along with the never-top vertex and its designated edge, interact. For instance, using a never-bottom with α where the edge is the designated edge of a never-top condition with face β provides overlap so 11 regions are eliminated. Combining a never-middle with a never-top condition where both designated regions for the never-top already have been eliminated leads to 13 dropped regions.

3.3 Combinations and Fishburn's Alternating Scheme

The calculus for three conditions is similar, so, instead of doing so, the above combinatoric rules are now used to obtain insight into what happens with the various combinations of conditions. The first result shows what can be obtained by using the same constraint with each triplet.

Theorem 2 *If the never-top or the never-bottom condition is used with each triplet, then the smallest associated Condorcet Domain is empty; the largest Condorcet Domain has 8 rankings. If the never-middle requirement is used with each triplet, then* *the resulting Condorcet Domain has either 4 or 8 rankings. The unique arrangement giving 8 rankings is equivalent to Fig. 4c. Each ranking in the never-middle Condorcet Domain has its reversal in the Domain.*

Proof. First consider the never-bottom condition. Use the never-bottom for the α and β faces as defined by the connecting edge; this eliminates all α, β rankings. Doing the same with the γ, δ faces means that all rankings have been eliminated, so the Condorcet Domain is empty. To have a minimum number of eliminated regions, select a face α; each edge of α connects to another face; using each of these edges to define the never-bottom condition for the connecting face means that each of these three conditions eliminate all α regions; in total 12 rankings have been dropped. It remains to use the never-bottom with α; the selected edge will eliminate the remaining four rankings from the connecting face, leaving the specified 8 rankings.

For the never-top condition, to eliminate all rankings, just use all four vertices. About each vertex, the condition eliminates all six rankings where that candidate is top-ranked, so all rankings are eliminated. At the other extreme, select a vertex; it connects three faces. For each face, select the never-top condition defined by that vertex. As the six rankings with that candidate top-ranked are eliminated three times, the total number of eliminated rankings is 12. The choice for the last face must be selected. The three conditions already selected define three designated legs. Select a vertex in this face so that it defines the same designated leg; only four more regions are eliminated. Hence the associated Condorcet Domain has 8 rankings. That this is best possible follows from the construction and the above combinatoric rules.

The never-middle conditions are left for last as they indicate a general strategy. For instance, to show that the never-middle conditions cannot have an empty Condorcet Domain, assume that it could; thus all rankings from each face must be eliminated. So we try to find what conditions permit this to obtain a contradiction. With the m's in Fig. 6a, the required conditions to eliminate all rankings in this α face defined by A_1, A_2, A_3 follow immediately: There is one "never-middle" condition from the β face of A_2, A_3, A_4 that never eliminates any regions from α; the other two never-middle choices from β leave two blank regions in α. A similar statement holds for any of the three faces bordering on α. Indeed, it is easy to see that the

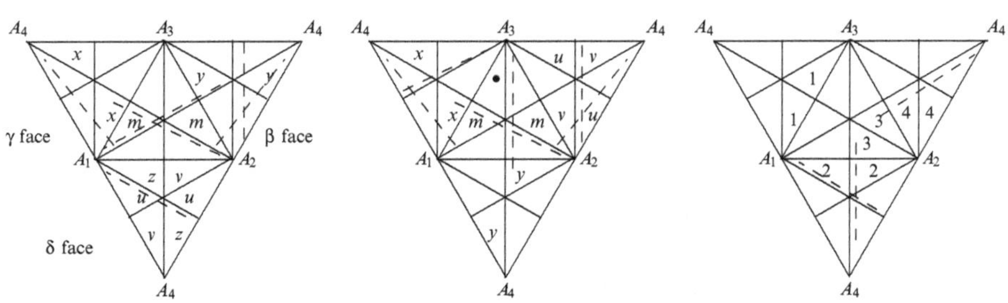

Fig. 6 Middle-ranked and alternating scheme

positioning of the x's in γ defined by A_1,A_3,A_4 and the y's in β, where the x is adjacent to an m, and the y is lifted a region, will eliminate all α face rankings. This choice is unique up to symmetry.

As the never-middle choices for three faces are uniquely specified to drop all α face rankings, it remains to find the never-middle choice for the bottom face δ given by A_1,A_2,A_4. As it is easy to check, each of the three choices of u, v, or z leaves four blank regions, so the associated Condorcet Domain has four rankings. Because this setting describes where all rankings from one face are eliminated, it follows in general that if all rankings from any face are not eliminated, then each face must have at least one blank region; i.e., with the never-middle conditions, the Condorcet Domain must always have at least four rankings.

If we do not want to have all rankings dropped from each face, then the next step is to determine how to select never-middle conditions so that a face has precisely one blank region. The two choices are where the blank region and one of the never-middle rankings are either the bottom two, or the top two, rankings for some alternative. For the first case, which is illustrated in Fig. 6b, the goal is to keep the $A_3 \succ A_1 \succ A_2 \succ A_4$ ranking; this ranking region is identified with the bullet. It is easy to see that the only choice for the x and y never-middle rankings are uniquely determined as illustrated. It remains to find the rankings for the β face. To keep the designated region blank, the only choices are denoted by u and v. If u is selected, all rankings in the side face are eliminated, which returns to the earlier case of four rankings in the Condorcet Domain. Selecting v is the Fig. 4c case of eight rankings in the Condorcet Domain.

The argument for the second case, where m and the blank region are the top two rankings for a candidate is essentially the same. This requirement uniquely defines the choices of never-middle for two faces. There are only two choices for the remaining face; one creates a face with all rankings removed, so it reduces to the earlier case having a Condorcet Domain of four rankings. The other choice leaves one blank ranking for each face; e.g., rankings of the $A_1 \succ A_2 \succ A_3 \succ A_4, A_4 \succ A_3 \succ A_2 \succ A_1, A_3 \succ A_1 \succ A_4 \succ A_2$, and $A_2 \succ A_4 \succ A_1 \succ A_3$, where each candidate is in each position once, emerge.

Finally, it is easy to show that it is impossible to have three blank rankings in a face. For four blank rankings, it is equally as easy to show that the situation is equivalent to that of Fig. 4c. This completes the proof. □

Before providing a geometric description of Fishburn's alternating scheme, notice how the above approach can be used to answer several other questions. For instance, is the set of rankings $\{A_1 \succ A_2 \succ A_3 \succ A_4, A_2 \succ A_1 \succ A_4 \succ A_3, A_4 \succ A_2 \succ A_1 \succ A_3\}$ a Condorcet Domain? If so, is it a complete Condorcet Domain? If not, how can it be completed? To find answers, use the above approach used to determine whether a face can have the specified blank regions. In the same way, it is possible to determine the associated Ward-Sen conditions. If such conditions can be found, the set is a Condorcet Domain. If additional blank regions emerge, then the set is not complete and the added regions define a completion.

This approach leads to a geometric description that is equivalent to Fishburn's alternating scheme. Start with face α. For each of the remaining three faces, use the never-bottom condition adjacent to the α edge. In this way, 12 rankings are eliminated; all six in the α face and two from each of the other three faces. But whatever Ward–Sen choice is made for α, never-top, never-middle, or never-bottom, it eliminates four more regions from other faces, which defines a Condorcet Domain of eight rankings. Alternatively, by using the never-top ranked choice with the same vertex, whatever choice is made for the remaining face, four more rankings are excluded.

The next natural approach coming from the calculus is to use two never-bottom conditions, say for faces γ and δ, where both eliminate all α face rankings, and two never-top conditions, for the remaining faces α and β, that use the same vertex. In Fig. 6c, the never-bottom choices are illustrated with the 1's and 2's; they eliminate all α rankings. The only two choices for the common vertex of the α and β faces are A_2 and A_3. Either works; I selected A_2 as given by the 3's and 4's. Observe how this construction creates overlaps with the never-bottom condition, which means that the Condorcet Domain has nine rankings –the nine blank regions in Fig. 6c outside of the α face.

Using the above machinery of computing when all rankings from a face are eliminated, etc., it is not difficult to show that this is the maximum, and it can be attained only in this manner. To recover Fishburn's alternating scheme, select the names of the vertices in an appropriate manner. Notice that while Fishburn proved that the alternating scheme does not hold for all values of n, the calculus of the geometric approach described above does apply to any number of alternatives.

3.4 More Candidates

The approach for $n \geq 4$ candidates is similar, but assistance coming from concrete geometric objects is missing for $n \geq 6$. (For $n = 5$, the simplex opens into a tetrahedron, which can be opened into a 96 region version of Fig. 3b plus another copy for 24 interior ranking regions.) Any Ward–Sen condition with triplet eliminates $\frac{n!}{3}$ rankings.

The structure remains similar; e.g., the "never-middle ranked" condition eliminates rankings and their reversals; these rankings lie along two "indifference ranking" surfaces. If a triplet includes two of the alternatives from the specified triplet, the never-middle condition eliminates half of them; if it has one or none, it eliminates all of the triplet rankings. The never-bottom ranked condition defines an edge and eliminates all rankings in $[\frac{n!}{3} - 3]/6$ triplets. The never-top condition defines a vertex; it eliminates all $(n-1)!$ rankings sharing this vertex (that is, all rankings where the candidate identified with the vertex is top-ranked) and $\frac{(n-1)!}{3}(n-3)$ other rankings which involve rankings on both sides of edges from the designated vertex to the other vertices not on this face. Again, if the triplet includes two alternatives from the specified triplet, the excluded rankings are along an edge; if it includes one or none, the triplet is eliminated.

In this manner, calculus rules for combining Ward–Sen conditions can be determined. The way to do so is to emphasize the interactions among edges, bases, and vertices. For instance, check whether any of the designated edges from the never-top condition coincide with the edge from the never-bottom condition. In this manner, the analysis to determine what happens with $n = 5$ turned out to be straightforward, and it is not overly difficult to find conditions leaving a fixed number of blank regions in a triangle. I have yet to examine what happens with $n \geq 6$.

4 Profile Coordinates

It is widely appreciated that a Condorcet Domain imposes a far too strict constraint to avoid cyclic behavior. This is illustrated with the Fig. 7a profile, which fails *all* of the Ward conditions. Nevertheless, the majority vote rankings are transitive, and, going far beyond what could ever be expected from a Condorcet Domain, the differences in tallies satisfy an extreme "tally consistent transitivity" whereby adding the difference in A, B tallies $(13-9=4)$ to the difference in B, C tallies $(13-9=4)$ equals the difference in the A, C tallies $(15-7=8)$! If we embrace the value of the Condorcet Domain problem, then it becomes necessary to understand why this example, which violates all of Ward conditions, enjoys a much stronger form of majority vote transitivity.

This example was constructed by adding multiples of Figs. 7b, c profiles with appropriate permutations of the A_j names; these component profiles do satisfy Ward conditions. Indeed, the Fig. 7a profile is two units of the Fig. 7b profile where $\{A_1,A_2,A_3\} = \{A,B,C\}$ plus one unit where $\{A_1,A_2,A_3\} = \{B,A,C\}$ plus two units of Fig. 7c where $\{A_1,A_2,A_3\} = \{C,A,B\}$.

The construction of this example suggests that, perhaps, a way to analyze voting rules is to use appropriate configurations of rankings rather than individual rankings. To make this comment more concrete, let me introduce what I call the "Generalized Condorcet Domain" problem; it is to determine how to replace "individual rankings" with specific "configurations of rankings" in a way so that any multiples of these configurations never allow cycles.

This Generalized Condorcet Domain problem can be completely solved for any n. For three candidates, not only can this generalized problem be solved, but the

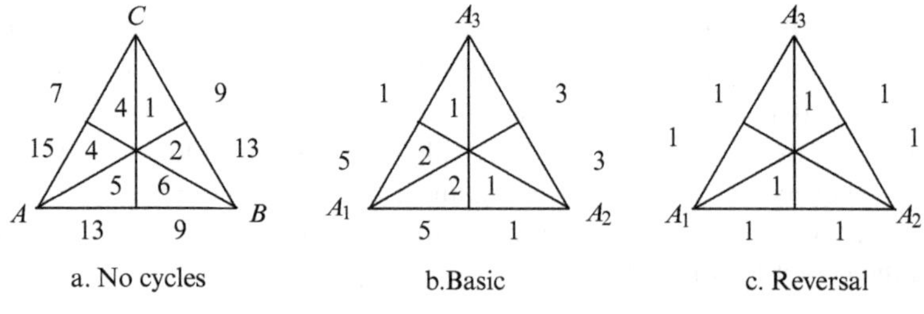

Fig. 7 A non-cylic example

tallies always satisfy the tally-consistent transitivity of majority votes if and only if the profile is a sum of multiples of permutations of the Fig. 7b, c configuration of rankings (Saari (1999))! Notice, these configurations of profiles define a five-dimensional subspace, which is a dimension larger than possible with any Condorcet Domain. Staying with the theme of the Condorcet Domain, it turns out that with these Fig. 7b. c, configurations, and only with these configurations, can any multiple of them be added without ever encountering a cyclic outcome, or without ever violating tally-consistency.

To generalize the discussion, recall that a "positional rule" tallies ballots by assigning specified number of points for candidates based on their position on a ballot. The plurality vote assigns one point to a voter's top-positioned candidate and zero to all others. The Borda Count for n candidates assigns $n-j$ points to a voter's j^{th} positioned candidate.

A recent approach (Saari (1999, 2000b,a, 2001b)), which currently is being refined, is to find appropriate profile coordinate systems that will handle all possible combinations of positional rules. The goal is similar to that of a Condorcet Domain; it is to find appropriate configurations of profiles – profile coordinates –so that when adding any multiple of a coordinate to a profile, we know in advance the effect it will have on all possible positional methods. As true with the Condorcet Domain, no restrictions are imposed on how much of a particular profile coordinate is added or subtracted. The difference is that the Condorcet Domain problem concentrates on individual rankings; the profile coordinate system concentrates on specified configurations of preferences.

As an illustration, the Fig. 7b, c configurations define certain three-alternative coordinate directions; it is easy to show that the Fig. 7b configurations never permit conflict among positional and binary rankings while Fig. 7c configurations, which consist of a ranking and its reversal, has no effect on binary rankings but change positional outcomes. To further illustrate this program while connecting it with Condorcet Domains, notice that to understand how and why positional outcomes over triplets differ from positional outcomes over all four candidates and over pairs, we need to find a coordinate direction that affects the positional election outcomes of triplets without ever affecting binary or four-candidate positional rankings. An example of how this can be done is with the earlier derived Condorcet Domain $\{A_1 \succ A_2 \succ A_3 \succ A_4, A_4 \succ A_3 \succ A_2 \succ A_1, A_3 \succ A_1 \succ A_4 \succ A_2, A_2 \succ A_4 \succ A_1 \succ A_3\}$ where each candidate is in each position precisely once (so all four-candidate positional outcomes end in a tie), and for each pair $\{A_j, A_k\}$, two rankings have $A_j \succ A_k$ while two others have $A_k \succ A_j$; i.e., all pairwise outcomes end in ties. But with this configuration of rankings, all non-Borda Count positional outcomes for any triplet never are ties. By discovering and using configurations of this type, it becomes possible to explain all differences among all positional elections of all possible subsets of candidates.

Of particular relevance for the current paper is that one part the emerging profile coordinate system identifies all profile configurations that cause pairwise voting cycles. As these coordinates are closely related to the Condorcet Domain problem, they are described in more detail.

5 The Source of All Pairwise Cycles

The Condorcet Domain problem searches for the maximal dimensional profile subspaces where cycles never occur. The approach can be described as finding a space of rankings by use of the Ward–Sen conditions that is orthogonal to Condorcet triplets. In this analysis, the coordinate directions are determined by individual rankings; this choice is what causes the inefficiencies of the Condorcet Domain in that these domains form overly strict conditions to avoid cycles. To avoid these inefficiencies, precisely the same program is carried out next except that specific "profile coordinate directions" replace individual rankings. Namely, the objective is to find appropriate profile coordinates, and the associated profile subspace, so that *any profile orthogonal to this subspace never allows a cycle with the majority vote.* The following theorem states the result.

Theorem 3 *(Saari (2000b, 1999)) For $n \geq 3$ alternatives, in the $n!$ dimensional profile space, there is a $\frac{(n-1)!}{2}$ dimensional subspace with the following property. Any profile that is orthogonal to this subspace can never have a majority vote cycle. Thus, this cycle-free subspace has dimension $n! - \frac{(n-1)!}{2} = \frac{(n-1)!}{2}(2n-1)$. Each triplet $\{A_j, A_k, A_s\}$ has the tally-consistent transitivity property where adding the difference of the majority vote tallies between A_j and A_k to the difference between A_k and A_s equals the difference between A_j and A_s.*

The last statement goes far beyond assuring non-cyclic outcomes to ensure the transitivity of pairwise rankings *and* tally-consistent transitivity. These results, then, are much stronger than possible with the Condorcet Domain. Also, the dimension of the orthogonal space, $\frac{(n-1)!}{2}$, is much smaller than the number of dimensions dismissed by *just one Ward-Sen condition applied to just one triplet,* which is $\frac{n!}{3}$. Thus, the cycle-free subspace ensured by Theorem 3 has a dimension significantly larger than that of any Condorcet Domain. For instance, the largest dimension of a subspace attached to a four-candidate Condorcet Domain is nine, while the subspace from Theorem 3 has dimension $24 - 3 = 21$, so it is more than twice as large. The largest dimension of a subspace attached to a five-candidate Condorcet Domain has dimension 20; the cycle-free subspace guaranteed by Theorem 3 for five candidates is $5! - \frac{4!}{2} = 120 - 12 = 108$, or a five fold increase.

The following theorem illustrates some positive consequences possible from this subspace.

Theorem 4 *(Saari (2001a)) For any number of candidates, if profiles are restricted to the $n! - \frac{(n-1)!}{2}$ dimensional subspace defined in Theorem 3, an admissible rule satisfying Arrow's assumptions Arrow (1951) is the Borda Count. In the same subspace, there exist rules where Sen's Paretian Liberal impossibility result Sen (1970) does not lead to a cycle.*

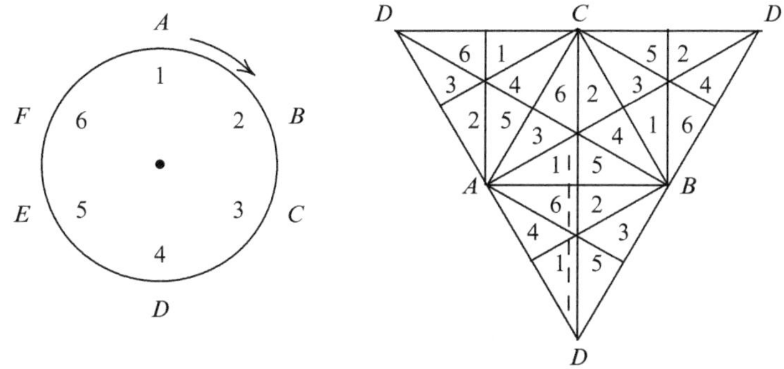

a. Ranking wheel b. Six Condorcet four-tuples

Fig. 8 Profile coordinates to Condorcet Domains

5.1 Coordinates

To find a profile coordinate system for the orthogonal subspace, use what I call a ranking wheel (Saari (2000b, 2001b)), which is a freely rotating wheel attached at its center to a wall. With n candidates, list the numbers from 1 to n in a uniform manner near the wheel's edge. In Fig. 8a, this is illustrated with $n = 6$. Next, select a ranking and list the names of the candidates on the wall next to the appropriate "ranking number." In Fig. 1, the generating ranking is $A \succ B \succ C \succ D \succ E \succ F$.

The first ranking is as given; for Fig. 8a it is the specified $A \succ B \succ C \succ D \succ E \succ F$. Next, rotate the ranking wheel so that the ranking number "1" is positioned next to the second candidate and read off the new ranking. Illustrating with Fig. 8a, the rotated ranking wheel now has "1" next to B, so the new ranking is $B \succ C \succ D \succ E \succ F \succ A$. Continue in this fashion until the ranking number "1" has been next to each candidate precisely once. I call this the "Condorcet n-tuple" generated by the starting ranking. With the Fig. 8a example, the "Condorcet six-tuple generated by $A \succ B \succ C \succ D \succ E \succ F$" is

$$\begin{array}{l} A \succ B \succ C \succ D \succ E \succ F,\ B \succ C \succ D \succ E \succ F \succ A,\ C \succ D \succ E \succ F \succ A \succ B, \\ D \succ E \succ F \succ A \succ B \succ C,\ E \succ F \succ A \succ B \succ C \succ D\ \ F \succ A \succ B \succ C \succ D \succ E \end{array} \tag{3}$$

A Condorcet n-tuple can be generated by any ranking, and each ranking is in precisely one Condorcet n-tuple. There are $n!$ possible rankings, so there are precisely $\frac{n!}{n} = (n-1)!$ Condorcet n-tuples. To illustrate with $n = 4$, the six Condorcet triplets are generated by

Name	Ranking	Name	Ranking
1	$A \succ B \succ C \succ D$	2	$D \succ C \succ B \succ A$
3	$A \succ C \succ B \succ D$	4	$D \succ B \succ C \succ A$
5	$B \succ A \succ C \succ D$	6	$D \succ C \succ A \succ B$

(4)

Each Condorcet four-tuple has four rankings; by using the Eq. (4) assigned names, the positioning of these rankings are located in Fig. 8b. Each face of the tetrahedron has precisely one representative from each Condorcet four-tuple. For n candidates, each of the n faces of the corresponding equilateral object has precisely one representative from each of the Condorcet n-tuples.

On each row of Eq. (4), each ranking is the reverse of the other. The same effect occurs for any n, a Condorcet n-tuple generated by a ranking can be associated with a Condorcet n-tuple generated by the reverse of the original ranking. Indeed, a coordinate direction in profile space is given by one unit of one of these Condorcet n-tuples and -1 units of the other. (To see the role of negative numbers in profiles, see Saari (1999). It just means that when adding such a profile to another profile, subtract voters from the specified rankings.) This defines the $\frac{(n-1)!}{2}$ orthogonal coordinate directions for the Theorem 3 subspace.

With three candidates, place a 1 in each Fig. 2 starred region, and a -1 in each of the bulleted regions. Listing the Fig. 7a profile coordinates in a counterclockwise manner starting from the lower left corner defines the vector $(5,4,4,1,2,6)$ while the Condorcet profile vector is $(1,-1,1,-1,1,-1)$. It now is trivial to show that the two vectors are orthogonal, as required by Theorem 3. However, using Fig. 1a, with the associated vector $(0,x,y,z,w,0)$, it follows that Ward's never-bottom condition satisfies the tally-consistent property if and only if the coordinates satisfy the added restriction $x+z=y+w$. A similar assertion holds for the other two Ward conditions. Namely, profiles associated with Condorcet Domains still have vestiges of the Condorcet n-tuples that the Ward-Sen approach tries to eliminate.

5.2 Condorcet Domains in Condorcet n-tuples

Central to the Ward-Sen condition is that any three rankings from a Condorcet n-tuple creates a cycle. (For an illustrating example, notice that selecting *any* three rankings from the six choices in Eq. (3) creates a cycle.) Consequently, a Condorcet Domain cannot include more than two rankings from any n-tuple, so at least $n-2$ of the rankings from each Condorcet n-tuple must be dropped. Thus a Condorcet Domain can have at most $2(n-1)!$ terms. The actual value is much smaller. The reason is that, as illustrated in Fig. 8b with projections, the rankings of the different Condorcet n-tuples are intimately intertwined. For instance, using a Ward–Sen condition with any triplet drops rankings from *each* of the six Condorcet four-tuples. Thus the choices of what rankings to eliminate from each four-tuple are closely interrelated. As an illustration, the dashed lines shows that associated with the α face and the $A \succ B \succ C \succ D$ region in Fig. 8b are two rankings from the number 1 Condorcet four-tuple and specific number 4 and 6 rankings. If the never-bottom, or never-middle, or never-top condition is used in this α face with $A \succ B \succ C \succ D$, then each choice eliminates at least one ranking from the remaining number 2, 3, and 5 Condorcet four-tuples. Indeed, by using geometry with Fig. 8b and the above conditions, it can be shown that the maximal number of rankings in a four-alternative

Condorcet Domain is less than ten. This interesting connection between the Condorcet four-tuples and the Ward–Sen conditions extends to any $n \geq 3$; i.e., the projection approach captures the weaving interactions needed to eliminate rankings from among the Condorcet n-tuples.

6 Summary

It is interesting how geometry can capture the intricacies of the combinatoric problem of finding and characterizing all Condorcet Domains. As a special case and as shown here, elementary geometry can be used to provide an alternative way to explain Fishburn's alternating scheme. Of course, for the practical issue of understanding and avoiding majority vote cycles, Theorem 3 is stronger and more useful than the overly strict conditions imposed by Condorcet Domains. Nevertheless, the Condorcet Domain problem remains an intriguing question in part because it uncovers valued structures about voting that should be more carefully examined. The projection approach introduced here is a new way to do so. Because this geometric approach identifies all Condorcet Domains for any number of alternatives, it would be interesting to carry out it out for $n \geq 6$; what needs to be done is to determine the calculus conditions for the different Ward–Sen conditions.

Even more, the symmetries disclosed by analyzing this Condorcet Domain issue most surely have other applications in understanding other complex problems that arise in social choice theory. As indicated above, for instance, such symmetries arise when examining positional methods. To explain another benefit of this approach, start with the fact that Fishburn was blessed with an intuitive insight about how to handle the associated and complex combinatorics that are characteristic of this research area. For those of us who are not gifted with such insight, it is important to create a systematic approach to uncover the source of fundamental problems in this area. My sense is that the appropriate tools involve mathematical symmetries, and a way to uncover the appropriate symmetries of social choice is to appeal to the underlying geometry.

Acknowledgements This research was supported by NSF grants DMI 0640817 and DMS 0631362. My thanks to Bernard Monjardet and two anonymous referees for their several useful suggestions.

References

Arrow, K. (1951). *Social choice and individual values*. New York: Wiley.
Black, D. (1948). On the rationale of group decision-making. *Journal of Political Economy, 56*, 23–34.
Condorcet, M. (1785). *Éssai sur l'application de l'analyse à la probabilité des décisions rendues à la pluralité des voix*, Paris.

Fishburn, P. (1997). Acyclic sets of linear orders. *Social Choice and Welfare, 14*, 113–124.
Fishburn, P. (2002). Acyclic sets of linear orders: A progress report. *Social Choice and Welfare, 19*, 431–447.
Gaertner, W. (2001). *Domain conditions in social choice theory*. New York: Cambridge University Press.
Gehrlein, W. (2006). *Condorcet's Paradox*. Berlin: Springer.
Li, L., & Saari, D. G. (2004). *Sen's theorem: Geometric Proof*. New Interpretations, IMBS working paper series. Social Choice Welfare (2008) *31*, 393–413.
Monjardet, B. (2006a). Condorcet domains and distributive lattices. *Workshop on Voting Theory and Preference Modeling*. DIMACS-LAMSADE, Paris, October 26, 2006.
Monjardet, B. (2006b). Condorcet domains and distributive lattices, pp. 285-301, *Annales du LAMSADE*, Université Paris, Daupine.
Monjardet, B. Acyclic domains of linear orders:a survey, in this volume.
Saari, D. G. (1999). Explaining all three-alternative voting outcomes. *Journal of Economic Theory, 87*, 313–355.
Saari, D. G. (2000a). Mathematical structure of voting paradoxes 1: pairwise vote. *Economic Theory, 15*, 1–53.
Saari, D. G. (2000b). Mathematical structure of voting paradoxes 2: positional voting. *Economic Theory, 15*, 55–101.
Saari, D. G. (2001a). *Decisions and elections, explaining the unexpected*. New York: Cambridge University Press.
Saari, D. G. (2001b)., 2001, *Chaotic elections! A mathematician looks at voting*. Providence, RI: American Mathematical Society.
Saari, D. G., & Petron, A. (2006). Negative externalities and Sen's liberalism theorem. *Economic Theory, 28*, 265–281.
Saari, D. G., & Sieberg, K. (2004). Are part wise comparisons reliable? *Research in Engineering Design, 15*, 62–71.
Sen, A. K. (1966). A possibility theorem on majority decisions, *Econometrica, 34*, 491–499.
Sen, A. K. (1970) The impossibility of a Paretian libera. *The Journal of Political Economy, 78*, 152–157.
Ward, B. (1965). Majority voting and the alternative forms of public enterprise. In J. Margolis (Ed.), *The Public Economy of Urban Communities* (pp. 112-126). Baltimore, MD: Johns Hopkins University Press.

Condorcet's Paradox with Three Candidates

William V. Gehrlein

Condorcet formally developed the notion of cyclical majorities over two centuries ago (Condorcet, 1785), and Peter Fishburn introduced me to that phenomenon in 1971. When Peter first described the idea behind Condorcet's Paradox during a course in Social Choice Theory at Pennsylvania State University, my response was that the phenomenon simply could not happen. When he reproduced the classic example of its existence with three voters and three candidates, my immediate response was that this phenomenon certainly could not be very likely to ever be observed in realistic situations. Peter quickly suggested that I should work on developing some estimates of the probability that the paradox might occur, and very soon afterward that pursuit began. We completed many co-authored papers on related topics over the following years, but it is only after more than 30 years of effort that I feel a good answer can be given to the challenge that Peter presented in that classroom in 1971. The following essay can be viewed as a long overdue course project report, and we can finally see a theoretical model that clearly explains why observations of Condorcet's Paradox are so rare in elections on a small number of candidates.

1 Introduction

We consider three-candidate elections in which each voter has a complete and transitive preference ranking on the candidates $\{A,B,C\}$. There are six possible preference rankings that each voter might have on the candidates, as shown in Fig. 1. Here, n_i denotes the number of voters with the corresponding preference ranking. The total

W.V. Gehrlein
Department of Business Administration, University of Delaware, Newark, DE 19716, USA
e-mail: wvg@udel.edu

S.J. Brams et al. (eds.), *The Mathematics of Preference, Choice and Order: Essays in Honor of Peter C. Fishburn*, Studies in Choice and Welfare,

Fig. 1 Possible linear preference rankings on three candidates

A	A	B	C	B	C
B	C	A	A	C	B
C	B	C	B	A	A
n_1	n_2	n_3	n_4	n_5	n_6

number of voters is $n = \sum_{i=1}^{6} n_i$, and any given combination of n_i's is referred to as a profile of voters' preferences for a specified value of n.

A *Pairwise Majority Rule Winner* (*PMRW*) exists for a given profile if some candidate can defeat each of the other two candidates by majority rule voting on the corresponding pairs. For example, A beats B by pairwise majority rule, denoted by $\boldsymbol{AMB}$, if $n_1 + n_2 + n_4 > n_3 + n_5 + n_6$. Then A is the PMRW in a profile if $\boldsymbol{AMB}$ and $\boldsymbol{AMC}$. Condorcet's Paradox occurs when a PMRW does not exist, and there is a majority-rule cycle like $\boldsymbol{AMB}$, $\boldsymbol{BMC}$ and $\boldsymbol{CMA}$. We assume that n is odd throughout this study to avoid having to deal with ties by pairwise majority rule.

When we began studying this phenomenon, few empirical studies had been conducted to attempt to find actual examples of Condorcet's Paradox in real-life situations. In proceeding to develop some basic theoretical models to estimate the probability that a PMRW exists in a random voting scenario, some elementary and predictable assumptions were made regarding the likelihood that various profiles of voters' preferences would be observed. As more and more empirical studies were performed to indicate that occurrences of Condorcet's Paradox are relatively rare with a small number of candidates, the basic theoretical models were modified to try to develop an explanation of what was being observed. The preliminary theoretical models appeared to be treating the procedure as to how voters formed their preferences as being much too random a process.

The general notion in work that followed was that as voters tend to have preferences that are more consistently in mutual agreement with some logical model to explain the process by which their preferences were formed, the probability that a PMRW exists should tend to increase. Stated in an alternative form, the probability that Condorcet's Paradox is observed should decrease as this happens. Many studies, including (Fishburn and Gehrlein, 1980a, b), have been performed to try to establish a relationship between the probability that a PMRW exists and various forms of the internal consistency of voters' preferences within a profile. A survey of these studies in (Gehrlein, 2004) indicated that there is unfortunately only a very weak relationship between the probability that a PMRW exists and most measures of internal consistency of voters' preferences that had been considered to that point. The most promising studies of this type were developed by (Fishburn, 1973) and by (Niemi, 1969). Fishburn (1973) measures the underlying consistency of voters' preferences with Kendall's Coefficient of Concordance and (Niemi, 1969) uses a measure of the proximity of voter preference profiles to the condition of perfectly single-peaked preferences.

2 Probability Representations

The logic of developing a basic representation for the probability that a PMRW exists is quite simple. We simply enumerate the subset of all profiles that have a PMRW and sum their respective probabilities. It is therefore necessary to establish some model of assigning probabilities to profiles. The model that is used in the current study is a variation of *Impartial Anonymous Culture* (*IAC*), which assumes that all combinations of n_i's that sum to a specified n are equally likely to be observed.

Gehrlein and Fishburn (1976) develop a representation for the probability, $P^{PMRW}(n,IAC)$, that a PMRW exists for n voters under IAC as

$$P^{PMRW}(n,IAC) = \frac{15(n+3)^2}{16(n+2)(n+4)}, \text{ for odd} \, n \geq 3 \tag{1}$$

Lepelley (1989) develops a similar representation for even n.

It is well known from (Black, 1958) that a PMRW must exist if voters' preferences are perfectly single-peaked, and (Arrow, 1963) shows that voters' preferences are perfectly single-peaked in three-candidate elections if and only if some candidate is never ranked last in the preference ranking of any voter. The measure proposed by (Niemi, 1969) is related to the minimum number, b, of times that any candidate is bottom ranked by voters in a given profile, with

$$b = Min\{n_5+n_6, n_2+n_4, n_1+n_3\} \tag{2}$$

If b equals zero for a profile, the associated profile is perfectly single-peaked, and profiles become more distant from the condition of perfect single-peakedness as b increases.

Gehrlein (2004) develops a representation to link the probability that a PMRW exists to b, with the expectation that this probability should decrease as b increases. This was done by using algebraic techniques to develop representations for the conditional probability, $P_b^{PMRW}(n,IAC|k)$, that a PMRW exists for n voters, given that b has a specified value k. The basic logic behind the notion of IAC is used here since all profiles with the specified conditional value k are assumed to be equally likely to be observed. The representations for odd n are given by

$$\begin{aligned} & P_b^{PMRW}(n,IAC|k) \\ &= \frac{\{k(-17+21k+11k^2)+(5-26k-4k^2)n+3(2-k)n^2+n^3\}}{(n-3k)\{(n+1)(n+5)-3k(2+k)\}} \\ & \quad \text{for } 0 \leq k \leq (n-1)/4, \\ &= \frac{\{(n+1)(9+2n+n^2)-6(1+n^2)k+18nk^2-18k^3\}}{2(k+1)\{(n+1)(n+5)-3k(2+k)\}} \\ & \quad \text{for } (n+1)/4 \leq k \leq (n-1)/3, \\ &= 3/4, \text{ for } k = n/3. \end{aligned} \tag{3}$$

The critical observation in (Gehrlein, 2004) is that $P_b^{PMRW}(n, IAC|k)$ generally decreases as k increases, in complete agreement with our intuition.

A PMRW must also exist for three-candidate elections if voters' preferences are perfectly single-dipped, and the proximity of a profile to perfect single-dippedness can be measured in a manner similar to that used in the definition of b. A profile will be perfectly single-dipped for three candidates if and only if some candidate is never ranked as most preferred by any voter, and we follow the logic of earlier discussion to define the proximity of a voter preference profile to perfectly single-dipped preferences by the minimum number of times, t, that any candidate is top ranked in voters' preferences, with

$$t = Min\{n_1 + n_2, n_3 + n_5, n_4 + n_6\} \tag{4}$$

We define the conditional probability $P_t^{PMRW}(n, IAC|k)$ following the logic of our definition of $P_b^{PMRW}(n, IAC|k)$, and it follows directly from symmetry arguments that $P_b^{PMRW}(n, IAC|k) = P_t^{PMRW}(n, IAC|k)$ for any k.

A PMRW must also exist with three candidates if some candidate is never middle ranked by any voter. This represents a perfectly polarized preference scenario since some candidate is either most preferred or least preferred by all voters. We measure the proximity of voters' preferences in a profile to perfect polarization with m, where

$$m = Min\{n_3 + n_4, n_1 + n_6, n_2 + n_5\} \tag{5}$$

The algebraic procedures that were used to obtain the representation in (3) are extremely cumbersome to implement, and (Gehrlein, 2005) develops a procedure, called EUPIA2, that makes it much easier to obtain such representations. This procedure is used here to directly obtain a representation for $P_m^{PMRW}(n, IAC|k)$ as

$$\begin{aligned} & P_m^{PMRW}(n, IAC|k) \\ &= \frac{\begin{bmatrix} (k+1)\{-3-169k+333k^2+139k^3+4(14-95k-7k^2)n+18(5-3k)n^2+16n^3\} \\ -3\delta_k^2\{(6k^2+24k-1)+4(k-2)n-2n^2\} \end{bmatrix}}{16(m+1)(n-3m)\{(n+1)(n+5)-3k(2+k)\}} \\ & \text{for } 0 \le k \le (n-1)/4, \\ &= \frac{\begin{bmatrix} (n-3k)\{(52-44k-72k^2+39k^3)+(88+48k-63k^2)n+(20+29k)n^2-n^3\} \\ -3\delta_k^2\{(6k^2+24k-1)+4(k-2)n-2n^2\} \end{bmatrix}}{16(k+1)(n-3k)\{(n+1)(n+5)-3k(2+k)\}} \\ & \text{for } (n+1)/4 \le k \le (n-1)/3, \\ &= \frac{27+42n+7n^2}{8(n+3)^2} \text{ for } k = n/3. \end{aligned} \tag{6}$$

Here, $\delta_k^2 = 1$ if k is an even number, otherwise $\delta_k^2 = 0$.

A critical observation that can now be made from computed values is that $P_m^{PMRW}(n, IAC|k)$ generally decreases as k increases, in complete agreement with our intuition. So, the probability that a PMRW exists tends to increase as voter preference profiles tend to become closer to perfectly single-peaked preferences, perfectly single-dipped preferences or perfectly polarized preferences.

Unfortunately, this observation could be very misleading. In particular, these probabilities remain quite large for relatively large values of k. However, this does not account for the proportions of all possible profiles that they represent. For example, $P_b^{PMRW}(n, IAC|k)$ could be quite large for a relatively wide range of k values, but the results are meaningless if this range of k only accounts for a small proportion of all possible profiles. In order to adequately address this issue, it is necessary to develop representations for the cumulative number of profiles that have specified parameter values in some given range.

3 Cumulative Probabilities

We begin this process by developing representations for the cumulative conditional probabilities $CP_b^{PMRW}(n, IAC|k)$, $CP_t^{PMRW}(n, IAC|k)$ and $CP_m^{PMRW}(n, IAC|k)$, as defined in the obvious fashion. All profiles with a specified parameter value $k*$ with $0 \le k* \le k$ are assumed to be equally likely to be observed in these representations. Using algebraic summations on the equations that led to the numerators and denominators in the representations above in (3) and (6), we find:

For parameters b and t:

$$CP_b^{PMRW}(n, IAC|k) = CP_t^{PMRW}(n, IAC|k)$$

$$= \frac{2\left\{\left(-41+69k+22k^2\right)k+5\left(5-18k-2k^2\right)n+10\left(3-k\right)n^2+5n^3\right\}}{\left\{\left(-73+117k+36k^2\right)k+5\left(10-33k-3k^2\right)n+20\left(3-k\right)n^2+10n^3\right\}}$$

$$\text{for } 0 \le k \le (n-1)/4,$$

$$= \frac{\left[\begin{array}{c} 195-1968k-720k^2+3840k^3+4320k^4+1728k^5 \\ +\left(1661-1680k-6000k^2-5760k^3-2880k^4\right)n+10\left(165+200k+216k^2+192k^3\right)n^2 \\ +30\left(9-8k-24k^2\right)n^3+5\left(15+32k\right)n^4-11n^5 \end{array}\right]}{16\left(k+1\right)\left(k+2\right)\left\{\left(-73+117k+36k^2\right)k+5\left(10-33k-3k^2\right)n+20\left(3-k\right)n^2+10n^3\right\}}$$

$$\text{for } (n+1)/4 \le k \le (n-1)/3,$$

$$= \frac{15\left(n+3\right)^2}{16\left(n+2\right)\left(n+4\right)} \text{ for } k = n/3. \qquad (7)$$

For parameter m:

$$CP_m^{PMRW}(n,IAC|k)$$

$$=\frac{\left[(k+1)\left[\begin{array}{c}165-783k+1743k^2+1597k^3+278k^4+10\left(71-233k-143k^2-7k^3\right)n\\+30\left(31+3k-6k^2\right)n^2+80(k+2)n^3\end{array}\right]\right.\\\left.-15\delta_k^2\left\{11+30k+6k^2-2(3-2k)n-2n^2\right\}\right]}{8(k+1)(k+2)\left\{(-73+117k+36k^2)k+5(10-33k-3k^2)n+20(3-k)n^2+10n^3\right\}}$$

for $0\leq k\leq(n-1)/4$,

$$=\frac{\left[\begin{array}{c}435-952k+480k^2+2200k^3-90k^4-468k^5\\+\left(1349-2520k-4160k^2+840k^3+1140k^4\right)n+10\left(177+120k-162k^2-100k^3\right)n^2\\+10\left(39+72k+32k^2\right)n^3-5(3+4k)n^4+n^5-30\delta_k^2\left\{11+30k+6k^2-2(3-2k)n-2n^2\right\}\end{array}\right]}{16(k+1)(k+2)\left\{(-73+117k+36k^2)k+5(10-33k-3k^2)n+20(3-k)n^2+10n^3\right\}}$$

for $(n+1)/4\leq k\leq(n-1)/3$,

$$=\frac{15(n+3)^2}{16(n+2)(n+4)},\ \text{for } k=n/3. \tag{8}$$

These representations have been verified by computer enumeration, but they are totally intractable for any type of useful analysis in their present form. By considering the limiting form of these representations as $n\to\infty$, they can be significantly simplified. The limiting case does not permit us to consider any specific finite values of k for any of the parameters b, t or m. Instead, we must use the minimum proportion,α_k, of the n voter preference rankings that have the associated parameters b, t or m. Based on the definitions of b, t and m it is obvious that $0\leq\alpha_k\leq 1/3$. To obtain the limiting representations, we replace k with $n\alpha_k$ in the representations above, and then let $n\to\infty$. The resulting limiting representations are:

For parameters b and t:

$$\begin{aligned}&CP_b^{PMRW}(\infty,IAC|\alpha_k)=CP_t^{PMRW}(\infty,IAC|\alpha_k)\\&=\frac{10-20\alpha_k-20\alpha_k^2+44\alpha_k^3}{10-20\alpha_k-15\alpha_k^2+36\alpha_k^3}\ \text{for } 0\leq\alpha_k\leq 1/4,\\&=\frac{-11+160\alpha_k-720\alpha_k^2+1920\alpha_k^3-2880\alpha_k^4+1728\alpha_k^5}{16\alpha_k^2\left(10-20\alpha_k-15\alpha_k^2+36\alpha_k^3\right)}\\&\qquad\text{for } 1/4\leq\alpha_k\leq 1/3.\end{aligned} \tag{9}$$

For parameter m:

$$\begin{aligned}&CP_m^{PMRW}(\infty,IAC|\alpha_k)\\&=\frac{40-90\alpha_k-35\alpha_k^2+139\alpha_k^3}{40-80\alpha_k-60\alpha_k^2+144\alpha_k^3}\ \text{for } 0\leq\alpha_k\leq 1/4,\\&=\frac{1-20\alpha_k+320\alpha_k^2-1000\alpha_k^3+1140\alpha_k^4-468\alpha_k^5}{16\alpha_k^2\left(10-20\alpha_k-15\alpha_k^2+36\alpha_k^3\right)}\\&\qquad\text{for } 1/4\leq\alpha_k\leq 1/3.\end{aligned} \tag{10}$$

These limiting representations are far easier to work with, and they represent the potentially most interesting case of large electorates. Following previous discussion, it is obvious from these limiting representations that

$$CP_b^{PMRW}(\infty, IAC|0) = CP_t^{PMRW}(\infty, IAC|0) = CP_m^{PMRW}(\infty, IAC|0) = 1, \quad (11)$$

and that

$$\begin{aligned} &CP_t^{PMRW}(\infty, IAC|1/3) = CP_b^{PMRW}(\infty, IAC|1/3) = \\ &CP_m^{PMRW}(\infty, IAC|1/3) = 15/16. \end{aligned} \quad (12)$$

4 Proportions of Profiles with Specified Parameters

A representation for the cumulative proportion, $CP_b^{\Pr ofiles}(n, IAC|k)$, of all possible voter preference profiles that have their parameter b equal to k^* in the specified range with $0 \leq k* \leq k$ was also obtained, and the result is given by

$$\begin{aligned} &CP_b^{\Pr ofiles}(n, IAC|k) \\ &= \frac{\left[3(k+1)(k+2)\left\{\begin{array}{c}(-73+117k+36k^2)k+ \\ 5(10-33k-3k^2)n+20(3-k)n^2+10n^3\end{array}\right\}\right]}{(n+1)(n+2)(n+3)(n+4)(n+5)} \\ &\quad \text{for } 0 \leq k \leq (n-1)/3, \\ &= 1 \text{ for } k = n/3. \end{aligned} \quad (13)$$

A representation for the limiting case, $CP_b^{\Pr ofiles}(\infty, IAC|\alpha_k)$, of all possible profiles that have parameter b in the specified range $0 \leq \alpha_{k^*} \leq \alpha_k$ as $n \to \infty$ follows from discussion above. It also follows from earlier work in (Gehrlein, 2004) that $CP_b^{\Pr ofiles}(\infty, IAC|\alpha_k) = CP_t^{\Pr ofiles}(\infty, IAC|\alpha_k) = CP_m^{\Pr ofiles}(\infty, IAC|\alpha_k)$. The resulting representations are given by:

$$\begin{aligned} CP_b^{\Pr ofiles}(\infty, IAC|\alpha_k) &= CP_t^{\Pr ofiles}(\infty, IAC|\alpha_k) \\ &= CP_m^{\Pr ofiles}(\infty, IAC|\alpha_k) \\ = 3\alpha_k^2&\left(10-20\alpha_k-15\alpha_k^2+36\alpha_k^3\right), \text{ for } 0 \leq \alpha_k \leq 1/3. \end{aligned} \quad (14)$$

A search procedure was used on the representation in (14) to obtain the specific β_b^p values of α_k that give $CP_b^{\Pr ofiles}(\infty, IAC|\beta_b^p) = p$ for each proportion $p = 0.00(.05)1.00$, and the results are summarized in Table 1. The results in Table 1 indicate for example that 65% of all possible voter preference profiles are included in the range of α_b parameter values that are within the range $0 \leq \alpha_b \leq .1924$. Based on discussion above, it follows that $\beta_b^p = \beta_t^p = \beta_m^p$ for all p.

It is now possible to use the results that are included in Table 1 along with the limiting representations from (9) to compute the limiting conditional cumulative

Table 1 Computed values of β_b^p, β_t^p, β_m^p, β_u^p, and β_ℓ^p for each proportion $p = 0.00(.05)1.00$

p	$\beta_b^p = \beta_t^p = \beta_m^p$	β_u^p	β_ℓ^p
.00	.0000	.0000	.0000
.05	.0428	.0308	.0256
.10	.0619	.0449	.0375
.15	.0772	.0564	.0473
.20	.0908	.0667	.0562
.25	.1033	.0763	.0646
.30	.1150	.0854	.0727
.35	.1264	.0943	.0806
.40	.1374	.1031	.0885
.45	.1483	.1118	.0965
.50	.1591	.1206	.1046
.55	.1700	.1296	.1130
.60	.1811	.1388	.1217
.65	.1924	.1484	.1308
.70	.2042	.1585	.1407
.75	.2166	.1695	.1514
.80	.2298	.1815	.1634
.85	.2445	.1951	.1774
.90	.2614	.2117	.1946
.95	.2829	.2344	.2191
1.00	.3333	.3333	.3333

probability $CP_b^{PMRW}\left(\infty, IAC|\beta_b^p\right)$ that a PMRW exists for the p percent of profiles that are closest to being perfectly single-peaked. For example, the limiting probability that a PMRW exists for the 65% of all voter preference profiles that are closest to being perfectly single-peaked is obtained by evaluating CP_b^{PMRW} (∞, $IAC|.1924$), given the results in Table 1. In the same fashion, it is also possible to obtain similar conditional probabilities for both $CP_t^{PMRW}\left(\infty, IAC|\beta_t^p\right)$ from (9) and for for $CP_m^{PMRW}\left(\infty, IAC|\beta_m^p\right)$ from (10). Computed results for all three are summarized in Table 2 for each proportion $p = 0.00(.05)1.00$.

The computed values that are given in Table 2 show some very interesting and compelling results. We see for example that the 10% of voter preference profiles that are closest to being perfectly single-peaked have a PMRW with a probability of .9980. An even more important observation is that the 50% of voter preference profiles that are closest to being perfectly single-peaked have a PMRW with a probability of .9857. Thus, the presence of any reasonable degree of internal consistency within voters' preferences that approaches perfectly single-peaked preferences clearly results in a high likelihood that a PMRW will exist. The impact of having voters' preferences that indicate the presence of a candidate approaching a perfectly polarizing candidate is also quite strong, but it is not as dramatic as the presence of some proximity to single-peakedness or single-dippedness since $CP_b^{PMRW}\left(\infty, IAC|\beta_b^p\right) > CP_m^{PMRW}\left(\infty, IAC|\beta_m^p\right)$ for all $0 < p < 1$.

Table 2 Computed Values of $CP_X^{PMRW}(\infty, IAC|\beta_X^p)$, for $\beta_X^p = \beta_b^p, \beta_t^p, \beta_m^p, \beta_u^p, \beta_\ell^p$ for each proportion $p = 0.00(.05)1.00$

p	b, t	m	u	ℓ
.00	1.0000	1.0000	1.0000	1.0000
.05	.9991	.9895	.9995	.9976
.10	.9980	.9850	.9989	.9963
.15	.9969	.9814	.9983	.9951
.20	.9956	.9782	.9975	.9940
.25	.9943	.9753	.9967	.9928
.30	.9929	.9726	.9958	.9916
.35	.9913	.9701	.9948	.9903
.40	.9896	.9676	.9936	.9890
.45	.9877	.9652	.9924	.9876
.50	.9857	.9628	.9910	.9860
.55	.9834	.9605	.9894	.9843
.60	.9809	.9582	.9876	.9825
.65	.9781	.9558	.9856	.9804
.70	.9749	.9535	.9832	.9781
.75	.9712	.9510	.9804	.9753
.80	.9669	.9486	.9770	.9721
.85	.9616	.9460	.9728	.9680
.90	.9548	.9433	.9671	.9628
.95	.9466	.9405	.9583	.9550
1.00	.9375	.9375	.9375	.9375

5 More Dramatic Results with Combinations of *b* and *t*

We have seen that values of parameters b, t or m that reflect any significant degree of proximity, respectively, to single-peakedness, single-dippedness or polarization have a dramatic effect on the probability that a PMRW exists. Even more dramatic results can be observed if various combinations of b, t and m are considered. We begin by considering an overall measure, u, of the presence of a *unifying candidate* where

$$u = Minimum\{b,t\}. \tag{15}$$

If b is a small number relative to n, then some candidate is viewed as the least preferred candidate by very few of the voters, so that particular candidate can be viewed as being *positively unifying* among the electorate. If t is small relative to n, then there is some candidate that is most preferred by very few of the voters. That particular candidate is *negatively unifying* for the electorate in the sense that the voters are generally in agreement in their opposition to having that candidate selected as the winner.

Using the EUPIA2 procedure that is developed in (Gehrlein, 2005), we are able to obtain a representation for $CP_u^{\mathrm{Profiles}}(n, IAC|k)$ as:

$$CP_u^{\mathrm{Pr}ofiles}(n,IAC|k)$$
$$= \frac{6(k+1)(k+2)\{2(15+56k+111k^2+13k^3)-5(2+27k-7k^2)n+10(3-4k)n^2+10n^3\}}{(n+1)(n+2)(n+3)(n+4)(n+5)}$$
$$\text{for } 0 \le k \le (n-1)/4,$$
$$= \frac{3(n-2k)\begin{bmatrix}18(k+1)(13+42k+63k^2+27k^3)-3(35+250k+360k^2+144k^3)n\\ +(25+24k)(5+6k)n^2-3(5+6k)n^3+n^4\end{bmatrix}}{(n+1)(n+2)(n+3)(n+4)(n+5)}$$
$$\text{for } (n+1)/4 \le k \le (n-1)/3,$$
$$= 1 \text{ for } k = n/3. \tag{16}$$

The same logic that was used in previous discussion is then used to obtain the limiting representation for $CP_u^{\mathrm{Pr}ofiles}(\infty,IAC|\alpha_k)$ as $n \to \infty$.

$$\begin{aligned} &CP_u^{\mathrm{Pr}ofiles}(\infty,IAC|\alpha_k)\\ &= 6\alpha_k^2(10-40\alpha_k+35\alpha_k^2+26\alpha_k^3) \text{ for } 0 \le \alpha_k \le 1/4,\\ &= 3(1-2\alpha_k)(1-18\alpha_k+144\alpha_k^2-432\alpha_k^3+486\alpha_k^4)\\ &\text{for } 1/4 \le \alpha_k \le 1/3. \end{aligned} \tag{17}$$

A search procedure was then used with the representation in (17) to obtain the values of β_u^p for which $CP_u^{\mathrm{Pr}ofiles}(\infty,IAC|\beta_u^p) = p$ for each proportion $p = 0.00(.05)1.00$, and the results are summarized in Table 1. As noted above, 65% of all possible profiles are included in the range of α_b parameter values within the range $0 \le \alpha_b \le .1924$. However, 65% of all possible profiles are included in a much smaller range for parameter u, with $0 \le \alpha_u \le .1484$.

Following the logic of earlier discussion, representations for the cumulative conditional probability $CP_u^{PMRW}(n,IAC|k)$ are obtained as

$$CP_u^{PMRW}(n,IAC|k)$$
$$= \frac{\{30+121k+261k^2+38k^3-10(1+15k-3k^2)n+10(3-4k)n^2+10n^3\}}{\{2(15+56k+111k^2+13k^3)-5(2+27k-7k^2)n+10(3-4k)n^2+10n^3\}}$$
$$\text{for } 0 \le k \le (n-1)/4,$$
$$= \frac{\begin{bmatrix}27(25+64k+480k^2+1280k^3+1440k^4+576k^5)\\ +9(101-960k-3840k^2-5760k^3-2880k^4)n+90(29+128k+288k^2+192k^3)n^2\\ -10(85+576k+576k^2)n^3+15(37+64k)n^4-59n^5\end{bmatrix}}{16(n-2u)\begin{bmatrix}18(k+1)(13+42k+63k^2+27k^3)-3(35+250k+360k^2+144k^3)n\\ +(25+24k)(5+6k)n^2-3(5+6k)n^3+n^4\end{bmatrix}}$$
$$\text{for } (n+1)/4 \le k \le (n-1)/3,$$
$$= \frac{15(n+3)^2}{16(n+2)(n+4)} \text{ for } k = n/3. \tag{18}$$

The resulting limiting cumulative conditional probability representations for $CP_u^{PMRW}(\infty, IAC|\alpha_k)$ as $n \to \infty$ are given by:

$$\begin{aligned} CP_u^{PMRW}&(\infty, IAC|\alpha_k) \\ &= \frac{10-40\alpha_k+30\alpha_k^2+38\alpha_k^3}{10-40\alpha_k+35\alpha_k^2+26\alpha_k^3} \text{ for } 0 \le \alpha_k \le 1/4, \\ &= \frac{-59+960\alpha_k-5760\alpha_k^2+17280\alpha_k^3-25920\alpha_k^4+15552\alpha_k^5}{16(1-2\alpha_k)(1-18\alpha_k+144\alpha_k^2-432\alpha_k^3+486\alpha_k^4)} \\ &\qquad \text{for } 1/4 \le \alpha_k \le 1/3. \end{aligned} \tag{19}$$

This representation is used with entries from Table 1 to compute numerical values of $CP_u^{PMRW}(\infty, IAC|\beta_u^p)$ for each $p = 0.00(.05)1.00$, and these values are shown in Table 2. It is clear that by considering the joint measure of voter preference unification, u, there is a much greater impact on the probability that a PMRW exists. The results indicate that the 50% of voter profiles that are most closely related to voter unification have a probability .9910 of having a PMRW and that the 65% of voter profiles that are most closely related to voter unification have a probability .9856 of having a PMRW. Any voter preference profiles that are at all close to representing unified preferences, as measured by u, will clearly have a very high probability of yielding a PMRW.

Ward (1965) defines a condition on profiles that requires the existence of a PMRW for three candidates. This condition requires that voters' preferences do not contain any *Latin Squares*. This is equivalent to the requirement that there is some candidate that is never ranked first, is never ranked last, or is never ranked in the middle by any voter. We define a parameter ℓ to measure the proximity of a profile to Ward's Condition, with

$$\ell = Minimum\{b, t, m\}. \tag{20}$$

If $\ell = 0$ for a profile, that the profile does not contain any Latin Squares. We then obtain a representation for $CP_\ell^{\text{Profiles}}(n, IAC|k)$ by using the EUPIA2 procedure described in (Gehrlein, 2005) as

$$\begin{aligned} &CP_\ell^{\text{Profiles}}(n, IAC|k) \\ &= \frac{9(k+1)(k+2)\{-3k(17+27k+36k^2)+15(2+3k+9k^2)n-60kn^2+10n^3\}}{(n+1)(n+2)(n+3)(n+4)(n+5)} \\ &\text{for } 0 \le k \le (n-1)/3 \\ &= 1 \text{ for } k = n/3. \end{aligned} \tag{21}$$

The limiting distribution $CP_\ell^{\text{Profiles}}(\infty, IAC|\alpha_k)$ as $n \to \infty$ is found to be

$$\begin{aligned} &CP_\ell^{\text{Profiles}}(\infty, IAC|\alpha_k) \\ &= 9\alpha_k^2(10-60\alpha_k+135\alpha_k^2-108\alpha_k^3) \text{ for } 0 \le \alpha_k \le 1/3 \end{aligned} \tag{22}$$

This representation is then used to obtain β_ℓ^p values for each proportion $p = 0.00(.05)1.00$, and the results are listed in Table 1. It was noted above that 65% of all profiles are included in the range for parameter u with $0 \le \alpha_u \le .1484$. Here, 65% of all possible profiles are contained in the range of the parameter ℓ with $0 \le \alpha_\ell \le .1308$. Based on definitions, we must have $\ell \le u$ for every profile so it follows that $\alpha_\ell^p \le \alpha_u^p$ for all p.

Following the logic of previous discussion, we develop a representation for $CP_\ell^{PMRW}(n, IAC|k)$ with the EUPIA2 procedure as

$$CP_\ell^{PMRW}(n,IAC|k)$$

$$= \frac{\left[\begin{array}{c}(k+1)\left[\begin{array}{c}-135-2547k-4293k^2-6687k^3-2538k^4+10\left(153+273k+759k^2+327k^3\right)n\\ -10\left(3+295k+146k^2\right)n^2+240\left(2+k\right)n^3\end{array}\right]\\ +15\delta_k^2\left\{9\left(1+2k+2k^2\right)-6\left(1+2k\right)n+2n^2\right\}\end{array}\right]}{24\left(k+1\right)\left(k+2\right)\left\{-3k\left(17+27k+36k^2\right)+15\left(2+3k+9k^2\right)n-60kn^2+10n^3\right\}}$$

for $0 \le k \le (n-1)/4$,

$$= \frac{\left[\begin{array}{c}27\left(25+96k+440k^2+840k^3+810k^4+324k^5\right)+9\left(69-880k-2520k^2-3240k^3-1620k^4\right)n\\ +30\left(83+252k+486k^2+324k^3\right)n^2-10\left(41+324k+324k^2\right)n^3+15\left(23+36k\right)n^4-31n^5\\ +30\delta_k^2\left\{9\left(1+2k+2k^2\right)-6\left(1+2k\right)n+2n^2\right\}\end{array}\right]}{48\left(k+1\right)\left(k+2\right)\left\{-3k\left(17+27k+36k^2\right)+15\left(2+3k+9k^2\right)n-60kn^2+10n^3\right\}}$$

for $(n+1)/4 \le k \le (n-1)/3$,

$$= \frac{15\left(n+3\right)^2}{16\left(n+2\right)\left(n+4\right)} \text{ for } k = n/3. \tag{23}$$

The limiting distribution $CP_\ell^{PMRW}(\infty, IAC|\alpha_k)$ as $n \to \infty$ is given by

$$\begin{aligned} &CP_\ell^{PMRW}(\infty,IAC|\alpha_k)\\ &= \frac{120-730\alpha_k+1635\alpha_k^2-1269\alpha_k^3}{12\left(10-60\alpha_k+135\alpha_k^2-108\alpha_k^3\right)} \; for\; 0 \le \alpha_k \le 1/4\\ &= \frac{-31+540\alpha_k-3240\alpha_k^2+9720\alpha_k^3-14580\alpha_k^4+8748\alpha_k^5}{48\alpha_k^2\left(10-60\alpha_k+135\alpha_k^2-108\alpha_k^3\right)}\\ &\quad \text{for } 1/4 \le \alpha_k \le 1/3. \end{aligned} \tag{24}$$

This representation is used with entries from Table 1 to compute numerical values of $CP_\ell^{PMRW}(\infty, IAC|\beta_\ell^p)$ for each $p = 0.00(.05)1.00$, and these resulting values are given in Table 2.

We noted above that $CP_b^{PMRW}(\infty, IAC|\beta_b^p) = CP_t^{PMRW}(\infty, IAC|\beta_t^p)$ and that $CP_b^{PMRW}(\infty, IAC|\beta_b^p) > CP_m^{PMRW}(\infty, IAC|\beta_m^p)$ for $0 < p < 1$. The impact of having a polarizing candidate is therefore not as strong as having a positively-unifying candidate or a negatively-unifying candidate. As a result, despite the fact that $\beta_\ell^p \le \beta_u^p$ for all p, we find $CP_\ell^{PMRW}(\infty, IAC|\beta_\ell^p) < CP_u^{PMRW}(\infty, IAC|\beta_u^p)$ for all $0 < p < 1$ in Table 2.

6 Conclusion

When voter preference profiles are at all close to being single-peaked, single-dipped, or completely polarized, the probability that a PMRW exists is quite high. When voters' preferences are at all close to being unified, the probability that a PMRW exists is very high. It must also be noted that the associated underlying models that lead to single-peaked, single-dipped, or completely polarized preferences do not actually have to be the basis of the formation of voters' preference rankings in a profile. We only require that the preferences in a profile could have been obtained by the associated models. As a result, Condorcet's Paradox should rarely be observed in any real three-candidate elections with large electorates, as long as voters' preferences reflect any significant degree of group coherence.

These observations are in general agreement with numerous empirical studies that are summarized in (Gehrlein, 2006). Only a few true examples of Condorcet's Paradox have been observed in results from real elections with large electorates on three candidates, despite many attempts to find them. Riker (1982) presents evidence that some other observations of Condorcet's Paradox have been contrived by politicians through the manipulation of voting situations by various means. Levmore (1999) suggests that such actions would only be taken by political interest groups for general election situations in which they would have the greatest likelihood of success. In our analysis, that would suggest situations in which voters' preferences do not reflect any significant degree of mutual consistency, where parameters b, t and m would have relatively large values.

Acknowledgement I appreciate the input from two anonymous reviewers who read the original draft of this paper.

References

Arrow, K. J. (1963). *Social choice and individual values* (2nd ed.). New Haven CT: Yale University Press.

Black, D. (1958). *The theory of committees and elections*. Cambridge: Cambridge University Press.

Condorcet, M. de. (1785). An essay on the application of probability theory to plurality decision making: An election between three candidates. In F. Sommerlad, & I. McLean (1989, Eds.), *The political theory of Condorcet* (pp. 69–80). Oxford: University of Oxford Working Paper.

Fishburn, P. C. (1973). Voter concordance, simple majority, and group decision methods. *Behavioral Science, 18*, 364–376.

Fishburn, P. C., & Gehrlein, W. V. (1980a). Social homogeneity and Condorcet's paradox. *Public Choice, 35*, 403–420.

Fishburn, P. C., & Gehrlein, W. V. (1980b). The paradox of voting: Effects of individual indifference and intransitivity. *Journal of Public Economics, 14*, 83–94.

Gehrlein, W. V. (2004). Consistency in measures of social homogeneity: A connection with proximity to single peaked preferences. *Quality and Quantity, 38*, 147–171.

Gehrlein, W. V. (2005). Probabilities of election outcomes with two parameters: The relative impact of unifying and polarizing candidates. *Review of Economic Design, 9*, 317–336.

Gehrlein, W. V. (2006). *Condorcet's paradox*. Berlin: Springer.
Gehrlein, W. V., & Fishburn, P. C. (1976). Condorcet's paradox and anonymous preference profiles. *Public Choice, 26*, 1–18.
Lepelley, D. (1989). *Contribution a l'analyse des procedures de decision collective*. Université de Caen, unpublished doctoral dissertation.
Levmore, S. (1999). Voting paradoxes and interest groups. *Journal of Legal Studies, 28*, 259–281.
Niemi, R. G. (1969). Majority decision-making under partial unidimensionality. *American Political Science Review, 63*, 488–497.
Riker, W. H. (1982). *Liberalism against populism: A confrontation between the theory of democracy and the theory of social choice*. New York: Freeman Press.
Ward, B. (1965). Majority voting and alternative forms of public enterprise. In J. Margolis (Ed.), *Public economy of urban communities* (pp. 112–126). Baltimore, MD: Johns Hopkins Press.

On the Probability to Act in the European Union

Marc R. Feix, Dominique Lepelley, Vincent Merlin, and Jean-Louis Rouet

1 Introduction

Since its foundation by Arrow in his seminal contribution (Arrow, 1963), one of the main merit of social choice theory has been to provide a coherent framework for the analysis and comparison of different voting rules. First, many normative requirements about voting rules can be expressed precisely in this framework. Then it is possible to check whether a given voting rule satisfies a given property. Ideally, this type of analysis may lead to the axiomatic characterization of a voting rule. At last the propensity of situations for which a voting rule fails to satisfy a condition can be evaluated.

Peter Fishburn's contributions to this research program have been extremely important. For example, he proposed many new normative conditions for the analysis of voting rules (see in particular Fishburn, 1974, 1977; Fishburn & Brams, 1983), and developed axiomatic analysis for binary voting (Fishburn, 1973) and approval voting (Fishburn, 1978). Together with Gehrlein, he launched an important research program on the probabilistic analysis of voting rules. After Guilbauld's paper (Guilbauld, 1952), the use of probability models in voting was limited to the evaluation of the majority voting paradox under the assumption that each voter would pick his preference independently from the others from a uniform distribution. This assumption, today called the Impartial Culture assumption, puts an equal weight on each profile. Fishburn and Gehrlein developed the use of probabilistic models in two directions. First, to analyze the occurrence of Condorcet cycles, they proposed in Gehrlein and Fishburn (1976) a new probability assumption, the Impartial Anonymous Culture assumption, which assumes that each anonymous profile is equally likely to appear. Secondly, they applied these two probability models to a wider range of problems, the relationships between the scoring rules and the Condorcet

V. Merlin (✉)
CREM, Faculté de Sciences Économiques et de Gestion, UMR6211 CNRS / Université de Caen, 19 rue Claude Bloch, 14032 Caen cedex, France
e-mail: vincent.merlin@unicaen.fr

S.J. Brams et al. (eds.), *The Mathematics of Preference, Choice and Order: Essays in Honor of Peter C. Fishburn*, Studies in Choice and Welfare,

principle being their favorite issue (see Fishburn & Gehrlein, 1976; Gehrlein & Fishburn, 1978a, 1978b). The results we will present in this paper are clearly a continuation of this research program, as we will compare voting rules suggested for the European Union on their propensity to fulfill a given property according to different probability assumptions.

Indeed, in the last 5 years, a considerable body of research on the choice of the best voting rules for federal unions have been inspired by the debates on the Treaty of Nice and the projects for an European Constitution. Without being exhaustive, we can mention the work by Baldwin, Berglof, Giavazzi, and Widgren (2001); Baldwin and Widgren (2004); Barberà and Jackson (2006); Bobay (2001); Beisbart, Bovens, and Hartmann (2005); Feix, Lepelley, Merlin, and Rouet (2007); Felsenthal and Machover (2001); Felsenthal and Machover (2004b); Laruelle and Widgren (1998). All these contributions share a common organization: the authors propose a voting model, including an a priori probabilistic description of the behavior of the voters, and then seek for the voting rule or the constitution that fits better according to some normative criteria.

In particular, a key parameter for the analysis of voting systems is the a priori probability by which a decision is taken against the status quo. This probability is called by Coleman (1971) "the power of a collectivity to act", and the "decision-making efficiency" or the "probability of passage" by Baldwin and Widgren (2004). There is clearly a trade off between a low and a high value of this probability. If the probability to act is too low, the political system may be inefficient in the sense that no decision, even those supported by a large majority of the voters, may ever be approved. On the other hand, when the protection of minority opinion matters, the probability of passage should decently stay below 50%. This criteria can be used to analyze the different decision making procedures of the European Union with 27 members (EU27 hereafter). Currently, the decision scheme of the Council of Ministers is the one described in the Treaty of Nice. First, each country is endowed with a certain number of mandates, ranging from 3 for Malta to 29 for Germany. A proposal must then receive 255 mandates out of 345.[1] It should also pass two extra conditions: it must be approved by a majority of states, gathering at least 62% of the population. Felsenthal and Machover (2001) have shown that there is only a handful of cases out of 227 where the second and third conditions are not met while the first one is satisfied, which justifies the fact that most of the time, the analysis only focuses on the first game. This simplification is no longer possible for the decision scheme outlined in the draft constitution proposed by the European Convention in 2003. The convention suggested that a decision would be adopted if it could be supported by 50% of the states gathering 60% of the total population. The

[1] The Treaty of Nice specified that when all candidate countries have acceded, the blocking minority in a Union of 27 will be raised to 91. Thus, the quota has been lowered to 255 instead of 258, which was first specified elsewhere in the treaty. This strange specification of the treaty explains why the 255 and 258 thresholds have both been studied in the literature. For a detailed analysis of the Treaty of Nice, see Felsenthal and Machover (2001).

constitutional treaty finally proposes a similar procedure: a proposal must receive the support of 55% of the states, representing at least 65% of the population.[2]

Recently a welcome and quite useful discussion between a Swedish diplomat (Axel Moberg) and scientists (Dan Felsenthal and Moshe Machover) has developed (Felsenthal & Machover, 2004a; Moberg, 2002). At the origin, the scientific analysis of the Treaty of Nice (Baldwin et al., 2001; Baldwin & Widgren, 2004; Felsenthal & Machover, 2001) claims that the need of 255 (resp. 258) votes on a total of 345 to approve a proposition at the council of ministers of the European Union will result in a serious deadlock with an a priori probability of passage of 2%. (resp. 1.7%). A. Moberg disagrees strongly, pointing out that the result ignores the "strong consensual culture of the EU". Who is right? In fact, the scientific analysis given in Baldwin et al. (2001) and Felsenthal and Machover (2001) is only a part of the full story: it is based on the use of the Impartial Culture (IC hereafter) model, which states that each country chooses to vote "yes" or "no" independently with equal probability. In other words, each country flips a fair coin to take a decision! But other models to describe the behavior of the voters exist. In particular, the Impartial Anonymous Culture (IAC) model, proposed by Gehrlein and Fishburn (1976) asserts that all the distributions of the votes at the Union level are equally likely.[3]

The aim of this note is to show that the use of a model related to the IAC one is able to give answers which are closer to the reality of the European Union with 27 members and, in some way, takes into account the consensual character of the vote. By departing from the common IC assumption, we obtain a theoretical probability of passing a motion that turns out to be higher. Our result concerns not only the Treaty of Nice with its famous 73.4% majority rule (one key vote), but also the double key vote decision schemes that have been suggested during the debates for the European Constitution. The position that has been defended during Spring 2007 by the Polish Government, i.e., attributing weights proportionally to the square root of the state population and using one key vote, will also be considered.

The paper is organized as follows. In Sect. 2, we present the voting models and the different probability assumptions, and we briefly discuss their adequacy to the vote at the council. In Sect. 3, we give the theoretical probability of approval under the Generalized Impartial Anonymous Culture assumption in the asymptotic limit, i.e., when the number of countries (denoted by N in what follows) goes to infinity. Section 4 checks the relevance of this asymptotic solution for an illustrative example and for EU27, by providing numerical simulations. We present our conclusions in Sect. 5.

[2] When the Council of Ministers is not acting on the basis of a proposal made by the Commission or on the initiative of the Union Minister for Foreign Affairs, this last quota is risen to 72% of the population.

[3] Notice that the widely used Banzhaf power index relies upon the IC probability assumption, which is known as the Independence assumption in the power index literature (Straffin, 1977). For its part, the IAC model can be associated to the Shapley–Shubik power index, and is then called the Homogeneity assumption (Straffin, 1977). The link between the probability models in social choice theory and power indices literature was first emphasized by Berg (1999).

2 The Model

2.1 The Voting Rules

We consider binary issue votes "yes" or "no" for the N states (elsewhere voters, MPs, etc.) of a federal union. The decisions are made by the conjunction of two weighted quota games. Each state has two mandates a_i and b_i, and his (her) vote ("yes" or "no") is used in two qualified majority games $\mathcal{A}$ and $\mathcal{B}$, the respective quotas being Q_A and Q_B. Notice that for each state i, it is the same vote ("yes" or "no") which is used to compute the number of mandates obtained by a motion respectively with keys $\mathcal{A}$ and $\mathcal{B}$. The two quotas must be reached for final approval. In the EU Constitution project, each key is related to a certain type of legitimacy. For country i, it proposes $a_i = 1$ and b_i equal to the population of state i. Let $A = \sum_{i=1}^{N} a_i$ and $B = \sum_{i=1}^{N} b_i$. We will denote the relative quotas by $q_A = Q_A/A$ and $q_B = Q_B/B$.

2.2 The Impartial Culture

In the IC model, each vote is independent of the others and each voter says "yes" or "no" with equal probability $p = 1/2$. IC has serious drawbacks. It describes a vote where everybody is undecided (no exchange of points of view allowing the emergence of a majority has taken place) which leads to the existence of two blocks of equal importance. When we consider one weighted quota game defined by a quota (Q_A) and a vector of weights $(a_i)_{i=1...N}$ with (1) a large number of voters, and (2) no dominant player in term of weight, a natural way to handle the IC case is to notice that the probability that a proposal receives between x and $x + \Delta x$ mandates (with Δx small) can be approximated by a normal distribution (see Feix et al. 2007 for example) with mean $m = \frac{1}{2}\sum_{i=1}^{N} a_i$ and variance $\sigma^2 = \frac{1}{4}\sum_{i=1}^{N} a_i^2$. Then the vote will be won by a margin in term of mandates going as σ as N grows. This explains the low probability of approval with a quota of 255/345, i.e., 73.9% in the Treaty of Nice decision scheme which is characterized by $m = 172.5$ and $\sigma = 39.84$.

2.3 Toward Homogeneity

A natural way to escape from the divided society described by the IC assumption has been, both in the game theory and in social choice literature, to consider that all the partitions with x states in favor of a proposal and $N - x$ against it should be equally likely. Thus, the equiprobability assumption is put on the results of the votes. This leads to the homogeneity assumption and the definition of the Shapley–Shubick index in the power indices literature, and to the so called IAC assumption in social choice theory.[4] For binary vote, a classical interpretation of the IAC model is to state that, before the vote, a probability p of voting for the proposal is drawn from the uniform distribution on $[0,1]$.

The idea is consequently to introduce a model where a probability p different from 1/2 has emerged. Moreover, our knowledge of p is itself of a probabilistic nature, it is mathematically described by the function $f(p)$ which is the density function of p. The emergence of a probability p different from 1/2 seems rather natural in an assembly where certainly long discussions, explanations, compromises, package deals, etc., precede each vote (the "consensual culture" of A. Moberg). Notice that, all these discussions result in a $p \neq 1/2$ and that the subsequent votes are independent. Then the Generalized Impartial Anonymous Culture (GIAC) model is characterized by a given $f(p)$ with $0 \leq p \leq 1$, $f(p) \geq 0$ and $\int_0^1 f(p)\mathrm{d}p = 1$. The function $f(p) = 1$ for all p gives back the IAC model. Strictly speaking, the "impartial" specification implies a symmetric density function $f(p)$ relatively to 0.5 (and consequently $\int_0^1 p f(p)\mathrm{d}p = 1/2$), nevertheless, we will here consider more general cases only constrained by the normalization of $f(p)$.

3 The Probability of Approval in the Asymptotic Limit

Till now, we have discussed the behavior of the voters and the distribution of their votes, without taking into account their number of mandates. Indeed, we need this information in order to evaluate the probability to act of the collectivity.[5]

Proposition 1. *One key-vote case. Let $(a_k)_{k=1}^{\infty}$ be a sequence of mandates, which are strictly positive numbers, chosen once for all.*

Let $p \in [0,1]$ be a fixed constant and $(U_k)_{k=1}^{\infty}$ a sequence of independent random variables distributed uniformly on $[0,1]$. Let X_N be the proportion of mandates brought by the states which, for this election, are in favor of a proposal, with

$$X_N = \frac{\sum_{k=1}^{N} a_k \mathbb{1}_{U_k<p}}{A}, \tag{1}$$

where $A = \sum_{k=1}^{N} a_k$ is the total number of mandates of the N states and $\mathbb{1}_{U_k<p}$ takes the value 1 if $U_k < p$ (the k-state votes in favor of the proposal) and 0 otherwise (the k-state does not vote for the proposal).

[4] The IAC assumption has been introduced in social choice theory by Gehrlein and Fishburn (1976) in order to compute a priori the likelihood of the Condorcet Paradox for three alternatives. Here, there are six possible preference types, and a probability p is now a vector $(p_1, p_2, p_3, p_4, p_5, p_6)$ in the unit simplex, where p_i is the probability of picking preference type i for each voter. The IC assumption is based upon the vector $p = (1/6, 1/6, 1/6, 1/6, 1/6, 1/6)$ while the IAC assumption assumes that p is drawn from a uniform distribution on the unit simplex. For more on the likelihood of the Condorcet paradox and the use of probability models in social choice, see the recent book by Gerhlein (2006).

[5] The number of mandates attributed to each state is also useful when one wants to evaluate their influence by the mean of a power index.

With the hypothesis: $\sum_{k=1}^{\infty} a_k = \infty$ *and* $\sum_{k=1}^{\infty} a_k^2 < \infty$, *or* $\sum_{k=1}^{\infty} a_k^2 = \infty$ *and* $0 < a_k \leq \alpha$ *for any k, where* α *is a constant, in the limit when N goes to infinity, we have for the single-key case*

$$X_N \stackrel{a.s.}{\rightarrow} p\,. \tag{2}$$

That is X_N *tends to* p *with probability 1. The same is true if* p *is random independent of the* (U_k). *If* p *has the density function* f, *we have for any* $q \in [0,1]$ *fixed:*

$$P(X_N < q) \rightarrow \int_0^q f(u)\,\mathrm{d}u\,. \tag{3}$$

Proof. For one election, the sketch of the proof is the following:

- With $\sum_{k=1}^{\infty} a_k^2 < \infty$, but $\sum_{k=1}^{\infty} a_k = \infty$ using martingale arguments it follows that $X_N \rightarrow p$ with probability 1 as N goes to infinity (see Theorem (4.8) p. 220 of Durrett 1991).
- With $\sum_{k=1}^{\infty} a_k^2 = \infty$ and $0 < a_k \leq \alpha$ for any k, where α is a constant, using Theorem (4.9) p. 220 of Durrett (1991), we still have $X_N \rightarrow p$ with probability 1 as N goes to infinity.

But almost surely convergence implies convergence in distribution, so this gives (3). □

Proposition 2. *Two key-vote case. Let* $(a_k)_{k=1}^{\infty}$ *and* $(b_k)_{k=1}^{\infty}$ *be two sequences of strictly positive numbers (mandates of key* $\mathcal{A}$ *and key* $\mathcal{B}$ *respectively) chosen once for all.*

For an election, let $p \in [0,1]$ *be a fixed constant and* $(U_k)_{k=1}^{\infty}$ *a sequence of independent random variables distributed uniformly on* $[0,1]$. *Let* X_N, *resp.* Y_N, *be the proportion of mandates of first key, resp. second key, brought by the states which are in favor of a proposal, with*

$$X_N = \frac{\sum_{k=1}^{N} a_k \mathbb{1}_{U_k<p}}{A}\,, \quad Y_N = \frac{\sum_{k=1}^{N} b_k \mathbb{1}_{U_k<p}}{B}\,, \tag{4}$$

where $A = \sum_{k=1}^{N} a_k$ *and* $B = \sum_{k=1}^{N} b_k$ *are the total number of mandates of first key, resp. second key, of the N states.*

With the same hypothesis on the a_k *and* b_k *as in Proposition 1, we have for the double-key case*

$$(X_N, Y_N) \stackrel{a.s.}{\rightarrow} (p,p)\,. \tag{5}$$

That is X_N *and* Y_N *tends to* p *with probability 1. The same is true if* p *is random independent of the* (U_k). *If* p *has the density function* f, *we have for any* $q \in [0,1]$ *and any* $r \in [0,1]$ *fixed:*

$$P(X_N < q, Y_N < r) \rightarrow \int_0^{\min(q,r)} f(u)\,\mathrm{d}u\,. \tag{6}$$

Proof. The proof is the same as for Proposition 1 for each of the two keys. □

Remarks:

- It is easy to extend these results for a three-key vote or more.
- It could be shown that the propositions remain true if the mandates a_k or b_k are themselves random. One could imagine that the mandates of the states vary from one election to the other, but in this work, the mandates are attributed once for all the elections.

From a practical point of view, Propositions 1 and 2 imply two limits. The first one is clearly given by a number N of states which goes to infinity to reach the limit p given by (2) or (5). Now as the number M of elections is going also to infinity, we will be able to perform the empirical density function of X_N which will converge toward f for the single-key vote case and the empirical density function of (X_N, Y_N) will converge toward $\delta(x = p, y = p)f(p)$ for the second key-vote case. This will be illustrated be the Monte Carlo method presented in the next section.

In the limit where both the number N of states and then the number M of elections go to infinity, the repartition of the proportion of mandates brought by the states in favor of a proposal is given by $f(p)$. For the double-key vote, the points are located on the segment joining $(0,0)$ and $(1,1)$ according to $f(p)$, then, as a consequence, in the case of unequal quota, the highest one will set up the frequency of "yes" votes. In the special case of the IAC model ($f(p) = 1$) and a single-key vote, Proposition 1 means a flat density and for the double-key vote, the points are located uniformly on the segment joining $(0,0)$ and $(1,1)$.

Note that Propositions 1 and 2 hold for N going to infinity. It can be shown that the first correction (N large but not infinite) provides a diffusion around these points of the order of $N^{-1/2}$. While this scattering slightly modifies the flatness of the density distribution of X_N for the one key vote, it transforms the segment of the two key vote into a long ellipse with a ratio long over small axes of the order of $N^{1/2}$. A simulation with 100 states will illustrate these facts in the next section. Now, from an operational point of view, how large should be N? We will tackle this question for the IAC model in the next section. First we will observe the convergence to the limit with an example where the a_i and the b_i will be drawn randomly and independently from a uniform distribution on the segment $[1,5]$. Next, we will study whether the specific distribution of the mandates in the EU27 affects the convergence to the limit.

4 Numerical Simulations Under the IAC Assumption

4.1 An Illustration of Propositions 1 and 2

In this section, the results of numerical simulations will be shown for the IAC case. Because we want to reach the asymptotic limit which supposes both an important number of elections and a large number of states, Monte Carlo method should be used. Actually, it is not possible, when the number of voters is large, to enumerate, stock and compute the 2^N configurations because of lack of memories and computation time. In addition, the Monte Carlo technique will illustrate clearly the double probabilistic character of the IAC model.

The method has two steps. First a probability p is chosen at random from the density function $f(p)$ and a vote configuration is determined according to this probability p: for each of the N voters (or states), a random number is taken in a uniform distribution, if this number is lower than p, the voter gives its mandates (it is a "yes" vote) while he does not if the number is higher. This is in fact an acceptation-rejection method and if the number of voters is large, the number of "yes" voters divided by N will tend toward p. Then, the number of "yes" mandates divided by A is also derived according to (1). Second, this process is repeated for a large number M of elections with, at each election, a choice of a new p into $f(p)$ and so on.[6]

We consider the results for $N = 10$ and $N = 100$ and $M = 50{,}000$ elections, both for a single key and a double key vote using the Monte Carlo technique in the IAC case ($f(p) = 1$). The mandates of the N states have been taken at random from a uniform distribution between 1 and 5, and the sum has been normalized to $A = B = 100$. We use the same set of mandates for the M elections. Notice that the draws of a_i and b_i are independent. As a consequence, it is possible to find a pair of states (i, j) such that $a_i > a_j$ while $b_i < b_j$.

For the single key case, Figs. 1 and 3 show the histogram of the number of configurations, as a function of the related number of mandates. This normalization does not change the ratio 5 between the highest value of the mandates and the smallest one. The histogram becomes flatter as N increases in agreement with Proposition 1 and the probability of approval tends to $(1 - q_A)$.

For the double key case, Figs. 2 and 4 give the results of the M elections in the plane (x, y), one point representing one election. Because all the points have

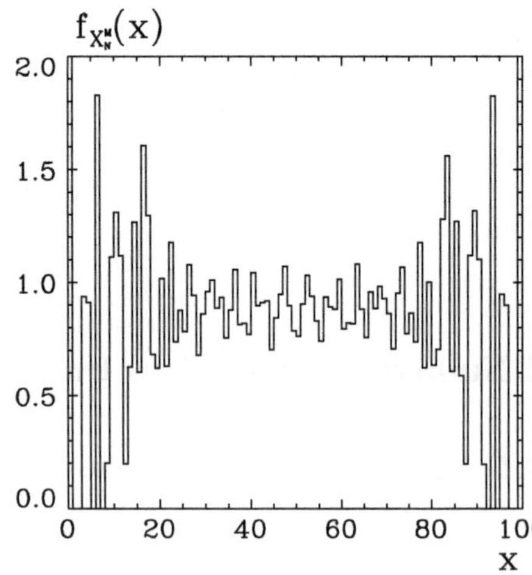

Fig. 1 One key vote. Distribution of the results of the votes for $N = 10$ voters and 50,000 elections using the Monte Carlo technique. The histogram is built with an increment $\Delta x = 1$

[6] Notice that the results of the IC model could also be obtained by this technique. The probability p of the N voters is then equal to $1/2$ which corresponds to $f(p) = \delta(p - 1/2)$.

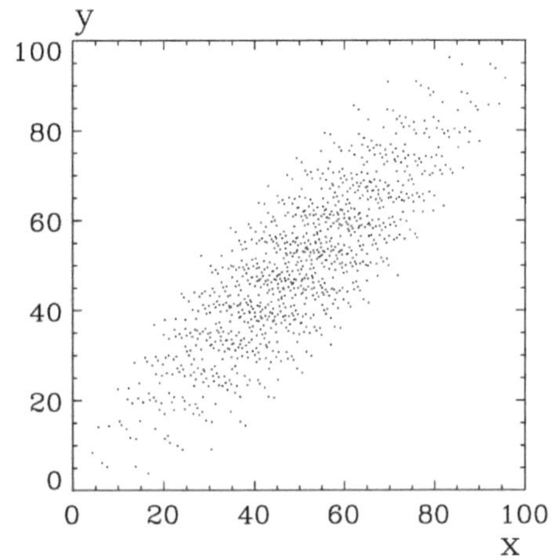

Fig. 2 Double key vote. Distribution of the results of the votes for $N = 10$ voters and 50,000 elections using the Monte Carlo technique. Each point represents the result of an election

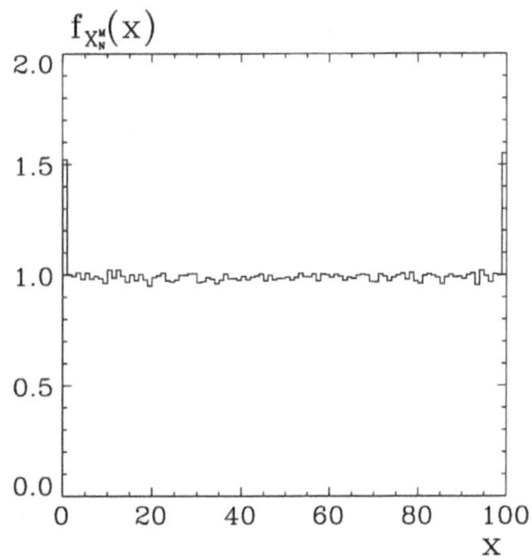

Fig. 3 Same as Fig. 1 but for $N = 100$ voters

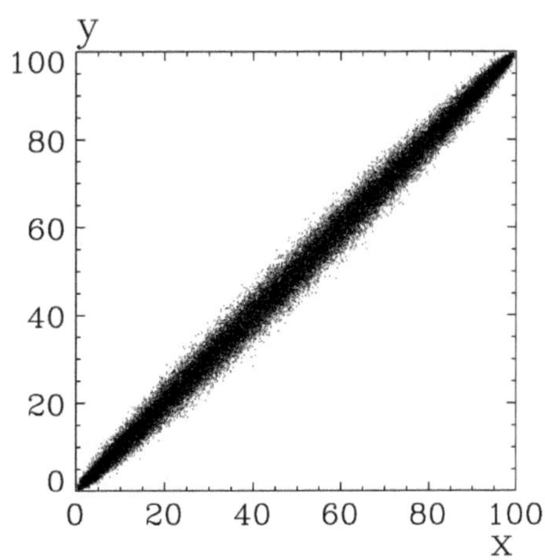

Fig. 4 Same as Fig. 2 but for $N = 100$ voters

Table 1 Double key vote. Probability of passage for the simulation presented Fig. 3 as a function of the two quotas q_A and q_B

		Key $\mathcal{A}$				
		50%	60%	70%	80%	90%
	50%	49.04	40.25	30.43	20.58	10.54
	60%	40.19	39.01	30.43	20.58	10.54
Key $\mathcal{B}$	70%	30.27	30.27	29.31	20.58	10.54
	80%	20.56	20.56	20.56	19.66	10.54
	90%	10.61	10.61	10.61	10.61	9.86

the same weight, their density reaches $\delta(x = p, y = p)f(p)$ in accordance with Proposition 2. As expected, the points are roughly distributed on the segment delimited by the two points $(0,0)$ and (A,B). In addition to this global behavior, the distribution shows a certain scattering, which is less pronounced for $N = 100$. We have checked that, for this case, the probability of approval is closely given by $1 - \max(q_A, q_B)$ as shown Table 1. With $N = 100$, for $q_A = q_B = 70\%$, we get 29.31% of approval and for $q_A = 50\%$ and $q_B = 80\%$, we get 20.56%.

4.2 The Probability to Act for EU27

We have just seen that going to $N = 100$ was already enough to apply Propositions 1 and 2 for an illustrative example where the weights of the states were drawn in the interval $[1,5]$. Now, the question is to know whether or not the asymptotic limit is also a good approximation for the EU27, where the number of states is smaller, and where the mandates, ranging from 3 to 29, are more dispersed.[7] Again, we focus on the IAC case only. It is now possible to enumerate the 2^{27} vote configurations (but taking care of their different weights).

For the single key case, Fig. 5 shows the histogram of the number of configurations as a function of the related number of mandates. The central part of the curve is flat, in accordance with Proposition 1 but we cannot avoid the effect of a finite number of states on the edges. For $Q_A = 255$, the probability of approval is 27.50%, rather close to the result predicted by the asymptotic limit, $(1 - q_A) = (1 - 255/245) = 26.08\%$. Also notice that we are far above the 2% level of approval predicted by the IC model!

Figure 6 shows the histogram when the number of mandates have been taken proportional to the square root of the state populations, according to the voting mechanism that was defended by the Polish government in Spring 2007. This case had been first considered by Sweden in early negotiations for the Nice Treaty as a compromise between the state legitimacy and the citizen legitimacy (see Moberg, 2002). In Spring 2007, the Polish government unearthed the Penrose Square root

[7] The number of mandates and the population data for 2003 can be found in Moberg (2002).

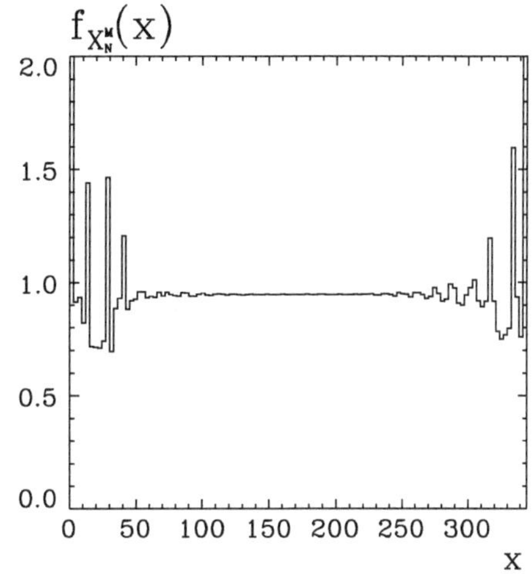

Fig. 5 One key vote. Distribution of the results of the votes for the EU27 for the Treaty of Nice. The histogram is built with a bin size equal to 3

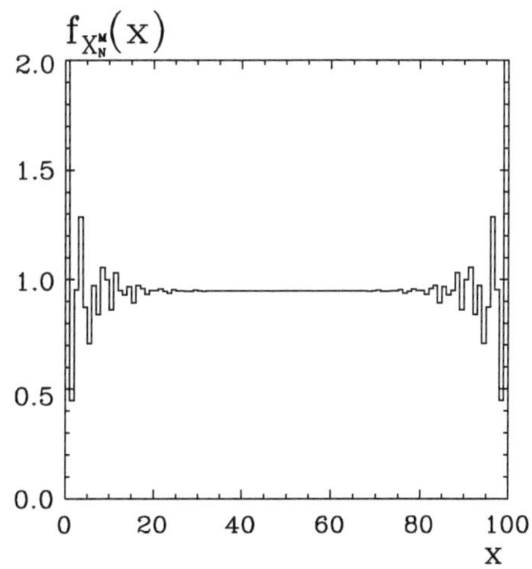

Fig. 6 One key vote. Distribution of the results of the votes for the EU27. The mandates are proportional to the square root of the populations of the states and the sum is normalized to 100

rule (Penrose, 1946, 1952) as a justification of its use. Again, the curve is flat, at least for q between 0.2 and 0.8, indicating that the asymptotic limit could be used for this single key vote.

We turn back to Monte Carlo simulations to analyze the dispersion of the votes for the two keys (although complete enumeration is possible) because each point has the same weight. Then, it is easier to interpret Fig. 7 which gives the distribution of 2,700 vote configurations in the plane (x,y) (one point represents result for an election) for the European Treaty voting procedure. For key $\mathcal{A}$, all the mandates are equal to 1 (state legitimacy) while for key $\mathcal{B}$, the number of mandates of a state is proportional to its population. The sum of the mandates of key $\mathcal{B}$ has been normalized to 100. Because of the discrete nature of the key $\mathcal{A}$ mandates, the points are aligned on vertical lines distant of 1. The scattering of the points, not negligible, is compatible with the $N^{-1/2}$ law as stated before.

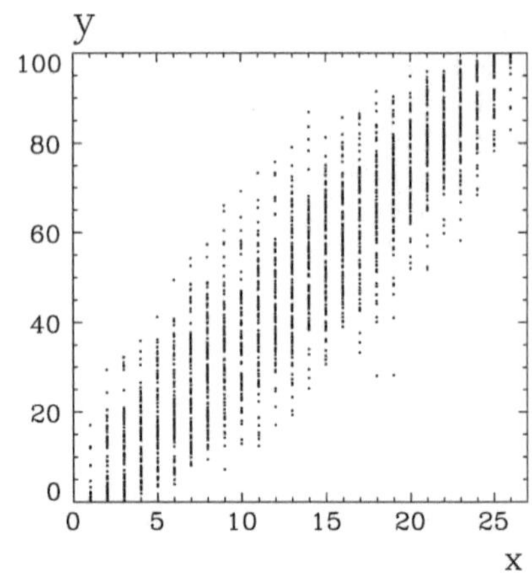

Fig. 7 Double key vote. Distribution of the results of the votes for the EU27. For key $\mathcal{A}$ (x variable) all the mandates are equal to 1, for key $\mathcal{B}$ (y variable) the mandates are proportional to the populations of the states and the sum is normalized to 100

Table 2 Double key vote. Percentage of approval for the EU27 with the constitutional treaty under IAC as a function of the two quotas Q_A and Q_B. The results have been obtained by complete enumeration of all the vote configurations

		Key $\mathcal{A}$					
		14/27 51.85%	15/27 55.56%	17/27 62.96%	19/27 70.37%	22/27 81.48%	25/27 92.59%
	50%	45.42	43.42	38.24	31.91	21.42	10.71
	60%	39.54	38.60	35.57	30.84	21.33	10.71
	65%	35.77	**35.24**	33.24	29.62	21.17	10.71
Key $\mathcal{B}$	70%	31.55	31.29	30.16	27.68	20.67	10.71
	72%	29.84	**29.66**	28.79	26.74	20.37	10.68
	80%	22.80	22.77	22.56	21.86	18.46	10.56
	90%	13.86	13.86	13.86	13.82	12.97	9.10

For this double key case, the probability to act is given for different values of the keys Q_A and Q_B in Table 2 which proves that the rule $1 - \max(Q_A, Q_B)$ for the approval is fairly satisfied. In particular, we observe that for 15 states gathering 65% of the population, the probability of passage is 35.24%, far above the probabilities obtained by Baldwin and Widgren (2004) for different two key decision method under the IC assumption.

5 Conclusion

In most of the applications of statistical models to voting theory, like the studies computing the Condorcet effect probability or evaluating the Condorcet efficiency of scoring rules, it was often found that the IC and IAC models were giving very similar results in terms of the magnitude of the paradoxes. It is with the study of binary votes that the fundamental differences between the two models become apparent.

The issue is of great importance if we remember that the two main power indices (Banhzaf and Shapley–Shubik) are respectively based on IC and IAC assumptions. In particular any recommendation on the number of mandates to attribute to a state in a federal union based upon a power index is contingent on the underlying probability model that we use.

Similarly, in our study on the probability to act, except if we take quotas close to $\frac{1}{2}$, IC and IAC give results which differ by a large factor. While many studies (Baldwin et al., 2001; Baldwin & Widgren, 2004) suggest a fast decline of the probability of passage under IC when the quota rises, the use of the IAC assumption to model a more consensual behavior of the states gives a different picture. We have shown, both for the IC case (see Feix et al. 2007) and for IAC (this paper) that the density function of probability to get x yes votes in the EU27 is already quite close to the results one would obtain in the asymptotic case. Thus, in the asymptotic limit, the probability of passage tends to $(1-q_A)$ for one key vote and $1-\max(q_A, q_B)$ for two key vote under IAC. As a consequence, quotas of 60–70% can still be considered as acceptable with the IAC model, while a similar study done with IC would lead to the opposite conclusion. To some extent, we have shown that the critics of A. Moberg were directed against the IC model but can be easily answered through the use of the IAC model or another GIAC model characterized by the adequate $f(p)$.

Thus, can we decide which model is the more appropriate? At this point, after years of studies of the voting rules with a priori models, to which Peter Fishburn greatly contributed, it is worth noticing that scientists are starting to look at the data or stylized facts. For example, the fact that most of the decisions are taken at the unanimity in the European Union have inspired Laruelle and Valenciano (2007) to design their model of bargaining in committees. In voting theory, a recent study by Gelman, Katz, and Bafumi (2004) gives first insights on the nature of the relevant probability models for two candidates. The chief merit of this study is that it analyzes data from American and European elections. For the US example, they show that margins between republicans and democrats measured in percent do not depend upon the size of the state, a clear contradiction of the IC assumption.[8] This confirms that the search for the adequate $f(p)$ (which must be reasonably stable from one election to the other) is of crucial importance. Similarly, Regenwetter, Grofman, Marley, and Tsetlin (2006) have started to analyze the repartition of the preferences among three or more candidates, and revised the common wisdom on the probability of voting paradoxes. Thus, after a first age, where the a priori assumption played a crucial role, it seems that the probabilistic analysis of voting rules is entering a new age, where the probability model must, in some way, be related to the observed behavior of the voters. Our results are a modest contribution to this approach, as they clearly state that the conclusions on the probability of passage of different decision schemes could be wrongly evaluated if one does not consider the right a priori probability model.

[8] More precisely, using statistical techniques, the authors test different values of n^{α} as a predictor of the difference of votes, n being the number of voters per state. They arrive at $\alpha = 0.9$, but themselves insist that this value must be taken with caution and that a n scale may be correct.

Acknowledgments We gratefully acknowledge the two referees for their comments and remarks. Special thanks are due to Dominique Lepingle for helpful discussions on Propositions 1 and 2.

References

Arrow, K. J. (1963). *Social choice and individual values* (2nd ed.). New York: Wiley.

Baldwin, R., Berglof, E., Giavazzi, F., & Widgren, M. (2001). *Nice try: Should the Treaty of Nice be ratified?* London: CEPR.

Baldwin, R., & Widgren, M. (2004). Winners and losers under various dual majority rules for the EU's Council of Ministers. In M. Wiberg (Ed.) *Reasoned choices* (pp. 42–92). Helsinki: The Finnish Political Science Association.

Barberà, S., & Jackson, M. O. (2006). On the weights of nations: Assigning voting weights in a heterogeneous union. *Journal of Political Economy, 114*, 317–339.

Beisbart, C., Bovens, L., & Hartmann, S. (2005). A utilitarian assessment of alternative decision rules in the council of ministers. *European Union Politics, 6*, 395–418.

Berg, S. (1999). On voting power indices and a class of probability distributions, with applications to EU data. *Group Decision and Negotiation, 8*, 17–31.

Bobay, F. (2001). La réforme du Conseil de l'Union Européenne à partir de la théorie des jeux. *Revue Française d'Économie, 16*, 3–58.

Coleman, J. S. (1971). Control of collectivities and the power of a collectivity to act. In B. Lieberman (Ed.) *Social choice* (pp. 269–300). New York: Gordon and Breach.

Durrett, R. (1991). *Probability: Theory and examples*. Pacific Grove, CA: Wadsworth & Brooks/Cole.

Feix, M. R., Lepelley, D., Merlin, V. R., & Rouet, J.-L. (2007). On the voting power of an alliance and the subsequent power of its members. *Social Choice and Welfare 28*, 181–207.

Felsenthal, D. S., & Machover, M. (2001). The Treaty of Nice and qualified majority voting. *Social Choice and Welfare, 18*, 431–465.

Felsenthal, D. S., & Machover, M. (2004a). The Nice Treaty and voting rules in the council: A reply to Moberg. *Journal of Common Market Studies, 40*, 259–282.

Felsenthal, D. S., & Machover, M. (2004b). An analysis of QM rules in the draft constitution for the Europe proposed by the European convention, 2003. *Social Choice and Welfare, 23*, 1–20.

Fishburn, P. C. (1973), *The theory of social choice*. Princeton: Princeton University Press.

Fishburn, P. C. (1974). Paradoxes of voting. *American Political Science Review, 68*, 537–546.

Fishburn, P. C. (1977). Condorcet social choice functions. *SIAM, Journal of Applied Mathematics, 33*, 469–489.

Fishburn, P. C. (1978). Axioms for approval voting: A direct proof. *Journal of Economic Theory, 19*, 180–185.

Fishburn, P. C., & Brams, S. (1983). Paradoxes of preferential voting. *Mathematics Magazine, 56*, 207–214.

Fishburn, P. C., & Gehrlein, W. V. (1976). Borda's rule, positional voting and condorcet's simple majority principle. *Public Choice, 28*, 79–88.

Gerhlein, W. V. (2006). *Condorcet's Paradox*. Berlin: Springer.

Gehrlein, W. V., & Fishburn, P. C. (1976). Condorcet's paradox and anonymous preference profiles. *Public Choice, 26*, 1–18.

Gehrlein, W. V., & Fishburn, P. C. (1978a). Coincidence probabilities for simple majority and positional voting rules. *Social Science Research, 7*, 272–283.

Gehrlein, W. V., & Fishburn, P. C. (1978b). Probabilities of election outcomes for large electorates. *Journal of Economic Theory, 19*, 38–49.

Gelman, A., Katz, J. N., & Bafumi, J. (2004). Standard voting power indexes don't work: An empirical analysis. *British Journal of Political Science, 34*, 657–674.

Guilbauld, G. T. (1952). Les théories de l'intérêt général et le problème logique de l'agrégation. *Economie Appliquée, 5*, 501–584.

Laruelle, A., & Valenciano, F. (2007). Bargaining in committees as an extension of Nash's bargaining theory. *Journal of Economic Theory, 132*, 291–305.

Laruelle, A., & Widgren, M. (1998). Is the allocation of voting power among the EU states fair? *Public Choice, 94*, 317–339.

Moberg, A. (2002). The Nice Treaty and voting rules in the council. *Journal of Common Market Studies, 40*, 259–282.

Penrose, L. S. (1946). The elementary statistics of majority voting. *Journal of the Royal Statistical Society, 109*, 53–57.

Penrose, L. S. (1952). *On the objective study of crowd behavior*, London: H.K. Lewis.

Regenwetter, M., Grofman, B., Marley, A., & Tsetlin, I. (2006). *Behavioral social choice*. Cambridge: Cambridge University Press.

Straffin, P. D. (1977). Homogeneity, independence and power indices. *Public Choice, 30*, 107–118.

Voting Rules

Voting Systems that Combine Approval and Preference

Steven J. Brams and M. Remzi Sanver

1 Introduction

Social choice theory, while postulating that voters have preferences over candidates, does not ask them to stipulate where, in their preference rankings, they would draw the line between acceptable and unacceptable candidates. Approval voting (AV) does ask voters to draw such a line, but it ignores rankings above and below this line.

Rankings and approval, though related, are fundamentally different kinds of information. They cannot necessarily be derived from one another. Both kinds of information are important in the determination of social choices. We propose a way of combining them in two hybrid voting systems, *preference approval voting* (PAV) and *fallback voting* (FV), that have several desirable properties.

Approving of a subset of candidates is generally not difficult, whereas ranking all candidates on a ballot, especially if the list is long, may be arduous. PAV asks for both kinds of information, whereas FV asks voters to rank only those candidates they approve of, making it simpler than systems that elicit complete rankings.

We describe, analyze, and compare each of these systems in tandem. In Sect. 2 we give definitions and assumptions. In Sect. 3 we describe PAV and analyze which candidates can and cannot win under this system. Although a PAV winner may not be a Condorcet winner or AV winner, PAV satisfies what we call the *strongest-majority principle for voters.* More specifically, if a majority-approved candidate is preferred by a majority to the AV winner and other majority-approved candidates, PAV "corrects" the AV result by electing the majority-preferred candidate. That is, PAV elects the majority-approved candidate who is most preferred.

A majority-preferred candidate is likely to have a more coherent point of view than an AV winner, who may be the most popular candidate because he or she is

---

S.J. Brams (✉)
Department of Politics, New York University, New York, NY 10012, USA
e-mail: steven.brams@nyu.edu

S.J. Brams et al. (eds.), *The Mathematics of Preference, Choice and Order: Essays in Honor of Peter C. Fishburn*, Studies in Choice and Welfare,

bland or inoffensive – a kind of lowest common denominator who tries to appease everybody. However, this problem does not seem to be a common one (Brams & Fishburn, 2005; Brams, 2008, Chap. 1). Sometimes *not* choosing such a candidate when two or more candidates receive majority approval makes PAV *coherence-inducing for candidates* by giving an advantage to candidates who are principled but, nevertheless, command broad support.

In Sect. 4 we describe FV and compare its properties with those of PAV. Like PAV, FV tends to help those candidates who are relatively highly ranked by a majority of voters. Both systems may give different winners from nonranking systems (e.g., plurality voting and AV), ranking systems (e.g., the Borda count and single transferable vote, or STV), and each other.

In Sect. 5 we show that PAV and FV are monotonic in two different senses: Voters, by either approving of a candidate or raising him or her in their rankings, can never hurt and may help this candidate get elected. The latter property (rank-monotonicity) is not satisfied by a number of ranking systems, including STV, whereas the former property (approval-monotonicity) is satisfied by AV.

Like all voting systems, PAV and FV are manipulable. In Sect. 6 we show that voters may induce preferred outcomes either by contracting or by expanding their approval sets. Because each voting system may give outcomes in equilibrium when the other does not, neither system is inherently more stable than the other.

In Sect. 7 we develop a dynamic model of voter responses to polls in 3-candidate elections, wherein voter preferences are either single-peaked or cyclic. If voters respond to successive polls by adjusting their approval strategies to try to prevent their worst choices from winning, they elect the Condorcet winner, though not necessarily in equilibrium, if their preferences are single-peaked. If their preferences are cyclical, the candidate ranked first or second by the most voters wins after voters respond to several polls. These outcomes are in equilibrium under both PAV and FV.

We conclude in Sect. 8 that PAV, and to a less extent FV, subtly interweave two different kinds of information: Approval information determines those candidates who are sufficiently popular to be serious contenders if not outright winners; ranking information enables voters to refine the set of potential winners if more than one candidate receives majority approval.

Together, these two kinds of information facilitate the election of majoritarian candidates with coherent positions. But more than abetting their election, PAV and FV may well have a salutary impact on which candidates choose to run – and how they choose to campaign – encouraging the entry of candidates who appeal to a broad segment of the electorate but do not promise them the moon.

2 Definitions and Assumptions

Consider a set of voters choosing among a set of candidates. We denote individual candidates by small letters a, b, c, ...

We assume that voters strictly rank the candidates from best to worst, so there is no indifference. Thus, for any candidates a and b, either a is preferred to b or b is preferred to a. This assumption simplifies the subsequent analysis but does not in any significant way affect our results, which can readily be extended to the case of nonstrict preferences.

We assume that rankings are transitive, so that for any candidates a, b and c, a is preferred to c whenever a is preferred to b and b is preferred to c. In addition, we assume that a voter evaluates each candidate as either acceptable or unacceptable, which we will refer to as *approved* and *disapproved* candidates.

The *preference-approval* of voters is based on both their rankings and their approval of candidates. Although different, these two types of information exhibit the following consistency: Given two candidates a and b, if a is approved and b is disapproved, then a is ranked above b.

We represent a voter's preference-approval by an ordering of candidates from left to right and a vertical bar, to the left of which candidates are approved and to the right of which candidates are disapproved. For example,

$$ab|cd$$

indicates that the voter's two top-ranked candidates, a and b, are approved, and the voter's two bottom-ranked candidates, c and d, are disapproved.

At one extreme, a voter may approve of all candidates, and at the other extreme of no candidates. As we discuss in Sect. 6, these extreme strategies are dominated strategies in a voting game in which voters have strict preferences, but these strategies are not illegal, as such, under PAV of FV.

Some voters will approve of a single favorite candidate, and some will approve of all except a worst choice. Many voters, however, are likely to select some middle ground, approving of two or three candidates in, say, a field of five (for empirical data on this question under AV, see Brams & Fishburn, 2005; Brams, 2008, Chap. 1).

A *preference-approval profile* is a list of preference-approvals of all voters. A *social-choice rule*, as we use the term here, aggregates preference-approval profiles into social choices. Thereby our framework generalizes the standard social-choice model – wherein a voter is characterized simply by his or her ranking of candidates – to one that adds a line in the ranking separating the voter's approvals from disapprovals.

In subsequent sections, we will use a number of examples to illustrate results as well as prove some propositions. Voters who have the same ranking of candidates will be put into classes, distinguished by Roman numerals I, II, III, ... For simplicity, we assume in the examples that all voters in a class draw the line separating approvals and disapprovals at the same point in their rankings, but none of our results depends on this assumption.

To describe PAV in the next section, we need two definitions. A *Condorcet winner* is a candidate who is preferred by a majority to every other candidate in pairwise comparisons. A *cycle* among 3 or more candidates a, b, c,... occurs if $a < b < c < \ldots < a$, where "$<$" indicates "is preferred by a majority to." (Notice that there can never be an "approval cycle" – approval is strictly ordered from candidates with the most approval to candidates with the least, except, of course, in the case of a tie.) The majority preference relation between any two candidates may lead to a tie if and only if there is an even number of voters, which we assume is broken by random tie-breaking.

3 Preference Approval Voting (PAV)

The winner under PAV is determined by two rules, the second comprising two cases:

1. If no candidate, or exactly one candidate, receives a majority of approval votes, the PAV winner is the AV winner – that is, the candidate who receives the most approval votes.
2. If two or more candidates receive a majority of approval votes, then (i) If one of these candidates is preferred by a majority to every other majority-approved candidate, then he or she is the PAV winner – even if not the AV or Condorcet winner among all candidates. (ii) If there is not one majority-preferred candidate because of a cycle among the majority-approved candidates, then the AV winner among them is the PAV winner – even if not the AV or Condorcet winner among all candidates.

It is rule 2 that distinguishes PAV from AV. It allows for the election of candidates who are not the most approved and, therefore, not AV winners. As we will see, a PAV winner may in fact be the least-approved candidate in a race.

Compared with preference-based voting systems, PAV is somewhat more demanding in the information that it requires of voters. Besides ranking candidates, voters must indicate where they draw the line between acceptable and unacceptable candidates, which is an issue we will return to when we compare the complexity of PAV and FV.

In the remainder of this section, we show what kinds of candidates PAV may and may not elect:

Proposition 1. *A Condorcet winner may not be a PAV winner under rule 1, rule 2(i), and rule 2(ii).*

Proof. **Rule 1.** Consider the following 3-voter, 3-candidate example, in which the voters divide into three preference classes:

Example 1.

I. 1 voter: $ab|c$
II. 1 voter: $b|ac$
III. 1 voter: $c|ab$

Candidate b is the AV winner, approved of by 2 of the 3 voters, whereas candidates a and c are approved of by only 1 voter each. Because candidate b is the only candidate approved of by a majority, b is the PAV winner under rule 1. But it is candidate a, who is preferred to candidates b and c by majorities of 2 votes to 1, that is the Condorcet winner.

Rule 2(i). Consider the following 3-voter, 4-candidate example:

Example 2.

I. 1 voter: $abc|d$
II. 1 voter: $bc|ad$
III. 1 voter: $d|acb$

Candidates b and c tie for AV winner with majorities of 2 votes each. Because candidate b is preferred to candidate c by 2 votes to 1, b is the PAV winner under rule 2(i). But it is candidate a, who is preferred to candidates b, c, and d by majorities of 2 votes to 1 (but who is not majority-approved), that is the Condorcet winner.

Rule 2(ii). Consider the following 5-voter, 5-candidate example:

Example 3.

I. 1 voter: $dabc|e$
II. 1 voter: $dbca|e$
III. 1 voter: $e|dcab$
IV. 1 voter: $abc|de$
V. 1 voter: $c|bade$

Candidates a (3 votes), b (3 votes), and c (4 votes) are all majority-approved and in a cycle as well: $a > b > c > a$. Because the Condorcet winner, candidate d (2 votes), is not majority-approved, he or she cannot be the PAV winner. Instead, the most approved candidate in the cycle, c, is the PAV winner. Q.E.D.

Not only may PAV fail to elect Condorcet winners when they exist, but it may also fail to elect unanimously approved candidates.

Proposition 2. *A unanimously approved AV winner may not be a PAV winner under either rule 2(i) or rule 2(ii).*

Proof. **Rule 2(i).** Consider the following 3-voter, 3-candidate example:

Example 4.

I. 2 voters: $ab|c$
II. 1 voter: $bc|a$

Candidate b is approved of by all 3 voters, whereas candidate a is approved of by 2 voters and candidate c by 1 voter. Nevertheless, candidate a is the PAV winner, because under rule 2(i) he or she is preferred by 2 votes to 1 to the other majority-approved candidate, b.

Rule 2(ii). Consider the following 8-voter, 4-candidate example:

Example 5.

I. 3 voters: $a\,b\,c|d$
II. 3 voters: $d\,a\,c|b$
III. 2 voters: $b\,d\,c|a$

Candidate c is approved of by all 8 voters, whereas candidates a, b, and d are approved of by majorities of either 5 or 6 voters. The latter three candidates are in a top cycle in which $a > b > d > a$; all are preferred by majorities to candidate c, the AV winner. But because candidate a receives more approvals (6) than candidates b and d (5 each), candidate a is the PAV winner under rule 2(ii). Q.E.D.

Proposition 2 shows how a unanimously approved AV winner may be displaced by a less approved majority winner under PAV. In fact, the conflict between AV and PAV winners may be even more extreme.

Proposition 3. *A least-approved candidate may be a PAV winner under rule 2(i).*

Proof. Consider the following 7-voter, 4-candidate example:

Example 6.

I. 2 voters: $a\,c\,b|d$
II. 2 voters: $a\,c\,d|b$
III. 3 voters: $b\,c\,d|a$

Candidate c is approved of by all 7 voters, candidates b and d by 5 voters each, and candidate a by 4 voters. While all candidates receive majority approval, candidate a is the PAV winner, because he or she is preferred by a majority (class I and II voters) to the AV winner (candidate c), as well as candidates b and d, under rule 2(i) Q.E.D.

When the PAV winner and the AV winner differ, as in Example 6, the PAV winner is arguably the more coherent majority choice. Two of the three classes of voters rank candidate a as their top choice in Example 6, whereas candidate c, the AV winner, is not the top choice of any class of voters.

Finally, we show that PAV may give winners different from the two-best known ranking systems (for more information on these and other voting systems, see Brams and Fishburn, 2002).

Proposition 4. *A PAV winner may be different from winners under the Borda count and single transferable vote (STV).*

Proof. If there are n candidates, the Borda count assigns $n - 1$ points to the first choice of a voter, $n - 2$ points to the second choice,..., and 0 points to the last choice; the candidate with the most points wins. In Example 6, candidate c wins with 14 points (2 points each from all 7 voters), whereas the PAV winner, candidate a, receives 12 points (3 points each from 4 voters and 0 points from 3 voters).

Under STV, only first-place votes are counted initially. In Example 5, candidates a, d, and b receive 3, 3, and 2 votes, respectively, from the voters who rank them first.

Because candidate b receives the fewest votes, the votes or his or her supporters are transferred to their second choice, candidate d, giving d a total of 5 votes, which is a majority and makes candidate d the winner. By contrast, candidate a is the PAV winner. Q.E.D.

In summary, we have shown that PAV may not elect Condorcet winners, or winners under AV, the Borda count, or STV. Nevertheless, PAV winners are strong contenders on grounds of both approval and preference, which we will say more about later.

We turn next to a voting system that asks less than PAV, requiring voters to rank only those candidates of whom they approve. It shares some properties of PAV but by no means all.

4 Fallback Voting (FV)

Fallback voting (FV) proceeds as follows:

1. Voters indicate all candidates of whom they approve, who may range from no candidate (which a voter does by abstaining from voting) to all candidates. Voters rank only those candidates of whom they approve.
2. The highest-ranked candidate of all voters is considered. If a majority of voters agree on one highest-ranked candidate, this candidate is the FV winner. The procedure stops, and we call this candidate a level 1 winner.
3. If there is no level 1 winner, the next-highest ranked candidate of all voters is considered. If a majority of voters agree on one candidate as *either* their highest or their next-highest ranked candidate, this candidate is the FV winner. If more than one candidate receives majority approval, then the candidate with the largest majority is the FV winner. The procedure stops, and we call this candidate a level 2 winner.
4. If there is no level 2 winner, the voters descend – one level at a time – to lower and lower ranks of *approved* candidates, stopping when, for the first time, one or more candidates are approved of by a majority of voters, or no more candidates are ranked. If exactly one candidate receives majority approval, this candidate is the FV winner. If more than one candidate receives majority approval, then the candidate with the largest majority is the FV winner. If the descent reaches the lowest rank of all voters and no candidate is approved of by a majority of voters, the candidate with the most approval is the FV winner.

The appellation "fallback" comes from the fact that FV successively falls back on lower-ranked approved candidates if no higher-ranked approved candidate receives majority approval. This nomenclature was first used in Brams & Kilgour (2001), but it was applied to bargaining rather than voting, in which the decision rule was assumed to be unanimity (the assent of all parties was necessary) rather than a simple majority.

Brams & Kilgour (2001), in what they called "fallback bargaining with impasse," did not require that the bargainers rank all alternatives. Rather, the bargainers ranked only those they considered better than "impasse," because impasse was preferable to any alternative ranked lower. Bargainers not ranking alternatives below impasse are analogous to voters not approving of candidates below a certain level, whom they do not rank.

Like FV, the "majoritarian compromise" proposed by Sertel and his colleagues (Sertel & Yilmaz, 1999; Sertel & Yilmaz, 1999; Hurwicz & Sertel, 1999) elects the first candidate approved of by a majority in the descent process. However, voters are assumed to rank all candidates – they do not stop their ranking at some point at which they consider candidates they rank lower unacceptable.

James W. Bucklin assumed, as we do with FV, that if a voter did not rank all candidates, he or she disapproved of those not ranked.[1] Thus, when the fallback process descends to a level at which a voter no longer ranks candidates, that voter is assumed to approve of no additional candidates should the process continue to descend for other voters because no candidate has yet reached majority approval. Bucklin's system is FV absent the designation of approved candidates, who are implicitly assumed to be only those candidates that voters rank.

In the analysis of FV that follows, we assume that voters have preferences for all candidates, though they reveal their rankings only for approved candidates. As we will see, the non-revealed information may lead to the election of different candidates from PAV. First, however, we indicate properties that FV shares with PAV.

Proposition 5. *Condorcet winners and unanimous AV winners may not be FV winners, whereas least-approved candidates may be FV winners.*

Proof. In Example 1, there is no level 1 winner. Because candidate *b* is the only candidate approved of by a majority (voters II and III) at level 2, *b* is the FV winner, whereas candidate *a* is the Condorcet winner.

In Example 4, candidate *a* is the FV winner at level 1, but candidate *b* is the unanimous AV winner. In Example 6, candidate *a* is the FV winner at level 1, but *a* is the least approved of the four candidates. Q.E.D.

While FV and PAV share the properties listed in Proposition 5, FV, unlike PAV, may fail to elect a majority-preferred candidate among the majority-approved candidates.

Proposition 6. *Suppose there are two or more majority-approved candidates. If one is majority-preferred among them, FV may not elect him or her.*

Proof. Consider the following 5-voter, 4-candidate example:

---

[1] Bucklin, a lawyer and founder of Grand Junction, Colorado, proposed his system for Grand Junction in the early twentieth century, where it was used from 1909 to 1922 – as well as in other cities – but it is no longer used today. See Hoag & Hallet (1926, pp. 485–491), http://www.gjhistory.org/cat/main.htm, http://en.wikipedia.org/wiki/Bucklin_voting, and http://wiki.electorama.com/wiki/ER-Bucklin.

Example 7.

I. 2 voters: $ab|cd$
II. 1 voter: $dca|b$
III. 2 voters: $ca|bd$

There is no level 1 majority-approved candidate with at least 3 votes. Because candidate a receives more approval (4 votes) than candidate c (3 votes) at level 2, a is the FV winner. But candidate c is majority-preferred to candidate a by 3 votes to 2. Q.E.D.

In fact, candidate c is the Condorcet winner among *all* candidates, defeating candidates b and d as well. PAV, because of rule 2(i), picks candidate c, even though candidate a is more approved at level 2 and is unanimously approved at level 3 (to which FV never descends).

A similar conflict between FV and PAV may occur when there is no Condorcet winner.

Proposition 7. *A unanimously approved candidate in a cycle may not be the FV winner.*

Proof. Consider the following 9-voter, 4-candidate example:

Example 8.

I. 2 voters: $abc|d$
II. 3 voters: $bdc|a$
III. 4 voters: $ca|db$

There is a cycle whereby $a > b > c > a$. Candidate c is the only candidate approved of by all 9 voters and so would be the PAV winner under rule 2(ii). Under FV, no candidate is majority-approved at level 1, but at level 2 candidate a receives 6 votes and candidate b receives 5 votes, making a the FV winner. Q.E.D.

Proposition 8. *FV, PAV, and AV may all give different winners for the same preference-approval profile.*

Proof. Consider the following 9-voter, 4-candidate example:

Example 9.

I. 4 voters: $abc|d$
II. 3 voters: $bc|ad$
III. 2 voters: $dac|b$

There is no level 1 majority-approved candidate, but candidates a and b each receive majority approval (6 and 7 votes, respectively) at level 2. Because candidate b (7 votes) is more approved of than candidate a (6 votes), FV elects candidate b. But candidate c is unanimously approved (9 votes) – at level 3 for the class I and III voters (to which FV never descends) – so AV elects candidate c. Finally, PAV elects candidate a, who is majority-preferred to the two other majority-approved candidates, b and c. Q.E.D.

Note in Example 9 that no class of voters ranks the unanimously approved AV winner (candidate c) first, so he or she is likely to be only a lukewarm choice of everybody. Neither FV nor PAV favors such candidates if there are majority-approved candidates ranked higher by the voters.

In Examples 7, 8, and 9, one can determine from the rankings of the approved candidates that candidate a is majority-preferred to candidate b. Thus in Example 9, even though the class II voters do not indicate that they prefer candidate b to candidate a when they rank their two approved candidates, b and c, the fact these voters do not approve of candidate a implies that candidate b, whom they do approve of, is ranked higher than candidate a. Similarly, one can ascertain from the ranking of the class III voters that they prefer candidate a to candidate b.

That PAV would have given a different outcome from FV may not always be revealed.

Proposition 9. *Information used to determine an FV winner may not reveal that PAV would have chosen a different winner.*

Proof. Consider the following 3-voter, 4-candidate example:

Example 10.

I. 1 voter: $abc|d$
II. 1 voter: $bda|c$
III. 1 voter: $c|abd$

There is no level 1 majority-approved candidate, but at level 2 candidate b receives majority approval (2 votes) and is, therefore, the FV winner. Because the class III voter does not rank candidates below candidate c under FV, it would not be known whether candidate a would defeat candidate b, or vice versa, in a pairwise contest between these two candidates (while candidate a is preferred by the class I voter, candidate b is preferred by the class II voter, leaving the contest undecided). But under PAV, wherein voters rank all candidates, the fact that the class III voter prefers a to b would not only be revealed but also would render candidate a the winner, because a is majority-preferred to b.[2] Q.E.D.

That FV ignores information on the lower-level preferences of voters is one reason why it gives different outcomes from PAV. Although we think information on nonapproved candidates should not be ignored, we recognize that it sometimes may be difficult for voters to provide it.

[2] To be sure, if the class III voter did not rank any candidates below candidate c, the outcome under PAV would, as under FV, be a tie between candidates a and b. While voters would be encouraged to rank all candidates under PAV, we do not think their ballots should be invalidated if they do not do so.

5 Monotonicity of PAV and FV

Such well-known voting systems as STV, also called "instant runoff voting" (IRV), do not satisfy a property called "monotonicity." This renders them vulnerable to what Brams and Fishburn (2002, p. 215) call "ranking paradoxes." As an example of such a paradox, a voter may, by ranking a candidate first, cause him or her to lose, whereas this voter, by ranking the candidate last, enable him or her to win – just the opposite effect of what one would expect a top ranking to have.

Because PAV and FV are hybrid voting systems, it is useful to define two kinds of monotonicity.

1. A voting system is *approval-monotonic* if a class of voters, by approving of a new candidate – without changing their approval of other candidates – never hurts and may help this candidate get elected.
2. A voting system is *rank-monotonic* if a class of voters, by raising a candidate in their ranking – without changing their ranking of other candidates – never hurts and may help this candidate get elected.

A *monotonicity paradox* occurs when a voting system is not approval-monotonic or rank-monotonic; violations of rank-monotonicity have been investigated by Fishburn (1982), among others.

Proposition 10. *PAV and FV are approval-monotonic.*

Proof. Consider PAV. Under rule 1, a class of voters, by approving of a candidate, helps him or her become the unique AV, and therefore the PAV, winner. Under rule 2(i), a class of voters, by approving of a candidate, helps him or her become one of the majority-approved candidates and, therefore, a possible PAV winner. Under rule 2(ii), a class of voters, by approving of a candidate, helps him or her become the AV, and therefore the PAV, winner among the majority-approved candidates in a cycle. Consider FV. Approving of a candidate allows him or her to be ranked and receive votes in the descent, thereby helping him or her become the FV winner. Q.E.D.

Proposition 11. *PAV and FV are rank-monotonic.*

Proof. Consider PAV. Under rule 1, ranks have no effect. Under rule 2(i), a class of voters, by raising a candidate in their ranking, helps that candidate defeat other majority-approved candidates in pairwise contests and thereby become the PAV winner. Under rule 2(ii), a class of voters, by raising a candidate in their ranking, helps that candidate be a member of the cycle – if there is no majority-preferred candidate among the majority-approved candidates – and thereby become a possible PAV winner. Consider FV. A class of voters, by raising a candidate in their ranking, helps that candidate become majority-approved at an earlier level, or receive the largest majority if two or more candidates are majority-approved at the same level, and thereby become the FV winner. Q.E.D.

Thus, a class of voters can rest assured that giving either approval or a higher ranking to a candidate can never hurt and may help him or her get elected under PAV and FV.

However, this may lead to the defeat of an already approved candidate that one prefers, which is illustrated by the following 7-voter, 4-candidate example:

Example 11.

I. 1 voter: $ab|cd$
II. 3 voters: $b|acd$
III. 2 voters: $ca|bd$
IV. 1 voter: $d|abc$

Under PAV, candidate b is the only candidate to be majority-approved (4 votes) and so is the PAV winner under rule 1.

But now assume that the 3 class II voters approve of candidate a as well as candidate b:

II$'$. 3 voters: $ba|cd$

Candidate a receives 5 votes and candidate b 4 votes, so both are majority-approved. But because candidate a is majority-preferred to candidate b by 4 votes to 3, candidate a is the PAV winner under rule 2(i), contrary to the interests of the class II voters who switched from strategy b to strategy *ba*.

Similarly, for the original approval strategies of the voters in Example 11, candidate b is the FV winner, picking up 4 votes at level 2. But when the class II voters switch from strategy b to strategy *ba*, candidate a wins with 5 votes at level 2. As under PAV, the strategy shift by the class II voters is detrimental to their interests.

In Sect. 7, we will show how information from polls may affect voters' calculations about how many candidates to approve of under PAV, and to approve of and rank under FV. As we will see, these calculations may or may not result in equilibrium outcomes.

The stability of outcomes under PAV and under FV reflects their robustness against manipulation, so it is important to assess its extent. Stability may be looked at in either static or dynamic terms. In Sect. 6 we view it statically – when will voters be motivated to try or not try to upset an outcome? – whereas in Sect. 7 we analyze how unstable outcomes, based on a dynamic poll model, evolve over time.

6 Nash Equilibria Under PAV and FV

Because PAV and FV give the same outcome as AV when either no candidate or one candidate receives the approval of a majority, they share many of the properties of AV. For example, in a field in which at most one candidate is likely to obtain majority approval, PAV and FV, like AV, give candidates an incentive to broaden their appeal to try to maximize their level of approval.

When candidates reach out to try to attract more votes, voters are likely to consider them acceptable and approve of more than one candidate. But if more than one candidate actually receives majority approval, the preferences of voters under

PAV and FV matter, so the most-approved candidate may not win, as we showed earlier. Thus, a key question that both PAV and FV raise is how many candidates a voter should approve of if he or she deems more than one acceptable. As we showed in Sect. 5, sometimes voting for additional candidates may sabotage the election of a preferred candidate.

In the analysis that follows, we assume that voters, in order to try to elect their preferred candidates, choose strategically where to draw the line between approved and disapproved candidates. But we assume that they are truthful in their rankings of candidates, which is equivalent to assuming that they choose from among their admissible and sincere AV strategies.[3]

An AV strategy *S* is *admissible* if it is not dominated in a game-theoretic sense – that is, there is no other strategy that in all contingencies leads to at least as good an outcome and in some contingency a better outcome. Admissible strategies under AV involve always approving of a most-preferred candidate and never approving of a least-preferred candidate (Brams & Fishburn, 1978, 1983, 2007).

An AV strategy *S* is *sincere* if, given the lowest-ranked candidate that a voter considers acceptable, he or she also approves of all candidates ranked higher. Thus, if *S* is sincere, there are no "holes" in a voter's approval set: Everybody ranked above a voter's lowest-ranked, but acceptable, candidate is also approved; and everybody ranked below this candidate is not approved.[4]

As we will illustrate shortly, voters may have multiple sincere strategies, which some analysts consider desirable but which others consider problematic; this clash has sparked considerable controversy about AV.[5] Given the multiplicity of sincere strategies, we are led to ask what, if any, strategies are stable under PAV and FV.

We define an outcome to be *in equilibrium* if the approval strategies of each preference class of voters that produce it constitute a Nash equilibrium. At such an equilibrium, no class of voters has an incentive to depart unilaterally from its approval strategy, because it would induce no better an outcome, and possibly a worse one, by doing so.

---

[3] In Sect. 7 we consider the possibility that voters may change their rankings as well as their approval in order to try to manipulate outcomes. For an excellent study of the manipulability of voting systems that focuses on manipulation through the misrepresentation of rankings, see Taylor (2005).

[4] Admissible strategies may be insincere if there are four or more candidates. For example, if there are exactly four candidates, it may be admissible for a voter to approve of a first and third choice without also approving of a second choice (see Brams & Fishburn 1983, 2007, pp. 25–26, for an example). However, the circumstances under which this happens are sufficiently rare and nonintuitive that we henceforth suppose that voters choose only sincere approval strategies under PAV and FV. Sincere strategies are always admissible if we exclude "vote for everybody," which we henceforth do.

[5] Saari & Van Newenhizen (1988) provoked an exchange with Brams, Fishburn, and Merrill (1988) over whether the plethora of AV outcomes that different sincere strategies may produce more reflected AV's "indeterminacy" (Saari and Van Newenhizen) or its "responsiveness" (Brams, Fishhburn, and Merrill); other critiques of AV are referenced in Brams & Fishburn (2005; Brams, 2008, Chap. 1). We view PAV and FV as ways to make AV more responsive to voter preferences.

Proposition 12. *Truth-telling strategies of voters under PAV and FV may not be in equilibrium. In particular, voters may induce a better outcome either by contracting or expanding their approval sets.*

Proof. We first prove this proposition for PAV using the following 7-voter, 4-candidate example:

Example 12.

I. 3 voters: $ab|cd$
II. 2 voters: $c|abd$
III. 2 voters: $db|ac$

Candidate *b*, approved of by 5 voters, is the only candidate approved of by a majority and so is the PAV winner.

To show the possible effects of contraction, assume that the 3 class I voters *contract* their approval set from strategy *ab* to strategy *a*:

I′. 3 voters: $a|bcd$

Then candidate *a*, who is preferred by the class I voters to candidate *b*, will win under PAV rule 1, receiving 3 votes to 2 votes each for candidates *b*, *c*, and *d*.

To show the possible effects of expansion in the original Example 12, assume the 2 class II voters *expand* their approval set from strategy *c* to strategy *ca*:

II′. 2 voters: $ca|bd$

Then candidates *a* and *b* tie with 5 votes each (candidates *c* and *d* each receive 2 votes). Because candidates *a* and *b* both receive majority approval, we apply PAV rule 2(i). Since candidate *a* is preferred to candidate *b* by a majority of 5 votes to 2, candidate *a*, whom the class II voters prefer to candidate *b*, is the winner.

Thereby both the contraction and the expansion of an approval set by a class of voters may induce a preferred outcome, rendering PAV strategies in Example 12 not in equilibrium. It is easy to show that the same contraction and expansion of approval sets induces preferred outcomes under FV (candidate *a* instead of candidate *b* in the case of contraction I′; a tie between candidates *a* and *b* in the case of expansion II′). Q.E.D.

We showed earlier that PAV, FV, and AV may lead to three different outcomes for the same preference-approval profile (Proposition 8). The fact that an outcome is in equilibrium under one system, however, does not imply that it is in equilibrium under another system.

Proposition 13. *When PAV and FV give different outcomes, one may be in equilibrium and the other not.*

Proof. In Example 9, we showed that candidate *a* (the Condorcet winner) wins under PAV and candidate *b* wins under FV. Candidate *a* is in equilibrium under PAV, because none of the three classes of voters, by switching to a different approval strategy, can induce an outcome they prefer to candidate *a*. On the other hand, candidate *b* is not in equilibrium under FV, because the 4 class I voters, by switching from strategy *abc* to *a*, can induce the election of candidate *a*, whom they prefer to candidate *b*. This example shows that PAV may give an equilibrium outcome when FV does not.

To show that FV may give an equilibrium outcome when PAV does not, consider the following example:

Example 13.

I. 1 voter: $ab|cd$
II. 1 voter: $ca|db$
III. 1 voter: $c|bad$
IV. 1 voter: $db|ac$
V. 1 voter: $db|ca$

Candidate *b* is the only candidate approved of by a majority of 3 voters. No voter, by switching to a different approval strategy under FV, can induce a preferred outcome to candidate *b* at level 2, making candidate *b* an equilibrium outcome. Candidate *b*, being the sole majority-approved candidate, is also the winner under PAV. But voter II, by switching from strategy *ca* to *cad*, can render both candidates *d* and *b* majority-approved (3 votes each). Since *d* is preferred to *b* by a majority of 3 voters, including voter II, voter II would have an incentive to induce this tied outcome under PAV, showing that FV may give an equilibrium outcome when PAV does not. Q.E.D.

The fact that equilibria under PAV do not imply equilibria under FV, or vice versa, indicates that one system is not inherently more stable than the other.[6]

7 The Effects of Polls in 3-Candidate Elections

In elections for major public office in the United States and other democracies, voters are not in the dark. Polls provide them with information about the relative standing of candidates and may also pinpoint their appeal, or lack thereof, to voters.

In this section, we focus on 3-candidate elections, because they are the simplest example in which information about the relative standing of candidates can affect the strategic choices of voters. Also, such elections are relatively common. We will show how voter responses to a sequence of polls may dynamically change outcomes under PAV and FV.[7]

To assess the effects of polls in 3-candidate elections, we make the following assumptions:

1. *No majority winner.* None of the three candidates, *a*, *b*, or *c*, is the top choice of a majority of voters.

---

[6] AV yields candidates *c* in Example 9, and candidate *b* in Example 13 – but neither in equilibrium – showing that equilibria under PAV and FV are not always the same as under AV. Merrill & Nagel (1987) suggest that outcomes under multistage systems like PAV and FV may be more manipulable than outcomes under single-stage systems like AV, but the manipulation of PAV and FV are computationally more demanding and, consequently, probably more impracticable.

[7] The effects of polls under plurality voting and AV were analyzed in Brams (1982) and Brams & Fishburn (1983, 2007, chap. 7) using a different dynamic model; see also Meirowitz (2004) and citations therein.

2. *Initial support of only top choice.* Before the poll, each voter approves of only his or her top choice.
3. *Poll information.* The poll indicates the relative standing of the candidates. For example, the ordering $n_a > n_b > n_c$ indicates that candidate a receives the most approval votes, candidate b the next most, and candidate c the fewest (for simplicity, we do not allow for ties).
4. *Strategy shifts.* After the results of the poll are announced, voters may shift strategies by approving of a second choice as well as a top choice. Voters will vote for their two top choices if and only if the poll indicates (i) the about-to-become-winner is their worst choice and (ii) they can prevent this outcome by approving of a second choice, too, given they did not previously approve of this choice.
5. *Repeated responses.* After voters respond to a poll, they respond to new information that is revealed in subsequent polls, as described in assumption 4 above.
6. *Termination.* Voters cease their strategy shifts when they cannot induce a preferred outcome.

We assume that voters truthfully rank the three candidates at the start and do not change these rankings in response to the initial poll or any subsequent poll. We next investigate what outcomes occur in response to polls under PAV for two different kinds of preferences.

1. *Single-peaked preferences.* Voters perceive the candidates to be arrayed along a left-right continuum, with candidate a on the left, candidate b in the middle, and candidate c on the right. Each voter most prefers one of these candidates, next most prefers an adjacent candidate, and least prefers the candidate farthest from his or her most-preferred candidate, who may or may not be adjacent.

More specifically, a-voters on the left with preference ranking abc may switch from strategy a to strategy ab, whereas c-voters on the right with preference ranking cba may switch from strategy c to strategy cb. The b-voters in the middle split into two groups: one group prefers candidate a over candidate $c(bac)$, and the other group prefers candidate c over candidate $a(bca)$. The former group may switch from strategy b to strategy ba, whereas the latter group may switch from strategy b to strategy bc.

Because no candidate is the first choice of a majority and preferences are single-peaked, the candidate preferred by the median voter, b, is the unique Condorcet winner – he or she is preferred by a majority to both candidate a and candidate c. We show in Table 1 the three qualitatively different poll rankings that the initial poll may give:

$$\text{(i) } n_a > n_b > n_c; \text{ (ii) } n_a > n_c > n_b; \text{ (iii) } n_b > n_a > n_c,$$

where n_i indicates the number of approval voters of candidate i. If the roles of candidates a and c are reversed, there are three analogous rankings, which we do not show in Table 1:

$$\text{(iv) } n_c > n_b > n_a; \text{ (v) } n_c > n_a > n_b; \text{ (vi) } n_b > n_c > n_a.$$

Table 1 Strategy switches of voters in response to a poll under PAV and FV: single-peaked preferences with three poll rankings (b Condorcet winner)

Poll ranking	(i) $n_a > n_b > n_c$	(ii) $n_a > n_c > n_b$	(iii) $n_b > n_a > n_c$
Initial strategies	$a\|bc$	$a\|bc$	$a\|bc$
	$b\|ac$	$b\|ac$	$b\|ac$
	$b\|ca$	$b\|ca$	$b\|ca$
	$c\|ba$	$c\|ba$	$c\|ba$
Outcome	a	a	b
Shift in strategies (if any)	$a\|bc$	$a\|bc$	
after initial poll	$b\|ac$	$b\|ac$	
	$bc\|a$	$bc\|a$	
	$cb\|a$	$cb\|a$	
Outcome	b	b	

For poll ranking (i) in Table 1, the voters with preference rankings *bca* and *cba* will switch from strategies *b* and *c*, respectively, to strategies *bc* and *cb* to try to prevent their worst choice, candidate *a*, from winning (assumption 4). This results in the election of candidate *b*, whether candidate *b* is the unique majority-approved candidate – with approval from three classes of voters – or candidate *c* also wins a majority – with approval from two classes of voters – in which case candidate *b* will defeat candidate *c* in a pairwise contest. Because no voters can effect a preferred outcome under PAV through any subsequent shifts in their strategies – in response to a poll that shows candidate *b* to be the unique or largest-majority winner – no voters will have an incentive to make further shifts.

The same shifts will occur for poll ranking (ii), again boosting candidate *b* to winning status. As for poll ranking (iii), no voters will have an incentive to shift in response to the initial poll, because the plurality winner, candidate *b*, is not the worst choice of any voters.

Under FV, candidate *b* will also prevail. In the case of poll rankings (i) and (ii), this occurs because candidate *b* is the unique or largest-majority winner after the shift. In the case of poll ranking (iii), candidate *b* is the initial plurality winner, after which the descent of voters ceases because no voter ranks *b* last.

In summary, whichever of the three qualitatively different poll rankings occurs when voter preferences are single-peaked, the responses of voters to an initial poll leads to the election of Condorcet winner *b* under both PAV and FV. But when preferences are cyclical and there is no Condorcet winner, the evolution of a winner is more drawn out, requiring up to three shifts rather than just one.

2. *Cyclical preferences.* We consider the simplest case of cyclical preferences, wherein three classes of voters, none with a majority of votes initially, have preferences *abc*, *bca*, and *cab*, so $a > b > c > a$. For simplicity, we exclude voters with preferences that do not contribute to the cyclic component of these voters (e.g., *acb*).

If, as assumed earlier, voters initially approve of only their top choices, there are two qualitatively different poll rankings that the initial poll may give:

$$\text{(i) } n_a > n_b > n_c; \text{ (ii) } n_a > n_c > n_b.$$

Table 2 Strategy switches of voters in response to a poll under PAV and FV: cyclic preferences with two poll rankings

Poll ranking	(i) $n_a > n_b > n_c$	(ii) $n_a > n_c > n_b$
Initial strategies	$a\|bc$	$a\|bc$
	$b\|ca$	$b\|ca$
	$c\|ab$	$c\|ab$
Outcome	a	a
Shift I (Initial Poll)	$a\|bc$	$a\|bc$
	$bc\|a$	$bc\|a$
	$c\|ab$	$c\|ab$
Outcome	c	c
Shift II (2nd Poll)	$ab\|c$	$ab\|c$
	$bc\|a$	$bc\|a$
	$c\|ab$	$c\|ab$
Outcome	b	b
Shift III (3rd Poll)		$ab\|c$
		$bc\|a$
		$ca\|b$
Outcome		a

The four other possible rankings are analogous, with candidate *b* ranked first in two cases and candidate *c* ranked first in the other two:

$$\text{(iii) } n_b > n_a > n_c;\ \text{(iv) } n_b > n_c > n_a;\ \text{(v) } n_c > n_a > n_b;\ \text{(vi) } n_c > n_b > n_a.$$

In Table 2, we show the strategy shifts that voters will make in response to poll rankings (i) and (ii). After an initial poll that shows candidate *a* to be in first place in each case, there will be one shift by the *bca* voters (Shift I) – and up to two additional shifts (Shift II and Shift III) in response to subsequent polls that show other candidates to be in first place – as voters try to prevent their worst choice from winning.

To illustrate for poll ranking (i), the *bca* voters will switch from strategy *b* to strategy *bc* in Shift I to try to prevent candidate *a* from winning with a plurality of votes. But when this shift leads to candidate *c*'s receiving a majority of votes, the *abc* voters will switch from strategy *a* to strategy *ab* in Shift II, giving candidates *b* and *c* each a majority.

Under PAV, candidate *b* will be majority-preferred to candidate *c* in the contest between these two majority-approved candidates after Shift II. Under FV, candidate *b*, with approval from both *abc* and *bca* voters at level 2, will receive a larger majority than candidate *c* – based on the initial poll ranking – with approval from *bca* and *cab* voters.

At this stage, even if the *cab* voters switched from strategy *c* to strategy *ca*, they could not induce the election of candidate *a*, who will get a smaller majority than candidate *b*, based on the initial poll ranking. Hence, the shifts will terminate after shift II, resulting in the election of candidate *b*, the candidate with more first and second-place approval than any other candidate.

For poll ranking (ii), three shifts are required to induce the election of candidate *a*. In the absence of a Condorcet winner, the most approved candidate in the cycle – when all voters support their two top candidates – emerges as the winner under PAV and FV.

In summary, when preferences are cyclical, the candidate who is ranked first or second by the most voters prevails after three shifts under both PAV and FV. Together with our results on single-peaked preferences, we have the following:

Proposition 14. *In the poll model for 3-candidate elections under PAV and FV, strategy shifts result in the election of (1) the Condorcet winner if preferences are single-peaked and (2) the candidate ranked first or second by the most voters if preferences are cyclical.*

These outcomes, however, may not be stable.

Proposition 15. *In the poll model for 3-candidate elections under PAV and FV, strategy shifts may result in outcomes that are not in equilibrium when there is a Condorcet winner.*

Proof. Assume that voter preferences are single-peaked (Table 1), and consider poll ranking (ii) after the shift. Assume that the *bca* and *cba* voters constitute a majority. Then the *cba* voters, by switching from strategy *cb* to strategy *c* (a contraction), will induce the election of candidate *c*, whom they prefer to candidate *b*. As the sole majority-approved candidate, candidate *c* wins under both PAV and FV, rendering candidate *b* not in equilibrium. Q.E.D.

Surprisingly, it is not the cyclical preferences of voters (in Table 2) that produce instability but the single-peaked preferences of voters (in Table 1) for poll ranking (ii) – and poll ranking (i) as well if the *bca* and *cba* voters constitute a majority in this situation – that produce instability. Thus, the strategy shifts of voters in response to polls, while leading to the outcomes indicated in Proposition 14, may not terminate at these outcomes because of the possible nonequilibrium status of candidate *b* for poll rankings (i) and (ii) in Table 1.

This is not to say that the Condorcet winner (in Table 1), candidate *b*, cannot be supported as a Nash equilibrium in this situation. It turns out that the "critical strategy profile" of candidate *b*,

$$ab|c; b|ac; b|ca; cb|a,$$

which maximizes *b*'s approval vis-à-vis the other candidates, supports *b* as a *strong* Nash equilibrium – no coalition of voter classes, by choosing different approval strategies, can induce an outcome they prefer to candidate *b*. Not only is it impossible for a coalition to replace *b* with a preferred candidate under PAV and FV, but this is also true of AV. In fact, under AV, candidates are strong Nash equilibria at their critical strategy profiles if and only if they are Condorcet winners (Brams & Sanver, 2006; Brams, 2008, Chap. 2).

We have assumed up until now that while voters may changes their levels of approval in order to try to induce preferred outcomes, they are steadfast in their rankings of candidates, which we assumed are truthful. But what if they can falsify their rankings? Then the candidates will be more vulnerable. But falsifying rankings, especially if information is incomplete, is a risky strategy that many voters are likely to shun.[8]

8 Conclusions

It is worth emphasizing that PAV and FV duplicate AV when at most one candidate receives a majority of approval votes. In such a situation, there seems good reason to elect the AV winner, because if there is a different Condorcet winner, he or she would not be majority-approved. If the AV winner also is not majority-approved, his or her election seems even more compelling, because this is the most acceptable candidate in a field in which nobody is approved of by a majority.

When two or more candidates are majority-approved, PAV and FV may elect different winners from AV, the Borda count, STV, and each other. PAV chooses the majority-preferred candidate, if there is one, among those who are majority-approved, whereas FV chooses the first candidate to receive a unique or largest majority in the descent.[9]

If there is no majority-preferred candidate among the majority-approved candidates, PAV chooses the most approved candidate in the cycle. FV does the same if this candidate is in the first set of candidates to receive majority approval in the descent; if not, a majority-approved candidate with less approval – but received earlier – will be the FV winner. PAV and FV winners, if different from the AV winner, are likely to have more coherent majoritarian positions, not just be the lukewarm choices of most voters.

Candidates with coherent positions are more likely to run if they believe, without egregious pandering, that they can win. Consequently, PAV and FV may well encourage candidates to enter the fray who might otherwise be deterred because they are unwilling to sacrifice their fundamental tenets in order to win.

PAV and FV afford voters the opportunity to approve of lower-ranked candidates without necessarily helping them to win. Unlike AV, in which voting for a less-preferred candidate can cause the displacement of a more-preferred candidate, PAV and FV impede this event, though they do not rule it out entirely.

---

[8] AV, of course, does not permit such falsification since voters do not rank candidates. While AV leads to the same outcomes as PAV and FV in the poll model, it may give very different outcomes in other situations, as we showed earlier.

[9] Majority approval may be too high a bar to impose if the field of candidates is large. This bar has been lowered in some plurality elections in the United States, wherein a candidate can win outright if he or she obtains at least 40% of the vote; otherwise, there is a runoff election between the two highest vote-getters. Our view is open about the amount of approval (1) that two or more candidates must receive in order that rule 2 take effect under PAV or (2) that one candidate must receive for the descent to stop under FV. Perhaps a simple majority should not be the *sine qua non*. A lower threshold may be appropriate in elections in which at most one candidate is likely to receive majority approval and, therefore, the winner will always be the AV winner, obviating the need for PAV and FV.

PAV, for example, takes into account voter preferences, which can override the greater approval a less-preferred candidate receives. Both PAV and FV are are approval-monotonic and rank-monotonic, so approving of a candidate or ranking him or her higher never hurts, and may help, this candidate to get elected.

PAV is more information demanding than FV, which asks voters to rank only their approved candidates. Without complete information on preference rankings, FV is less able to ensure the election of a majority-preferred – or the most approved if there is no majority-preferred – candidate among the majority-approved candidates.

PAV and FV may elect different candidates in equilibrium if voters contract or expand their approval sets; neither system is inherently more stable than the other. In the 3-candidate dynamic poll model, Condorcet winners are elected after one shift when voter preferences are single-peaked – though not always in equilibrium – whereas candidates ranked first or second by the most voters are equilibrium choices after several shifts when voter preferences are cyclic.

By combining information on approval and preferences, PAV and FV may yield outcomes that neither kind of information, by itself, produces. Although PAV is more likely to lead to majority-preferred winners among the majority-approved, its greater information demands of voters may make FV a better practical choice. Such trade-offs require careful consideration, as do other ways of mixing approval and preferences to coax better social choices out of a voting system.[10]

Finally, it is worth mentioning a situation in which PAV was recently adopted by the New York University politics department because of a failure, at least initially, of plurality voting (PV) to choose a candidate for a faculty position. Two candidates, A and B, were vying for that position, with almost two-thirds of department members favoring one or the other.

But the department split almost evenly over which candidate members preferred. Because the more than one-third who favored neither candidate won under PV, it seemed that neither candidate would be hired, though a substantial majority preferred either A or B over no hire. In the end, however, the majority prevailed in a second vote over hiring one or the other, with a third vote showing which one of the two candidates was preferred.

Under PAV, there would have been three options: Hire A, hire B, or hire neither (the position did not have to be filled). The nearly two-thirds who favored either A or B over no hire presumably would have approved of both, at which point their preferences for either A or B would have elected one of the two candidates (except in the case of a tie).

---

[10] Ossipoff & Smith (2005) survey a number of such voting methods, several of which disqualify candidates if another candidate is ranked over them on more than half the ballots. Thus, if there is a Condorcet winner, this candidate will disqualify all others and will, therefore, be elected, independent of how approved he or she is. In our view, a Condorcet winner who receives less than majority approval – as we showed can happen under very different circumstances in Examples 1, 2, and 3 – should *not* be elected when there are other candidates who receive majority approval. Both PAV and FV give precedence to majority-approved candidates over Condorcet winners when there is a conflict. But *among* majority-approved candidates, Condorcet winners take precedence under PAV.

Note that AV might not have succeeded, because some of the A and B supporters might have approved of only their favorite, which could have prevented either from winning. But under PAV, there is no good strategic reason for A and B supporters not to approve of both, knowing that their preferences will determine a winner between the two if both are majority-approved. Thus PAV mitigates, if not prevents, certain kinds of strategizing to which AV may be vulnerable, including what Nagel (2006, 2007) calls the "Burr dilemma."[11]

Acknowledgement This chapter appears in slightly different form in Brams (2008, chap. 3). We thank an anonymous referee for helpful suggestions.

References

Brams, S. J. (1982). Strategic information and voting behavior. *Society, 19*(6), 4–11.

Brams, S. J. (2008). *Mathematics and democracy: Designing better voting and fair-division procedures*. Princeton, NJ: Princeton University Press.

Brams, S. J., & Fishburn, P. C. (1978). Approval voting. *American Political Science Review, 72*(3), 831–857.

Brams, S. J., & Fishburn, P. C. (1983, 2007). *Approval voting*. Cambridge, MA: Birkhäuser Boston 2nd ed.; New York: Springer.

Brams, S. J., & Fishburn, P. C. (2002). Voting procedures. In K. Arrow, A. Sen, and K. Suzumura (Eds.), *Handbook of social choice and welfare*, vol. 1 (pp. 175–326). Amsterdam: Elsevier Science.

Brams, S. J., & Fishburn, P. C. (2005). Going from theory to practice: The mixed success of approval voting. *Social Choice and Welfare*, 25(2), 457–474.

Brams, S. J., Fishburn, P. C., & Merrill, S. III. (1988). The responsiveness of approval voting: Comments on Saari and Van Newenhizen and rejoinder to Saari and Van Newenhizen. *Public Choice*, 59, 121–131, 149.

Brams, S. J., & Kilgour, D. M. (2001). Fallback bargaining. *Group Decision and Negotiation* 10(4), 287–316

Brams, S. J., & Sanver, M. R. (2006). Critical strategies under approval voting: Who gets ruled in and ruled out. *Electoral Studies*, 25, 287–305.

Fishburn, P. C. (1982). Monotonicity paradoxes in the theory of elections. *Discrete Applied Mathematics*, 4(2), 119–134.

Hoag, C. G., & Hallet, G. H., Jr. (1926). *Proportional representation*. New York: Macmillan.

[11] This dilemma occurred in the U.S. presidential election of 1800, in which Republicans Thomas Jefferson and Aaron Burr tied in electoral votes when Republicans voted for both of them. (AV was used at that time: Electors could vote for more than one candidate; the candidate with the most votes became president, and the candidate with the second-most votes became vice-president.) The tie in the Electoral College sent the election to the House of Representatives, wherein each state had one vote. After 35 ballots cast over six days, Jefferson finally won. But if PAV had been used in the Electoral College, Republicans could, as they did, approve of both Jefferson and Burr with impunity and thereby defeat the opposition candidates. In addition, by being able to express their preferences for either Jefferson or Burr, the electors would have elected their preferred candidate as president and their non-preferred candidate as vice-president, except in the unlikely event of a tie.

Hurwicz, L., & Sertel M. R. (1999). Designing mechanisms, in particular for electoral systems: The majoritarian compromise. In M. R. Sertel (Ed.), *Economic Design and Behaviour*. London: Macmillan.

Meirowitz, A. (2004). Polling games and information revelation in the Downsian framework. *Games and Economic Behavior*, 51, 464–489.

Merrill, S., III., & Nagel, J. H. (1987). The effect of approval balloting on strategic voting under alternative decision rules. *American Political Science Review*, 81(2), 509–524.

Nagel, J. (2006). A strategic problem in approval voting. In B. Simeone, & F. Pukelsheim (Eds.), *Mathematics and democracy: Recent advances in voting systems and collective choice* (pp. 133–150). Berlin: Springer.

Nagel, J. H. (2007). The Burr dilemma in approval voting. *Journal of Politics*, 69(1), 43–58.

Ossipoff, M., & Smith, W. D. (2005). *Survey of voting methods that avoid favorite-betrayal.* Preprint available at http://rangevoting.org/FBCsurvey.html.

Saari, D. G., & Van Newenhizen J. (1988). The problem of indeterminacy in approval, multiple, and truncated voting systems and Is approval voting an unmitigated evil?: A response to Brams, Fishburn, and Merrill. *Public Choice*, 59, 101–120, 133–147.

Sertel, M., & Sanver, M. R. (1999). Designing public choice mechanisms. In I. Limam (Ed.), *Institutional reform and development in the MENA region* (pp. 129–148). Cairo, Egypt: Arab Planning Institute.

Sertel, M., & Yilmaz, B. (1999). The majoritarian compromise is majoritarian-optimal and subgame-perfect implementable. *Social Choice and Welfare,* 16(4), 615–627.

Taylor, A. D. (2005). *Social choice and the mathematics of manipulation*. New York: Cambridge University Press.

Anonymous Voting Rules with Abstention: Weighted Voting

William S. Zwicker

1 Introduction

We consider legislative voting rules that govern collective approval or disapproval of a bill or a motion, and that allow abstention (or absence) as a "middle option" distinct from a *yes* or *no* vote. In contrast with Peter Fishburn's work on *representative systems*, or *RS*s, we do not treat collective approval and disapproval symmetrically; a voting rule may have a built-in bias against passing motions, for example. In this asymmetric case, the additional assumption that a rule is anonymous (all votes count equally) still allows for a significant variety of rules, a number of which are used by real voting bodies (see Freixas & Zwicker (2003)). We provide three characterizations of weighted voting in this context, and discuss potential applications.

In real legislative voting bodies an abstention or absence often does have an effect different from a voter's *yes* or *no* vote. Yet since the publication of *Theory of Games and Economic Behavior* von Neumann & Morgenstern (1949) the standard mathematical model for a legislative voting system has been the *simple game*, which by virtue of its structure treats any non-*yes* vote as a *no*. Peter Fishburns 1973 work seems to be the earliest to have taken abstention seriously, but others followed: Rubenstein (1980); Bolger (1986, 1993a,b); Felsenthal & Machover (1997, 1998); Amer et al. (1998); Freixas & Zwicker (2003); Côrte-Real & Pereira (2004); Dougherty & Edward (2004); Bilbao, Fernández, Jiménez, & López (2005a,b).

Distinguishing features of a RS, as defined in Fishburn (1973), include:

- It is constant-sum (treats outcomes symmetrically): if each vote is flipped (from *no* to *yes*, *yes* to *no*, and *abstain* to *abstain*), the outcome is flipped,[1]
- It admits ties in the outcome: a motion may neither pass nor fail, but be on the border,

[1] Fishburn refers to this property as "duality".

W.S. Zwicker
Department of Mathematics, Union College, Schenectady, NY
e-mail: zwickerw@union.edu

S.J. Brams et al. (eds.), *The Mathematics of Preference, Choice and Order: Essays in Honor of Peter C. Fishburn*, Studies in Choice and Welfare,

- Such a tie is "knife-edge": a change in vote by any one non-dummy voter breaks the tie,
- And voters may have different influence: the rule is not required to be *anonymous*.

Any constant-sum voting rule that allows abstention must admit ties,[2] so the second requirement is forced. However, a number of real voting rules that allow abstention are not constant sum and do not admit ties (Freixas & Zwicker (2003, 2008)). We argue in Freixas & Zwicker (2003) that the appropriate model for such a rule is a $(3,2)$ *game*, and show that *grade trade robustness* (a generalization of Elgot's *asummability* Elgot (1960/ 1961), or of *trade robustness* in Taylor & Zwicker (1992); Taylor & Zwicker (1999), for simple games) characterizes weighted voting for $(3,2)$ *games*. Without the constant sum property, a weighted rule that assigns weight 0 to each abstention may, for example, assign more (negative) weight to a *no* than (positive) weight to a *yes*. In fact, this must be the case for a permanent member of the UN Security Council (Freixas & Zwicker (2003)). A weighted *RS* is quite different: the constant-sum property effectively requires us to assign to a voter the same positive weight for her *yes* as negative weight for her *no*.

What happens when we impose anonymity on an *RS*? The constant-sum condition (together with a monotonicity requirement saying that more *yes* votes never cause a motion to fail) implies that a motion must pass with strictly more *yes* than *no* votes, fail with strictly more *no* than *yes* votes, and tie with equal numbers of each. So this version of *majority rule with abstention* – a weighted rule (as we'll see) – is the only anonymous *RS*. This assertion can be thought of as *May's Theorem for Representative Systems*. May's original version of this theorem was in the "no abstentions, no ties" setting; it asserts that the only anonymous, monotonic, voting rule that is constant-sum (equivalently, that is neutral for two alternatives) is majority rule with an odd number of voters (May (1952); Taylor (1995)). This is a very special instance of weighted voting and the following restatement is helpful for our purposes:

May's Theorem (Recast) In the no ties, no abstentions, constant-sum setting, anonymity + monotonicity implies weighted voting.

We show in Freixas & Zwicker (2008) that the situation is more interesting for *anonymous* $(3,2)$ *games*: many rules are possible, and not all are weighted (as we will soon see). In fact, each such rule for n voters corresponds to a *quota function* q that assigns, to each integer a in $[0,n]$, the minimum number $q(a)$ of *yes* votes required for collective approval, given that there are a abstentions. Figures 1 and 2 show the quota function for *relative majority rule* (also called *simple majority rule*, in which approval of a motion requires *strictly* more *yes* votes than *no*), and for the *majority threshold* system used in certain referenda in Hungary (see Côrte-Real & Pereira (2004)), wherein passage requires more *yes* votes than *no*, subject to the requirement that at least 25% of all registered voters vote *yes*.

---

[2] For example, when everyone abstains. But for most realistic constant sum voting rules, abstentions force many other ties, as well.

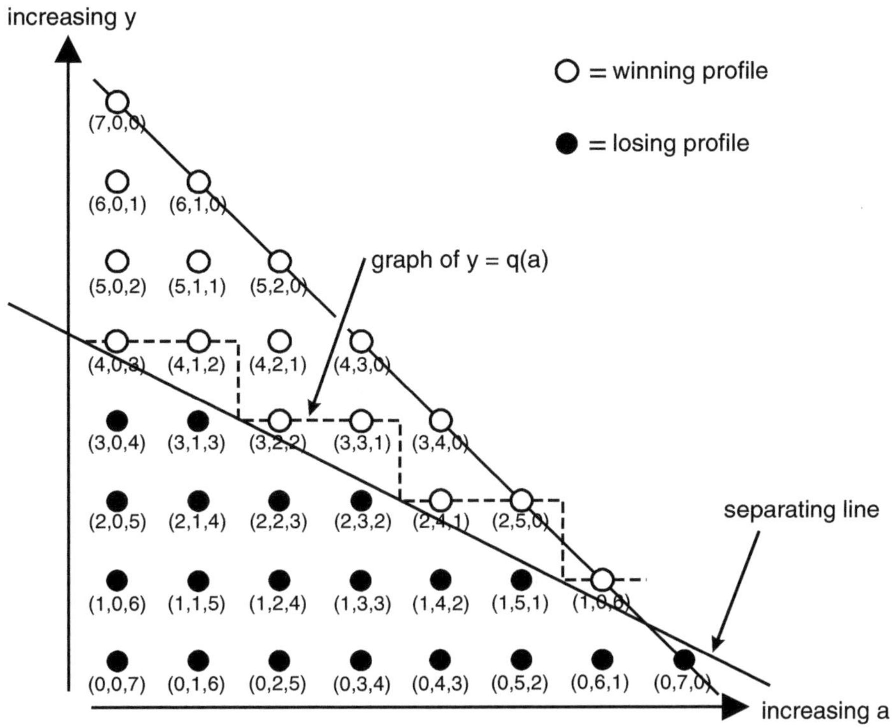

Fig. 1 Quota function diagram: relative majority rule, with abstention, for 7 voters

In fact the monotonicity requirement for $(3,2)$ *games* (see Definition 2.2) implies that the graph of $y = q(a)$ must share the following features with Fig. 1:

- It must be a step function
- The graph can never step up, and whenever the graph steps down, it can only step down by one unit
- If the graph "runs off the edge" by crossing the hypotenuse of the triangular grid of points, then it becomes undefined, and it remains undefined for all larger values of a – that is, if it is possible for the number a of abstentions to become so high that the bill fails to pass even when each non-abstainer votes *yes* then any number of abstentions greater than a must also preclude passage.

These conditions are stated precisely in Freixas & Zwicker (2008), where we show that they are necessary and sufficient for the function q to correspond to an anonymous $(3,2)$ *game*.

Theorem 1. *Quota Function Characterization Theorem*

Every anonymous $(3,2)$ game *corresponds to a quota function satisfying the properties specified above, and every such quota function induces a unique anonymous* $(3,2)$ game.

This result can be thought of as *May's Theorem for (3,2) games*. As a corollary, in Freixas & Zwicker (2008) we showed that for n voters the number of distinct anonymous voting rules with abstention (anonymous $(3,2)$ games) is 2^{n+1} – significantly greater than the number of anonymous *RS*s!

Our purpose in this note is to provide three characterizations of weighted voting for the context of anonymous voting rules that allow abstention, but disallow ties:

Theorem 2. *Main Theorem on Weighted Anonymous Voting with Abstention*[3] *Let G be an anonymous* $(3,2)$ *game. Then the following are equivalent:*

1. *G is* weighted
2. *G is* bimonotonic
3. *G's quota-function diagram is* linearly separable
4. *G satisfies the* almost equal plateau sum condition, *or* a.e.p.s.

We'll discuss weighted voting shortly. Bimonotonicity has two interpretations. It is a very weak form of *grade-trade robustness*, the condition that characterizes weightedness for (j,k) games in general. Alternately, it is a strong form of monotonicity. Keeping in mind this second interpretation, we now restate part of the theorem:

> Main Theorem (part (2) $\Rightarrow$ (1), recast) In the no ties, abstentions allowed setting, anonymity + bimonotonicity implies weighted voting.

A comparison with the previously recast version of May's Theorem suggests that this fragment of the main theorem may also lay claim to being *May's Theorem for* $(3,2)$ *games*, although it is not the same as the version we mentioned earlier.

Linear separability and a.e.p.s. are conditions on the shape of the graph of the quota function. The first asserts that the same separation accomplished by the step function can be achieved by a straight line (as, in Fig. 1, we observe to be the case for relative majority rule but not for the majority threshold rule of Fig. 2). The second is a more constructive condition that puts precise limits on the amount by which the plateaus (steps) in the step function q_G can vary in length. It can be seen as a version of the requirement that the convex hull of all losing profiles be disjoint from that of all winning coalitions.

2 Anonymous (3,2) *Games*, Weighted Voting, and Linear Separability

If we allow each voter in a yes-no voting system the additional option of abstaining, then the profile of their individual decisions can be represented as an ordered triple (Y,A,H) consisting of the sets of voters who choose *yes, abstain*, and *no*, respectively. In Freixas & Zwicker (2003) we define a $(3,2)$ game in terms of a value

[3] In the body of the paper, Theorem 2 is separated into Proposition 1, Theorem 3, and Theorem 4.

function that assigns, to each such ordered 3-partition (Y,A,H), a value in the set {win, lose}. More generally, a (j,k) game employs an ordered j-partition to profile voters' choices from among j ordered levels of input approval, and a value function that assigns to each ordered j-partition an element from a value set containing k ordered levels of output approval. We impose a certain monotonicity condition on these structures, and argue that the resulting level of generality is the appropriate one to model a broad variety of real decision rules. In the anonymous $(3,2)$ context, however, things can be kept more simple; we'll record the profile of an election in the form of three numbers y, a, and h representing the number of *yes* voters, abstainers, and *no* voters, respectively.

Definition 1. Given a natural number n, *an anonymous profile for n voters* is an ordered triple (y,a,h) of nonnegative integers with sum n. A *value function V* for n voters assigns to each anonymous profile p for n voters a single value $V(p)$ in the set $\{win, lose\}$.

Of course, not every value function corresponds to a reasonable voting rule.

Definition 2. Given two anonymous profiles $p = (y,a,h)$ and $q = (y',a',h')$ for n voters we will write $p <_1 q$ if either $y' = y+1, a' = a-1$, and $h' = h$, or if $y' = y, a' = a+1$, and $h' = h-1$. The *left-shift order* $<_{LS}$ is the transitive closure of $<_1$, and a value function V is *monotonic* if the winning profiles are closed upwards in the left-shift order: whenever $V(p) = win$ with $p <_{LS} q$, it follows that $V(q) = win$.

Notice that $p <_{LS} q$ holds precisely when p can be transformed into q via a series of switches in vote by individual voters, each of which is in the direction of greater approval (from *no* to *abstain* or to *yes*, or from *abstain* to *yes*).

Definition 3. An *anonymous* $(3,2)$ *game for n voters* is a pair $G = (n,V)$ in which n is a natural number and V is a monotonic value function for n voters. We'll say that a profile p is *winning* if $V(p) = win$, and *losing* if $V(p) = lose$.[4]

Each anonymous profile (y,a,h) for 7 voters corresponds to a node in the quota function diagram of Fig. 1. Figure 2 is a "population generic" diagram, in which the black (losing) and white (winning) nodes have merged into darker and lighter regions, respectively. We might imagine that the number n of voters is too great for the nodes to resolve as individual points; alternately, such a diagram corresponds to a sequence of voting rules, one for each positive integer n.

Definition 4. An anonymous $(3,2)$ *game* $G = (n,V)$ is *weighted* if there exists a *weighted representation*, consisting of a *weight vector*

$$w = (w_{yes}, w_{abstain}, w_{no})$$

[4] Notice that we do allow the $(3,2)$ *game* for which every profile is winning, as well as that for which every profile is losing; this simplifies some theorem statements at the cost of admitting two games that are of little use as real voting rules.

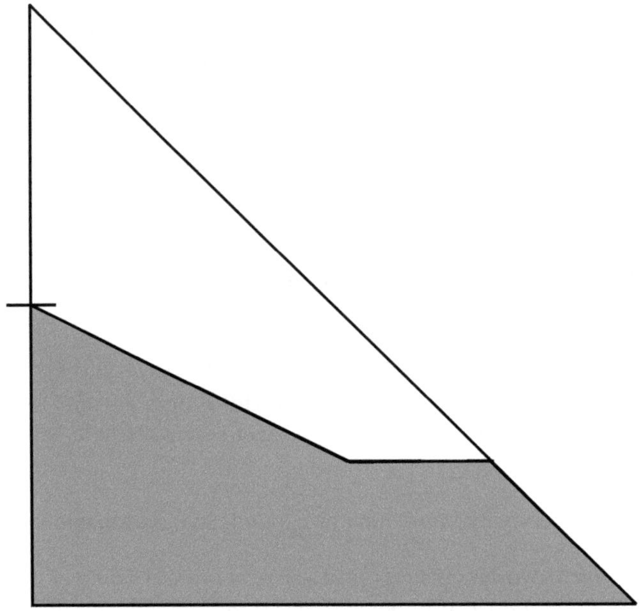

Fig. 2 Population generic quota function diagram: Hungarian "majority threshold" referendum rule

with real number components satisfying $w_{yes} \geq w_{abstain} \geq w_{no}$ together with a *threshold* or *quota* t such that for every anonymous profile $p = (y, a, h)$ for n voters we have

$$V(y,a,h) = win \Leftrightarrow p \cdot w \geq t.$$

Thus *yes, abstain*, and *no* votes are each assigned a fixed weight. The dot product represents the total weight $W_w(p) = yw_{yes} + aw_{abstain} + hw_{no}$ cast by all voters, and the motion carries if and only if this total meets or exceeds the preset threshold. Notice that by subtracting the constant vector $(w_{abstain}, w_{abstain}, w_{abstain})$ from the weight vector $(w_{yes}, w_{abstain}, w_{no})$, while subtracting the product $nw_{abstain}$ from t, we can obtain an equivalent weighted representation in which the abstainers cast no weight, *yes*-voters cast nonnegative weight, and *no*-voters cast non-positive weight, which some readers may find more palatable.

But is this definition congruent with one's naive notion of what *weighted voting with abstention* ought to mean in the anonymous setting? Initially, one might guess that each voter should receive a *single* vote, which is cast either against or for the proposal (or not cast at all, in the case of an abstaining or absent voter). Perhaps collective approval should require that the *yes* votes exceed the *no* votes by some fixed *margin m*. Or perhaps it should require that some minimum *ratio r* of *yes* votes to *no* votes be achieved, e.g. $r = 2$ if one requires for collective approval that of the *active voters* (those present and not abstaining), at least two-thirds should vote *yes*. It is easy to see that the first proposal is tantamount to restricting Definition 4 by requiring that $w = (1, 0, -1)$ with $t = m$, while the second is equivalent to the restriction $w = (1, 0, -r)$ with $t = 0$. Proposition 1, which follows, provides some justification for Definition 4, by suggesting that it represents a sort of minimal extension encompassing both the margin and the ratio approaches. However, we would argue that a stronger justification is provided by the combinatorial characterization via grade trade robustness of the non-anonymous version of Definition 4; we refer the reader to Freixas & Zwicker (2003) for details.

With the help of three adjustments (none of which change the underlying $(3,2)$ game G) we can now put any weight vector $w = (w_{yes}, w_{abstain}, w_{no})$ into a standard form. First, we arrange, as described above, $w_{yes} \geq 0, w_{abstain} = 0$, and $w_{no} \leq 0$. Next, we exploit the "wiggle room" extant in any weighted rule with finitely many voters by adding some small positive value to w_{yes}. If this increment is sufficiently small, then we preserve the requirement that $W_w(p) > W_w(q)$ hold whenever p is winning and q is losing, so that a threshold t can separate by squeezing strictly between the weights of all winning profiles and those of all losing ones. Finally by multiplying our transformed vector w through by the reciprocal of w_{yes} (while multiplying the threshold by the same factor) we obtain the *standard form weight vector*

$$w = (1, 0, s); s \leq 0.$$

Thus we can specify any weighted rule via the two parameters $s \leq 0$ and $t \geq ns$, where s denotes the weight of a *no* vote and t is the threshold for passage. (Note that any rule with $t < ns$ produces the same outcomes as $t = ns$: all profiles are winning.)

In the rule with parameters s and t, a profile $p = (y, a, h)$ is winning if and only if

$$y + hs \geq t. \tag{1}$$

After substituting $n - y - a$ for h in this inequality, it is easy to see that it is equivalent to

$$y \geq \left(\frac{s}{1-s}\right)a + \left(\frac{t - ns}{1-s}\right). \tag{2}$$

Now if we define new parameters

$$m = \frac{s}{1-s}, b = \frac{t - ns}{1-s}; -1 < m \leq 0, b \geq 0$$

(where the limits on m and b correspond to those on s and t), then inequality (2) becomes

$$y \geq ma + b. \tag{3}$$

These new parameters have a simple geometric interpretation. When the equation $y = ma + b$ is graphed on the quota function diagram of G (in which y is the vertical axis, and a is the horizontal) we obtain a straight line L with slope m and y-intercept b.

Definition 5. An anonymous $(3,2)$ game G is *linearly separable* if there exists a straight line L with slope m satisfying $-1 < m \leq 0$ and with y intercept $b \geq 0$, such that all winning nodes on the diagram lie on or above L, and all losing nodes lie strictly below L.

The previous discussion constitutes a proof of the following:

Proposition 1. *Let* $G=(n,V)$ *be an anonymous* $(3,2)$ game. *Then the following are equivalent:*

1. G is weighted
2. There are real number parameters $s \leq 0$ and $t \geq ns$ such that G's winning profiles $p=(y,a,h)$ are precisely those satisfying $y+hs \geq t$.
3. G is linearly separable.

Relative majority rule (Fig. 1) is clearly linearly separable, while the majority threshold rule of Fig. 2 is not. It is well-known (see, for example, Taylor & Zwicker (1999)) that an ordinary simple game is weighted if and only if the winning coalitions (suitably plotted as a set of points in $\mathfrak{R}^n$) are separable from the losing coalitions via a hyperplane. The line L can be thought of as a projection, onto a subspace of small dimension, of a separating hyperplane.[5]

Note that in the majority threshold rule, the graph of the quota function $q(a)$ runs into the hypotenuse of the triangle and stops. For any rule sharing this feature, the last (rightmost) node on the graph of $q(a)$ represents a profile $p=(K,n-K,0)$ in which

- There are a certain number $n-K$ of abstentions,
- Of the active voters, all K of them vote *yes* (where, in the case of the Hungarian Referendum rule, $K=\frac{1}{4}$ of the registered voters),
- And the rule grants collective approval for p.

Because this node is the lowest white (passing) node on the diagram, we know that K is the absolute minimum number of *yes* votes that can ever achieve collective approval. Establishing some type of floor for collective approval is not uncommon among real voting rules, with the goal of avoiding situations wherein a tiny handful of active voters can change the *status quo* body of law. However, the more typical approach is to impose a *quorum*, which is a floor on the number of active voters, rather than a majority threshold, where the floor is on the number of *yes* voters. These two approaches have dramatically different effects, because a quorum typically violates monotonicity. The effect is to give an odd incentive to voters opposed to the motion under consideration: in many cases, they have greater influence by staying home than by voting *no*. In Italy the law governing abrogative referenda has just such a quorum provision and according to Uleri (2002) the effect on participatory democracy in Italy has been perverse – see further discussion in Côrte-Real & Pereira (2004), and Axtman (2003) makes amusing reading for the US context. A majority threshold has no such perverse effect, so in this respect it seems much preferable to a quorum.

But if we are designing some voting rule that will allow abstention or absence, imposing a majority threshold is not the only approach that simultaneously preserves monotonicity and establishes a participation floor – these ends are met by

---

[5] In this connection, see Remark after Definition 9.

any quota function that both "crosses the hypotenuse" at some point $(K, n-K, 0)$ and also satisfies the conditions of Theorem 1.

In particular, one can build any desired floor into a *weighted* rule. In Freixas & Zwicker (2008) we refer to such rules as *soft quorum* weighted rules. An example would be the rule that requires, for collective approval, that the number of *yes* votes be at least $\frac{1}{3}$ of the assembly together with at least $\frac{2}{5}$ of any active voters beyond this $\frac{1}{3}$:

$$y \geq \frac{n}{3} + \frac{2}{5}\left(y + h - \frac{n}{3}\right).$$

Note that this rule can be thought of as a sort of sliding ratio quota. With all voters active a 60% approval rate is required for passage, but as the number of active voters gradually falls, the required fraction of votes in favor gradually increases, until it reaches 100% when only $\frac{1}{3}$ of voters are active. With fewer than $\frac{1}{3}$ of the voters active the percentage in effect rises above 100%, rendering collective approval impossible and establishing the floor. The example suggests yet a third parameterization of weighted rules,[6] using fractions r_1 and r_2 corresponding to the $\frac{1}{3}$ and $\frac{2}{5}$ of the example.

As far as we know, no such rule has been implemented in practice. This is a bit surprising, as the principle seems to be simple.

3 Grade Trade Robustness and Bimonotonicity

In the non-anonymous context, a profile for a $(3,2)$ *game G* consists of a vector $p_i = ({}_{yes}A_i,\ {}_{abstain}A_i,\ {}_{no}A_i)$ in which ${}_xA_i$ denotes the *set* of x-voters for p_i. Given a vector $P = (p_1, p_2, \ldots, p_k)$ of such profiles, a *migration* consists of a change $p_i{}', p_j{}'$ in exactly two of the profiles p_i, p_j $(i \neq j)$ from P, of the following kind: there exists some individual voter s and some pair $x, y \in \{yes, abstain, no\}$ of possible votes such that

- $s \in {}_xA_i$ and $s \in {}_yA_j$
- $s \in {}_yA_i{}'$ and $s \in {}_xA_j{}'$
- there are no other differences between p_i and p_i', or p_j and p_j',

with $P' = (p_1, p_2, \ldots, p_i', \ldots, p_j', \ldots p_k)$. A *k-grade-trade* consists of a finite sequence

$$P, P', P'', \ldots, P''\cdots' = P^*$$

of migrations that convert a *pre-trade vector* $P = (p_1, p_2, \ldots, p_k)$ into a *post-trade vector* $P^* = (p_1*, p_2*, \ldots, p_k*)$. We say that G is *k-grade-trade robust* if no such trade can convert a vector P of winning profiles into a vector P^* of losing profiles,

[6] However, as described here r_1 and r_2 only parameterize that subclass of weighted rules for which a participation floor exists.

and is *grade-trade robust* if it is *k-grade-trade robust* for every integer $k \geq 2$. We show, in Freixas & Zwicker (2003), that G is grade-trade robust if and only if it is weighted.[7]

We might expect that in the context of anonymous voting rules, something considerably less than the full power of grade-trade robustness would suffice to guarantee weightedness.

Definition 6. A *shift* is an ordered pair (u,v) of nonnegative integers. An anonymous profile (y,a,h) and a *shift* (u,v) are *compatible* if $a \geq u+v$, and the (u,v) *shift* of a compatible anonymous profile (y,a,h) is the anonymous profile $(y+u, a-u-v, h+v)$.

Definition 7. Let $G=(n,V)$ be an anonymous $(3,2)$ game. We say that the shift (u,v)

- *Never hurts* if the (u,v) shift of every winning compatible anonymous profile for G is winning
- *Sometimes helps* if the (u,v) shift of at least one compatible losing profile is winning
- Is *big* if it both never hurts and sometimes helps.

Similarly, the shift (u,v)

- *Never helps* if the (u,v) shift of every losing compatible anonymous profile for G is losing
- *Sometimes hurts* if the (u,v) shift of at least one compatible winning profile is losing
- Is *small* if it both never helps and sometimes hurts.

These definitions were suggested by the following observation. For any weighted, anonymous $(3,2)$ game $G=(n,V)$, consider a standard form weight vector $w=(1,0,s)$, $s \leq 0$. Consider any (u,v) shift $(y+u, a-u-v, h+v)$ of a compatible (y,a,h). The effect of this shift is to increment the total weight cast by the amount $u+vs$, which is nonnegative when $\frac{u}{v} \geq |s|$. In the weighted case, then, any (u,v) shift with $\frac{u}{v} \geq |s|$ never hurts, while any with $\frac{u}{v} \leq |s|$ never helps.

Definition 8. A symmetric $(3,2)$ game is *bimonotonic* if every shift (u,v) either never hurts or never helps.

Remark 1. (i) Notice that bimonotonicity is a strong form of monotonicity, in the following sense: ordinary monotonicity is equivalent to the statement that shifts of the form $(u,0)$ never hurt, while those of form $(0,v)$ never help.[8]

[7] In fact, this characterization extends to (non-anonymous) $(j,2)$ games. An elaborated version characterizes weighted voting for (j,k) games.

[8] As bimonotonicity does not actually imply monotonicity, it would be more accurate to say that this property is a strong form of *unateness* (see Taylor & Zwicker (1999)) as recast in the $(3,2)$ setting.

(ii) If bimonotonicity fails, there is a (u,v) that sometimes hurts, as witnessed by a winning compatible (y,a,h) with $(y+u,a-u-v,h+v)$ losing, and also sometimes helps, as witnessed by a losing compatible $(y^{\#},a^{\#},h^{\#})$ with $(y^{\#}+u,a^{\#}-u-v,h^{\#}+v)$ winning. It is straightforward to then see that a grade-trade can convert the two winning profiles (y,a,h), $(y^{\#}+u,a^{\#}-u-v,h^{\#}+v)$ into the two losing profiles $(y+u,a-u-v,h+v)$, $(y^{\#},a^{\#},h^{\#})$.[9] Thus a failure of bimonotonicity is equivalent to a certain particular type of failure of 2-grade-trade robustness. In fact, for anonymous $(3,2)$ games *every* failure of grade trade robustness between two non-anonymous profiles can easily be shown to induce a failure of bimonotonicity. Thus, in the anonymous $(3,2)$ setting, bimonotonicity is equivalent to 2-grade trade robustness.

Theorem 3. *An anonymous* $(3,2)$ *game is weighted if and only if it is bimonotonic.*

Proof. The earlier remarks of this section establish that weightedness implies bimonotonicity. For the other direction, let $G=(n,V)$ be a bimonotonic anonymous $(3,2)$ game. The following claim is key to the proof that G is weighted.

Claim. 1 If (u_1,v_1) is a big shift and (u_2,v_2) is a small shift, then $u_1v_2 > u_2v_1$. Of course, when $v_1,v_2 \neq 0$ this inequality asserts $\frac{u_1}{v_1} > \frac{u_2}{v_2}$.

Proof of claim An ordered quadruple (u_1,v_1,u_2,v_2) of nonnegative integers represents a *counterexample of size* $u_1+v_1+u_2+v_2$ if (u_1,v_1) is a big shift, (u_2,v_2) is a small shift, and

$$u_1v_2 \leq u_2v_1. \tag{4}$$

If a counterexample exists, then there is one of minimal size. Hence, to prove the claim it suffices to show that from any purported counterexample C we can construct a strictly smaller counterexample $C^{\dagger}$. Before the construction, it helps to establish the following:

Claim. 2 Every counterexample $C=(u_1,v_1,u_2,v_2)$ satisfies $u_1>0, v_1>0, u_2>0$, and $v_2>0$.

To establish claim 2, note that ordinary monotonicity implies that $(u_2,0)$ is not small and that $(0,v_1)$ not big, whence $u_1>0$ and $v_2>0$. Thus $u_1v_2>0$ and condition (4) implies $u_2>0$ and $v_1>0$.

The construction of $C\dagger$ proceeds by cases.

<u>Case 1</u> Assume that $u_1 \leq u_2$ and $v_1 \leq v_2$. Let $u_2\dagger = u_2-u_1$ and $v_2\dagger = v_2-v_1$. Let $C\dagger = (u_1,v_1,u_2\dagger,v_2\dagger)$. Clearly, $C\dagger$ consists of nonnegative integers and $C\dagger$ is strictly smaller than C. As condition (4) holds of C, we obtain

$0 \leq v_1u_2-u_1v_2 = v_1(u_1+u_2\dagger)-u_1(v_1+v_2\dagger) = v_1u_1+v_1u_2\dagger-u_1v_1-u_1v_2\dagger =$
$*v_1u_2\dagger-u_1v_2\dagger$

[9] Literally, of course, the trade takes place between non-anonymous profiles, with the anonymous profiles being vectors of cardinalities of the corresponding sets.

whence $C\dagger$ also satisfies (4). To finish case 1 we need only show that $(u_2\dagger, v_2\dagger)$ sometimes hurts; our assumption of bimonotonicity then implies that $(u_2\dagger, v_2\dagger)$ is small. We know that (u_2, v_2) sometimes hurts, so we may pick a winning anonymous profile $p = (y, a, h)$ with $a \geq u_2 + v_2$ such that $p' = (y + u_2, a - u_2 - v_2, h + v_2)$ is losing. Consider the sequence obtained by applying first a (u_1, v_1) shift to p, and then a $(u_2\dagger, v_2\dagger)$ shift to the result:

$$p = (y, a, h)$$
$$p_1 = (y + u_1, a - u_1 - v_1, h + v_1)$$
$$p_2 = (y + u_1 + u_2\dagger, a - u_1 - u_2\dagger - v_1 - v_2\dagger, h + v_1 + v_2\dagger)$$

As (u_1, v_1) never hurts and p is winning, p_1 is winning. As p_1 is winning and p_2 is losing, it follows that $(u_2\dagger, v_2\dagger)$ sometimes hurts.

Case 2 Assume $u_1 \geq u_2$ and $v_1 \geq v_2$. Let $u_1\dagger = u_1 - u_2$ and $v_1\dagger = v_1 - v_2$. Using arguments similar to those in case 1, we can show that $C\dagger = (u_1\dagger, v_1\dagger, u_2, v_2)$ is a smaller counterexample than C.

Case 3 Assume $u_1 > u_2$ and $v_1 < v_2$. This case assumption violates (4), so the case does not occur.

Case 4 Assume $u_1 < u_2$ and $v_1 > v_2$. We claim that by setting $u_2\dagger = u_1$ we obtain a smaller counterexample $C\dagger = (u_1, v_1, u_2\dagger, v_2)$. As $u_2\dagger = u_1$ and $v_1 > v_2$, (4) holds for $C\dagger$. By ordinary monotonicity, any anonymous profile (y, a, h) that is hurt by a (u_2, v_2) shift is also hurt by any (u, v_2) shift with $u < u_2$, and from bimonotonicity it now follows that $(u_2\dagger, v_2)$ is small.

This completes Case 4 and Claim 1. Returning to the proof of Theorem 3, we note that there are but finitely many small shifts (u_S, v_S) and finitely many big shifts (u_B, v_B), so we may choose a real number d such that

$$d > \frac{u_S}{v_S} \text{ and } d < \frac{u_B}{v_B}$$

holds for each small and big shift, respectively. Note that the first fraction $\frac{u_S}{v_S}$ is always well defined, as v_S is never zero for a small shift (by ordinary monotonicity). While $v_B = 0$ is indeed possible, we'll simply declare "$d < \frac{u_B}{v_B}$" true in this event.

We claim there exists a weighted representation of G that assigns each voter the standard form weight vector $(1, 0, -d)$. It suffices to prove that the weight of an arbitrary winning profile (y_W, a_W, h_W) is strictly greater than that of an arbitrary losing profile (y_L, a_L, h_L), i.e., that

$$y_W - dh_W > y_L - dh_L$$

for then we can insert some real number threshold t between the greatest weight attained by any losing profile and the least weight achieved by any winning profile.

Assume first that $y_L \geq y_W$. Then ordinary monotonicity implies that $h_L > h_W$ so that if we set $u = y_L - y_W$, $v = h_L - h_W$, then a (u, v) shift converts the winning profile (y_W, a_W, h_W) into the losing profile (y_L, a_L, h_L). So (u, v) is too small, whence

$$d > \frac{u}{v} = \frac{y_L - y_W}{h_L - h_W}$$

so $d(h_L - h_W) > y_L - y_W$, whence $y_W - dh_W > y_L - dh_L$, as desired.

The other possibility is that $y_L < y_W$. Then if $h_L \geq h_W$, the desired inequality $y_W - dh_W > y_L - dh_L$ follows immediately. So let us assume $h_L < h_W$. Set $u = y_W - y_L, v = h_W - h_L$. Then a (u,v) shift converts the losing profile (y_L, a_L, h_L) into the winning profile (y_W, a_W, h_W). Thus (u,v) is big, whence

$$d < \frac{u}{v} = \frac{y_W - y_L}{h_W - h_L}$$

from which $y_W - dh_W > y_L - dh_L$, as desired.

4 The Almost Equal Plateau Sum Condition

For a small value of n, we can tell by inspection of the quota function diagram for G whether or not the game is linearly separable, but with more voters the issue is not so obvious. Can we find a more constructive condition on the degree of straightness of q_G's graph that is equivalent to weightedness for a $(3,2)$ game G? To motivate the answer, consider 3, which depicts two plateaus of some quota function graph, whose widths differ by 2. As can be seen in the figure, this leads to a failure of bimonotonicity, because the $(1,3)$ shift of the winning profile (y,a,h) is the losing profile $(y+1, a-4, h+3)$, which in turn has as its $(1,3)$ shift the winning profile $(y+2, a-8, h+6)$. However, the failure would evaporate if the shorter plateau owed its reduced length to the fact that q's graph had run across the hypotenuse of the triangular array of nodes (as in Fig. 4), or across the vertical (left) leg, and the shorter plateau were thus "incomplete." Similarly, the failure would disappear if the longer plateau were a "bottom plateau" – owing its length to the fact that q's graph had bottomed out along the horizontal leg of the triangle, as in Fig. 5. Providing that we set aside incomplete shorter plateaus, and bottomed out longer ones, it is easy to

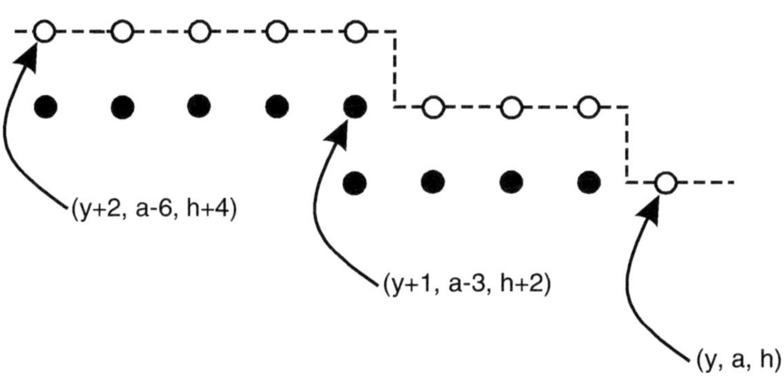

Fig. 3 A plateau of width 5 and a plateau of width 3 result in a failure of bimonotonicity

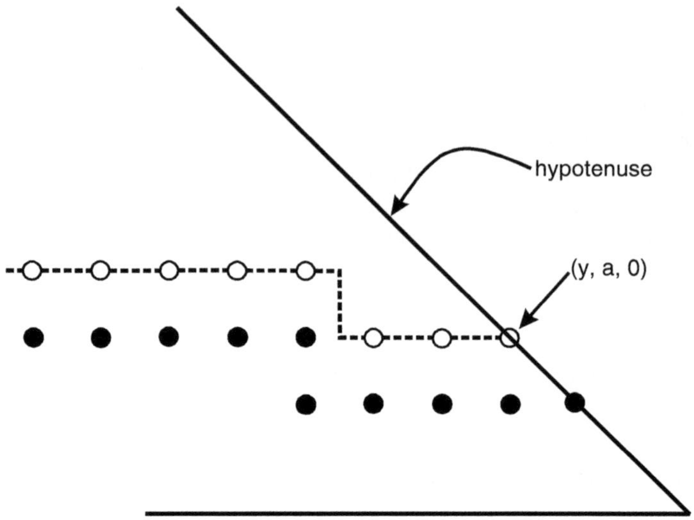

Fig. 4 The shorter plateau does not violate bimonotonicity, because it is *incomplete* (truncated by the hypotenuse)

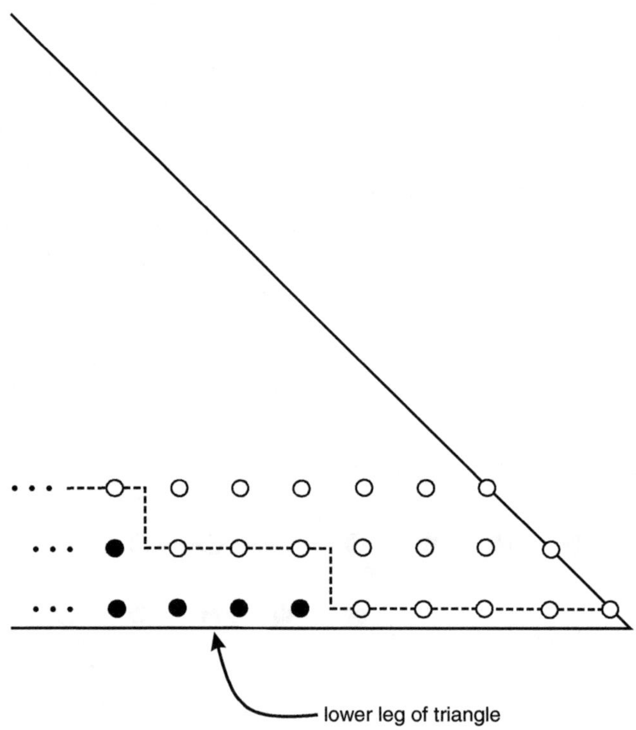

Fig. 5 The longer plateau does not violate bimonotonicity, because it is *bottom* (runs along the lower leg of the triangle)

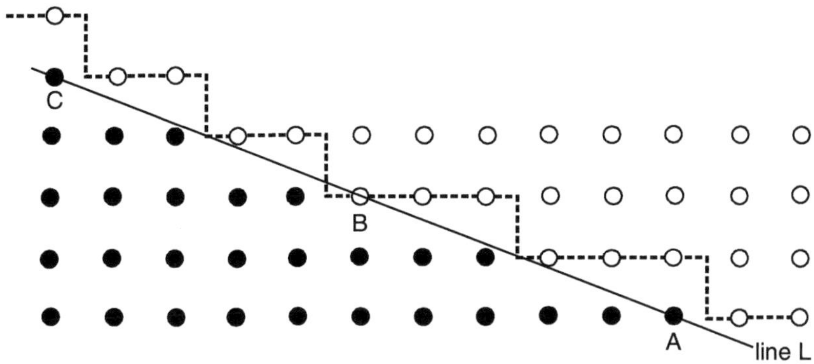

Fig. 6 A failure of the almost equal plateau sum condition leads to a failure of bimonotonicity

see that a failure of bimonotonicity results whenever any plateau exceeds the width of some shorter plateau by 2 or more (and we'll see later that this is the case even when the plateaus in question are not adjacent).

This leads us to the following necessary condition for weightedness of G: there exists some positive integer j such that every complete non-bottom plateau has width j or $j+1$, and every incomplete non-bottom plateau has width $j+1$ or less.

However, this condition is not sufficient. In Fig. 6 we see a (partial) quota function diagram for which each complete plateau has width 2 or 3, yet the nodes labeled A, B, and C show that there is a failure of bimonotonicity $\big(C$ is the $(2,3)$ shift of B, which is the $(2,3)$ shift of $A\big)$ arising from two adjacent plateaus with total width $3+3=6$, in combination with another pair of adjacent (complete) plateaus with total width of $4=2+2$, which is 2 less.

Definition 9. A $(3,2)$ game $G=(n,V)$ satisfies the *almost equal plateau sum condition* (or *a.e.p.s.* condition) if for every $k \leq n$, the sum of the widths of any k consecutive non-bottom plateaus of q_G never exceeds by more than 1 the sum of the widths of any (other) k consecutive complete plateaus.

Remark 2. In fact, the line L of Fig. 6 shows that B is the midpoint of A and C; in particular this reveals that the winning node B lies in the convex hull of the losing nodes. Thus the equivalence of linear separability and the a.e.p.s. condition can be interpreted as a special case of the well known fact that two closed compact sets can be (strictly) separated by a hyperplane if and only if their convex hulls are disjoint.[10]

The intuitive meanings of the terms used in Definition 9 may already be clear, but we'll begin by being more precise.

Definition 10. If G is an anonymous $(3,2)$ game and $q=q_G$ is the induced quota function, then a *plateau of* q is a nonempty interval $I=[j, j+1, \ldots, j+r]$, (with

[10] On the face of it, of course, the a.e.p.s. condition looks weaker than the assertion that the convex hulls are disjoint ... which is why there is something left to be proved.

$j, r \geq 0$) of integers in $Dom(q)$ such that $q(j) = q(j+1) = \ldots = q(j+r)$. A plateau I is *complete* if conditions (i) and (ii) below are both met, and *incomplete* otherwise:

(i) $j > 0$ $(so\, j-1 \in Dom(q))$ and $q(j-1) = q(j)+1$.
(ii) $j+r+1 \in Dom(q)$ and $q(j+r+1) = q(j+r)-1$.

The *width* of the plateau $I = [j, j+1, \ldots, j+r]$ is the number $r+1$ of integers in I, the *height* of I is the common value of q on the integers in I, and I is *bottom* if its height is 0. A sequence $I_1, I_2, \ldots, I_k$ of k plateaus is consecutive if the greatest integer in I_{s-1} is one less that the least integer in I_s, for each $s = 2, 3, \ldots, k$.

According to this definition, any nonempty, proper subinterval of a plateau is also a plateau – albeit an incomplete one; this feature eases the phrasing in the subsequent proof. However, for a moment let us only consider *maximal* plateaus – those not properly contained in other plateaus. Then, with the possible exception of the first and last plateaus, it is easy to see that each is both non-bottom and complete. The a.e.p.s. condition thus implies that these "middle" maximal plateaus come in either one width, or two – a shorter width j and a longer one $j+1$. When there are two, the condition implies that the ratio of shorts to longs is roughly the same in any two regions of q's graph.

Theorem 4. *An anonymous* $(3,2)$ *game G is weighted if and only if its associated quota function* $q = q_G$ *satisfies the almost equal plateau sum condition.*

Proof. : $[\Rightarrow]$ We'll start by showing that if G is weighted then it satisfies the a.e.p.s. condition. With the help of Theorem 3, it suffices to demonstrate that any failure of the a.e.p.s. condition yields a failure of bimonotonicity. Assume there is an a.e.p.s. failure, and choose a failure of minimal total width w – that is, choose an integer $k \geq 1$ together with k consecutive non-bottom plateaus $I_1, I_2, \ldots, I_k$ of G's quota function q whose widths sum to w, and k consecutive complete plateaus $I'_1, I'_2, \ldots, I'_k$ having total width w' with $w \geq w'+2$, for which w is as small as possible.

Claim. In any such minimal failure, $w = w'+2$. To establish the claim, it suffices to assume that $w > w'+2$, and construct a failure of a.e.p.s. having a smaller value of w. This is easy to do if the width of I_k is at least 2; just modify the original failure by deleting the largest integer of the last plateau I_k and the result is to lower w by 1. If the width of I_k is 1, then $w > w'+2$ certainly implies that $k \geq 2$. In this case, delete the last intervals I_k and I'_k of each consecutive sequence, reducing k to $k-1$. It is easy to see that this decreases w' by at least as much as it does w, preserving $w > w'+2$.

Now, let a denote the largest member of I_k, and note that $q(a) \geq 1$, as I_k is non-bottom. Let y denote $q(a)-1$, and $h = n-a-q(a)+1$ (so that $y+a+h = n$). Then (y, a, h) is a losing profile (and its location in the quota function diagram is one node directly beneath the right-most node of the right-most interval I_k). Also, $a-w+1$ is the first member of I_1, and as there are k consecutive plateaus in $I_1, I_2, I_3, \ldots I_k$, the difference in height between I_1 and I_k is $k-1$, so that $q(a-w+1) = q(a)+k-1 =$

$y+k$. It follows that $\left(y+k, a-w+1, h+(w-1-k)\right)$ is a winning profile (and its location in the quota function diagram is the left-most node of the left-most interval I_1), so that the shift $(k, w-1-k)$ sometimes helps.

We'll complete this direction of the proof by using $I'_1, I'_2, \ldots, I'_k$ to establish that this same shift sometimes hurts. Let e' be the largest member of I'_k, and set $a' = e'+1$. As I'_k is complete, we know that a' is in q's domain, and $q(a') = q(e')-1$. Let $y' = q(a')$ and $h' = n-y'-a'$. Then (y', a', h') is a winning profile (and its position in the quota function diagram is one below and one to the right of the right-most node in the right-most plateau I'_1). Now the first member of I'_1 is

$$e' - w' + 1 = (a'-1) - (w-2) + 1 = a' - w + 2$$

and as there are k steps in the sequence, we know that

$$q(a'-w+2) = q(e'-w'+1) = q(e') + k - 1 = q(a') + k = y' + k.$$

As I'_1 is complete, we know that $a'-w+1$ is in q's domain, with

$$q(a'-w+1) = q(a'-w+2) + 1 = q(a') + k + 1 = y' + k + 1.$$

It follows that $(y'+k, a'-w+1, h'+(w-1-k))$ is a losing profile (and its position in the quota-function diagram is one to the left of the left-most node of the left-most plateau I'_1), whence $(k, w-1-k)$ sometimes hurts. This provides the desired violation of bimonotonicity.

($\Leftarrow$) To show that the a.e.p.s. condition implies weightedness, we again appeal to Theorem 3.5 and show that any failure of bimonotonicity triggers a failure of a.e.p.s. Assume that (u, v) is a shift that sometimes helps and sometimes hurts, and further that among such shifts it is minimal in the following sense: the value of u is as small as it can be (for the game G at hand) and the value of v is as small as it can be for this smallest value of u. Choose a losing profile (y, a, h) such that $(y+u, a-u-v, h+v)$ is winning, and a winning profile (y', a', h') such that $(y'+u, a'-u-v, h'-v)$ is losing. Note that

(i) $q(a) \geq y+1$
(ii) $q(a-u-v) \leq y+u$,
(iii) $q(a') \leq y'$ and
(iv) $q(a'-u-v) \geq y'+u+1$.

From (iii) plus the fact that q never steps down by more than 1, it follows that

(v) $q(a'-1) \leq y'+1$,

and from (iv) and the fact that q never steps down by more than 1, it similarly follows that

(vi) $q(a'-u-v+1) \geq y'+u$.

Claim.

(i)* $q(a) = y + 1$
(ii)* $q(a - u - v) = y + u$,
(v)* $q(a' - 1) = y' + 1$, and
(vi)* $q(a' - u - v + 1) = y' + u$.

Proof of claim It suffices to show that, were any of the inequalities (i), (ii), (v), or (vi) strict, we could contradict the presumed minimality of (u,v) by constructing a shift (u^*, v^*) with $u^* < u$, or with $u^* = u$ and $v^* < v$, such that (u^*, v^*) sometimes helps and sometimes hurts.

Case 1 Assume $q(a) > y+1$. Let $u^* = u-1$, and $v^* = v+1$. Let $(y^*, a^*, h^*) = (y+1, a, h-1)$, so that $(y^* + u^*, a - u^* - v^*, h^* + v^*) = (y, a - u + v, h + v)$. Then by the case assumption, (y^*, a^*, h^*) is still losing, while clearly $(y^* + u^*, a - u^* - v^*, h^* - v^*)$ is still winning, so (u^*, v^*) sometimes helps. Let $(y'^*, a'^*, h'^*) = (y', a', h')$, which is winning. Then, as $(y' + u, a' - u - v, h' + v)$ was losing, monotonicity tells us that $(y'^* + u^*, a'^* - u^* - v^*, h'^* + v^*) = (y' + u - 1, a' - u - v, h' + v + 1)$ is also losing, so (u^*, v^*) sometimes hurts.

Case 2 Assume that $q(a - u - v) < y + u$. Let $u^* = u - 1, v^* = v + 1$, and $(y^*, a^*, h^*) = (y, a, h)$. Then (y^*, a^*, h^*) is still losing, while the case assumption tells us that $(y^* + u^*, a - u^* - v^*, h^* + v^*) = (y + u - 1, a - u - v, h + v + 1)$ is winning, so (u^*, v^*) sometimes helps. The argument that (u^*, v^*) sometimes hurts is exactly as in case 1.

Case 3 Assume that $q(a' - 1) < y' + 1$. Let $(u^*, v^*) = (u, v - 1)$, and $(y^*, a^*, h^*) = (y, a, h)$, so that (y^*, a^*, h^*) is still losing. As $(y + u, a - u - v, h + v)$ was winning, monotonicity tells us that $(y^* + u^*, a - u^* - v^*, h^* + v^*) = (y + u, a - u - v + 1, h + v - 1)$ is still winning, and thus (u^*, v^*) sometimes helps. Let $(y'^*, a'^*, h'^*) = (y', a' - 1, h' + 1)$. Then by our case assumption, (y'^*, a'^*, h'^*) is winning, whereas $(y'^* + u^*, a'^* - u^* - v^*, h'^* + v^*) = (y' + u, a' - u - v, h' + v)$ remains losing, so that (u^*, v^*) sometimes hurts.

Case 4 Assume that $q(a' - u - v + 1) > y' + u$. Let $(u^*, v^*) = (u, v - 1)$, and $(y^*, a^*, h^*) = (y, a, h)$. Exactly as in case 3, we can show that (u^*, v^*) sometimes helps. Let $(y'^*, a'^*, h'^*) = (y', a', h')$, so (y'^*, a'^*, h'^*) remains winning, but $(y'^* + u^*, a'^* - u^* - v^*, h'^* + v^*) = (y' + u, a' - u - v + 1, h' + v - 1)$ is losing by our case assumption, so that (u^*, v^*) sometimes hurts.

This completes proof of the claim, establishing that (u,v) satisfies equations $(i)^*, (ii)^*, (v)^*, and (vi)^*$. Our u plateaus $I_1, I_2, \ldots, I_u$ will now be defined as follows: recalling that $q(a - v - u) = y + u$, we will set I_1 equal to the integer interval $[a - v - u, j_{y+u}]$, where j_m is defined to be the greatest integer j for which $q(j) = m$. In general, for $k = 2, 3, \ldots, u - 1$, I_k will equal the integer interval $[j_{y+u-k+2} + 1, j_{y+u-k+1}]$, so that$I_{u-1} = [j_{y+3} + 1, j_{y+2}]$, and we will set $I_u = [j_{y+2} + 1, a]$; here, note that $a \in [j_{y+2} + 1, j_{y+1}]$, as we know $q(a) = y + 1$ (condition $(i)^*$). Note that the sum of the widths of these plateaus is $(a) - (a - v - u) + 1 = u + v + 1$, and the last (and lowest) interval I_u is non-bottom, as $q(a) = y + 1 \geq 1$, so all the plateaus are non-bottom.

Finally, we define the sequence $I'_1, I'_2, \ldots, I'_u$. Recalling that $q(a' - u - v + 1) = y' + u$ (condition $(vi)^*$) and that $q(a' - u - v) \geq y' + u + 1$ (condition (iv)), we can conclude, using the fact that q's graph never steps down by more than 1, that $q(a' - u - v) = y' + u + 1$. Let I'_1 be the integer interval $[a' - u - v + 1, j_{y'+u}]$, and notice that we have shown, in the previous sentence, that I'_1 is complete. For $k = 2, 3, \ldots, u-1$, I'_k will equal the integer interval $[j_{y'+u-k+2} + 1, j_{y'+u-k+1}]$, so that $I_{u-1} = [j_{y'+3} + 1, j_{y'+2}]$, and we will set $I_u = [j_{y'+2} + 1, a' - 1]$; here, note that $a' - 1 \in [j_{y'+2} + 1, j_{y'+1}]$, as we know $q(a' - 1) = y' + 1$. Furthermore, as a' is in $q's$ domain, and $q(a') \leq y'$ (condition (iii)) the fact that q's graph never steps down by more than 1 tells us that $q(a') = y'$, so that I'_u is complete. The construction of the I' sequence then guarantees that all the intervening intervals are also complete. Finally, the sum of the widths of the I' intervals is $(a' - 1) - (a' - u - v + 1) + 1 = u + v - 1$. Thus we have constructed u sequential non-bottom plateaus $I_1, \ldots, I_u$ and u sequential complete plateaus $I'_1, \ldots, I'_u$, such that the total width of the first sequence is 2 greater than that of the second – a failure of the a.e.p.s. condition.

Acknowledgements I'd like to thank my occasional co-author, Josep Freixas. Without his original insight, this paper would not exist. The comments of two anonymous referees significantly improved the exposition.

References

Amer, R., Carreras, F., & Magaña, A. (1998). The Banzhaf-Coleman index for games with r alternatives, *Optimization, 44*, 175–198.

Axtman, K. (2003). *As redistricting tale ends, fight echoes beyond Texas, newspaper article in the September 19 edition of The Christian Science Monitor*. page 2.

Bilbao, J. M., Fernández, J. R., Jiménez, N., & López, J. J. (2005a). *Probabilistic values for bicooperative games*, preprint. Available online at: http://econpapers.repec.org/paper/ceadoctra/

Bilbao, J. M., Fernández, J. R., Jiménez, N., & López, J. J. (2005b). *The Shapley value for bicooperative games*, preprint: University of Seville.

Bolger, E. M. (1986). Power indices for multicandidate voting games. *International Journal of Game Theory, 14*, 175–186.

Bolger, E. M. (1993a). A value for games with n players and r alternatives. *International Journal of Game Theory, 22*, 319–334.

Bolger, E. M. (1993b). A consistent value for games with n players and r alternatives. *International Journal of Game Theory, 29*, 93–99.

Côrte-Real, P. P., & Pereira P. T. (2004). The voter who wasn't there: referenda, representation and abstention. *Social Choice and Welfare, 22*, 349–369.

Dougherty, K., & Edward, J. (2004). *Simple vs. absolute majority rule*, preprint. Available online at: http://www.arches.uga.edu/dougherk./

Elgot, C. C. (1960,1961). Truth functions realizable by single threshold organs, AIEE Conference Paper 60-1311 (October); revised November 1960; paper presented at *IEEE Symposium on Switching Circuit Theory and Logical Design*, September 1961.

Felsenthal, D., & Machover, M. (1997). Ternary voting games. *International Journal of Game Theory, 26*, 335–351.

Felsenthal, D., & Machover, M. (1998). *The measurement of voting power: Theory and practice, problems and paradoxes*, Cheltenham: Edward Elgar Press.

Fishburn, P. C. (1973). *The theory of social choice*. Princeton: Princeton University Press.
Freixas, J., & Zwicker, W. S. (2003). Weighted voting, abstention, and multiple levels of approval. *Social Choice and Welfare, 21*, 399–431.
Freixas, J., & Zwicker, W. S. (2008). Anonymous yes-no voting with abstention and multiple levels of approval: to appear in *Games and Economic Behavior*.
May, K. (1952). A set of independent, necessary and sufficient conditions for simple majority decision. *Econometrica, 20*, 680–684.
Rubenstein, A. (1980). Stability of decision systems under majority rule. *Journal of Economic Theory, 23*, 150–159.
Taylor, A. D. (1995). *Mathematics and politics*. Berlin: Springer.
Taylor, A. D. & Zwicker, W. S. (1992). A characterization of weighted voting. *Proceedings of the American Mathematical Society, 115*, 1089–1094.
Taylor, A. D., & Zwicker, W. S. (1999). *Simple games: Desirability relations, trading, and pseudoweightings*. Princeton: Princeton University Press.
Uleri, P. V. (2002). On referendum voting in Italy: YES, NO, or nonvote? How Italian parties learned to control referendums. *European Journal of Political Research, 41*, 863–883.
von Neumann, J., & Morgenstern, O. (1949). *Theory of games and economic behavior*. Princeton: Princeton University Press.

Social Choice

Pareto, Anonymity or Neutrality, but Not IIA: Countably Many Alternatives

Donald E. Campbell and Jerry S. Kelly

1 Introduction

We would like to express our indebtedness to Peter for his pioneering work in social choice theory and our pleasure in co-authoring with him.[1] More specific to this paper, Peter had early doubts about rules that required preference information on all alternatives in order to socially rank two alternatives. Addressing independence he writes (see Fishburn, 1973):

> If in fact the social choice can depend on infeasibles, which infeasibles should be used? For with one set of infeasibles, feasible x might be the social choice, whereas feasible $y \neq x$ might be the social choice if some other infeasible set were adjoined to [feasible set] Y. Hence, the idea of allowing infeasible alternatives to influence the social choice introduces a potential ambiguity into the choice process that can be at least alleviated by insisting on the independence condition.

Campbell and Kelly (2000) provides a formal answer to Fishburn's question by defining and exploring "relevance sets." Fishburn continues:

> This obviously ties into the choice of the universal set X of alternatives in a particular situation. If independence is adopted, then the contents of X are not especially important as long as they include, at least conceptually, anything that might qualify as a feasible candidate or alternative. If independence is not adopted, the ambiguity noted in the preceding paragraph may cause significant problems in attempting to justify just what should and should not be included in X.

For a finite number of alternatives, Campbell and Kelly (2007) have shown that in the presence of Pareto, non-dictatorship, full domain, and transitivity, an extremely weak interprofile condition (see Fishburn, 1987) is incompatible with each

[1] Campbell and Fishburn (1980); Fishburn and Kelly (1997).

J.S. Kelly (✉)
Department of Economics, Syracuse University, Syracuse, NY 13244-1090, USA
e-mail: jskelly@maxwell.syr.edu

S.J. Brams et al. (eds.), *The Mathematics of Preference, Choice and Order: Essays in Honor of Peter C. Fishburn*, Studies in Choice and Welfare,

of anonymity and neutrality. This paper explores how those results are affected when there are countably many alternatives.

We will show that there *do* exist neutral rules that satisfy all of Arrow's conditions except IIA and also anonymous rules that satisfy all of Arrow's conditions except IIA. There also exist anonymous rules that satisfy all of Arrow's conditions except IIA, and for which the social ordering on a pair depends only on the individual preferences restricted to a finite set. We show that there do not exist neutral rules for which the social ordering on even one pair depends only on the individual preferences restricted to a proper subset of the outcome space. And that result will not require either transitivity of the social ordering or a Pareto condition.

2 Framework[2]

X is the set of all *alternatives* or *outcomes*. In this paper, we assume X has countably many elements. The binary relation $\succeq$ on X is read "x is weakly preferred to y" or "x is preferred or indifferent to y."

A binary relation $\succeq$ on X is *complete* if for all $x, y \in X$, either $x \succeq y$ or $y \succeq x$ holds. Note that a complete relation is *reflexive*, which means that $x \succeq x$ holds for each $x \in X$. The *asymmetric* part of $\succeq$ is denoted by $\succ$, so $x \succ y$ if and only if $x \succeq y$ holds but $y \succeq x$ does not. When $x \succ y$ we often say that x is strongly preferred to y, or that x ranks strictly above y in $\succeq$. Relation $\succeq$ is *transitive* if for all x, y, and z in X, if $x \succeq y$ and $y \succeq z$ then $x \succeq z$; a complete and transitive relation $\succeq$ is an *ordering*.

In this paper, we will simplify our analysis by assuming that an individual is never indifferent between distinct alternatives, in which case we say that the preference ordering is *strong*. Formally, we say that the complete binary relation $\succeq$ is *antisymmetric* if for all $x, y \in X$, $x \succeq y$ and $y \succeq x$ imply $x = y$. A binary relation is a *strong ordering* if it is complete, transitive, and antisymmetric. Let $L(X)$ denote the set of strong orderings on X.

The set N of individuals whose preferences are to be consulted is the (finite) set $\{1, 2, \ldots, n\}$ with $n > 1$. A *domain* is some nonempty subset $\wp$ of $L(X)^N$. A member p of $L(X)^N$ is called a *profile*, and it assigns the ordering $p(i)$ to individual $i \in N$. We typically write $x \succ_i^p y$ to indicate that individual i strictly prefers x to y in ordering $p(i)$. When p is understood, we sometimes write $\succ_i$ for $\succ_i^p$. A *social welfare function* for outcome set X and domain $\wp$ is a function f from $\wp$ into the set of complete binary relations on X. Social welfare functions are often called "rules." We say that rule f has *full domain* if $\wp = L(X)^N$. If x is ranked higher than y at the image $f(p)$ of f at profile p we write $x \succ_{f(p)} y$.

We next introduce some restrictions on the social welfare function f on domain $\wp$: If $f(p)$ is transitive for each $p \in \wp$ we say that f is *transitive-valued* or satisfies *transitivity*. The rule f satisfies *nondictatorship* if there is no individual i such that for every p in $\wp$ and every x and y in X, $x \succ_i^p y$ implies $x \succ_{f(p)} y$. Rule f satisfies

---

[2] Much of this section is drawn from Campbell and Kelly (2007).

the *Pareto criterion* if for every $p \in \wp$ and all $x, y \in X$, we have $x \succ_{f(p)} y$ if $x \succ_i^p y$ for all $i \in N$. Rule f satisfies *weak unanimity* if for every $p \in \wp$ and all $x \in X$, we have $x \succ_{f(p)} y$ for all $y \neq x$ *if* $x \succ_i^p y$ for all $y \neq x$ and all $i \in N$. In words, if x is at the top of everyone's ordering at p, then it is at the top of $f(p)$.

The *independence of irrelevant alternatives* (IIA) condition is quite different in spirit from the Pareto criterion or nondictatorship, each of which requires a kind of responsiveness to individual preferences on the part of the social welfare function. IIA requires the social ordering of x and y to be the same at two profiles if the restrictions of those profiles to $\{x, y\}$ are the same: Formally, rule f satisfies *IIA* if for all $p, q \in \wp$ and all $x, y \in X$, $p|_{\{x,y\}} = q|_{\{x,y\}}$ implies $f(p)|_{\{x,y\}} = f(q)|_{\{x,y\}}$, where $p|_{\{x,y\}}$ and $f(p)|_{\{x,y\}}$ are the restrictions to $\{x, y\}$ of profile p and social ranking $f(p)$ respectively.

We will also need some weaker versions of independence:

Independence of Some Alternative (ISA): For *every pair* of alternatives x and y in X there is a proper subset Y of X such that for any two profiles p and p' in the domain, if $p|_Y = p'|_Y$ then $f(p)|_{\{x,y\}} = f(p')|_{\{x,y\}}$.

Weakest Independence (WI): For *at least one pair* of alternatives x and y in X there is a proper subset Y of X such that for any two profiles p and p' in the domain, if $p|_Y = p'|_Y$ then $f(p)|_{\{x,y\}} = f(p')|_{\{x,y\}}$.

The modified independence conditions suggest some new terminology: Given a rule f and a subset Y of X, we say that Y is *sufficient* for $\{x, y\}$ if for any two profiles p and p' in the domain, $f(p)|_{\{x,y\}} = f(p')|_{\{x,y\}}$ if $p|_Y = p'|_Y$. If Y is sufficient for $\{x, y\}$ and $Y \subseteq Z \subseteq X$, then clearly Z is also sufficient for $\{x, y\}$. The family of sufficient sets can place substantial restrictions on the possible departures from IIA, as the following *intersection principle*, proven in Campbell and Kelly (2000), shows. It is important to note that it does not assume finiteness of X, or the Pareto criterion, or any independence condition, or any type of transitivity property for $f(p)$.

Intersection principle: If the domain of f is $L(X)^N$, and Y and Z are each sufficient for $\{x, y\}$ then $Y \cap Z$ is sufficient for $\{x, y\}$.

For the case of finite X and for each pair $\{x, y\}$ of distinct alternatives, the intersection principle ensures the existence of a smallest set sufficient for $\{x, y\}$ – a sufficient set that is a subset of every set sufficient for $\{x, y\}$. Such a smallest set sufficient for $\{x, y\}$ is the *relevant set* for $\{x, y\}$ and is denoted by $\Psi(\{x, y\})$ or $\Psi(x, y)$. Thus IIA is equivalent to $\Psi(x, y) \subseteq \{x, y\}$ for all x, y in X. With countable X, some pairs may not have a relevant set:

Example 1. For any profile $p \in L(X)^N$ set $x \succ_{f(p)} y$ if $x \succ_1^p y$ unless individual 2 has both (1) $y \succ_2^p x$ and (2) infinitely many alternatives between y and x in $p(2)$ in which case set $y \succ_{f(p)} x$. It is easy to confirm that f is transitive-valued and satisfies Pareto, neutrality, and non-dictatorship. It is also easy to check that $Y \subset X$ is sufficient for a pair $\{x, y\}$ if and only if $\{x, y\} \subseteq Y$ and $X \setminus Y$ is finite. Therefore, there is no relevant set for any pair: If Y is sufficient for $\{x, y\}$ then so is $Y \setminus \{z\}$ for any $z \in Y \setminus \{x, y\}$.

Arrow (1963) has shown that for $|X| \geq 3$ there does not exist any transitive-valued social welfare function satisfying full domain, the Pareto condition, nondictatorship, and IIA. But if we simply delete the requirement of IIA, there are many rules satisfying the rest of Arrow's conditions plus the interprofile conditions (see Fishburn, 1987) of *neutrality* and *anonymity* which we define next.

Suppose that σ is a permutation of N. Such a permutation induces a map σ on profiles where $\sigma(p)$ assigns ordering $p(\sigma(i))$ to individual i. A rule f is *anonymous* if for every permutation σ on N and for every profile p in the domain of f, $\sigma(p)$ is also in the domain and $f(\sigma(p)) = f(p)$.

Turn now from individuals to alternatives. Any permutation μ of X, the set of alternatives, induces a permutation on preference orders where $\mu(R)$ is defined by

$$\mu(x)\mu(R)\mu(y) \quad \text{if and only if } xRy.$$

In turn, this induces a permutation on profiles where $\mu(p)$ assigns ordering $\mu(p(i))$ to individual i. A rule f is *neutral* if for every profile p in the domain of f and every permutation μ on X, $\mu(p)$ is also in the domain and $f(\mu(p)) = \mu(f(p))$.

We clarify these symmetry conditions by contrasting them with other possible versions. Several authors, e.g., Sen (1970, p. 72) and Fishburn (1973, p. 161), define neutrality in such a way as to incorporate considerable independence. An informal version of this is given by Rae and Schickler (1997, p. 167):

> *Neutrality*: Suppose that all individual ordinal preferences over (x,y) are the same as they are over (w,z), then the collective outcomes over the two pairs of options must be the same.

Because we want to work in weak independence contexts, we do not use their definition.

It is also helpful to contrast our definitions with conditional versions from Campbell and Fishburn (1980):

Conditional anonymity: For every permutation σ on N and for every profile p in the domain $\wp$ of f, *if* $\sigma(p)$ is also in $\wp$ then $f(\sigma(p)) = f(p)$;

Conditional neutrality: For every profile p in the domain $\wp$ of f and every permutation μ on X, *if* $\mu(p)$ is also in $\wp$ then $f(\mu(p)) = \mu(f(p))$.

Our (unconditional) anonymity requires that $\wp$ be closed under permutations of individuals; (unconditional) neutrality requires that $\wp$ be closed under permutations of alternatives.

An impossibility result for infinite X does not follow immediately from an impossibility theorem for the finite case. Consider the following result from Campbell and Kelly (2007):

Theorem 1. *If X is finite with $|X| \geq 3$, there does not exist a social welfare function satisfying full domain, transitivity, Pareto, nondictatorship, weakest independence, and neutrality.*

Suppose X is infinite and there exists a social welfare function f on X satisfying full domain, transitive-valuedness, Pareto, nondictatorship, weakest independence, and neutrality. Pick a finite subset Y of X with $|Y| \geq 3$ and select an ordering Q on $X \setminus Y$. Define g on Y as follows, for each profile q on Y, extend each $q(i)$ to all of X by appending Q below $q(i)$ to create $p(i)$. Then $g(q)$ is defined to be $f(p)|_Y$.

This g function inherits from f the full domain condition, Pareto, and neutrality. But it need not inherit either nondictatorship or weakest independence (see Example 5 below). So we cannot use the nonexistence of a rule satisfying full domain, transitive-valuedness, Pareto, nondictatorship, weakest independence, and neutrality for finite X to rule out the existence of such a rule when X is countably infinite.

3 Examples

For finite X, a paradigm example of a neutral and anonymous rule satisfying all of Arrow's conditions except IIA is Borda's rule. Borda's rule violates even weakest independence. As commonly defined this rule does not work in the countably infinite context as there need not be a "first" (topmost) element, or "second," etc. However, there is an alternative definition that works for both finite and countably infinite X (with a somewhat restricted domain). The resulting rule satisfies transitive-valuedness, Pareto, non-dictatorship, anonymity, and neutrality (but violates both full domain and weakest independence):

Example 2. Let X be countable and let S be the largest subset of $L(X)$ such that for every ordering in S and every x and y in X, there are at most finitely many alternatives between x and y. Then the following is a transitive-valued, Paretian, neutral and anonymous social welfare function on $\wp = S^N$. At profile p, for any pair x,y of distinct alternatives in X, let $A(i,x,y)$ be defined as follows:

(1) If $x \succ_i y$, then $A(i,x,y)$ is (1 + the number of alternatives between x and y in $\succ_i$).
(2) If $y \succ_i x$, then $A(i,x,y)$ is the negative of (1 + the number of alternatives between x and y in $\succ_i$).
Then $x \succ_{f(p)} y$ if and only if $\sum_{i \in N} A(i,x,y) > 0$.

This example can be extended to all of $L(X)^N$ so as to still satisfy neutrality (but not anonymity):

Example 3. Let S be as in Example 2. For p in S^N, let $f(p)$ also be determined as in Example 2. If p is in $L(X)^N \setminus S^N$, let i be the individual with the lowest label such that $p(i) \notin S$ and set $f(p) = p(i)$. Because S is closed under permutations of alternatives, this rule is neutral.

Hence, even with countably infinite X, there do exist neutral rules that satisfy all of Arrow's conditions except IIA. We'll see in the next section how far we will have to deviate from IIA.

Turning from neutrality to anonymity, if X is countably infinite, there *do* exist social welfare functions satisfying full domain, transitivity, the Pareto criterion, nondictatorship, and anonymity. Many examples below are, like Example 4, weighted-scoring rules that employ "utility" representations. We are, of course, aware of the problems presented *in some contexts* by utilitarianism.

Example 4. Since X is countable we can write it as a list: $X = < x(1), x(2), \dots >$. Let $\succ$ be an arbitrary strong ordering on X; there is a numerical representation of $\succ$ (in fact, one where all images are rational), i.e., there is a rational-valued function u on X such that $u(x) > u(y)$ just when $x \succ y$. See Birkhoff (1940, p. 200); also Fishburn (1970), and Rader (1972). For each $\succ$ select one such representation. Now, given a profile $r = (r(1), r(2), \dots, r(n))$ in $L(X)^N$, let u_i be the chosen representation of $r(i)$ and set $x \succeq$ y in the social ranking just when

$$\sum_{i=1}^{n} u_i(x) \geq \sum_{i=1}^{n} u_i(y).$$

This rule is defined on all of $L(X)^N$ and satisfies Pareto and anonymity. However, other properties of the rule depend on the choice of representations. This is true for neutrality and for independence conditions. For some choices, the rule will satisfy a strong form of independence (relevant sets are all finite); for other choices, X is the relevant set for every pair, in which case we might say that the rule is nowhere independent. To illustrate, we present a rule where no finite set is sufficient for any pair $\{x, y\}$.

Example 5. Start with any set of representations; they may, for example, yield small relevant sets. We use these representations u to define a new representation v on $L(X)$. Partition $L(X)$ into $L_1 \cup L_2$, where L_1 consists of all those orderings with a minimal element and $L_2 = L(X) \setminus L_1$. Given $\succ$ in $L(X)$, define v by

$$v(x(i)) = \begin{cases} u(x(i)) & \text{if} \succ \in L_1, \\ 2u(x(i)) & \text{if} \succ \in L(X) \setminus L_1. \end{cases}$$

The rule is given by setting $x \succeq y$ in the social ranking just when

$$\sum_{i=1}^{n} v_i(x) \geq \sum_{i=1}^{n} v_i(y).$$

Then no finite subset Y of X that contains $\{x, y\}$ is a sufficient set for $\{x, y\}$ because the ordering restricted to any finite subset of X cannot be used to determine if the ordering is in L_1 or L_2.

In the next example, representations are chosen so that all relevant sets are finite.

Example 6. Since X is countable we can write it as a list: $X = < x(1), x(2), \dots >$. Let $\succ$ be an arbitrary strong ordering on X. We will describe a particular numerical representation of $\succ$. Let $u(x(1)) = 1$. If $x(2) \succ x(1)$, assign $u(x(2)) = 2$; if $x(1) \succ x(2)$, assign $u(x(2)) = 0$. Proceeding inductively, suppose that we have defined $u(x(i))$ for all $i < n$. Then $u(x(n))$ is specified by one of the following three statements:

1. If $x(n) \succ x(i)$ for all $i < n$, let θ be given by

   $$u(x(\theta)) = \max u(x(1)), u(x(2)), \dots, u(x(n-1)),$$

   then $u(x(n)) = u(x(\theta)) + 1$.

2. If $x(i) \succ x(n)$ for all $i < n$, let θ be given by

$$u(x(\theta)) = \min u(x(1)), u(x(2)), \ldots, u(x(n-1)),$$

then $u(x(n)) = u(x(\theta)) - 1$.

3. Otherwise, consider $\succ$ restricted to $\{x(1), x(2), \ldots, x(n)\}$; let $x(s)$ and $x(p)$ be the immediate successor and predecessor respectively of $x(n)$; i.e., $x(s) \succ x(n) \succ x(p)$ and there are no alternatives in $\{s(1), s(2), \ldots, s(n-1)\}$ ranked between $x(s)$ and $x(p)$. Then set

$$u(x(n)) = 0.5[u(x(s)) + u(x(p))].$$

Now, given a profile $r = (r(1), r(2), \ldots, r(n))$ in $L(X)^N$, let u_i be the above representation of $r(i)$ and set $x \succeq y$ in the social ranking just when

$$\sum_{i=1}^{n} u_i(x) \geq \sum_{i=1}^{n} u_i(y).$$

This rule is defined on all of $L(X)^N$, satisfies Pareto and anonymity, and for every pair of alternatives, the relevant set is finite: given $x(i)$ and $x(j)$ in X, with $i > j$, the relevant set for this pair is $\{x(1), x(2), \ldots, x(i)\}$. This rule is not neutral as it depends on the initial listing of the elements of X.

If every pair in X has a finite sufficient set , we say the rule satisfies *finite dependence*.

4 An Impossibility Result

Now that we know there is a rule, defined on all of $L(X)^N$, that satisfies Pareto, anonymity, and satisfies finite dependence we ask what happens when we substitute neutrality for anonymity. We will show that no such rules exist, for any countably infinite set of alternatives – whether or not we impose transitivity of social preference or the Pareto criterion. Recall that for x, y in X and a given rule $f, \Psi(\{x,y\})$ denotes the smallest sufficient set for $\{x,y\}$, *if there is such a set.*[3]

Theorem 2. *For X infinite, there does not exist a social welfare function on* $L(X)^N$ *that is neutral but where there is even a single pair* $\{x,y\}$ *such that*

(1) $\Psi(\{x,y\}) \setminus \{x,y\} \neq \emptyset$.
(2) $\Psi(\{x,y\})$ *is a proper subset of* X.

Proof. Suppose for social welfare function f on $L(X)^N$ there does exist a pair $\{x,y\}$ such that for some a and z,

[3] Note, for example, that the rule of Example 1 has no relevance map.

$$a \in \Psi(\{x,y\}) \setminus \{x,y\}; \quad \text{and}$$
$$z \in X \setminus \Psi(\{x,y\}).$$

Since $a \in \Psi(\{x,y\})$, there exist profiles u and $u*$ such that

$$u|_{\Psi(\{x,y\})\setminus\{a\}} = u * |_{\Psi(\{x,y\})\setminus\{a\}}$$

with $x \succeq_{f(u)} y$ but $y \succ_{f(u*)} x$. We construct a profile u' from u by taking each $u(i)$ and moving only z to the "same position" alternative a has in $u^*(i)$.

To be more explicit:

(1) If $u(i)|_{\Psi(\{x,y\})} = u^*(i)|_{\Psi(\{x,y\})}$, insert z just above a.
(2) If $u(i)|_{\Psi(\{x,y\})} \neq u^*(i)|_{\Psi(\{x,y\})}$, insert z according to the following rules:

  (a) $u'(i)|_{X\setminus\{z\}} = u(i)|_{X\setminus\{z\}}$.
  (b) $(z,w) \in u'(i)$ for all $w \in X \setminus \{z\}$ such that $(a,w) \in u^*(i)$ while $(w,z) \in u'(i)$ for all $w \in X \setminus \{z\}$ such that $(w,a) \in u*(i)$.
  (c) Since $u(i)|_{\Psi(\{x,y\})} \neq u^*(i)|_{\Psi(\{x,y\})}$, either in $u(i)$ alternative a is preferred to some w with $(w,a) \in u^*(i)$ – in which case $(a,z) \in u'(i)$ – or, in $u(i)$ alternative a is below some w with $(a,w) \in u^*(i)$ – in which case $(z,a) \in u'(i)$. The resulting $u'(i)$ relation is transitive.

Since u and u' agree on $\Psi(\{x,y\})$, we have $x \succeq_{f(u')} y$. Finally, construct profile u'' from u' by interchanging a and z. If f satisfied neutrality, the social ranking at u'' would be obtained from the social ranking at u' by just interchanging a and z. In particular, $x \succeq_{f(u'')} y$. But that contradicts

$$u''(i)|_{\Psi(\{x,y\})} = u^*(i)|_{\Psi(\{x,y\})}$$

and $y \succ_{f(u*)} x$. Therefore f must not satisfy neutrality. □

Note that the proof doesn't use anything close to a full domain. The proof is valid for any domain that is closed with respect to the operations employed to switch the positions of alternatives.

Neutrality is one of the virtues of the Borda rule. A serious liability of that rule is that a pair of alternatives typically cannot be socially ordered without obtaining information about each individuals preference relation over the entire set X. The theorem tells us that this drawback applies generally to neutral rules. Given *any* neutral rule on a full domain, for every pair $\{x,y\}$ that *has* a relevant set – i.e., a smallest sufficient set – either $\Psi(\{x,y\}) = \{x,y\}$, or X is its only sufficient set. Pairs without relevant sets have as sufficient sets only infinitely large sets; in fact, there must be infinitely many infinitely large sufficient sets. All neutral rules that violate IIA have extremely demanding information processing requirements.

A rule exhibits *finite neutrality* if for every profile p in the domain and every permutation μ on X that moves only finitely many alternatives, $\mu(p)$ is also in the domain and $f(\mu(p)) = \mu(f(p))$. Much of what we intend regarding equal treatment of alternatives is captured by finite neutrality.

There does exist a way of choosing utility representations that *can* lead to finite neutrality.

Example 7. Let $\sim$ be the relation on $L(X)$ that, for arbitrary orderings R and Q in $L(X)$, sets $R \sim Q$ just when $Q = \sigma(R)$ for some permutation σ on X that moves only finitely many alternatives of X. This is an equivalence relation that induces a partition on $L(X)$. For each partition component, select one ordering, R, from the component and then select the representation $u(R)$ given by Example 6. For any Q in the same partition component, there is a unique σ with $Q = \sigma(R)$. Permute the representation in the same way to arrive at the utility representation of Q and then set $x \succeq y$ in the social ranking just when

$$\sum_{i=1}^{n} u_i(x) \geq \sum_{i=1}^{n} u_i(y).$$

This rule satisfies Pareto, anonymity, and finite neutrality on $L(X)^N$. Unfortunately it violates finite dependence, and the need to elicit preference information over an infinite set may make information processing prohibitively expensive.

Example 7 cannot be extended to include all infinite permutations. Given a Q in the same partition component as R, there may be many different σ's with $Q = \sigma(R)$. For example, let R be an ordering with x and y on top and then a double-ended ordering below:

$$xy \ldots a_3 a_2 a_1 b_1 b_2 b_3 \ldots.$$

Next let Q reverse x and y:

$$yx \ldots a_3 a_2 a_1 b_1 b_2 b_3 \ldots.$$

Q is in the same component of R, but there are infinitely many permutations mapping R to Q. Besides the obvious transposition $(x\, y)$, there is transposition-plus-a-shift:

$$\begin{array}{ccccccccc} x & y & \ldots & a_3 & a_2 & a_1 & b_1 & b_2 & b_3 \ \ldots \\ \downarrow & \downarrow & & \downarrow & \downarrow & \downarrow & \downarrow & \downarrow & \downarrow \\ y & x & \ldots & a_4 & a_3 & a_2 & a_1 & b_1 & b_2 \ \ldots \end{array}$$

Accordingly, choosing a representation for R does not induce a unique representation for Q.

Finally, we observe that for countable X, there is a fundamental design problem in implementing rules with countably many alternatives. How does an individual submit her preference ordering to the social choice procedure? Obviously it is not possible to write down an infinite ordering on a ballot. Of course, some orderings have a finite description; the preference

$$\ldots x(6) \succ x(4) \succ x(2) \succ \cdots \succ x(5) \succ x(3) \succ x(1)$$

could be: "I prefer all even numbered alternatives to all odd-numbered alternatives; within each of those two groups I prefer higher-numbered alternatives to lower-numbered alternatives."

The problem with this is that while there are only countably many orderings with finite descriptions, the set $L(X)$ is uncountable. So there are orderings with no finite description and how are such orderings to be submitted?

Acknowledgements We would like to thank Lauren Merrill for technical assistance in preparing this manuscript for publication.

References

Arrow, K. J. (1963). *Social choice and individual values* (2nd ed.). New York: Wiley.
Birkhoff, G. (1940). *Lattice theory*. Providence: American Mathematical Society.
Campbell, D. E. & Fishburn, P. C. (1980). Anonymity conditions in social choice theory. *Theory and Decision, 12*, 21–39.
Campbell, D. E., & Kelly, J. S. (2000). Information and preference aggregation. *Social Choice and Welfare, 17*, 85–93.
Campbell, D. E., & Kelly, J. S. (2007). Social welfare functions that satisfy Pareto, anonymity, and neutrality, but not IIA. *Social Choice and Welfare, 29*, 69–82.
Fishburn, P. C. (1970). *Utility theory for decision making*. New York: Wiley.
Fishburn, P. C. (1973). *The theory of social choice*. Princeton: Princeton University Press.
Fishburn, P. C. (1987). *Interprofile conditions and impossibility*. Chur, Switzerland: Harwood.
Fishburn, P. C. & Kelly, J. S. (1997). Super-Arrovian domains with strict preferences. *SIAM Journal of Discrete Mathematics, 10*. 83–95.
Rader, T., (1972). *Theory of microeconomics*. New York: Academic.
Rae, D. W., & Schickler, E. (1997). Majority rule. In D. C. Mueller (Ed.) *Perspectives on public choice* (pp. 163–180). Cambridge: Cambridge University Press.
Sen, A. K. (1970). *Collective choice and social welfare*. San Francisco: Holden Day.

Fair Division

Bruhat Orders and the Sequential Selection of Indivisible Items

Brian Hopkins and Michael A. Jones

1 Introduction

For two players with identical preferences, cake-cutting procedures, such as Cut-and-Choose (Brams & Taylor, 1996) and the Surplus Procedure (Brams, Jones, & Klamler, 2006), guarantee that each player receives exactly half of the cake, according to their preferences. In essence, receiving exactly half is a worst-case scenario because when their preferences are not identical, the opportunity often exists for both players to receive more than half of the cake, measured by their preferences. This potential reward is balanced by risk, as these differences in preferences provide an incentive for players to misrepresent their preferences in an effort to gain a more valuable piece. In contrast, players may not be able to exploit information about an opponent's preferences when indivisible objects are allocated to two players, even when the players' preferences are different. Our purpose is to determine the structure of, relationship between, and frequency of two players' preferences for which players receive their worst or best possible outcomes when dividing a finite set of indivisible goods, independent of strategic behavior.

Kohler and Chandrasekaharan (1971) pose and solve three optimization problems in which a finite set of players, with linear preference orders over the items, alternate taking turns selecting a number of items from a set of indivisible items. We adopt their framework, as Brams and Straffin (1979) do, to the case when two players alternate selecting a single item from a set of indivisible items. Although Kohler and Chandrasekaharan (1971) assume that players have values associated with each item and subsets are valued according to the sum of the values of its objects, like Brams and Straffin (1979), we assume that the players' preferences for subsets of items are partially ordered, induced by the linear orders.

M.A. Jones (✉)
Department of Mathematical Sciences, Montclair State University, Montclair, NJ, USA
e-mail: jonesm@mail.montclair.edu

S.J. Brams et al. (eds.), *The Mathematics of Preference, Choice and Order: Essays in Honor of Peter C. Fishburn*, Studies in Choice and Welfare,

Brams and Fishburn (2000) consider fair divisions of a finite set of indivisible items between two players with identical linear orders over the individual items, but with possibly different preferences over subsets of items. They do not require the players to alternate selecting items, but instead consider alternative procedures that award possibly different-sized subsets of items to the players. Edelman and Fishburn (2001) extend the work to three or more players. Their focus on identical linear orders highlights potentially contentious divisions. When two players with identical preferences alternate taking turns selecting an item, the division is no longer contentious in that the subsets of items the players receive are the same, whether or not either player acts strategically. In this instance, both players receive their worst possible subsets of items.

Brams, Edelman, and Fishburn (2003) also require linear orders over items, but allow for players to have different preferences. As in Brams and Fishburn (2000) and Edelman and Fishburn (2001), they examine the compatibility of various fairness criteria, including an evenness criterion, in which no player receives more than one item more than any other player. By having the players select alternately, our outcomes are guaranteed to satisfy the evenness property. For players with reversed rankings over items, once again, as in the case in which players have identical preferences, sequentially selecting items results in the same outcomes, whether or not either player acts strategically. Notably, players with reversed linear orders both receive their respective best outcomes. We determine necessary and sufficient conditions for players to receive their worst and best outcomes.

In Sect. 2, we formulate and extend algorithms from Kohler and Chandrasekaharan (1971) to our context. In Sect. 3, we survey background material on the combinatorics of permutations that we will use. In Sect. 4, we apply the combinatorics of permutations to sequential selection procedures, focusing on preferences where strategy has no impact and the outcomes are the best or worst possible for either of the players, or both.

We write $\lceil x \rceil$ for the least integer greater than or equal to x (the integer ceiling), and $\lfloor x \rfloor$ for the greatest integer less than or equal to x (the integer floor). We write $|A|$ for the cardinality of a set A.

2 Strategies for Sequential Selection

Assume two players, Left and Right, have linear preference orders over n indivisible items. Label the items so that Left's preferences are given by the ordering $1 \ldots n$. Right's preferences are given by some permutation $\pi \in S_n$ where $\pi(i) = j$ indicates that item j is ranked in the ith position by Right. Starting with Left, the two players alternate selecting an unclaimed item. Notice that if n is odd, Left will finish with one more item than Right. Let ℓ_i denote the item selected by Left in round i. Define r_i similarly.

Each player's linear order induces a partial ordering $\leq$ on sets of k items. Assume that the elements in different sets of items $S = \{s_1, \ldots, s_k\}$ and $T = \{t_1, \ldots, t_k\}$ are arranged in decreasing preference to a player. Then a player prefers S to T if the player prefers s_i to t_i or $s_i = t_i$ for all i. As an example, Left prefers $\{1,2,4\}$ to both $\{1,3,4\}$ and $\{1,2,5\}$. Yet, Left's linear order does not indicate a preference between $\{1,3,4\}$ and $\{1,2,5\}$. See Fig. 3 for Left's partial order on subsets of three elements from $\{1,2,3,4,5\}$.

There are four strategy combinations that we analyze, as either player may use a naïve approach or be strategic. The details of each combination are described as an algorithm and worked out for Right's linear order $\pi = 231645$.

Definition 1 (Naïve, Naïve). Both players follow the "top-down" strategy of selecting the most preferred remaining item, according to his or her linear order.

For $\pi = 231645$, Left and Right select items in the following sequence $\ell_1 = 1$, $r_1 = 2$; $\ell_2 = 3$, $r_2 = 6$; and $\ell_3 = 4$, $r_3 = 5$. We write $L_{NN}(\pi) = \{1,3,4\}$ and $R_{NN}(\pi) = \{2,5,6\}$ to denote this outcome.

Definition 2 (Strategic, Naïve). Left selects item 1 in round $\lceil \frac{j}{2} \rceil$ where $\pi(j) = 1$. For $i = 2, \ldots, n$, let $\pi(j) = i$. Left selects item i in round k where k is the greatest positive integer less than or equal to $\lceil \frac{j}{2} \rceil$ for which an item has not been determined. If no such k exists, then Left would only be able to select item i at the expense of a preferred item (less than i). Left continues to fill items for the rounds until she has an item for each round. Right selects naïvely, as above.

The strategy uses Left's knowledge of Right's linear order to determine the latest round that an item will be available. For $\pi = 231645$, Left does not select item 1 first, but in round $\lceil \frac{3}{2} \rceil = 2$ because $\pi(3) = 1$. Left selects item 2 in round $\lceil \frac{1}{2} \rceil = 1$ because $\pi(1) = 2$. For ℓ_3, Left would have to select item 3 in round $\lceil \frac{2}{2} \rceil = 1$ because $\pi(2) = 3$, but prefers $\ell_1 = 2$. Left selects item 4 in round $\lceil \frac{5}{2} \rceil = 3$ because $\pi(5) = 4$, completing all three rounds. This gives $L_{SN}(\pi) = \{1,2,3\}$, the best possible outcome for Left.

This is an application of a more general result of Kohler and Chandrasekaharan (1971). They solve a one-sided optimization problem in which Left selects items with complete information about Right's preferences and in which Right has no knowledge of Left's preferences, thereby selecting items naïvely. Their set up is more general, allowing for players to select more than one item and a different number of items each round.

Definition 3 (Naïve, Strategic). For $i = 1, \ldots, n$, Right selects item $\pi(i)$ in round k where k is the greatest positive integer less than or equal to $\lfloor \frac{\pi(i)}{2} \rfloor$, for which an item has not been determined. If no such k exists, then either Right would have to select item $\pi(i)$ at the expense of an item ranked higher than $\pi(i)$ or Right would have to select before Left, in the event that $\pi(1) = 1$. Right continues to fill rounds with items to select until he has an item for each round. Left selects naïvely.

This is the other possibility of one-sided information. After Left takes item 1 in the first round, this is the (Strategic, Naïve) algorithm with the roles reversed. For $\pi = 231645$. Right selects item $\pi(1) = 2$ in round $\lfloor \frac{2}{2} \rfloor = 1$. Right could only select item $\pi(2) = 3$ in round $\lfloor \frac{3}{2} \rfloor = 1$, but he prefers $\pi(1)$ to $\pi(2)$. Right cannot select item $\pi(3) = 1$, as $\ell_1 = 1$ (the algorithm would require round $\lfloor \frac{1}{2} \rfloor = 0$). Finally, Right selects item $\pi(4) = 6$ in round $\lfloor \frac{6}{2} \rfloor = 3$. Left selects naïvely in her linear order. The optimality of this approach for Right is basically the same argument for (Strategic, Naïve).

Definition 4 (Strategic, Strategic). For $n = 2k$, consider the rounds in reverse order, Right first. Each player selects the least-preferred item remaining on the other's linear order after items selected in later rounds have been deleted. So $r_k = 2k$, Left selects the last remaining item in Right's linear order for ℓ_k, etc. For $n = 2k-1$, start with $\ell_k = \pi(2k)$ and proceed as above. Actual selection begins ℓ_1, r_1.

Kohler and Chandrasekaharan (1971) show this procedure is optimal for both players when each player is aware of the other's linear order. Brams and Straffin (1979) refer to this algorithm as "bottom-up," because players fill the bottom round selections first. Essentially, this algorithm determines which item a player gets left with in a particular round. For $\pi = 231645$, start from the bottom with $r_3 = 6$ and $\ell_3 = 5$. Next, $r_2 = 4$ (as 5 and 6 are removed from Left's linear ordering) and $\ell_2 = 1$ (as 4, 5, and 6 are removed). Finally, $r_1 = 3$ and $\ell_1 = 2$.

Table 1 summarizes the examples of selecting items under the four strategy combinations. The items selected by the players, according to the column-heading strategies, are in numerical order. The ordered pairs below the columns indicate the order by rounds in which the items were selected.

The examples collected in Table 1 show that the same preferences for Right can lead to four different outcomes depending on the approaches taken by the two players. At the other extreme, there are also rankings for which strategy makes no difference. The following statements follow from more general proofs in Sect. 4, but these two specific cases may be self evident.

If Right's preferences are identical to Left's, i.e., if $\pi = 1 \ldots n$, then all possible strategy combinations lead to the same result. We write $L_{XY}(\pi) = \{1, 3, \ldots\}$ and $R_{XY}(\pi) = \{2, 4, \ldots\}$ to denote that the outcomes are the same for all strategy combinations (X, Y). This is the worst possible outcome for Left, and also the worst possible outcome for Right.

Table 1 Outcomes under different strategy pairs for Right's preferences $\pi = 231645$ followed by $(\ell_1, r_1)(\ell_2, r_2)(\ell_3, r_3)$

N	N	S	N	N	S	S	S
1	2	1	3	1	2	1	3
3	5	2	5	3	4	2	4
4	6	4	6	5	6	5	6
(1,2) (3,6) (4,5)		(2,3) (1,6) (4,5)		(1,2) (3,4) (5,6)		(2,3) (1,4) (5,6)	

If Right's preferences are the reverse of Left's, i.e., if $\pi = n \ldots 1$, then again all possible strategy combinations lead to the same result: $L_{XY}(\pi) = \{1, \ldots, \lceil n/2 \rceil\}$ and $R_{XY}(\pi) = \{\lceil n/2 \rceil + 1, \ldots, n\}$. This yields the best possible outcome for Left and also the best possible outcome for Right.

We are interested in analyzing the preferences for which strategy has no effect. In particular, the preferences that lead to the worst and best possible cases for Left, and then for Right. We conclude with a description of the preferences, like $1 \ldots n$ and $n \ldots 1$, for which *both* Left and Right receive the worst or best possible outcomes.

3 Partial Orders on Permutations

In this section, we define two partial orders on permutations and collect various properties that will be used later. See the references for further information. We denote a permutation $\pi \in S_n$ by $\pi(1) \ldots \pi(n)$, the one-line notation. Write e for the identity permutation $1 \ldots n$ and w_0 for $n \ldots 1$. To each permutation, we associate the following sets of ordered pairs.

Definition 5. For $\pi \in S_n$, let $LI(\pi) = \{(i, j) \mid i < j \text{ and } \pi(\text{i}) > \pi(\text{j})\}$ and $RI(\pi) = \{(\pi(i), \pi(j)) \mid i < j \text{ and } \pi(\text{i}) > \pi(\text{j})\}$.

Note that $LI(\pi)$ records the positions of decreasing pairs, while $RI(\pi)$ records the decreasing pairs themselves. Examples are given in the Table 2.

Each of these sets is called the inversions of a permutation by various authors; see Björner and Brenti (2005) and Bóna (2004). These sets are the basis for two partial orders on partitions.

Definition 6. Let $\pi, \sigma \in S_n$. If $LI(\pi) \subseteq LI(\sigma)$, then $\pi \leq_L \sigma$. If $RI(\pi) \subseteq RI(\sigma)$, then $\pi \leq_R \sigma$.

Referring to Table 2, $142536 \leq_L 142635$ but $142536 \not\leq_R 142563$ because $(5,3) \notin RI(142635)$. Similarly, $142536 \leq_R 142563$ but $142536 \not\leq_L 142563$ because $(4,5) \notin LI(142563)$. Also note that 142563 and 142635 are incommensurate under each partial order.

The partial orders $\leq_L$ and $\leq_R$ are called the left and right weak Bruhat orders. The names come from alternate definitions of the orders, involving left or right multiplication of a permutation by neighborly transpositions. We will write the permutation 123465 with the standard transposition notation $(5,6)$. Notice that $(5,6)$

Table 2 Left and right inversions for three elements of S_6

π	$LI(\pi)$	$RI(\pi)$
142536	$\{(2,3),(2,5),(4,5)\}$	$\{(4,2),(4,3),(5,3)\}$
142563	$\{(2,3),(2,6),(4,6),(5,6)\}$	$\{(4,2),(4,3),(5,3),(6,3)\}$
142635	$\{(2,3),(2,5),(4,5),(4,6)\}$	$\{(4,2),(4,3),(6,3),(6,5)\}$

composed with 142536 gives 142635 (the 5 and 6 are swapped), while 142536 composed with $(5,6)$ gives 142563 (the fifth and sixth entries are swapped). See Fig. 1 for the elements of S_4 under the right weak Bruhat order (ignore the circles at this point).

Proposition 1. *Each of the weak Bruhat orders on S_n is an ortholattice, with rank $|LI(\pi)| = |RI(\pi)|$ which is also the standard length of permutation π. In addition, the left and right weak Bruhat orders are related by $\pi \leq_L \sigma$ if and only if $\pi^{-1} \leq_R \sigma^{-1}$, because $LI(\pi) = RI(\pi^{-1})$.*

Proofs of these results, including the equivalence of the different definitions, may be found in Björner and Brenti (2005) and Aguiar and Mahajan (2006). There is one additional definition we will use.

Definition 7. Let $\pi, \sigma \in S_n$. The interval $[\pi, \sigma]_L = \{\rho \in S_n \mid \pi \leq_L \rho \leq_L \sigma\}$. Define $[\pi, \sigma]_R$ similarly.

In Fig. 1, the permutations surrounded by four closed circles constitute $[1324, 4321]_R$.

4 Results

We now apply the notions from algebraic combinatorics to the analysis of linear preference orders. Throughout, we identify Right's linear order and its corresponding permutation.

Theorem 1. *For $n = 2k-1$ or $2k$, there are C_k possible outcomes for Left, where $C_k = \frac{1}{k+1}\binom{2k}{k}$ is the kth Catalan number.*

Proof. Let $n = 2k$. Notice that the tables showing the results of the various strategy combinations in Table 1 may be considered as Young tableaux with two columns having length 3. For $n = 2k$, there would be two length k columns filled with the labels $1, \ldots, 2k$ such that labels increase down columns and across rows. Conversely, a Young tableau having two length k columns with first column $\ell_1 < \cdots < \ell_k$ comes from the (Naïve, Naïve) situation when Right's preferences are $\ell_1 r_1 \ldots \ell_k r_k$ where $\{r_1, \ldots, r_k\}$ is the complement of $\{\ell_1, \ldots \ell_k\}$ listed in increasing order. (E.g., to have $L_{NN}(\pi) = \{1,2,3\}$, use $\pi = 142536$). Young tableaux of this shape are among the many items counted by the Catalan numbers; see Stanley (1999).

For $n = 2k-1$, put an additional item $2k$ at the end of Right's list, which adds $2k$ to the bottom of the second column, and the same arguments apply.

The following theorem establishes the connection between strategies for sequential selection and the weak right Bruhat order.

Theorem 2. *Let $\pi, \sigma \in S_n$ with $\pi \leq_R \sigma$, the right weak Bruhat order. In the partial order of Left's outcomes, $L_{XY}(\pi) \leq L_{XY}(\sigma)$ for each of the four strategy combinations XY.*

Proof. It is sufficient to show that $L_{XY}(\pi) \leq L_{XY}(\sigma)$ for $\sigma = \pi\rho$ where ρ is the transposition $(i, i+1)$ with $|RI(\pi\rho)| = |RI(\pi)| + 1$. That is,

$$\sigma = \pi(1)\cdots\pi(i-1)\pi(i+1)\pi(i)\pi(i+2)\cdots\pi(n),$$

where $\pi(i) < \pi(i+1)$. The relationship between $L_{XY}(\pi)$ and $L_{XY}(\sigma)$ does not depend on the strategy pair XY, but on which players receive $\pi(i)$ and $\pi(i+1)$.

Assume $\pi(i)$ and $\pi(i+1)$ are both in $L_{XY}(\pi)$ or $R_{XY}(\pi)$. Under σ, $\pi(i)$ and $\pi(i+1)$ are selected by the same player in the same rounds, only the order of their selection may change. Hence, $L_{XY}(\pi) = L_{XY}(\sigma)$.

Assume $\pi(i) \in L_{XY}(\pi)$ and $\pi(i+1) \in R_{XY}(\pi)$. Under σ, Left selects $\pi(i)$ and Right selects $\pi(i+1)$ in the same rounds as under π. Hence, $L_{XY}(\pi) = L_{XY}(\sigma)$.

Assume $\pi(i) \in R_{XY}(\pi)$ and $\pi(i+1) \in L_{XY}(\pi)$. Under π for $XY = NN$, SS, or SN, if Right selects $\pi(i)$ in round r, then Left must select $\pi(i+1)$ in round $r+1$. Under σ, Right selects $\pi(i+1)$ in round r, allowing Left to take $\pi(i)$ in round $r+1$. Under π for $XY = NS$, from the floor function calculation, no round exists for Right to select $\pi(i+1)$ without giving up a preferred item. Under σ, $\pi(i+1)$ is selected instead of $\pi(i)$. And, no round exists for Right to select $\pi(i)$. Because Left now receives $\pi(i)$ instead of $\pi(i+1)$, $L_{XY}(\pi) \leq L_{XY}(\sigma)$.

Figure 1 shows S_4 under the weak right Bruhat order and gives the results of all four strategy combinations for each linear order (see caption for details). Notice that, moving left to right along edges, i.e., from π to σ with $\pi \leq_R \sigma$, there is never a transition in a given position from a filled circle to an open circle, illustrating Theorem 2.

We provide complete descriptions for linear orders that lead to the worst and best outcomes for Left.

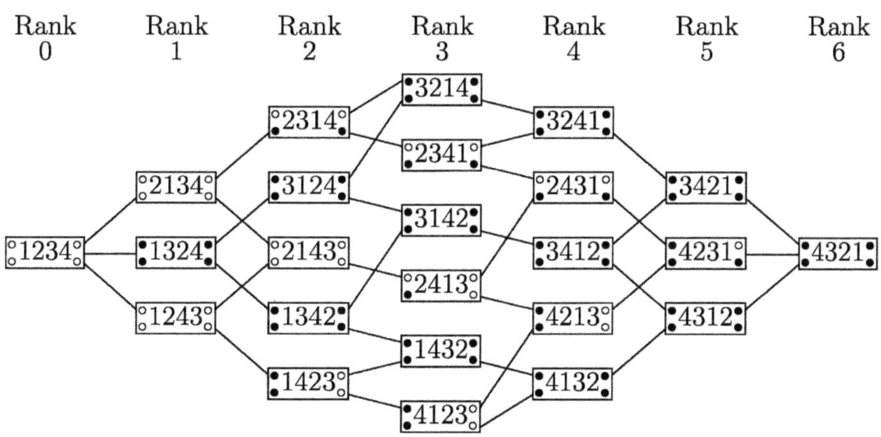

Fig. 1 S_4 under the right weak Bruhat order with outcomes for the four strategy combinations: *open circle* for Left outcome $\{1,3\}$, *closed circle* for Left outcome $\{1,2\}$ with *NN* to the *upper left* of the permutation, *NS* to the *upper right*, *SN* to the *lower left*, and *SS* to the *lower right*

Theorem 3. *The $\pi \in S_{2k}$ for which $L_{XY}(\pi) = \{1,3,\ldots,2k-1\}$, the worst possible outcome for Left, are the 2^k permutations $\pi \in [e, 2143\ldots(2k)(2k-1)]_R$.*

The $\pi \in S_{2k+1}$ with $L_{XY}(\pi) = \{1,3,\ldots,2k+1\}$ are the 2^k permutations $\pi \in [e, 2143\ldots(2k)(2k-1)(2k+1)]_R$.

Proof. Since the best of the four possible outcomes for Left occurs when Left is strategic and Right is naïve, it suffices to show that $L_{SN}(\pi)$ is the worst possible outcome for Left.

Assume first that $n = 2k$. The interval $[e, 2143\ldots(2k)(2k-1)]_R$ comprises permutations that are the product of any of the neighborly transpositions $(1,2),(3,4),\ldots,(2k-1,2k)$ (because these transpositions are disjoint, they commute and there is no need to distinguish left and right multiplication). I.e., these are the permutations for which $\{\pi(2i-1),\pi(2i)\} = \{2i-1,2i\}$ for all $i = 1,\ldots,k$.

Recall the (Strategic, Naïve) algorithm. Left can delay selecting item i until round $\lceil \pi(i)/2 \rceil$. But for the permutations in $[e, 2143\ldots(2k)(2k-1)]_R$, this is not different from the naïve approach, because $\lceil \pi(i)/2 \rceil = i$. On turn i, Left will choose $2i-1$ and Right, choosing the highest remaining item on his list, will choose $2i$. This results in $L_{SN}(\pi) = \{1,3,\ldots,2k-1\}$.

A permutation $\sigma \notin [e, 2143\ldots(2k)(2k-1)]_R$ is characterized by having $\sigma(2i) \notin \{2i-1,2i\}$ for some $i = 1,\ldots,k$. If $\sigma(2i) > 2i$, then Left's strategy can delay her choice of $2i$ to a later turn, allowing her to chose an item with lower label in round i. If $\sigma(2i) < 2i-1$, then Right will select $2i$ before round i, say in round j, which allows Left to take item $2j$.

For $n = 2k+1$, the permutations arise from the same transpositions and all have $\pi(2k+1) = 2k+1$; the same arguments apply. The count follows from the description of the permutations as products of k disjoint transpositions; the cases may be unified by writing $2^{\lfloor n/2 \rfloor}$. Also, the terminal element of each interval has rank $\lfloor n/2 \rfloor$.

We mention that $[e, 2143\ldots(2k)(2k-1)]_R$ (also an interval under $\leq_L$) is a Boolean sublattice, isomorphic to the k-dimensional hypercube.

Theorem 4. *The $\pi \in S_{2k-1}$ for which $L_{XY}(\pi) = \{1,\ldots,k\}$, the best possible outcome for Left, are the $(2k-1)[(k-1)!]^2$ permutations $\pi \in [1(k+1)(k+2)\ldots(2k-1)\,23\ldots k,\, w_0]_R$.*

The $\pi \in S_{2k}$ with $L_{XY}(\pi) = \{1,\ldots,k\}$ are the $2k(2k-1)[(k-1)!]^2$ permutations $\pi \in [1(k+1)(k+2)\ldots(2k-1)23\ldots k(2k),\, w_0]_R$.

Proof. Since the worst of the four possible outcomes for Left occurs when Left is naïve and Right is strategic, it will suffice to show that $L_{NS}(\pi)$ is the best possible outcome for Left.

Assume first that $n = 2k-1$. The specified interval comprises permutations π for which

$$\{(k+1,2),\ldots,(k+1,k),\ldots,(2k-1,2),\ldots,(2k-1,k)\} \subseteq RI(\pi). \qquad (1)$$

Recall the algorithm for (Naïve, Strategic). Notice, for the linear orders in this interval, that $2,\ldots,k$ are all preceded by at least $k-1$ items greater than 1, in particular, $k+1,\ldots,2k-1$. That means that Right will use all his rounds selecting items $k+1,\ldots,2k-1$, along with the guaranteed $2k$. Thus $L_{NS}(\pi)=\{1,\ldots,k\}$.

A permutation σ not in the interval has some $(i,j)\notin RI(\sigma)$ where $k+1\leq i\leq 2k-1$ and $2\leq j\leq k$. That is, j appears before i in σ and Right will choose it in round $\lfloor j/2\rfloor$ before Left does, so that $j\notin L_{NS}(\sigma)$.

For the size of the interval, consider the equivalent formulation in terms of $RI(\pi)$. There are $(k-1)!$ ways to order $\{k+1,\ldots,2k-1\}$ in the one-line presentation of π before the set $\{2,\ldots,k\}$, which can also be ordered in $(k-1)!$ ways. Item 1 can be listed in any of the $2k-1$ positions among these $2k-2$ numbers, giving a total of $(2k-1)[(k-1)!]^2$ permutations.

For $n=2k$, Right is guaranteed item $2k$. The same description of $RI(\pi)$ applies. The count is adjusted by the $2k$ positions where item $2k$ can be added into the linear order.

We note that, for both S_{2k-1} and S_{2k}, the initial elements of the intervals have rank $(k-1)^2$.

All five possible outcomes for linear orders on five items are given in Fig. 2. The interval $[12345,21435]_R$ giving $L_{XY}(\pi)=\{1,3,5\}$ is an example of Theorem 3, and the interval $[14523,54321]_R$ giving $L_{XY}(\pi)=\{1,2,3\}$ is an example of Theorem 4. Notice that $[13245,43215]_R$ giving $L_{XY}(\pi)=\{1,2,5\}$ is isomorphic to $[1324,4321]_R$ in Fig. 1, and that the linear orders for which $L_{XY}(\sigma)=\{1,2,4\}$ do not constitute an interval in the lattice.

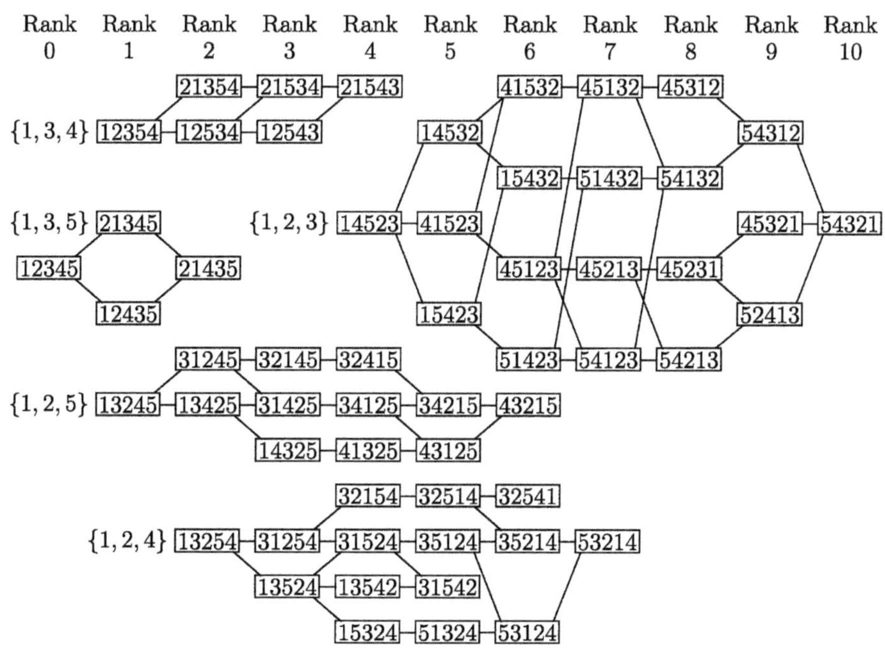

Fig. 2 Subset of the S_5 lattice under $\leq_R$ showing the permutations for which the four $L_{XY}(\pi)$ are the same. Permutations are aligned horizontally by rank

Table 3 Right's five possible outcomes for S_5 (omit 0), followed by a permutation giving each outcome

R	L	R	L	R	L	R	L	R	L
0	$\pi(3)$	0	$\pi(2)$	0	$\pi(2)$	0	$\pi(1)$	0	$\pi(1)$
$\pi(1)$	$\pi(4)$	$\pi(1)$	$\pi(4)$	$\pi(1)$	$\pi(3)$	$\pi(2)$	$\pi(4)$	$\pi(2)$	$\pi(3)$
$\pi(2)$	$\pi(5)$	$\pi(3)$	$\pi(5)$	$\pi(4)$	$\pi(5)$	$\pi(3)$	$\pi(5)$	$\pi(4)$	$\pi(5)$
34125		31425		31245		13425		13245	

We also analyze the outcomes of the various sequential selection procedures from the perspective of the second player, Right. From Table 1, we know $R_{NN}(231645) = \{2,5,6\}$, Right's first, fourth, and sixth items. To emphasize this perspective, we write $R_{NN}(231645) = \{\pi(1),\pi(4),\pi(6)\}$.

Theorem 5. *For $n = 2k-2$ or $2k-1$, there are C_k possible outcomes for Right, where $C_k = \frac{1}{k+1}\binom{2k}{k}$ is the kth Catalan number.*

Proof. Let $n = 2k-1$. We modify the proof of Theorem 1 by placing an item labeled 0 at the beginning of Right's preferences and allowing him to go first, which will have no effect on the (Naïve, Naïve) algorithm after 0 is removed. The resulting tables are then Young tableaux having two length k columns with labels $0,\pi(1),\ldots,\pi(2k-1)$ ordered by their input (see Table 3 for the $n = 5$ possibilities). Conversely, a Young tableau having two length k columns with first column $0,\pi(i_1),\ldots,\pi(i_{k-1})$ where $i_1 < \cdots < i_{k-1}$ arises from the modified (Naïve, Naïve) algorithm with 0 when Right's preferences have $\pi(i_1) = k,\ldots,\pi(i_{k-1}) = 2k-2$ and the complementary numbers are placed in the remaining positions in increasing order; see Table 3 for examples. For $n = 2k-2$, put an additional item $\pi(2k-1)$ at the end of Left's list, which adds $\pi(2k-1)$ to the bottom of the second column.

As Left's outcomes are related to the weak right Bruhat order, Right's outcomes are related to the weak left Bruhat order. The next three theorems are analogous to Theorems 2, 3, and 4 for Left's outcomes. Proofs only note variations from the previous arguments.

Theorem 6. *Let $\pi,\sigma \in S_n$ with $\pi \leq_L \sigma$, the left weak Bruhat order. In the partial order of Left's outcomes, $R_{XY}(\pi) \leq R_{XY}(\sigma)$ for all four strategy combinations XY.*

Proof. It is sufficient to show that $R_{XY}(\pi) \leq R_{XY}(\sigma)$ for $\sigma = \rho\pi$ where ρ is the transposition $(i,i+1)$ with $|LI(\pi\rho)| = |LI(\pi)|+1$. That is,

$$\sigma = \pi(1)\cdots\pi(j-1)\pi(k)\pi(j+1)\cdots\pi(k-1)\pi(j)\pi(k+1)\cdots\pi(n),$$

where $\pi(j) = i$, $\pi(k) = i+1$, and $j < k$.

A similar analysis as the proof of Theorem 2 holds by considering the four cases for which i or $i+1$ may be in $R_{XY}(\pi)$. The result hinges on i and $i+1$ being consecutive in Left's linear order, as $\pi(i)$ and $\pi(i+1)$ are consecutive in Right's linear order.

Theorem 7. *The* $\pi \in S_{2k+1}$ *for which* $R_{XY}(\pi) = \{\pi(2), \pi(4), \ldots, \pi(2k)\}$, *the worst possible outcome for Right, are the* 2^k *permutations* $\pi \in [e, 132\ldots(2k+1)(2k)]_L$.

The $\pi \in S_{2k+2}$ *with* $R_{XY}(\pi) = \{\pi(2), \pi(4), \ldots, \pi(2k+2)\}$ *are the* 2^k *permutations* $\pi \in [e, 132\ldots(2k+1)(2k)(2k+2)]_L$.

Proof. The intervals comprise permutations that are the product of any of the commuting neighborly transpositions $(2,3), (4,5), \ldots, (2k,\ 2k\ +\ 1)$. The cases may be unified by writing $2^{\lfloor (n-1)/2 \rfloor}$. Also, the terminal element of each interval has rank $\lfloor (n-1)/2 \rfloor$.

Theorem 8. *The* $\pi \in S_{2k-2}$ *for which* $R_{XY}(\pi) = \{\pi(1), \ldots, \pi(k-1)\}$, *the best possible outcome for Right, are the* $[(k-1)!]^2$ *permutations* $\pi \in [k\ldots(2k-2)1\ldots(k-1), w_0]_L$.

The $\pi \in S_{2k-1}$ *with* $R_{XY}(\pi) = \{\pi(1), \ldots, \pi(k-1)\}$ *are the* $(2k\ -\ 1)[(k-1)!]^2$ *orders* $\pi \in [k\ldots(2k-2)1\ldots(k-1)(2k-1), w_0]_L$.

Proof. The intervals comprise permutations π for which

$$\{(1,k), \ldots, (1,2k-2), \ldots, (k-1,k), \ldots, (k-1,2k-2)\} \subseteq LI(\pi). \quad (2)$$

For $n = 2k-2$, there are $(k-1)!$ ways to order $\{k, \ldots, 2k-2\}$ in the one-line presentation of π before the set $\{1, \ldots, k-1\}$, which can also be ordered in $(k-1)!$ ways. For $n = 2k-1$, the count is adjusted by the $2k-1$ positions where item $2k-1$ can be added into the linear order. The initial elements of the intervals have rank $(k-1)^2$.

For S_{2k-1}, the sizes of the intervals for the permutations giving the worst possible case for Left and Right match, similarly for the best possible case, and both players have C_k possible outcomes. In fact, their situations are completely analogous.

Proposition 2. *For* $\pi \in S_n$, *let* $\sigma = w_0 \pi w_0$. *There is a bijection between the left inversions of* π *and* σ, *similarly the right inversions.*

Proof. The mapping $\pi \mapsto w_0 \pi w_0$ is an automorphism whose effect is to reverse the order of the entries in the one-line presentation and to replace x with $n+1-x$. Suppose, for positions $i < j$, that $\pi(i) = x$ and $\pi(j) = y$ with $x > y$. That is, $(i,j) \in LI(\pi)$ and $(x,y) \in RI(\pi)$. Then $\sigma(n+1-j) = n+1-y$ and $\sigma(n+1-i) = n+1-x$, so that $(n+1-j, n+1-i) \in LI(\sigma)$ and $(n+1-y, n+1-x) \in RI(\sigma)$.

This automorphism preserves rank, inclusion, and all lattice structure, switching between the weak left and weak right Bruhat orders; see Björner and Brenti (2005) for more details. Under the connections to Left and Right outcomes established in Theorems 2 and 6, it follows that the Left outcome structure of S_{2k-1} under the weak right Bruhat order is isomorphic to the Right outcome structure of S_{2k-1} under the left weak Bruhat order. For example, the correspondences between the five sets of permutations in S_5 that give various outcomes for Left and Right regardless of strategy are given in Fig. 3; this is equivalent to applying the automorphism to the permutations in Fig. 2.

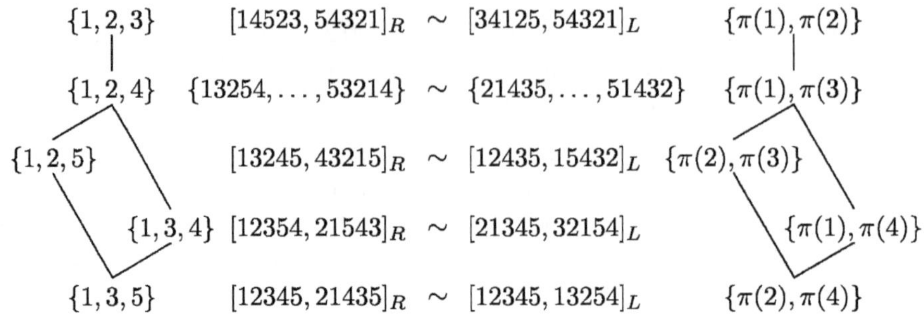

Fig. 3 Permutations giving the specified Left outcome for S_5 with corresponding data for Right outcomes, along with the partial order on outcomes

The two intervals in the first row of Fig. 3 share 12 elements, $[45123, 54321]_R$ (also an interval in the left weak Bruhat order). That is, there are 12 permutations for which Left and Right both receive the best possible outcome under all strategy combinations. The two intervals in the last row of Fig. 3 have only e in common, the one permutation in S_5 for which Left and Right both receive the worst possible outcome under all strategy combinations.

The connections are not so direct for S_{2k}, as there are C_k outcomes for Left and C_{k+1} outcomes for Right. For example, Fig. 1 shows occurrences of the two possible outcomes for Left, while the analogous figure for Right would require five symbols to denote his outcomes. Nonetheless, we can completely characterize the permutations in arbitrary S_n for which Left and Right both receive the worst possible outcome under all strategy combinations, and the same for best possible outcome.

Theorem 9.

(a) Only $e = 1\dots n \in S_n$ *gives* $L_{XY}(\pi) = \{1,3,\dots\}$ *and* $R_{XY}(\pi) = \{\pi(2), \pi(4), \dots\}$, *the worst possible outcomes for both Left and Right.*

(b) The $\pi \in S_{2k-2}$ *for which* $L_{XY} = \{1,\dots,k-1\}$ *and* $R_{XY}(\pi) = \{\pi(1),\dots,\pi(k-1)\}$, *the best possible outcomes for Left and Right, are the* $[(k-1)!]^2$ *permutations*

$$\pi \in [k\dots(2k-2)1\dots(k-1), w_0]_L.$$

(c) The $\pi \in S_{2k-1}$ *for which* $L_{XY} = \{1,\dots,k\}$ *and* $R_{XY}(\pi) = \{\pi(1),\dots,\pi(k-1)\}$ *are the* $k!(k-1)!$ *permutations*

$$\pi \in [(k+1)\dots(2k-1)1\dots k, w_0]_L.$$

Proof. (a) The permutations that give the worst possible outcome for Left, described in Theorem 3, are generated by the transpositions $(1,2), (3,4), \dots$ The permutations that give the worst possible outcome for Right, described in Theorem 7, are generated by the transpositions $(2,3), (4,5), \dots$ Therefore, e is the only permutation in both intervals.

(b) For $n = 2k-2$, the initial permutation of the interval from Theorem 8 giving the best outcomes for Right is in the interval from Theorem 4 giving the

best outcomes for Left. That is, the right inversions of $k\ldots(2k-2)1\ldots(k-1)$ include everything listed in (1) (with the index shifted down one) along with $(k,1),\ldots,(2k-2,1)$. So the intersection of the two intervals is exactly the interval from Theorem 8.

(c) For $n=2k-1$, the intersection of the intervals from Theorems 4 and 8 comprises permutations having both the right inversions listed in (1) and the left inversions listed in (2). The lowest rank permutation satisfying both sets of conditions is $(k+1)\ldots(2k-1)1\ldots k$, rank $k(k-1)$. For the size of the interval, there are $(k-1)!$ ways to order $\{k+1,\ldots,2k-1\}$ before the set $\{1,\ldots k\}$, which can be ordered in $k!$ ways.

We believe that these are only initial steps in the application of the combinatorics of permutations to the analysis of sequential selection, and we look forward to further investigations along these lines.

We would like to thank Steven J. Brams for the invitation to contribute to this volume. Extensive computation by Carl E. Bredlau of Montclair State University was foundational for this research.

References

Aguiar, M., & Mahajan, S. (2006). *Coxeter groups and hopf algebras*. Providence: American Mathematical Society.

Björner, A., & Brenti, F. (2005). *Combinatorics of coxeter groups*. New York: Springer.

Bóna, M. (2004). Combinatorics of permutations. Boca Raton: Chapman & Hall / CRC.

Brams, S. J., Edelman, P. H., & Fishburn, P. C. (2003). Fair division of indivisible items. *Theory and Decision, 55*, 147–180.

Brams, S. J., & Fishburn, P. C. (2000). Fair division of indivisible items between two people with identical preferences: Envy-freeness, Pareto-optimality, and equity. *Social Choice and Welfare, 17*, 247–267.

Brams, S. J., Jones, M. A., & Klamler, C. (2006). Better ways to cut a cake. *Notices of the American Mathematical Society, 53*, 1314–1321.

Brams, S. J., & Straffin, P. D. (1979). Prisoners' dilemma and professional sports drafts. *The American Mathematical Monthly, 86*, 80–88.

Brams, S. J., & Taylor, A. D. (1996). Fair division: From cake-cutting to dispute resolution. Cambridge, UK: Cambridge University Press.

Edelman, P., & Fishburn, P. (2001) Fair division of indivisible items among people with similar preferences. *Mathematical Social Sciences, 41*, 327–347.

Kohler, D. A., & Chandrasekaharan, R. (1971). A class of sequential games. *Operations Research, 19*, 270–277.

Stanley, R. P. (1999). *Enumerative combinatorics* (Vol. 2). New York: Cambridge University Press.

Part III
Posets, Graphs, Combinatorics, and Related Applied and Mathematical Topics

Partial Orders and Interval Orders

Fractional Weak Discrepancy of Posets and Certain Forbidden Configurations

Alan Shuchat, Randy Shull, and Ann N. Trenk

1 Introduction

A *weak order* is a poset $P = (V, \prec)$ that can be assigned a real-valued function $f : V \to \mathbf{R}$ so that $a \prec b$ in P if and only if $f(a) < f(b)$ Bogart (1990). Thus, the elements of a weak order can be ranked by a function that respects the ordering $\prec$ and issues a tie in ranking between incomparable elements ($a \parallel b$). When P is not a weak order, it is not possible to resolve ties as fairly. The *weak discrepancy* of a poset, introduced in Trenk (1998) as the *weakness* of a poset, is a measure of how far a poset is from being a weak order [Gimbel and Trenk (1998); Tanenbaum, Trenk, & Fishburn (2001)]. In Shuchat, Shull, and Trenk (2007), the problem of determining the weak discrepancy of a poset was formulated as an integer program whose linear relaxation yields a fractional version of weak discrepancy given in Definition 1 below.

Definition 1. The *fractional weak discrepancy* $wd_F(P)$ of a poset $P = (V, \prec)$ is the minimum nonnegative real number k for which there exists a function $f : V \to \mathbf{R}$ satisfying

(i) if $a \prec b$ then $f(a) + 1 \leq f(b)$ ("up" constraints)
(ii) if $a \parallel b$ then $|f(a) - f(b)| \leq k$. ("side" constraints)

Such a function f is called an *optimal fractional weak labeling* of P (or of V).

As an example, consider the salary assignment problem described in Shuchat, Shull, & Trenk (2006). A manager wishes to assign a salary $f(a)$ to each employee a in her division in a fair way. She can partially order the employees in her division based on their value to the company. The "up" constraints ensure that a more valuable employee receives a higher salary. The "side" constraints are fairness

A. Shuchat (✉)
Department of Mathematics, Wellesley College, Wellesley, MA 02481, USA
e-mail: ashuchat@wellesley.edu

S.J. Brams et al. (eds.), *The Mathematics of Preference, Choice and Order: Essays in Honor of Peter C. Fishburn*, Studies in Choice and Welfare,

conditions that restrict the salary discrepancies between incomparable employees. For a weak order, sets of pairwise incomparable employees (antichains) are assigned the same salary level and the fractional weak discrepancy is zero ($k = 0$ satisfies the definition). In general, with the appropriate choice of unit the manager can assign $f(a)$ according to Definition 1. The k in this definition is a measure of the fairness of the assignment.

Denote the disjoint union of two chains with r and s elements, respectively, by $\mathbf{r}+\mathbf{s}$. A number of important classes of posets can be characterized in terms of forbidden $\mathbf{r}+\mathbf{s}$ configurations. For example, *linear orders* are posets with no induced $\mathbf{1}+\mathbf{1}$, and it is not hard to show that weak orders are posets with no induced $\mathbf{2}+\mathbf{1}$ [Bogart (1990)]. Posets with no induced $\mathbf{2}+\mathbf{2}$ and no induced $\mathbf{3}+\mathbf{1}$ are known as *semiorders*. By a theorem of Scott and Suppes (1958), this class is equivalent to the class of *unit interval orders*, that is, posets which can be represented as follows: each element x of the ground set V is assigned a *unit* length interval I_x on the real number line so that $x \prec y$ if and only if the interval I_x is completely to the left of I_y. In Shuchat et al. (2006) we show how we can use fractional weak discrepancy to characterize the class of semiorders. In particular we establish the following two results.

Theorem 1 (Shuchat et al. (2006)). *If P is a semiorder then $wd_F(P) = \frac{r}{r+1}$ for some integer $r \geq 0$. Furthermore, for each integer $r \geq 0$, there exists a semiorder P with $wd_F(P) = \frac{r}{r+1}$. Equivalently, $\{wd_F(P) : P \text{ a semiorder}\} = \{0, \frac{1}{2}, \frac{2}{3}, \frac{3}{4}, \frac{4}{5}, \ldots\}$.*

Theorem 2 (Shuchat et al. (2006)). *If P is a poset that is not a semiorder then $wd_F(P)$ is a rational number that is at least one. Furthermore, for each rational number $q \geq 1$, there exists a poset P (that is not a semiorder) with $wd_F(P) = q$. Equivalently, $\{wd_F(P) : P \text{ a poset that is not a semiorder}\} = \{q \geq 1 : q \in \mathbf{Q}\}$.*

Combining the results of Theorems 1 and 2, we obtain the following characterization of semiorders.

Corollary 1 (Shuchat et al. (2006)). *A poset P is a semiorder if and only if $wd_F(P) = \frac{r}{r+1}$ for some integer $r \geq 0$.*

Posets possessing no induced $\mathbf{2}+\mathbf{2}$ and/or no $\mathbf{3}+\mathbf{1}$ have been studied extensively beyond the class of semiorders. Relaxing the requirement that the poset contain no $\mathbf{3}+\mathbf{1}$, but retaining our restriction on no induced $\mathbf{2}+\mathbf{2}$, yields the well-known class of *interval orders*. These are, by definition, posets in which each element x can be assigned an interval I_x on the real line so that $x \prec y$ if and only if I_x lies completely to the left of I_y [Fishburn (1985)]. Posets that are $\mathbf{3}+\mathbf{1}$ free but may or may not contain a $\mathbf{2}+\mathbf{2}$ are not as well known as either semiorders or interval orders, but have a number of important properties nevertheless. For example, Stanley's generalization of the chromatic polynomial is s-positive for the incomparability graph of such a poset [Gasharov (1996); Stanley (1995)]. Skandera characterized posets containing no induced $\mathbf{3}+\mathbf{1}$ in terms of their antiadjacency matrices and used this characterization to give a simple proof that the chain polynomial of such posets has only real zeros [Skandera (2001)].

Table 1 Summary of results for the range of wd_F

	No **3+1**	Yes **3+1**
	semiorders	
No **2+2**	$\{wd_F(P)\} = \{\frac{r}{r+1}\}$	$\{wd_F(P)\} = \{\text{rationals} \geq 1\}$
interval orders	(Corollary 1)	(Theorems 2, 5)
Yes **2+2**	$wd_F(P) = 1$	$\{wd_F(P)\} = \{\text{rationals} \geq 1\}$
	(Corollary 4)	(Theorem 4; Corollary 3; Fig. 3)

In this paper we study the fractional weak discrepancy of posets obtained by selectively relaxing the restrictions on induced $\mathbf{2}+\mathbf{2}$ and induced $\mathbf{3}+\mathbf{1}$. Together with Theorem 2, Theorem 5 will imply that the range of the fractional weak discrepancy function for interval orders (no induced $\mathbf{2}+\mathbf{2}$) that are not semiorders (contain an induced $\mathbf{3}+\mathbf{1}$) is precisely the set of all rational numbers greater than or equal to 1. Indeed, Corollary 3 states that any poset with fractional weak discrepancy greater than 1 must contain a $\mathbf{3}+\mathbf{1}$. We also show that the range of wd_F when an induced $\mathbf{2}+\mathbf{2}$ is present also depends on the presence of a $\mathbf{3}+\mathbf{1}$: when P contains no induced $\mathbf{3}+\mathbf{1}$ then $wd_F(P) = 1$ and when it does contain a $\mathbf{3}+\mathbf{1}$ then the range is again the set of rationals that are at least 1. These results are summarized in Table 1.

2 Forcing Cycles

We begin with some definitions and preliminary results.

Definition 2. A *forcing cycle* C of poset $P = (V, \prec)$ is a sequence $C : x_0, x_1, \ldots, x_m = x_0$ of $m \geq 2$ elements of V for which $x_i \prec x_{i+1}$ or $x_i \parallel x_{i+1}$ for each $i : 0 \leq i < m$. If C is a forcing cycle, we write $\text{up}(C) = |\{i : x_i \prec x_{i+1}\}|$ and $\text{side}(C) = |\{i : x_i \parallel x_{i+1}\}|$.

In Fishburn (1985), forcing cycles are called *picycles* (preference-indifference cycles).

Let C be a forcing cycle as in Definition 2. We may choose to start the cycle at an element x_0 that is the beginning of a sequence of "up" steps, i.e., $x_0 \prec x_1$ and $x_{m-1} \parallel x_m = x_0$. We call x_0 an *upward starting point* of C. In this case, C consists of s successive chains of $a_i \geq 1$ elements each followed by an incomparability, $i = 1, 2, \ldots, s$, where $\sum_{i=1}^{s} a_i = m$. We write $\text{type}(C) = [a_1, a_2, \ldots, a_s]$.

For example, the poset P in Fig. 1 has forcing cycle $C : a \prec b \parallel c \prec d \parallel e \prec f \prec g \parallel a$ with $\text{up}(C) = 4, \text{side}(C) = 3$ and $\text{type}(C) = [2, 2, 3]$. In general, a forcing cycle C with $\text{type}(C) = [a_1, a_2, \ldots, a_s]$ has $\text{up}(C) = \sum_{j=1}^{s}(a_j - 1)$ and $\text{side}(C) = s$. Note that $a_i = 1$ corresponds to two consecutive incomparabilities in the forcing cycle.

Given a forcing cycle C, we can obtain a closely related forcing cycle C' by choosing a different upward starting point. For example, in Fig. 1 we can start the

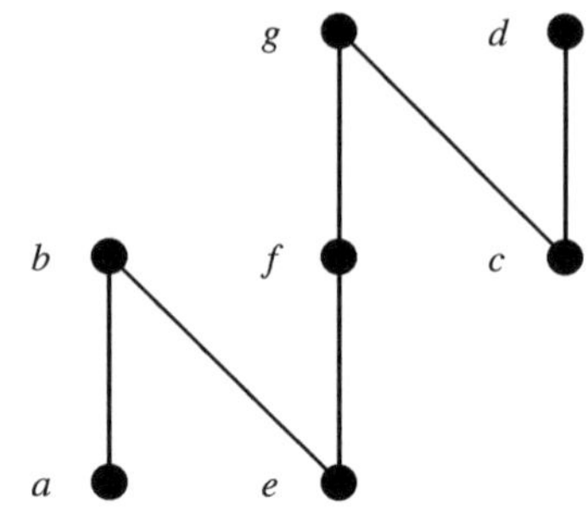

Fig. 1 A poset with similar forcing cycles starting at a with type $[2,2,3]$ and at e with type $[3,2,2]$

forcing cycle at e instead of a. Then we obtain $C' : e \prec f \prec g \parallel a \prec b \parallel c \prec d \parallel e$, which has $\text{type}(C') = [3,2,2]$.

Note that if P has no incomparable pair then it is a linear order, has no forcing cycle, and $wd_F(P) = 0$. The following result characterizes fractional weak discrepancy in terms of forcing cycles when P has an incomparable pair. The analogous result for weak discrepancy appears in Gimbel and Trenk (1998).

Theorem 3 (Shuchat et al. (2007)). *Let $P = (V, \prec)$ be a poset with at least one incomparable pair. Then $wd_F(P) = \max_C \frac{\text{up}(C)}{\text{side}(C)}$, taken over all forcing cycles C in P.*

The proof of Theorem 2 shows that for integers $r \geq s \geq 2$, if $q = \frac{r}{s}$ (not necessarily in lowest terms) then there exist a non-semiorder P with $wd_F(P) = q$ and a forcing cycle C in P with $\text{up}(C) = r, \text{side}(C) = s$. It is thus natural to conjecture that for integers $r \geq s \geq 2$, every poset P with $wd_F(P) = \frac{r}{s}$ has a forcing cycle C with $\text{up}(C) = r$ and $\text{side}(C) = s$. This is not the case even if $\frac{r}{s}$ is in lowest terms, as the following proposition shows.

Proposition 1. *There exists a poset P with $wd_F(P) = \frac{3}{2}$ but no forcing cycle C with $\text{up}(C) = 3$ and $\text{side}(C) = 2$.*

Proof. We show that the poset P in Fig. 2 has the desired property. By Definition 1, the labeling function shown there implies that $wd_F(P) \leq \frac{3}{2}$. By Theorem 3 the forcing cycle

$$x_1 \prec y_1 \parallel x_2 \prec y_2 \parallel x_3 \prec y_3 \parallel z_1 \prec z_2 \prec z_3 \prec z_4 \parallel x_1$$

shows that $wd_F(P) \geq \frac{6}{4} = \frac{3}{2}$, thus $wd_F(P) = \frac{3}{2}$.

It is easy to see that there is no $\mathbf{4+1}$ in P because there is only one chain of four elements, $z_1 \prec z_2 \prec z_3 \prec z_4$, and every other element in P is comparable to some z_i. Similarly, one can check that there is no $\mathbf{3+2}$ in P by considering all possible chains of three elements. This implies that P cannot contain a forcing cycle C with $\text{up}(C) = 3$ and $\text{side}(C) = 2$ because we could choose an upward starting point for such a cycle to yield one of type $[4,1]$, a $\mathbf{4+1}$, or of type $[3,2]$, a $\mathbf{3+2}$. □

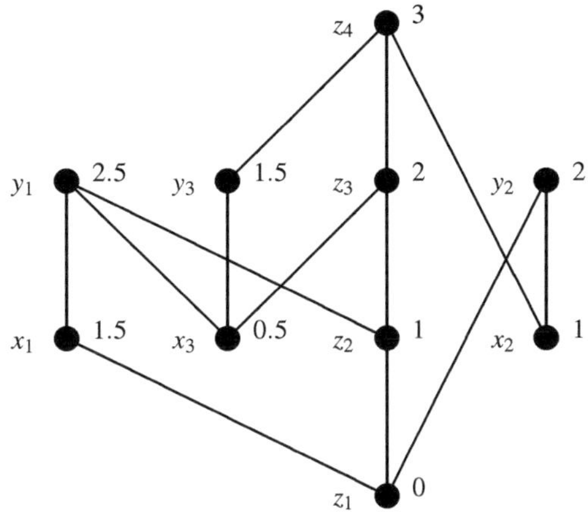

Fig. 2 A poset with $wd_F(P) = \frac{3}{2}$ but no forcing cycle with $\text{up}(C) = 3, \text{side}(C) = 2$

Lemma 1, which appears as Proposition 9 of Shuchat et al. (2006), allows us to describe optimal fractional weak labelings for forcing cycles whose "up" to "side" ratios achieve the maximum value of $wd_F(P)$. In particular, every optimal labeling is *tight* on such a forcing cycle in the following sense.

Lemma 1 (Shuchat et al. (2006)). *Let $C : x_0, x_1, \ldots, x_{m-1}, x_m = x_0$ be a forcing cycle for poset $P = (V, \prec)$ such that $k = wd_F(P) = \frac{\text{up}(C)}{\text{side}(C)}$ and let $f : V \to \mathbf{R}$ be an optimal fractional weak labeling of P. For each $i \in \{0, 1, \ldots, m-1\}$*
(i) if $x_i \prec x_{i+1}$ then $f(x_i) + 1 = f(x_{i+1})$
(ii) if $x_i \parallel x_{i+1}$ then $f(x_{i+1}) - f(x_i) = -k$.

For example, the labeling shown in Fig. 2 is tight on the forcing cycle given in the proof of Proposition 1.

3 The Range of wd_F and Interval Orders

In Theorem 2 we find the range of the fractional weak discrepancy function for posets that are not semiorders. In this section we divide the non-semiorders into two types and find the range for each: non-interval orders and interval orders that are not unit interval orders.

Theorem 4. *If P is a non-interval order, then $wd_F(P) \geq 1$. Furthermore, for any rational number $q \geq 1$, there exists a non-interval order P with $wd_F(P) = q$. Thus for the class of non-interval orders, the range of wd_F is $\{q \in \mathbf{Q} : q \geq 1\}$.*

Proof. If P is not an interval order (i.e., possesses an induced $\mathbf{2}+\mathbf{2}$) then P is not a semiorder, so $wd_F(P) \geq 1$ by Theorem 2.

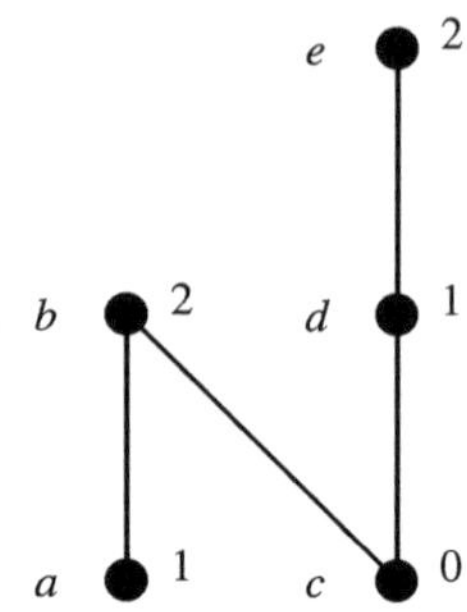

Fig. 3 A non-interval order P with $wd_F(P) = 1$

Now let $q > 1$ be rational. The proof of Theorem 2 includes the construction of a poset P with $wd_F(P) = q$. This construction, which appears in Proposition 14 of Shuchat et al. (2006), contains an induced $\mathbf{2}+\mathbf{2}$, so P is not an interval order.[1]

For the case of $q = 1$ we consider Fig. 3, which gives a poset P containing an induced $\mathbf{2}+\mathbf{2}$, so again P is not an interval order. By the first sentence of the proof $wd_F(P) \geq 1$, and the labeling of P shown in the figure demonstrates that $wd_F(P) \leq 1$. So $wd_F(P) = 1 = q$. □

We now establish a similar result for interval orders. In particular, we show how to achieve any rational number that is at least one as the fractional weak discrepancy of some interval order, which by Theorem 2 is necessarily not a semiorder. The proof is constructive.

Theorem 5. *For any rational number $q \geq 1$, there exists an interval order P with $wd_F(P) = q$.*

Proof. We write the given rational q as $q = r/s$ with integers $r \geq s \geq 2$. We will construct an interval representation of an order $P = (V, \prec)$ with $V = \{x_0, x_1, \ldots, x_r, y_1, y_2, \ldots, y_{s-1}\}$. For $0 \leq i \leq r$, let $I(x_i) = [is, is]$, that is, each of these intervals is a point. Let $I(y_{s-1}) = [(s-2)r, sr]$ and if $s > 2$, then for $1 \leq j \leq s-2$, let $I(y_j) = [(j-1)r, (j+1)r - \frac{1}{2}]$. Figure 4 shows the representation in the case $r = 7$ and $s = 4$. By construction, $x_i \prec x_{i+1}$ for $0 \leq i < r$ and $y_j \parallel y_{j+1}$ for $1 \leq j \leq s-2$. Furthermore, $x_r \parallel y_{s-1}$ and $y_1 \parallel x_0$. Thus P contains the forcing cycle $C : x_0 \prec x_1 \prec x_2 \prec \cdots \prec x_r \parallel y_{s-1} \parallel y_{s-2} \parallel \cdots \parallel y_1 \parallel x_0$ with $\mathrm{up}(C) = r$ and side $(C) = s$. Thus $wd_F(P) \geq r/s$ by Theorem 3.

It remains to show $wd_F(P) \leq r/s$. Define the labeling function $g : V \to \mathbf{Z}$ by setting $g(x_i) = is$ for $i = 0, 1, \ldots, r$ and setting $g(y_j) = jr$ for $j = 1, 2, \ldots, s-1$. (See the example in Fig. 4.) We show

[1] The $\mathbf{2}+\mathbf{2}$ in that construction is formed by the chains $x_{n-1} \prec y_{n-1}$ and $z_1 \prec z_2$.

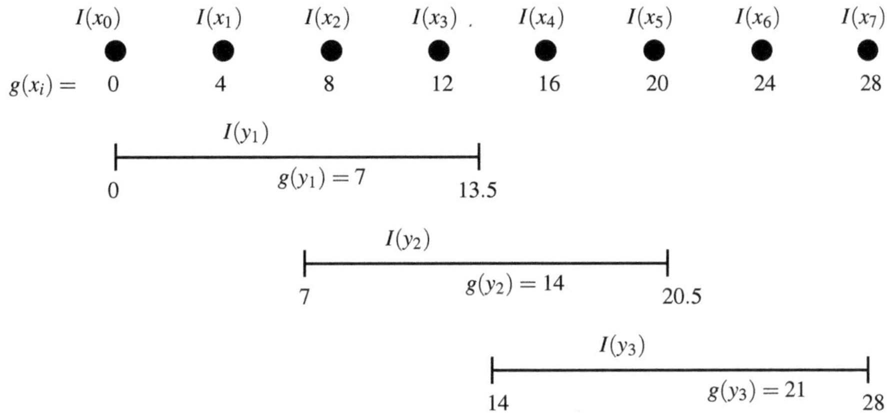

Fig. 4 An interval order P with $wd_F(P) = 7/4$ and labeling function g

(i) if $a \prec b$ then $g(a) + s \leq g(b)$ ("up" constraints)
(ii) if $a \parallel b$ then $|g(a) - g(b)| \leq r$. ("side" constraints)

Then it will follow that the function $f : V \to \mathbf{Q}$ defined by $f(x) = g(x)/s$ is an optimal fractional weak labeling of P satisfying Definition 1.

We will consider all pairs (a,b) of elements of V, classify their relation in the poset, and prove that the corresponding constraints are satisfied. First take $x_i, x_j \in V$ with $i < j$. By construction, $x_i \prec x_j$ and $g(x_i) + s = is + s = (i+1)s \leq js = g(x_j)$, satisfying (i) for this pair of elements.

Next consider $y_i, y_j \in V$ with $i < j$. If $j = i+1$, then $I(y_i) \cap I(y_j) \neq \emptyset$ so $y_i \parallel y_j$ and $|g(y_i) - g(y_j)| = |ir - jr| = r$, satisfying (ii). Otherwise, $j \geq i+2$. Let R be the right endpoint of the interval $I(y_i)$ and L be the left endpoint of the interval $I(y_j)$. Then $R = (i+1)r - \frac{1}{2} \leq (j-1)r - \frac{1}{2} < L$. Thus $y_i \prec y_j$ and $g(y_i) + s = ir + s \leq ir + r = (i+1)r < jr = g(y_j)$, satisfying (i).

Lastly, consider $x_i, y_j \in V$. By construction, $x_i \parallel y_j$ precisely when the point $I(x_i)$ is contained in the interval $I(y_j)$. In this case, for $1 \leq j \leq s-1$ we have $(j-1)r \leq is \leq (j+1)r$. Subtracting jr yields $-r \leq is - jr \leq r$. Thus $|g(x_i) - g(y_j)| = |is - jr| \leq r$, satisfying (ii).

If instead $x_i \prec y_j$ then the point $I(x_i)$ lies strictly to the left of $I(y_j)$, so $is < (j-1)r$. In this case, $g(x_i) + s = is + s < (j-1)r + s \leq (j-1)r + r = jr = g(y_j)$, satisfying (i). Finally, if $y_j \prec x_i$ then $j < s-1$ and the point $I(x_i)$ lies strictly to the right of the interval $I(y_j)$, and thus $(j+1)r - \frac{1}{2} < is$. Since all the parameters are integers, in fact, $(j+1)r \leq is$. In this case $g(y_j) + s = jr + s \leq jr + r = (j+1)r \leq is = g(x_i)$, satisfying (i). □

We can now fill in the top-right entry of Table 1. If P contains an induced $\mathbf{3+1}$ but no $\mathbf{2+2}$, then Theorem 2 implies $wd_F(P) \geq 1$. Conversely, by Theorem 5 any rational $q \geq 1$ equals $wd_F(P)$ for some interval order P (contains no $\mathbf{2+2}$), but Theorem 1 implies P is not a semiorder (thus contains a $\mathbf{3+1}$). We have shown the following.

Corollary 2. *For the class of posets that are interval orders but not semiorders (contain an induced* $\mathbf{3}+\mathbf{1}$ *but no* $\mathbf{2}+\mathbf{2}$*), the range of* wd_F *is* $\{q \in \mathbf{Q} : q \geq 1\}$.

4 An Upper Bound on wd_F for Posets with no $\mathbf{n}+\mathbf{1}$

As in Shuchat et al. (2007); Shuchat et al. (2006), we define the (*integer*) *weak discrepancy* $wd(P)$ of a poset $P = (V, \prec)$ as the minimum nonnegative integer k for which there exists a function $f : V \to \mathbf{Z}$ satisfying (i) and (ii) of Definition 1. This is equivalent to the concept of *weakness* first introduced in Trenk (1998). The following theorem (Proposition 7 of Shuchat et al. (2007)) allows us to calculate the weak discrepancy of a poset from its fractional weak discrepancy.

Theorem 6 (Shuchat et al. (2007)). *For any poset P we have* $wd(P) = \lceil wd_F(P) \rceil$.

In Trenk (1998), the author proved a result giving an upper bound on $wd(P)$ for posets with no induced $\mathbf{n}+\mathbf{1}$. We state the result in its contrapositive form.

Theorem 7 (Trenk (1998)). *Let n be an integer,* $n \geq 2$. *Every poset P with* $wd(P) > n-2$ *contains an induced* $\mathbf{n}+\mathbf{1}$.

Neither forcing cycles nor fractional weak discrepancy had been defined when Theorem 7 was first presented. In this section, we give a substantially simpler proof of the analogous theorem for fractional weak discrepancy and show the two results are in fact equivalent.

Theorem 8. *Let n be an integer,* $n \geq 2$. *Every poset P with* $wd_F(P) > n-2$ *contains an induced* $\mathbf{n}+\mathbf{1}$.

Proof. Let $P = (V, \prec)$ be a poset with $k = wd_F(P) > n-2$ and let $f : V \to \mathbf{R}$ be an optimal fractional weak labeling of P. By Theorem 3 there exists a forcing cycle $C : x_0, x_1, \ldots, x_m = x_0$ such that $k = wd_F(P) = \frac{\mathrm{up}(C)}{\mathrm{side}(C)}$. Without loss of generality, suppose that $x_{m-1} \parallel x_m = x_0$, i.e., the cycle closes with an incomparability. By Lemma 1, the labeling f is tight on C. In particular, $f(x_{m-1}) - f(x_0) = k$.

Consider the sequence S of differences

$$f(x_1) - f(x_0), f(x_2) - f(x_1), \ldots, f(x_{m-1}) - f(x_{m-2}).$$

Note that the sum of the elements of S is $f(x_{m-1}) - f(x_0) = k$. By Lemma 1, each term of S is either $+1$ or $-k$. Let t be the largest number of consecutive $+1$'s in S. If $t < k$ then every partial sum of S, and in particular the sum of all the terms, is less than k, a contradiction. Thus, $t \geq k$ and since t is an integer, $t \geq \lceil k \rceil$. By the definition of t, there is a longest chain in C containing t elements. Let x_j be its starting point and consider its subchain $x_j \prec x_{j+1} \prec \cdots \prec x_{j+\lceil k \rceil}$ of length $\lceil k \rceil$. By the maximality of t, $x_{j-1} \parallel x_j$ (if $j = 0$ we replace j by m) and thus $f(x_{j-1}) - k = f(x_j)$. Now

$$f(x_{j+\lceil k\rceil}) = f(x_j) + \lceil k\rceil = f(x_{j-1}) - k + \lceil k\rceil < f(x_{j-1}) + 1. \tag{1}$$

If $x_{j-1} \prec x_{j+\lceil k\rceil}$ then (1) contradicts the "up" constraint in Definition 1. If $x_{j+\lceil k\rceil} \prec x_{j-1}$ then $x_j \prec x_{j+\lceil k\rceil} \prec x_{j-1}$, contradicting $x_{j-1} \parallel x_j$. Thus, $x_{j-1} \parallel x_{j+\lceil k\rceil}$. We conclude that $x_j \prec x_{j+1} \prec \cdots \prec x_{j+\lceil k\rceil} \parallel x_{j-1}$ is a $(\lceil \mathbf{k}\rceil + \mathbf{1}) + \mathbf{1}$. Since $k > n-2$ and $n-2$ is an integer, $\lceil k\rceil \geq n-1$ and P contains an induced $\mathbf{n}+\mathbf{1}$. □

The bound given in Theorem 8 is the best possible, since $P = (\mathbf{n}-\mathbf{1})+(\mathbf{n}-\mathbf{1})$ has no induced $\mathbf{n}+\mathbf{1}$ but $wd_F(P) = \frac{2(n-2)}{2} = n-2$.

The hypotheses of Theorems 7 and 8 are equivalent because n is an integer and $wd(P) = \lceil wd_F(P)\rceil$. Thus our proof of Theorem 8 gives a shorter proof of Theorem 7 as well. Notice also that the proof of Theorem 8 relies on the existence of a forcing cycle and an optimal labeling that is tight on that cycle. This same argument cannot be used to prove Theorem 7 directly since the tightness condition need not hold for forcing cycles whose "up" to "side" ratios achieve the (integer) weak discrepancy of the poset. For example, let P be a $\mathbf{3}+\mathbf{2}$ with chains $a_0 \prec a_1 \prec a_2$ and $a_3 \prec a_4$. An optimal integer labeling is $f(a_0) = 0, f(a_1) = 1, f(a_2) = 2, f(a_3) = 1, f(a_4) = 2$ so $wd(P) = 2$, but the labeling is not tight on the forcing cycle $a_0 \prec a_1 \prec a_2 \parallel a_3 \prec a_4 \parallel a_0$.

5 The Range of wd_F and Non-interval Orders

In Theorem 4 we found the range of the fractional weak discrepancy function for non-interval orders. In this section, we divide these orders into two types and find the range for each: orders that contain an induced $\mathbf{3}+\mathbf{1}$ and those that do not. This will justify the entries at the bottom of Table 1.

The poset P in Fig. 2 has $wd_F(P) = 3/2$ and contains no induced $\mathbf{4}+\mathbf{1}$ but it does have a $\mathbf{3}+\mathbf{1}$, e.g., the elements of $z_1 \prec z_2 \prec z_3$ are all incomparable to x_2. Indeed, all posets with fractional weak discrepancy greater than one must contain a $\mathbf{3}+\mathbf{1}$ by Theorem 8, with $n = 3$. We state that specific case so we can refer to it more easily.

Corollary 3. *Every poset P with* $wd_F(P) > 1$ *contains an induced* $\mathbf{3}+\mathbf{1}$.

This result is best possible since, by Theorem 2, if $wd_F(P) < 1$ then P must be a semiorder and thus does not contain a $\mathbf{3}+\mathbf{1}$. On the other hand, Shuchat et al. (2007) show that $wd_F(\mathbf{3}+\mathbf{1}) = wd_F(\mathbf{2}+\mathbf{2}) = 1$ so if $wd_F(P) = 1$ then P may or may not contain a $\mathbf{3}+\mathbf{1}$.

Note that Theorem 4 implies that the range of wd_F for posets possessing an induced $\mathbf{2}+\mathbf{2}$ is the set of all rational numbers greater than or equal to 1. In the case of strict inequality, Corollary 3 implies all such posets must also have an induced $\mathbf{3}+\mathbf{1}$. The poset P given in Fig. 3 possesses both a $\mathbf{2}+\mathbf{2}$ and a $\mathbf{3}+\mathbf{1}$ and has $wd_F(P) = 1$. We conclude that the range of wd_F for posets possessing both an induced $\mathbf{2}+\mathbf{2}$ and an induced $\mathbf{3}+\mathbf{1}$ is also $\{q \in \mathbf{Q} : q \geq 1\}$, as indicated in the lower-right entry in Table 1.

By Corollary 3 a poset P with no induced $\mathbf{3}+\mathbf{1}$ must satisfy $wd_F(P) \leq 1$. Also, if P has an induced $\mathbf{2}+\mathbf{2}$ then it has a forcing cycle C with $\frac{\text{up}(C)}{\text{side}(C)} = 1$ and thus by Theorem 3, $wd_F(P) \geq 1$. So $wd_F(P) = 1$, which fills in the lower-left entry of Table 1 and which we state as an additional corollary. The converse is clearly false since $wd_F(\mathbf{3}+\mathbf{1}) = 1$. We have

Corollary 4. *Every poset P with an induced $\mathbf{2}+\mathbf{2}$ but no induced $\mathbf{3}+\mathbf{1}$ satisfies $wd_F(P) = 1$.*

Although Corollary 3 gives the best possible bound for $wd_F(P)$ over the class of all posets, the upper row of Table 1 suggests a slightly better bound when P is restricted to the class of interval orders. In particular, if C is a forcing cycle for P with $\text{up}(C) > r$ and $\text{side}(C) = r+1$ (so $wd_F(P) > \frac{r}{r+1}$), then C must contain a $\mathbf{3}+\mathbf{1}$. Furthermore, the proof of Theorem 1 given in Proposition 16 of Shuchat et al. (2006) shows how to construct, for each $r > 0$, an interval order P possessing an optimal forcing cycle C with $\text{up}(C) = r$ and $\text{side}(C) = r+1$ but no induced $\mathbf{3}+\mathbf{1}$. In the case $n = 3$, we can express the upper row as saying that if $\text{up}(C) > (n-2)r$ and $\text{side}(C) = r+1$, then P must contain an $\mathbf{n}+\mathbf{1}$. In Shuchat, Shull, and Trenk (in press) we extend this result to the case where $n \geq 3$.

Acknowledgements The third author feels very privileged to have collaborated with Peter Fishburn on papers Fishburn, Tanenbaum, and Trenk (2001) and Tanenbaum et al. (2001). Her work on the present article was supported in part by a Wellesley College Brachman Hoffman Fellowship.

References

Bogart, K. P. (1990). *Introductory Combinatorics*. New York: Harcourt Brace Jovanovich.

Fishburn, P. C. (1985). *Interval orders and interval graphs: A study of partially ordered sets*. New York: Wiley.

Fishburn, P. C., Tanenbaum, P. J., & Trenk, A. N. (2001). Linear discrepancy and bandwidth. *Order, 18*, 237–245.

Gasharov, V. (1996). Incomparability graphs of ($\mathbf{3}+\mathbf{1}$)-free posets are s-positive. *Discrete Mathematics, 157*, 211–215.

Gimbel, J. G., & Trenk, A. N. (1998). On the weakness of an ordered set. *SIAM Journal on Discrete Mathematics, 11*, 655–663.

Scott, D., & Suppes, P. (1958). Foundational aspects of theories of measurement. *Journal of Symbolic Logic, 23*, 113–128.

Shuchat, A., Shull, R., & Trenk, A. N. (2007). The fractional weak discrepancy of a partially ordered set. *Discrete Applied Mathematics, 155*, 2227–2235.

Shuchat, A., Shull, R., & Trenk, A. N. (2006). Range of the fractional weak discrepancy function. *Order , 23*, 51–63.

Shuchat, A., Shull, R., & Trenk, A. N. (in press). Fractional weak discrepancy and interval orders.

Skandera, M. (2001). A characterization of ($\mathbf{3}+\mathbf{1}$)-free posets. *Journal of Combinatorial Theory - Series A , 93*, 231–241.

Stanley R. (1995). A symmetric function generalization of the chromatic polynomial of a graph. *Advances in Mathematics, 111*, 166–194.

Tanenbaum, P. J., Trenk, A. N., & Fishburn, P. C. (2001). Linear discrepancy and weak discrepancy of partially ordered sets. *Order, 18*, 201–225.
Trenk, A. N. (1998). On k-weak orders: Recognition and a tolerance result. *Discrete Mathematics, 181*, 223–237.

Interval Order Representation via Shortest Paths

Garth Isaak

1 Introduction

Our goal in this paper is to illustrate how the representation theorems for finite interval orders and semiorders can be seen as special instances of existence results for potentials in digraphs. This viewpoint yields short proofs of the representation theorems and provides a framework for certain types of additional constraints on the intervals. We also use it to obtain a minimax theorem for the minimum number of endpoints in a representation. The techniques are based on techniques used by Peter Fishburn in proving results about bounded representations of interval orders.

Interval orders represent the order structure of a collection of intervals. For example, this can be used to model the relations between a set of events each of which occurs over some time interval. Semiorders are a special case where the intervals have the same length. These can be viewed as representing comparisons of values where a relation is noted only if the difference of values is above a certain threshold. We will not go into more detail here as there are many good references describing the various applications of interval orders and semiorders. See for example Fishburn (1985); Luce, Krantz, and Suppes (1971, 1989, 1990); Pirlot and Vincke (1997). See Fishburn (1997) for a good description of some more general models based on intervals.

Recall that an interval order is an asymmetric binary relation $\prec$ on a set U that satisfies ($a \prec x$ and $b \prec y$ implies $a \prec y$ or $b \prec x$) for all $a, b, x, y \in U$. These are (strict partial) orders as transitivity is implied by this definition. The name interval applies because these orders can be represented by a set of intervals in a linear order with the natural relation "less than" for the intervals. A (closed real) interval representation of a strict order $(U, \prec)$ is a set of closed real intervals $[l_x, r_x]$ for $x \in U$ such that $x \prec y$ if and only if $r_x < l_y$. A $\mathbf{2}+\mathbf{2}$ in an order is the disjoint union of two chains

G. Isaak
Department of Mathematics, Lehigh University, Bethlehem, PA 18015, USA
e-mail: gisaak@lehigh.edu

S.J. Brams et al. (eds.), *The Mathematics of Preference, Choice and Order: Essays in Honor of Peter C. Fishburn*, Studies in Choice and Welfare,

each with two elements. That is, a $\mathbf{2}+\mathbf{2}$ is a set of four elements x,y,a,b such that $a \prec x, b \prec y, a \sim y, b \sim x$ (with also $x \sim y$, $a \sim b$ implied by transitivity). Here we use the notation $\sim$ to represent incomparability, $x \sim y$ if and only if $x \not\prec y$ and $y \not\prec x$. A $\mathbf{2}+\mathbf{2}$ corresponds to a violation of the interval order condition above so we can take as our definition that an interval order is a strict partial order with no $\mathbf{2}+\mathbf{2}$.

This result that an order can be represented by intervals if and only if it has no $\mathbf{2}+\mathbf{2}$ was anticipated by Wiener in 1914 (see Fishburn & Monjardet, 1992) and shown by Fishburn (1970a, 1970b). Following Bogart (1993) we will refer to this as Fishburn's Theorem. When the ground set is finite we can use either open or closed intervals in the real numbers for the interval representation. When the ground set is infinite and in particular uncountable, things are a little more complicated. A linear order other than the reals may be required. For more on this see Fishburn's book (Fishburn, 1985). For this paper we will stick to the finite version and consider real number representations with the goal of seeing a connection to potentials, shortest paths and negative cycles in digraphs. We will assume that all orders considered in this paper are finite. We will not attempt to survey the many different proofs that have been given for these theorems nor the various related results. See Fishburn's book (Fishburn, 1985) for a very nice description of results on interval orders.

Fishburn's Theorem (for finite interval orders): A finite strict partial order has a closed real interval representation if and only if it has no $\mathbf{2}+\mathbf{2}$.

A finite unit interval order (also called a semiorder), introduced by Luce (1956) is an interval order that has a real representation in which all of the intervals have length 1. Such a representation will be called a unit interval representation. See Fishburn (1985) for a discussion of the infinite case. A $\mathbf{3}+\mathbf{1}$ is the disjoint union of a chain with 3 elements and a chain with 1 element. That is, r,s,t,u such that $r \prec s \prec t$ and $u \sim r$, $u \sim t$ (with $u \sim s$ implied by transitivity). This is the extra condition for unit interval representations as shown by Scott and Suppes (1958).

Scott–Suppes' Theorem (for finite unit interval orders): A finite strict partial order has a closed real unit interval representation if and only if it has no $\mathbf{2}+\mathbf{2}$ or $\mathbf{3}+\mathbf{1}$.

We will give short proofs of these results using shortest paths in an associated digraph and also show how this framework can be used to get results about representations with certain side constraints on the intervals. This technique was first used by Fishburn (1983) for representations with bounds on the interval lengths and later by the author (Isaak, 1990, 1993) for representations with bounds and integral endpoints. A similar technique was used by Pirlot (1990) for semiorder representation questions although with a different set of inequalities. It was also used by Doignon (1988a). The books Fishburn (1985) and Pirlot and Vincke (1997) have more details of these techniques in various settings. The representation theorems are implicit in these results. However, without the bounds we can get short proofs of the representation theorems. Finally we will use the framework to obtain a minimax theorem for the minimum number of endpoints in a representation.

2 Shortest Paths and Potentials

We first give a brief review of basic results on shortest paths and potentials which will provide the framework for our proofs. This can be found, for example in Schrijver (2003, p. 108).

A weighted digraph is a set V of vertices along with a set A of ordered pairs (x,y) of vertices called arcs. Each arc (x,y) has an associated length $w(xy)$. A potential function $p(x)$ defined on the vertices is a function satisfying $p(y) - p(x) \leq w(xy)$ for all arcs (x,y). That is, the potential value at y is at most the potential at x plus the length of arc (x,y). An $x-y$ path in a digraph is a sequence of distinct vertices $x = x_1, x_2, \ldots, x_t = y$ such that (x_i, x_{i+1}) is an arc for $i = 1,2,\ldots,t-1$. A cycle is the same as a path except that (x_t, x_1) is also an arc. The length of a path (cycle) is the sum of the arc lengths along the path (cycle). Let $\overline{p}(y)$ denote the shortest length of a path ending at y. It is a basic result that these shortest path lengths are defined for all vertices if and only if the graph has no negative cycle (i.e., a cycle with negative length). It is easy to see that if $\overline{p}(x)$ is the length of a shortest path ending at x for each x (assuming that these are well defined) then this is a potential. If a digraph has a negative cycle with length $-c < 0$ then "adding" the inequalities for the arcs on the cycle produces the inconsistency $0 \leq -c$ showing that there is no potential.

Hence a digraph has a potential function if and only if it has no negative cycle. Furthermore, if there is no negative cycle then the lengths of shortest paths ending at each vertex yield a potential.

3 Interval Orders

That an order with an interval representation has no $\mathbf{2}+\mathbf{2}$ is easy to check. Our goal is to show the converse: a finite order with no $\mathbf{2}+\mathbf{2}$ has a closed interval representation. For each element $x \in U$ create two variables $p(r_x)$ and $p(l_x)$ (which will correspond to left and right endpoints of intervals in a representation). Consider the p as representing placement of the endpoints and in what follows (with a slight abuse of notation) a potential function.

Let γ be a positive number. Consider the following inequalities:
(C) For all $x \prec y$, $p(r_x) \leq p(l_y) - \gamma$, equivalently $p(r_x) - p(l_y) \leq -\gamma$.
(I) For all $x \sim y$, $p(r_x) \geq p(l_y)$, equivalently $p(l_y) - p(r_x) \leq 0$.

The intervals would be $[p(l_x), p(r_x)]$. The inequalities C enforce "less than" for intervals of comparable elements and the inequalities I enforce "not less than" for incomparable elements. It is not difficult to check that $(U, \prec)$ has an interval representation using intervals $[p(l_x), p(r_x)]$ if and only if the $p(r_x), p(l_x)$ are a solution to the system of inequalities above. The γ is just a convenience to avoid writing strict inequalities. Note that when $x = y$ in I we have $p(r_x) \geq p(l_x)$, ensuring that these really are intervals, with the right endpoints at least as large as the left endpoints. Also, when $x \sim y$, switching the roles of x and y we see that we have both $p(l_y) - p(r_x) \leq 0$ and $p(l_x) - p(r_y) \leq 0$.

Each inequality has two variables, one with coefficient $+1$ and one with coefficient -1. We then recognize the inequalities as those for a potential function on a particular digraph.

For a given order $(U,\prec)$ we define (with a slight abuse of notation) a digraph D_U with vertices $\{l_x, r_x | x \in U\}$ and arcs $C \cup I$ where $C = \{(l_y, r_x) | x \prec y\}$ with length $-\gamma$ for some positive number γ and $I = \{(r_x, l_y) | x \sim y\}$ with length 0. Then, from the preceding section, $(U,\prec)$ has an interval representation if and only if D_U has no negative cycles. Furthermore the length of a shortest path ending at r_x can be used for the right endpoints $p(r_x)$ and similarly the length of a shortest path ending at l_y can be used for the left endpoints $p(l_y)$. We note that directly writing down D_U one can fairly easily show this claim without going through the idea of potentials. Looking at shortest paths, the length 0 on an arc (r_x, l_y) for $x \sim y$ forces a shortest path ending at l_y to have length no more than a shortest path ending at r_x. That is $p(l_y)$, would be at most $p(r_x)$ and so $x \not\prec y$. Similarly the length $-\gamma$ on arc (l_y, r_x) for $x \prec y$ forces a shortest path ending at r_x to have length strictly less a shortest path ending at l_y. That is, $p(r_x)$ would be less than $p(l_y)$ and so $x \prec y$. The framework of potentials is useful as motivation for why we construct the digraph in this manner and for easily yielding a proof that the technique works.

Proof of Fishburn's Theorem for Finite Orders

From the discussion above we need to show that $(U,\prec)$ has a $\mathbf{2+2}$ if and only if D_U contains a negative cycle.

We have already noted that it is easy to check directly that an order with a $\mathbf{2+2}$ has no interval representation. We can also show this using the digraph: If $a \prec x, b \prec y$, $a \sim y, b \sim x$ is a $\mathbf{2+2}$ then l_x, r_a, l_y, r_b is a cycle with length $-2\gamma < 0$.

We now need to show that if D_U contains a negative cycle then $(U,\prec)$ has a $\mathbf{2+2}$. Observe that D_U is bipartite. Partition the vertex set into $R = \{r_x | x \in U\}$ and $L = \{l_y | y \in U\}$. Then there are two types of arcs, each with one end in R and the other in L. Arcs from L to R have length $-\gamma$ and arcs from R to L have length 0. Any cycle must alternate between these two types of arcs and thus any cycle is negative. So it is enough to show that if D_U contains a cycle then $(U,\prec)$ has a $\mathbf{2+2}$.

Consider a cycle with the minimum number of vertices. It is easy to see that it cannot have exactly two vertices. Since the digraph is bipartite the cycle contains l_x, r_a, l_y, r_b for some x, a, y, b (not necessarily distinct). The arcs imply that $a \prec x$, $b \prec y$ and $a \sim y$. If $b \prec x$ then (l_x, r_b) is an arc and replace the segment with this arc for a negative cycle with fewer vertices. If $x \prec b$ then in the order $a \prec x \prec b \prec y$ and by transitivity $a \prec y$ contradicting $a \sim y$. So $x \sim b$. Thus we have $y \sim a$, $a \prec x$, $b \prec y$ and $x \sim b$. Using transitivity it is easy to see that x, y, a, b are distinct and so x, y, a, b induce a $\mathbf{2+2}$. □

To prove Fishburn's Theorem with open intervals in the representation we would give an almost identical proof except that the arcs in I would have length $-\varepsilon$ for some positive ε and the arcs in C would have length 0 (and the corresponding changes in the inequalities C and I).

4 Unit Interval Orders

That an order with a unit interval representation has no **2**+**2** and no **3**+**1** is easy to check. Our goal is to show the converse: a finite order with no **2**+**2** and no **3**+**1** has a closed unit interval representation. One approach is to use the same model as for interval orders and replace the constraints $p(r_x) \geq p(l_x)$ that right endpoints are at least as large as left endpoints with constraints $p(r_x) = p(l_x) + 1$ written as $p(r_x) - p(l_x) \leq 1$ and $p(l_x) - p(r_x) \leq -1$. Then appropriately adjust the corresponding digraph. Instead we will use a single variable for left endpoints, setting the right endpoints to be 1 more than the left.

We will use the same notation for D_U as the previous section however the construction here is different.

For each element $x \in U$ create a variable $p(l_x)$ (which will correspond to left endpoint of intervals in a representation). Consider the p as representing placement of the left endpoint and in what follows a potential function.

Let γ be a positive number. Consider the following inequalities:
(C) For all $x \prec y$, $p(l_x) + 1 \leq p(l_y) - \gamma$,
equivalently $p(l_x) - p(l_y) \leq -(1+\gamma)$.
(I) For all $x \sim y$ with $x \neq y$, $p(l_x) + 1 \geq p(l_y)$,
equivalently $p(l_y) - p(l_x) \leq 1$.

The intervals would be $[p(l_x), p(l_x)+1]$. The inequalities C enforce "less than" for intervals of comparable elements and the inequalities I enforce "not less than" for incomparable elements. Note that we do not need the condition $x \neq y$. However we include it to avoid these trivial inequalities which would add loops to the digraph. It is not difficult to check that $(U, \prec)$ has an unit interval representation using intervals $[p(l_x), p(l_x)+1]$ if and only if the $p(l_x)$ are a solution to the system of inequalities above. The γ is just a convenience to avoid writing strict inequalities.

As with the interval order case we recognize the system of inequalities as corresponding to those for a potential in a particular digraph.

For a given order $(U, \prec)$ we define (with a slight abuse of notation) a digraph D_U with vertices $\{l_x | x \in U\}$ and arcs $C \cup I$ where $C = \{(l_y, l_x) | x \prec y\}$ with length $-(1+\gamma)$ for some positive number γ and $I = \{(l_x, l_y) | x \sim y\}$ with length 1. Observe that if $x \sim y$ we have both arcs (l_x, l_y) and (l_y, l_x) with length 1. From the connection to potentials $(U, \prec)$ has a unit interval representation if and only if D_U has no negative cycles. Furthermore length of a shortest path ending at l_y can be used for the left endpoints $p(l_x)$. We note that directly writing down D_U one can fairly easily show this claim without going through the idea of potentials. Looking at shortest paths, the length 1 on an arc (l_x, l_y) for $x \sim y$ forces a shortest path ending at l_y to have length at most 1 more than a shortest path ending at l_x. That is the right endpoint $p(l_x) + 1$, would be at most $p(l_y)$ and so $x \not\prec y$. Similarly the length $-(1+\gamma)$ on arc (l_y, l_x) for $x \prec y$ forces a shortest path ending at l_x to have length more than 1 less a shortest path ending at l_y. That is, the right endpoint $p(l_x) + 1$ would be less than $p(l_y)$ and so $x \prec y$.

Proof of Scott–Suppes' Theorem for Finite Orders

From the discussion on potentials we need to show that $(U,\prec)$ has a $\mathbf{2}+\mathbf{2}$ or a $\mathbf{3}+\mathbf{1}$ if and only if D_U contains a negative cycle.

We have already noted that it is easy to check directly that an order with a $\mathbf{2}+\mathbf{2}$ or a $\mathbf{3}+\mathbf{1}$ has no unit interval representation. We can also show this using the digraph: If $a \prec x, b \prec y, a \sim y, b \sim x$ is a $\mathbf{2}+\mathbf{2}$ then l_x, l_a, l_y, l_b is a cycle with length $-2\gamma < 0$. If $r \prec s \prec t, u \sim r, u \sim t$ is a $\mathbf{3}+\mathbf{1}$ then l_t, l_s, l_r, l_u is a cycle with length $-2\gamma < 0$.

We now need to show that if D_U contains a negative cycle then $(U,\prec)$ has a $\mathbf{2}+\mathbf{2}$ or a $\mathbf{3}+\mathbf{1}$.

Observe that since arcs in C have length $-(1+\gamma)$ and arcs in I have length $+1$ there are at least as many arcs from C as from I in any negative cycle. Consider a negative cycle with the minimum number of vertices. Using transitivity in the order, it is easy to see that it cannot have 3 vertices. Since there are at least as many C arcs as I arcs, the cycle either alternates between C arcs and I arcs or contains 2 consecutive arcs from C.

In the first case consider l_x, l_a, l_y, l_b along the cycle with $(l_x, l_a), (l_y, l_b)$ in C and (l_a, l_y) in I. The arcs imply that $a \prec x$, $b \prec y$ and $a \sim y$. If $b \prec x$ then (l_x, l_b) is an arc in C and replace the segment with this arc for a cycle with fewer vertices. If $x \prec b$ then in the order $a \prec x \prec b \prec y$ and by transitivity $a \prec y$ contradicting $a \sim y$. So $x \sim b$. Thus we have $y \sim a$, $a \prec x$, $b \prec y$ and $x \sim b$. So x, y, a, b induce a $\mathbf{2}+\mathbf{2}$.

In the second case some pair of consecutive arcs from C is followed by an arc from I. (If all arcs are from C we get a violation of transitivity in the order). Consider l_t, l_s, l_r, l_u along the cycle with (l_t, l_s) and (l_s, l_r) in C and (l_r, l_u) in I. The arcs imply that $s \prec t$, $r \prec s$ and $r \sim u$. If $u \prec t$ then (l_t, l_u) is an arc from C and replace the segment by this arc for a cycle with fewer vertices. If $t \prec u$ then in the order $r \prec s \prec t \prec u$ and by transitivity $r \prec u$ contradicting $r \sim u$. So $t \sim u$. Thus we have $s \prec t$, $r \prec s$ and $r \sim u$ and $t \sim u$. So r, s, t, u induce a $\mathbf{3}+\mathbf{1}$. □

To prove the Scott–Suppes' Theorem with open intervals in the representation we would give an almost identical proof except that the arcs in I would have length $1-\varepsilon$ for some positive ε and the arcs in C would have length -1 (and the corresponding changes in the inequalities C and I).

5 Side Constraints

We next briefly note how the framework of inequalities and shortest paths can be used in a more general setting. We can, for example, place lower and upper bounds on the interval lengths in a representation. That is, given numbers $0 \leq \alpha(x) \leq \beta(x)$ for each element $x \in U$ we add the constraints $p(r_x) - p(l_x) \geq \alpha(x)$ and $p(r_x) - p(l_x) \leq \beta(x)$ for all x. These specify lower and upper bounds on each interval length. The additional constraints add additional arcs to the digraph D_U. These representations have been examined by Fishburn (1985, 1983) and with the additional requirement that the endpoints be integral in Isaak (1990, 1993). The use of shortest paths to construct an interval representation does imply an efficient algorithm for determining when an order has a representation subject to the additional constraints on interval length. However, while the result that there is a representation if and only if D_U contains no negative cycle is in a sense a characterization theorem it would be good to translate this to minimal forbidden suborders as negative cycles implied a $\mathbf{2}+\mathbf{2}$ in the proof of Fishburn's Theorem. Unfortunately this seems to be fairly complicated. The minimal forbidden suborders for a representation with integral endpoints and interval lengths between 0 and some positive integer α is given in Isaak (1990) but it is very messy. The same situation with a lower bound of 1 on interval lengths appears even messier although a possible infinite list is suggested in Isaak (1990).

Looking at the inequalities and not using the digraph model one can consider interval representations that are "optimal" by some other measure. For example, when there are lower bound on interval lengths, one could specify a utility for each element and seek an interval representation that minimizes the sums of the interval lengths weighted by the utilities. Linear programming algorithms and linear programming duality immediately give an efficient algorithm to find such a "weighted least length" interval representation as well as minimax theorem. It may be interesting to investigate if this can be translated to a more direct statement in terms of the order.

6 Magnitude

In this section we will show that the representation for interval orders constructed with interval endpoints determined by the values of shortest paths in D_U uses the minimum number of distinct endpoint values among all interval representations. This minimum value is called the magnitude and discussed in Sect. 2.3 of Fishburn (1985). Magnitude is presented from a slightly different perspective here. In particular we obtain a minimax result equating the minimum number of endpoints to the maximum "size" of a certain suborder.

Since the arc lengths in D_U are 0 and $-\gamma$, the shortest path lengths $p(r_x), p(l_x)$ take on values $0, -\gamma, -2\gamma, \ldots, -m\gamma$ for some nonnegative integer m. The interval representation constructed from D_U uses $m+1$ distinct endpoints. We will show that the magnitude is $m+1$.

In this section we will assume that all orders are interval orders. That is, D_U does not have a negative cycle.

We use a particular class of orders, called sequences of linked chains to obtain a lower bound on the magnitude and show that this bound matches the number of endpoints used in the shortest paths construction. These orders are special cases of the picycles introduced by Fishburn (1983) and what are called sequences of linked chains in Isaak (1990).

A *sequence of linked chains* in an interval order $(U, \prec)$ is a sequence of chains $C_i = u_{i1} \succ u_{i2} \succ \cdots \succ u_{in_i}$ for $i = 1, 2, \ldots, k$ such that the chains are nontrivial ($n_i \geq 2$)

and for $i = 1, 2, \ldots, k-1$ we have $u_{in_i} \sim u_{(i+1)1}$. In addition, elements of the chains are distinct except that possibly an element can appear on two chains as $u_{in_i} = u_{j1}$ for some $j > i+1$. The *size* of the sequence of linked chains is $1 + \sum_{i=1}^{k}(n_i - 1)$.

Observe that we have not specified all the relations between the elements so that different orders on the same set of elements can be sequences of linked chains. The relations that have been specified by the chains are enough for our proof.

Theorem 1. *For a finite interval order $(U, \prec)$, the minimum number of distinct endpoints in an interval representation is equal to the maximum size of a sequence of linked chains in $(U\ \prec)$. Furthermore, the representation constructed from shortest paths in D_U uses this many endpoints.*

Proof. Consider any interval representation of an order that contains a sequence of linked chains with size $m+1$. Use the notation $[l(u_{ij}), r(u_{ij})]$ for the intervals. From $u_{i(n_i-1)} \succ u_{in_i}$ and $u_{in_i} \sim u_{(i+1)1}$ we get $l(u_{i(n_i-1)}) > r(u_{in_i}) \geq l(u_{(i+1)1})$. Also, from the chains, the left endpoints satisfy $l(u_{i1}) > \cdots > l(u_{in_i})$. Thus the endpoints $l(u_{ij})$ for $i = 1, 2, \ldots, k$ and $j = 1, 2, \ldots, n_i - 1$ are distinct. (It is possible that $l(u_{in_i}) = l(u_{(i+1)1})$). Finally, $l(u_{kn_k})$ is distinct from any of these endpoints. Thus any representation requires at least $1 + \sum_{i=1}^{k}(n_i - 1) = m+1$ endpoints.

As noted above, the length of a shortest path in D_U will be $-m\gamma$ for some nonnegative integer m. Partition the I arcs in D_U into $I_1 = \{(r_x, l_x) | x \in U\}$ and $I_2 = \{(r_x, l_y | x \sim y, x \neq y\}$. Since arcs in I have length 0 we can assume that a shortest path is of the form $l_{u_1}, r_{u_2}, \ldots, r_{u_t}$. Deleting the arcs from I_2 leaves subpaths which alternate between C arcs and I_1 arcs. That is, subpaths $l_{u_{i1}}, r_{u_{i2}}, l_{u_{i2}}, \ldots, r_{u_{i(n_i-1)}}, l_{u_{i(n_i-1)}}, r_{u_{in_i}}$ for some $i = 1, 2, \ldots, k$ (where we assume the ith subpath appears before the $(i+1)$st). These correspond to chains $u_{i1} \succ u_{i2} \succ \cdots \succ u_{un_i}$ in the order. From the deleted I_2 arcs we have $u_{in_i} \sim u_{(i+1)1}$ and hence the order contains a sequence of linked chains. For each of the chains there are $n_i - 1$ corresponding arcs from C in the path in D_U and these are the only arcs from C. Hence the path length satisfies $m = \sum_{i=1}^{k}(n_i - 1)$ and the sequence of linked chains has size $m+1$.

The first paragraph shows that the minimum number of endpoints in an interval representation of $(U, \prec)$ is at least the maximum size of a sequence of linked chains. If the shortest path length in D_U is $-m\gamma$ then the construction in the proof of Fishburn's theorem yields a representation with exactly $m+1$ distinct endpoints. The second paragraph shows that $(U, \prec)$ contains a sequence of linked chains of size $m+1$. □

7 Conclusion

The technique of using systems of linear inequalities and shortest paths in digraphs naturally arising from these systems has been used to prove a variety of results about interval orders and semiorders. These have included representations with additional constraints on interval lengths and endpoints and also investigation of minimal representations. See Doignon(1988a, 1988b); Fishburn (1985, 1983); Isaak (1990, 1993); Pirlot (1990); Pirlot and Vincke (1997) for some of these. In this paper we have shown how this technique that has been used in more general settings can also be used to give short proofs of basic representation theorems including a structural perspective on minimal representations.

Acknowledgements Thanks to the referees for useful comments. I would also like to thank Peter Fishburn for his valuable support and encouragement early in my career.

References

Bogart, K. P. (1993). An obvious proof of Fishburn's interval order theorem. *Discrete Mathematics, 118*, 239–242.

Doignon, J.-P. (1988a). Sur les reprsentations minimales des semiordres et des ordres d'intervalles. *Mathématiques et Sciences Humaines, 101*, 49–59.

Doignon, J.-P. (1988b). Threshold representations of multiple semiorders. *SIAM Journal on Algebraic and Discrete Methods, 8*, 77–84.

Fishburn, P. C. (1970a). Intransitive indifference with unequal indifference intervals. *Journal of Mathematical Psychology, 7*, 144–149.

Fishburn, P. C. (1970b). *Utility theory for decision making*. New York: Wiley.

Fishburn, P. C. (1983). Threshold-bounded interval orders and a theory of picycles. *SIAM Journal on Algebraic and Discrete Methods, 4*, 290–305.

Fishburn, P. C. (1985). *Interval orders and interval graphs*. New York: Wiley.

Fishburn, P. C. (1997). Generalizations of semiorders: A review note. *Journal of Mathematical Psychology, 41*, 357–366.

Fishburn, P. C., & Monjardet, B. (1992). Norbert Wiener on the theory of measurement (1914, 1915, 1921). *Journal of Mathematical Psychology, 36*, 165–184.

Isaak, G. (1990). *Odd forests, reversing numbers and discrete representations of interval orders*. Ph.D. Thesis, Rutgers University, New brunswick, NJ.

Isaak, G. (1993). Bounded discrete representations of interval orders. *Discrete Applied Mathematics, 44*, 157–183.

Luce, R. D. (1956). Semiorders and a theory of utility discrimination. *Econometrica, 24*, 178–191.

Luce, R. D., Krantz, D. H., & Suppes, P. (1971). *Foundations of measurement, Vol. I: Additive and polynomial representations*. New York: Academic.

Luce, R. D., Krantz, D. H., & Suppes, P. (1989). *Foundations of measurement, Vol. II: Geometrical, threshold and probabilistic representations*. San Diego: Academic.

Luce, R. D., Krantz, D. H., & Suppes, P. (1990). *Foundations of measurement, Vol. III: Representation, axiomatization, and invariance*. San Diego: Academic.

Pirlot, M. (1990). Minimal representation of a semiorder. *Theory and Decision, 28*, 109–141.

Pirlot, M., & Vincke, P. (1997). *Semiorders*. Dordrecht: Kluwer.

Schrijver, A. (2003). *Combinatorial optimization: Polyhedra and efficiency*. Berlin: Springer.

Scott, D., & Suppes, P. (1958). Foundational aspects of theories of measurement. *Journal of Symbolic logic, 23*, 113–128.

Probe Interval Orders

David E. Brown and Larry J. Langley

1 Introduction

A graph G is a *probe interval graph* if there is a partition of $V(G)$ into P and N and a collection $\{I_v : v \in V(G)\}$ of intervals of $\mathbb{R}$ in one-to-one correspondence with $V(G)$ such that $uv \in E(G)$ if and only if $I_u \cap I_v \neq \emptyset$ and at least one of u, v belongs to P. The sets P and N are called the *probes* and *nonprobes*, respectively, and $\{I_v = [l(v), r(v)] : v \in V(G)\}$ together with the partition will be referred to as a *representation*. An interval graph is a probe interval graph with $N = \emptyset$ and this class of graphs has been studied extensively; see the texts Fishburn (1985), Golumbic (1980), and Roberts (1976) for introductions and other references.

The probe interval graph model was invented in connection with the task called *physical mapping* faced in connection with the human genome project (Zhang, 1994, 1997; Zhang et al., 1994). In DNA sequencing projects, a *contig* is a set of overlapping DNA segments derived from a single genetic source. In order for DNA to be more easily studied, small fragments of it, called clones, are taken from multiple copies of the same genome. Physical mapping is the process of determining how DNA contained in a group of clones overlap without having to sequence all the DNA in the clones. Once the map is determined, the clones can be used as a resource to efficiently contain stretches of genome. If we are interested in overlap information between each pair of clones, we can use an interval graph to model this problem: vertices are clones and adjacency represents overlap. Using the probe interval graph model, we can use any subset of clones, label them as probes, and test for overlap between a pair of clones if and only if at least one of them is a probe. This way there is flexibility, in contrast to the interval graph model, since all DNA fragments need not be known at time of construction of the probe interval graph model. Consequently, the size of the data set, which by nature can be quite large, is reduced.

L.J. Langley (✉)
Department of Mathematics, University of the Pacific, Stockton, CA 95211, USA
e-mail: llangley@pacific.edu

S.J. Brams et al. (eds.), *The Mathematics of Preference, Choice and Order: Essays in Honor of Peter C. Fishburn*, Studies in Choice and Welfare,

We now mention some of the recent results on probe interval graphs. The paper (McMorris, Wang, & Zhang, 1998) has results similar to those for interval graphs found in Fulkerson and Gross (1965) and Golumbic (1980); e.g., probe interval graphs are weakly chordal, in analogue to interval graphs being chordal, and, as maximal cliques are consecutively orderable in interval graphs, so-called quasi-maximal cliques are in probe interval graphs. The neighborhood of graph classes surrounding probe interval graphs has begun to be described in Brown and Lundgren (2006), Brown, Lundgren, and Flink (2002), and Golumbic and Lipshteyn (2001). Relationships between bipartite probe interval graphs, interval bigraphs and the complements of circular arc graphs are presented in Brown and Lundgren (2006). In Golumbic and Lipshteyn (2004) chordal graphs have been generalized to what are called "chordal probe graphs" in a way analogous to how probe interval graphs generalize interval graphs.

There are two recognition problems for probe interval graphs: partitioned and non-partitioned. The former entails recognizing, finding, and representing possible layouts of intervals for probe interval graphs with a prior specification of partition of vertices into probes and nonprobes. The latter problem entails determining whether a given graph is a probe interval graph with no partition specified. An $O(n^2)$ recognition algorithm for the partitioned case is described in Johnson and Spinrad (2001), and an $O(n+m\log n)$ algorithm is given in McConnell and Lundgren (2002), where m is the number of edges and n the number of vertices of the graph under consideration. An application of an algorithm for constructing probe interval graph models occurred in recognizing circular arc graphs (McConnell, 2001). The results on chordal probe graphs in Berry, Golumbic, and Lipshteyn (2004) have been useful for developing a recognition algorithm for the class of probe interval graphs we study, see Sect. 4.

A class of graphs is *hereditary* if any induced subgraph from that class is a member of that class. It is easy to see the class of probe interval graphs is *hereditary*. One way to describe the structure of a hereditary class of graphs is through a characterization via forbidden induced subgraphs. As far as this task goes for probe interval graphs there are the following results. In Sheng (1999) the first step was taken and cycle-free probe interval graphs were characterized by two forbidden induced subgraphs for the case with no specified probe/nonprobe partition, and by six for the case where probes are specified. A *unit probe interval graph* is a probe interval graph such that a representation exists where all intervals are of identical length. And in Golumbic and Trenk (2004) and Lipshteyn (2001) unit probe interval graphs were shown to be identical to *proper probe interval graphs*, which are probe interval graphs that admit a representation in which no interval contains another properly. In Brown, Lundgren, and Sheng (2008) the unit probe interval graphs that are cycle-free were characterized by two forbidden induced subgraphs and one infinite family in the non-partitioned case, and for the case where probes are specified, there are five forbidden induced subgraphs. Bipartite unit probe interval graphs are characterized in Brown and Langley (2006) by five forbidden induced subgraphs that are in addition to those in Brown et al. (2008) for trees. For more general classes of probe interval graphs, the task of characterization in this way appears difficult. To wit, in

Corneil and Pržulj (2005) it is shown that there are at least 62 forbidden induced subgraphs for 2-trees that are probe interval graphs.

We now give a few definitions and some notation we will use in the sequel, and which will give this paper more context. $\overline{G}$ denotes the complement of graph G, and $u \to v$ means that there is a directed edge from u to v. A *cocomparability graph* is a graph whose complement has a transitive orientation, and hence yields an order on its vertices via this orientation; hence the graph can be thought of as the incomparability graph of the vertices with respect to the order given by the orientation. If G is an interval graph, then its representation $\{I_v : v \in V(G)\}$ gives a natural transitive orientation to its complement: put $u \to v$ in $\overline{G}$ if I_u lies entirely to the left of I_v. An order with such a representation is an *interval order*; that is, $(V, \prec)$ is an interval order if to each $x \in V$ an interval I_x can be assigned so that $x \prec y$ if and only if I_x is entirely to the left of I_y. One notable characterization of interval orders, first proven by Fishburn (1970) (and independently by Mirkin, 1972, 1970) is that an ordered set is an interval order if and only if it contains no $\underline{2}+\underline{2}$. A $\underline{2}+\underline{2}$ is a set of four elements $\{w,x,y,z\}$ where x is comparable to w, y is comparable to z and all other pairs are incomparable. One consequence of this theorem is that a cocomparability graph is an interval graph if and only if it contains no 4-cycle. For additional discussion of the relationship between interval orders and interval graphs, see Bogart, Rabinovitch, and Trotter (1976); Fishburn (1985); Greenough (1974).

Here is a summary of this paper's structure and content. We give several characterizations for probe interval graphs that are cocomparability graphs, and hence give rise to a *probe interval order*. One characterization specifies conditions on the vertex set's partition into probes and nonprobes when restricted to 4-cycles in the graph. Another gives conditions on the part of a probe interval graph's representation restricted to the intervals for the nonprobes; specifically, the set of intervals corresponding to nonprobes must be such that no interval properly contains another, see Sects. 2 and 3. A probe interval graph with a representation in which no nonprobe interval contains another nonprobe interval properly will be called a *nonprobe–proper probe interval graph*. We adapt recognition algorithms of chordal probe graphs to recognize nonprobe–proper probe interval graphs, both in the partitioned and the non-partitioned case; see Sect. 4. Also in Sect. 4, we describe the graph classes in the neighborhood of nonprobe–proper probe interval graphs.

2 Nonprobe–Proper Probe Interval Graphs

An *asteroidal triple* in a graph is a set of three vertices with a path between each pair that avoids the neighborhood of the third. Interval graphs and, more generally, cocomparability graphs are asteroidal triple-free (Boland & Lekkerkerker, 1962; Gallai, 1967), whereas probe interval graphs are not. Thus, in order to identify the probe interval graphs that are cocomparability graphs we investigated the mechanism by which a probe interval graph can contain an asteroidal

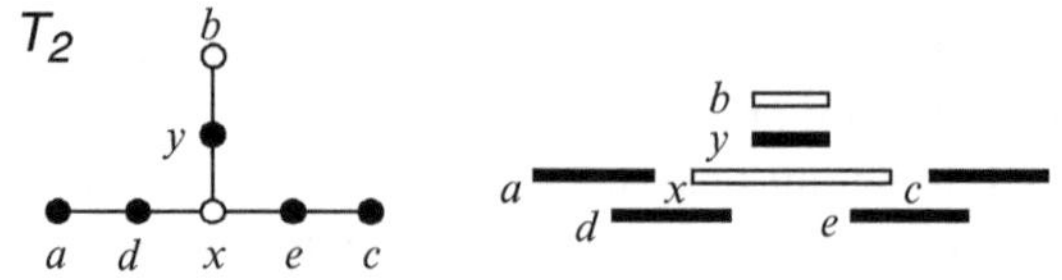

Fig. 1 A probe interval graph with an asteroidal triple and representation that exemplifies the general case of an asteroidal triple's presence in a probe interval graph. *Darkened vertices* are probes. Intervals are displaced vertically for easier visualization

triple. The observations we made lead us to define a nonprobe–proper probe interval graph. For example, consider the graph T_2 in Fig. 1 which has an asteroidal triple on $\{a,b,c\}$. The reader can verify that a representation of T_2 must have x a nonprobe and, up to symmetry, I_x must contain I_b properly.

Theorem 1 precisely identifies the class of probe interval graphs that has each member possessing a transitive orientation on its complement as the nonprobe–proper probe interval graphs. We prove only one direction of Theorem 1 here and defer the rest of the proof to Sect. 3.

Theorem 1. *A probe interval graph is a cocomparability graph if and only if it is a nonprobe–proper probe interval graph.*

Proof $\Leftarrow$: It turns out that the representation of a nonprobe–proper probe interval graph naturally yields a transitive orientation of its complement. Let G be a nonprobe–proper probe interval graph with $V(G)$ partitioned into $P \cup N$ and representation $\{I(v) = [l(v), r(v)] : v \in V\}$. In $\overline{G}$, put $u \to v$ if and only if $I(u)$ is entirely to the left of $I(v)$, or $l(u) < l(v)$ and $u, v \in N$. It is easy to check that this orientation is transitive, and hence G is a cocomparability graph.

Alternative proof $\Leftarrow$: To show that a nonprobe–proper probe interval graph is a cocomparability graph, we could also show that the former has a representation as a parallelogram graph. Since parallelogram graphs are cocomparability graphs (Golumbic & Trenk, 2004), the result then follows. Let G be a nonprobe–proper probe interval graph with partition $V(G) = P \cup N$, L_1 and L_2 be two parallel horizontal lines some vertical distance apart, and $\{I(v) : v \in V\}$ the representation for G. Place a copy of the representation on each of the lines L_1 and L_2. Supposing $P = \{p_1, \ldots, p_r\}$ and $N = \{n_1, \ldots, n_s\}$ let the parallelogram for p_i be defined by the region bounded by $I(p_i)$ on L_1 and on L_2 and the line from $l(p_i)$ on L_1 to $l(p_i)$ on L_2 and the line from $r(p_i)$ on L_1 to $r(p_i)$ on L_2. Then let the parallelogram for n_i be the line segment from $l(n_i)$ on L_1 to $r(n_i)$ on L_2. Since no nonprobe interval properly contains another, the line segments now representing nonprobes never cross, while $I(p_i) \cap I(p_j) \neq \emptyset$ if and only if regions now representing p_i and p_j intersect; similarly $I(p_i) \cap I(n_j) \neq \emptyset$ if and only if the region for p_i intersects the line segment for n_j; cf. Fig. 2. $\square$

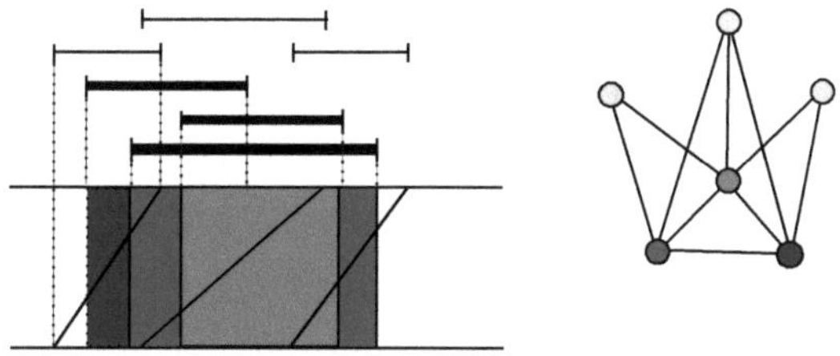

Fig. 2 A parallelogram representation of a nonprobe–proper probe interval graph. The *thick intervals* correspond to probes

3 Probe Interval Orders

Above we described how a representation of a probe interval graph G with proper containment among nonprobe intervals leads naturally to a transitive orientation of $\overline{G}$. This orientation in turn gives an ordering, say $\ll$, of $V(G)$ by putting $u \ll v$ if and only if $u \to v$ in $\overline{G}$. So, we define a *probe interval order* by, for $x, y \in V(G)$, $x \ll y$ if and only if either $r(x) < l(y)$ or both x and y are in N and $l(x) < l(y)$. Notice that this order restricted to N is a total ordering.

To complete the proof of Theorem 1 we will change our perspective to orders, and prove that an ordered set in which the $\underline{2}+\underline{2}$'s can be partitioned in a certain way is conducive to constructing a nonprobe–proper probe interval graph. But for now we will keep to the graph perspective and prove the following lemma for the sake of exposition, but note that it follows from a result in McMorris et al. (1998). Lemma 1 translates into a corresponding lemma about $\underline{2}+\underline{2}$'s in a probe interval order.

Lemma 1. *In any induced 4-cycle of a probe interval graph, two nonadjacent vertices must be nonprobes.*

Proof. Let $\langle a,b,c,d,a \rangle$ be an induced 4-cycle in probe interval graph G. Since a 4-cycle is not an interval graph, at least one of a,b,c,d must be a nonprobe, while no three can be all nonprobes, since nonprobes induce an independent set in G. Relabeling if necessary, assume $l(a) < l(b) < l(c)$, $l(b) < r(a)$, $l(c) < r(b)$, and that d is the only nonprobe. Since $I(d)$ must intersect both $I(a)$ and $I(c)$, we have $r(a), l(c) \in I(d)$, and hence $I(b) \cap I(d) \neq \emptyset$, a contradiction. Therefore exactly two vertices are nonprobes, and they are nonadjacent by definition. □

Corollary 1. *In a probe interval order, every* $\underline{2}+\underline{2}$ *has one 2-chain in P and the other in N.*

We are now ready to prove the other direction of Theorem 1, if G is a cocomparability probe interval graph, then it has a representation in which no nonprobe interval properly contains another nonprobe interval. We will begin with a partially ordered set having 2-chains partitioned according to Corollary 1, and construct a nonprobe–proper probe interval graph. Since we will not specify a transitive orientation, the proof also shows that a probe interval order is a cocomparability invariant. The methods of this construction closely follow those of Greenough (1974) and Langley (1995), using the notation from the latter. Let $D = (V, \ll)$ be a partially ordered set, with $\ll$ strict. If neither $a \ll b$ nor $b \ll a$, then a and b are incomparable and we write $a \sim b$. Define the predecessor set of x, $pred(x) = \{y \in V : y \ll x\}$. For a set $S \in V$, define $pred(S) = \bigcap_{x \in S} pred(x) = \{y \in V : \forall x \in S, y \ll x\}$. Similarly define the successor sets, $succ(x)$ and $succ(S)$. Note that for any $x \in V$, $x \in pred(succ(x))$.

Theorem 2. *Suppose D is a partially ordered set, with V partitioned into two sets N and P, such that D restricted to N is a total order, and for any* $\underline{2}+\underline{2}$ *in D, one 2-chain is in N and one is in P. Then D corresponds to a nonprobe–proper probe interval graph, with probes P and nonprobes N.*

To prove Theorem 2, we first note that the order restricted to P is an interval order since P contains no $\underline{2}+\underline{2}$, and prove the following lemmas. For these lemmas and the proof of Theorem 2 we define $T = \emptyset \cup V \cup \{pred(v) | v \in P\} \cup \{pred(succ(v)) | v \in P\}$. We will show that this set T is linearly ordered, and then use this order as a guide for constructing intervals for the representation.

Lemma 2. *Let A, B be in T. Let* $a \in A$ *and* $b \in B$*. If* $a \ll b$*, then* $a \in B$.

Proof. Let A, B be in T. Let $a \in A$ and $b \in B$ with $a \ll b$. Clearly $B \neq \emptyset$. If $B = V$, then $a \in B$. If $B = pred(x)$, then $b \ll x$ and by transitivity $a \ll x$, so $a \in pred(x) = B$. If $B = pred(succ(x))$, then, for all $y \in succ(x)$, we have $b \ll y$. By transitivity $a \ll y$ as well, so $a \in B$. □

Lemma 3. *T is totally ordered by set inclusion.*

Proof. Let A, $B \in T$. We will show that one set must be contained in the other by contradiction. Suppose there exists $a \in A$, $a \notin B$ and $b \in B, b \notin A$. Clearly neither A nor $B = \emptyset$, nor is either equal to V. By Lemma 2 $a \sim b$. We consider three cases:

(1) $A = pred(x)$, $B = pred(y)$ for some $x, y \in P$. By definition $a \ll x$ and $b \ll y$. Since $b \notin A$, $b \not\ll x$. Likewise $x \not\ll b$, since by transitivity, if $x \ll b$ then $a \ll b$ which is not possible by Lemma 2. So we have $x \sim b$ and by similar arguments $y \sim a$. Therefore a, x and b, y form a $\underline{2}+\underline{2}$ in D. Since both x and y are in P, this is a contradiction.

(2) $A = pred(x)$, $B = pred(succ(y))$ for some $x, y \in P$. Consider the vertex y, $y \in B$. If $y \in A$, then $y \ll x$, so $x \in succ(y)$. However, then $b \ll x$ and, consequently, $b \in A$. Therefore $y \notin A$. By the arguments for case 1, $y \sim x$. Since $a \notin B$, there must be some z in $succ(y)$, with $z \sim a$. So we have $a \ll x$ and $y \ll z$ form a $\underline{2}+\underline{2}$ in D, but x and y are both in P, a contradiction.

(3) $A = pred(succ(x))$, $B = pred(succ(y))$. If $succ(x) \subseteq succ(y)$, then $A \subseteq B$, so there is some vertex w, $x \ll w$, $y \sim w$. Likewise there is some vertex z, $y \ll z$, $x \sim z$. x, w and y, z form a $\underline{2}+\underline{2}$ with $x, y \in P$, a contradiction. □

We are now ready to constructively prove Theorem 2 and hence complete the proof of Theorem 1.

Proof. Order the elements of T by set inclusion and let $r(A)$ be the rank of set A in T. Label the vertices of N, $x_1, \ldots, x_n$, so $x_i \ll x_j$ if and only if $i < j$.

We will assign an interval to each element of V as follows:

If $v \in P$ let $I(v) = [r(pred(v)), r(pred(succ(v))) - 0.5]$. Since $v \in pred(succ(v))$ but $v \notin pred(v)$, $r(pred(succ(v))) > r(pred(v))$, so this interval is well defined.

If $v = x_i \in N$, let $A(v) \in T$ be the largest set contained within $pred(v)$, and $B(v) \in T$ be the smallest set containing $pred(succ(v))$. Let x_i be the label of v, let $I(v) = [r(A(v)) - (n-i)\varepsilon, r(B(v)) - 0.5 + i\varepsilon]$. Since $v \in B(v)$ and $v \notin A(v)$, this interval is well defined.

Let $x, y \in P$. $x \ll y$ if and only if $pred(succ(x)) \subseteq pred(y)$, so if $x \ll y$, $I(x)$ is completely to the left of $I(y)$ and if $x \sim y$, then $pred(y) \subset pred(succ(x))$ and $pred(x) \subset pred(succ(y))$ so the intervals have non trivial overlap.

Let $x \in P$, $y \in N$. We know $x \ll y$ if and only if $pred(succ(x)) \subseteq pred(y)$. Therefore, $pred(succ(x)) \subseteq A(y)$, and the interval for x is completely to the left of the interval for y. If $y \ll x$, then $pred(succ(y)) \subseteq pred(x)$ and hence $B(y) \subseteq pred(x)$, so the interval for y is completely to the left of the interval for x. If $x \sim y$, then $A(y) \subseteq pred(y) \subset pred(succ(x))$ and $pred(x) \subseteq pred(succ(y)) \subseteq B(y)$, so the intervals have nonempty intersection.

Finally Let x, y both be in N. Without loss of generality, assume $x \ll y$. Then $A(x) \subseteq pred(x) \subseteq pred(y)$. So $A(x) \subseteq A(y)$, and thus the left hand end point of the interval for x is to the left of the left hand endpoint of the interval of y. Likewise $pred(succ(x)) \subseteq pred(succ(y)) \subseteq B(y)$, so $B(x) \subseteq B(y)$, and the right hand end point of the interval for x is to the left of the right hand end point of the interval for y. □

Corollary 2. *Probe interval order is comparability invariant.*

Proof. Reorienting the underlying graph of D will not change the structure of a $\underline{2} + \underline{2}$ except perhaps reordering the vertices within a 2-chain. Thus, the conditions that decide if an order is a probe interval order will not change upon reorienting the underlying graph. □

4 Recognition Algorithms and a Hierarchy

We will adapt algorithms to recognize cocomparability probe interval graphs, either given a fixed partition of the vertices or not. Currently the fastest known algorithm for recognizing if G is a comparability graph or a cocomparability graph is $O(n^{2.38})$ (McConnell & Spinrad, 1999).

If G is partitioned, we may construct the proper non-probe interval representation using the methods of the proof of Theorem 2 in $O(n^3)$ time. However there are two faster algorithms to determine if G is a partitioned probe interval graph in general, Johnson and Spinrad (2001) of $O(n^2)$ and McConnell and Lundgren (2002) of $O(n + m\log n)$, as mentioned in the introduction. This reduces the speed of recognition of a partitioned cocomparability probe interval graph to the speed of recognizing whether it is a cocomparability graph.

In the non-partitioned case we will adapt a recognition algorithm for chordal probe graphs described by Golumbic and Lipshteyn (2004). Recall that a *stable set* in a graph G is a set of vertices S that induces a subgraph with no edges, so, in a probe interval graph, the non-probes must form a stable set. An equivalent definition for a probe interval graph, is any graph G with a vertex partition into a stable set N and a set P such that by adding edges between vertices in N we can create an interval graph. Golumbic and Lipshteyn (2004) generalize the notion of probe interval graphs to chordal probe graphs. A *chordal* graph G has no induced chordless cycle on more than two vertices. A *chordal probe graph* has a vertex partition into sets N and P where it is possible to create a chordal graph by adding edges between vertices of N. Since interval graphs are chordal, it follows that probe interval graphs are also chordal probe graphs. If a graph G is *weakly chordal* neither G nor its complement contain a chordless cycle of length greater than or equal to 5.

Theorem 3. *Let G be a graph. The following are equivalent:*

1. *G is a cocomparability graph and has a partition into P and a stable set N so that every 4-cycle alternates between N and P.*
2. *G is a probe interval graph with a proper nonprobe representation.*
3. *G is a bounded tolerance graph and a probe interval graph.*
4. *G is a cocomparability probe interval graph.*
5. *G is a cocomparability chordal probe graph and a weakly chordal graph.*
6. *G is a cocomparability chordal probe graph.*

Proof. The implication $1 \Rightarrow 2$ follows immediately from Theorem 2. That $2 \Rightarrow 3$ follows from the construction in Sect. 2. All bounded tolerance graphs are cocomparability graphs (Golumbic, Monma, & Totter, 1984), so $3 \Rightarrow 4$. For the implication $4 \Rightarrow 5$, note that probe interval graphs have no chordless cycles of length greater than 5, so they are weakly chordal; this was mentioned in Golumbic and Lipshteyn (2004). The implication $5 \Rightarrow 6$ is obvious. Chordal probe graphs by definition have a partition into P and a stable set N, and according to Golumbic and Lipshteyn (2004) every chordless even cycle in a chordal probe graph alternates between N and P. Cocomparability graphs have no chordless cycles of length greater than four, thus $6 \Rightarrow 1$. □

Golumbic and Lipshteyn provide an algorithm to recognize graphs that are both weakly chordal and chordal probe in the nonpartitioned case. The complexity of this algorithm is $O(m^2)$. The first step of the algorithm is to check to see if the graph is weakly chordal. By substituting a check for cocomparability instead, we may identify nonprobe–proper probe interval graphs.

The containment relationships identified in Theorem 3, and some others that are known, for classes of graphs in the neighborhood of cocomparability probe interval graphs are summarized in Fig. 3.

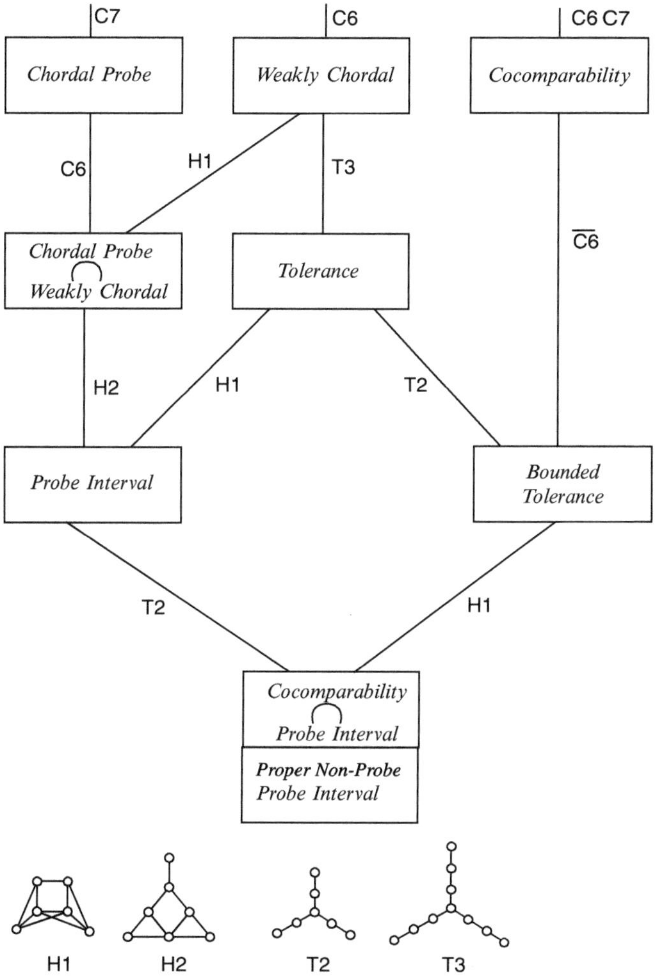

Fig. 3 Containment diagram including cocomparability probe interval graphs with separating examples

References

Berry, A., Golumbic, M., & Lipshteyn, M. (2004). Two tricks to triangulate chordal probe graphs in polynomial time. In *Proc. of the 12th ACM-SIAM symp. on discrete algorithms (SODA04)*. Association for Computing Machinery.

Bogart, K., Rabinovitch, I., & Trotter, W.T. (1976). A bound on the dimension of interval orders. *Journal of Combinatorial Theory A, 21*, 219–238.

Boland, J. & Lekkerkerker, C. (1962). Representation of a finite graph by a set of intervals on the real line. *Fundamenta Mathematicae, 51*, 45–64.

Brown, D. & Langley, L. (2006). Probe interval graphs and orders. In *Thirty-Sixth International Southeastern Conference on Combinatorics, Graph Theory, and Computing.*

Brown, D. & Lundgren, J. (2006). Bipartite probe interval graphs, interval point bigraphs, and circular arc graphs. *Australasian Journal of Combinatorics, 35*, 221–236.

Brown, D., Lundgren, J., & Flink, S. (2002). Interval *k*-graphs. *Congressus Numerantium, 156*, 5–16.

Brown, D., Lundgren, J., & Sheng, L. (2008). Cycle-free unit and proper probe interval graphs. *submitted to Discrete Applied Math.*

Brown D. E., Lundgren, J. R., & Sheng, L., A characterization of cycle-free unit probe interval graphs, Discrete Applied Mathematics (2008), doi:10.1016/j.dam.2008.07.004

Corneil, D. & Pržulj, N. (2005). 2-tree probe interval graphs have a large obstruction set. *Discrete Applied Mathematics, 150*(1), 216–231

Fishburn, P. C. (1970). Intransitive indifference with unequal indifference intervals. *Journal of Mathematical Psychology, 7*, 144-149.

Fishburn, P. C. (1985). *Interval orders and interval graphs*. New York: Wiley.

Fulkerson, D. & Gross, O. (1965). Incidence matrices and interval graphs. *Pacific Journal of Mathematics, 15*, 835–855.

Gallai, T. (1967). Transitiv orientbare graphen. *Acta Mathematica Academiae Scientiarum Hungaricae, 18*, 25–66.

Golumbic, M. (1980). *Algorithmic graph theory and perfect graphs*. New York: Academic.

Golumbic, M. & Lipshteyn, M. (2001). On the hierarchy of interval, probe, and tolerance graphs. *Congressus Numerantium, 153*, 97–106.

Golumbic, M. & Lipshteyn, M. (2004). Chordal probe graphs. *Discrete Applied Mathematics, 143*, 221–237.

Golumbic, M., Monma, C., & Totter, W. (1984). Tolerance graphs. *Discrete Applied Mathematics, 9*, 157–170.

Golumbic, M. & Trenk, A. (2004). *Tolerance graphs*. Cambridge: Cambridge University Press.

Greenough, T. (1974). *Representation and enumeration of interval orders and semiorders*. Dartmouth College.

Johnson, J. & Spinrad, J. (2001). A polynomial time recognition algorithm for probe interval graphs. *Proc. of the 12th ACM-SIAM symp. on discrete algorithms (SODA01)* (pp. 477–486). New York, NY: Association for Computing Machinery.

Langley, L. (1995). A recognition algorithm for orders of interval dimension two. *Discrete Applied Mathematics, 60*, 257–266.

Lipshteyn, M. (2001). *Probe graphs*. Unpublished master's thesis, Bar-Ilan University.

McConnell, M. (2001). Linear-time recognition of circular-arc graphs. In *Proc. of the 42nd annual IEEE symposium on foundations of computer science (FOCS 01)* (pp. 386–394).

McConnell, R. & Lundgren, J. (2002). Construction of probe interval models. *Proc. of the 13th ACM-SIAM symposium on discrete algs. (SODA02)* (pp. 866–875). San Francisco, CA: Association for Computing Machinery.

McConnell, R. & Spinrad, J. (1999). Modular decomposition and transitive orientation. *Discrete Mathematics, 201*, 189–241.

McMorris, F., Wang, C., & Zhang, P. (1998). On probe interval graphs. *Discrete Applied Mathematics, 88*, 315–324.

Mirkin, B. G. (1970). Ob odnom klasse otnoshenij predpochtenija. *Matematitcheskije woprosy formirovanija economitcheskich modelei.*

Mirkin, B. G. (1972). Description of some relations on the set of real-line intervals. *Journal of Mathematical Psychology, 9*, 243-252.

Roberts, F. (1976). *Discrete mathematical models*. Upper Saddle River, NJ: Prentice-Hall.

Sheng, L. (1999). Cycle-free probe interval graphs. *Congressus Numerantium, 88*, 33–42.

Zhang, P. (1994). *Probe interval graph and its applications to physical mapping of DNA*. (Manuscript)

Zhang, P. (1997). *Methods of mapping DNA fragments.* (United States Patent. http://www.cc.columbia.edu/cu/cie/techlists/patents/5667970.htm)

Zhang, P., Schon, E., Fischer, S., Cayanis, E., Weiss, J., Kistler, S., et al. (1994). An algorithm based on graph theory for the assembly of contigs in physical mapping of DNA. *CABIOS, 10*(3), 309–317.

Graphs

Mediatic Graphs

Jean-Claude Falmagne and Sergei Ovchinnikov

1 Background and Introduction

The core concept of this paper can occur in the guise of various representations. Four of them are relevant here, the last one being new:

1. A MEDIUM, that is, a semigroup of transformations on a set of states, constrained by strong axioms (see Eppstein, Falmagne, & Ovchinnikov, 2008; Falmagne, 1997; Falmagne & Ovchinnikov, 2002).
2. An ISOMETRIC SUBGRAPH OF THE HYPERCUBE, OR "PARTIAL CUBE." By "isometric", we mean that the distance between any two vertices of the subgraph is identical to the distance between the same two vertices in the hypercube (Djoković, 1973; Graham & Pollak, 1971). Each state of the medium is mapped to a vertex of the graph, and each transformation corresponds to an equivalence class of its arcs. Note that, as will become clear later on, no assumption of finiteness is made in this or in any of the other representation.
3. An ISOMETRIC SUBGRAPH OF THE INTEGER LATTICE. This representation is not exactly interchangeable with the preceding one. While it is true that any isometric subgraph of the hypercube is representable as an isometric subgraph of the integer lattice and vice versa, the latter representation lands in a space equipped with a considerable amount of structure. Notions of "lines", "hyperplanes", or "parallelism" can be legitimately defined if one wishes. Moreover, the dimension of the lattice representation is typically much smaller than that of the partial cube representing the same medium and so can be valuable in the representation of large media (see, in particular, Eppstein, 2005, in which an algorithm is described for finding the minimum dimension of a lattice representation of a partial cube).

J.-Cl. Falmagne (✉)
Department of Cognitive Sciences, University of California, Irvine, CA 92697, USA
e-mail: jcf@uci.edu

S.J. Brams et al. (eds.), *The Mathematics of Preference, Choice and Order: Essays in Honor of Peter C. Fishburn*, Studies in Choice and Welfare,

4. A MEDIATIC GRAPH. Axiomatic definitions are usually regarded as preferable whenever feasible, and that is what is given here.

The definition of a medium is recalled in the next section, together with some key concepts and the consequences of the axioms that are useful for this paper. Note that two axioms are used, which are equivalent to the original four used by Falmagne (1997) (see also Falmagne & Ovchinnikov, 2002; Eppstein & Falmagne, 2008). The graph of a medium and those graphs that induce media, called "mediatic graphs" are defined and studied in the following two sections. The last two sections of the paper are devoted to specifying the correspondence between mediatic graphs and media, for a given possibly infinity set – of vertices or states depending on the case.

The subject of this paper may at first seem to be singularly ill chosen for a volume honoring Peter Fishburn's, as its topic does not readily evoke any of Peter's favorite concepts. But the enormously rich span of his accomplishment is not so easily escaped: indeed, the set of all interval orders (Fishburn, 1971; Brightwell, Fishburn, & Winkler, 1993) on any finite set is representable as a mediatic graph, and so is the set of all semiorders (Fishburn, 1985; Fishburn & Trotter, 1999) on the same set, these few citations heading a list far too long to be included here. For the representability of families of interval orders or semiorders by mediatic graphs, see the concluding paragraph of this paper.

2 The Concept of a Medium

We begin with the terminology of "token systems" which provides a convenient framework.

Definition 1. Let $\mathcal{S}$ be a set of *states*. A *token* is a function $\tau : S \mapsto S\tau$ mapping $\mathcal{S}$ into itself. We shall use the abbreviations $S\tau = \tau(S)$, and $S\tau_1\tau_2\cdots\tau_n = \tau_n[\cdots\tau_2[\tau_1(S)]\cdots]$ for the function composition. By definition, the identity function τ_0 on $\mathcal{S}$ is not a token. Let $\mathcal{T}$ be a set of tokens on $\mathcal{S}$. The pair $(\mathcal{S},\mathcal{T})$ is called a *token system*. We suppose that $|\mathcal{S}| \geq 2$ and $\mathcal{T} \neq \varnothing$.

Let V and S be two distinct states. Then V is *adjacent* to S if $S\tau = V$ for some token τ. A token $\tilde{\tau}$ is a *reverse* of a token τ if, for any two adjacent states S and V, we have

$$S\tau = V \quad \Longleftrightarrow \quad V\tilde{\tau} = S, \tag{1}$$

and thus $S\tau\tilde{\tau} = S$. It is clear that a token has at most one reverse. If the reverse $\tilde{\tau}$ of a token τ exists, then $\tilde{\tilde{\tau}} = \tau$; that is, τ and $\tilde{\tau}$ are mutual reverses. If every token has a reverse, then adjacency is a symmetric relation on $\mathcal{S}$.

Definition 2. A *message* is a string of symbols representing tokens in the set $\mathcal{T}$. The message $\tau_1 \ldots \tau_n$ defines a function $S \mapsto S\tau_1\cdots\tau_n$ on the set of states $\mathcal{S}$. If $\boldsymbol{m} = \tau_1 \ldots \tau_n$ denotes a message, we also (by abuse of notation) write $\boldsymbol{m} = \tau_1\cdots\tau_n$ for the corresponding function. No ambiguity will arise from this double usage.

A message may consist in (the symbol representing) a single token. The *content* of a message $\boldsymbol{m} = \tau_1 \dots \tau_n$ is the set $\mathcal{C}(\boldsymbol{m}) = \{\tau_1, \dots, \tau_n\}$ of its tokens. We write $\ell(\boldsymbol{m}) = n$ to denote the *length* of the message $\boldsymbol{m}$. (We have thus $|\mathcal{C}(\boldsymbol{m})| \leq \ell(\boldsymbol{m})$.) A message $\boldsymbol{m}$ is *effective* (resp. *ineffective*) for a state S if $S\boldsymbol{m} \neq S$ (resp. $S\boldsymbol{m} = S$). A message $\boldsymbol{m} = \tau_1 \dots \tau_n$ is *stepwise effective* for S if $S\tau_1 \cdots \tau_k \neq S\tau_0 \cdots \tau_{k-1}$, $1 \leq k \leq n$. A message which is both stepwise effective and ineffective for some state is called a *return message* or, more brief, a *return* (for that state).

A message is *consistent* if it does not contain both a token and its reverse, and *inconsistent* otherwise. Two messages $\boldsymbol{m}$ and $\boldsymbol{n}$ are *jointly consistent* if $\boldsymbol{mn}$ (or, equivalently, $\boldsymbol{nm}$) is consistent. A consistent message which is stepwise effective for some state S and does not have any of its token occurring more than once is said to be *concise* (for S). A message $\boldsymbol{m} = \tau_1 \dots \tau_n$ is *vacuous* if the set of indices $\{1, \dots, n\}$ can be partitioned into pairs $\{i, j\}$, such that τ_i and τ_j are mutual reverses. By abuse of language, we sometimes call "empty" a place holder symbol that can be deleted, as in: "let $\boldsymbol{mn}$ be a message in which $\boldsymbol{n}$ is either a concise message or is empty" (that is, in the latter case, $\boldsymbol{mn} = \boldsymbol{m}$). If $\boldsymbol{m} = \tau_1 \dots \tau_n$ is a stepwise effective message producing a state V from a state S, then the *reverse* of $\boldsymbol{m}$ is defined by $\widetilde{\boldsymbol{m}} = \tilde{\tau}_n \dots \tilde{\tau}_1$. We then have clearly $V\widetilde{\boldsymbol{m}} = S$ and moreover $\tau \in \mathcal{C}(\boldsymbol{m})$ if and only if $\tilde{\tau} \in \mathcal{C}(\widetilde{\boldsymbol{m}})$.

Axioms for Medium 3 A token system $(\mathcal{S}, \mathcal{T})$ is called a *medium (on $\mathcal{S}$)* if the two following axioms are satisfied.

[Ma] For any two distinct states S, V in $\mathcal{S}$, there is a concise message producing V from S.
[Mb] Any return message is vacuous.

A medium $(\mathcal{S}, \mathcal{T})$ is *finite* if $\mathcal{S}$ is a finite set. The concept of a medium was proposed by Falmagne (1997) who proved various basic facts about media. Other results were obtained by Falmagne and Ovchinnikov (2002) (see also Eppstein & Falmagne, 2008; Ovchinnikov, 2008; Ovchinnikov & Dukhovny, 2000).

3 Some Basic Results

The material in this section, only part of which is new, is instrumental for the graph-theoretical results presented in this paper. We omit the proofs of straightforward or previously published results (see Falmagne, 1997; Falmagne & Ovchinnikov, 2002).

Lemma 4. *(i) No token can be identical to its own reverse.*
(ii) Let $\boldsymbol{m}$ be a message that is concise for some state; we have then $l(m) = |C(m)|$ and $\mathcal{C}(\boldsymbol{m}) \cap \mathcal{C}(\widetilde{\boldsymbol{m}}) = \varnothing$.
(iii) For any two adjacent (thus, distinct) states S and V, there is exactly one token producing V from S.
(iv) No token can be a 1–1 function.
(v) Suppose that $\boldsymbol{m}$ and $\boldsymbol{n}$ are stepwise effective for S and V, respectively, with $S\boldsymbol{m} = V$ and $V\boldsymbol{n} = W$. Then $\boldsymbol{mn}$ is stepwise effective for S, with $S\boldsymbol{mn} = W$.

(vi) Let $\boldsymbol{m}$ and $\boldsymbol{n}$ be two distinct concise messages transforming some state S. Then

$$S\boldsymbol{m} = S\boldsymbol{n} \quad\Longleftrightarrow\quad \mathcal{C}(\boldsymbol{m}) = \mathcal{C}(\boldsymbol{n}).$$

(vii) Any two consistent messages producing the same state and stepwise effective for two not necessarily distinct states are jointly consistent.

Lemma 4(vi) suggests an important concept.

Definition 5. Let $(\mathcal{S},\mathcal{T})$ be a medium. For any state S, define the *(token) content* of S as the set $\widehat{S}$ of all tokens each of which is contained in at least one concise message producing S; formally:

$$\widehat{S} = \{\tau \in \mathcal{T} \mid \exists V \in \mathcal{S}, V\boldsymbol{m} = S, \text{ for } \boldsymbol{m} \text{ concise with } \tau \in \mathcal{C}(\boldsymbol{m})\}.$$

We refer to the family $\widehat{\mathcal{S}}$ of all the contents of the states in $\mathcal{S}$ as the *content family* of the medium $(\mathcal{S},\mathcal{T})$.

Remark 6. In view of Condition (vii) of Lemma 4, the content of a state cannot contain both a token and its reverse.

Writing $\triangle$ for the symmetric set difference, and $+$ for the disjoint union, we have:

Theorem 7. *If $S\boldsymbol{m} = V$ for some concise message $\boldsymbol{m}$ (thus $S \neq V$), then $\widehat{V} \setminus \widehat{S} = \mathcal{C}(\boldsymbol{m})$, and so $\widehat{V} \triangle \widehat{S} = \mathcal{C}(\boldsymbol{m}) + \mathcal{C}(\widetilde{\boldsymbol{m}})$.*

Theorem 8. *For any token τ and any state S, we have either $\tau \in \widehat{S}$ or $\tilde{\tau} \in \widehat{S}$; so, $|\widehat{S}| = |\widehat{V}|$ for any two states S and V with $S = V$ if and only if $\widehat{S} = \widehat{V}$. Moreover, if $\mathcal{S}$ is finite, then $|\widehat{S}| = |\mathcal{T}|/2$ for any $S \in \mathcal{S}$.*

Definition 9. If $\boldsymbol{m}$ and $\boldsymbol{n}$ are two concise messages producing, from a state S, the same state $V \neq S$, we call $\boldsymbol{m}\widetilde{\boldsymbol{n}}$ an *orderly circuit* for S.

By Axiom [Mb], an orderly circuit is vacuous; therefore its length must be even. The following result is of general interest for orderly circuits.

Theorem 10. *Let S, N, Q and W be four distinct states of a medium and suppose that*

$$N\tau = S, \quad W\mu = Q, \quad S\boldsymbol{q} = N\boldsymbol{q}' = Q, \quad S\boldsymbol{w}' = N\boldsymbol{w} = W \tag{2}$$

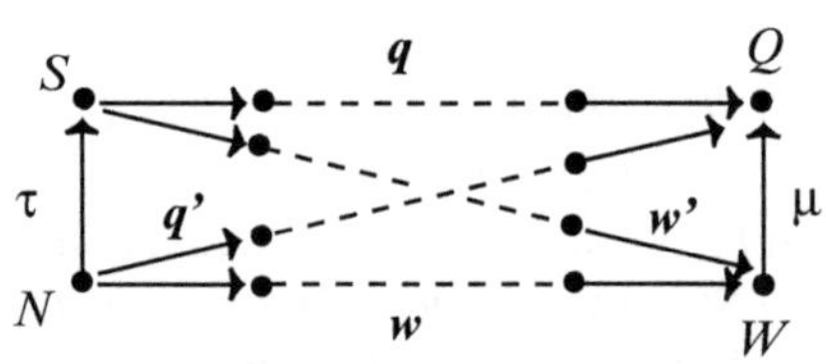

Fig. 1 For Theorem 10. Illustration of the conditions listed in Eq. (2).

for some tokens τ *and* μ *and some concise messages* $\boldsymbol{q}$, $\boldsymbol{q}'$, $\boldsymbol{w}$ *and* $\boldsymbol{w}'$ *(see Fig. 1). Then, the four following conditions are equivalent:*

(i) $\ell(\boldsymbol{q})+\ell(\boldsymbol{w}) \neq \ell(\boldsymbol{q}')+\ell(\boldsymbol{w}')$ and $\mu \neq \tilde{\tau}$.
(ii) $\tau = \mu$.
(iii) $\mathcal{C}(\boldsymbol{q}) = \mathcal{C}(\boldsymbol{w})$ and $\ell(\boldsymbol{q}) = \ell(\boldsymbol{w})$.
(iv) $\ell(\boldsymbol{q})+\ell(\boldsymbol{w})+2 = \ell(\boldsymbol{q}')+\ell(\boldsymbol{w}')$.

Moreover, any of these conditions implies that $\boldsymbol{q}\tilde{\mu}\widetilde{\boldsymbol{w}}\tau$ *is an orderly circuit for S with* $S\boldsymbol{q}\tilde{\mu} = S\tilde{\tau}\boldsymbol{w} = W$*. The converse does not hold.*

Proof. We prove (i) $\Rightarrow$ (ii) $\Leftrightarrow$ (iii) $\Rightarrow$ (iv) $\Rightarrow$ (i).

(i) $\Rightarrow$ (ii). Suppose that $\tau \neq \mu$. The token $\tilde{\tau}$ must occur exactly once in either $\boldsymbol{q}$ or in $\widetilde{\boldsymbol{w}}$. Indeed, we have $\mu \neq \tilde{\tau}$, both $\boldsymbol{q}$ and $\boldsymbol{w}$ are concise, and the message $\tau\boldsymbol{q}\tilde{\mu}\widetilde{\boldsymbol{w}}$ is a return for N, and so is vacuous by [Ma]. It can be verified that each of the two mutually exclusive, exhaustive cases: (a) $\tilde{\tau} \in \mathcal{C}(\boldsymbol{q}) \cap \mathcal{C}(\boldsymbol{w}')$; and (b) $\tilde{\tau} \in \mathcal{C}(\widetilde{\boldsymbol{w}}) \cap \mathcal{C}(\widetilde{\boldsymbol{q}'})$ lead to

$$\ell(\boldsymbol{q})+\ell(\boldsymbol{w}) = \ell(\boldsymbol{q}')+\ell(\boldsymbol{w}'), \tag{3}$$

contradicting (i). Thus, we must have $\tau = \mu$.

We only prove Case (a). The other case is treated similarly. Since $\tilde{\tau}$ is in $\mathcal{C}(\boldsymbol{q})$, neither τ nor $\tilde{\tau}$ can be in $\mathcal{C}(\boldsymbol{q}')$. Indeed, both $\boldsymbol{q}$ and $\boldsymbol{q}'$ are concise and $\boldsymbol{q}\widetilde{\boldsymbol{q}'}\tau$ is a return for S. It follows that both $\tilde{\tau}\boldsymbol{q}'$ and $\boldsymbol{q}$ are concise messages producing Q from S. By Theorem 7, we must have $\mathcal{C}(\tilde{\tau}\boldsymbol{q}') = \mathcal{C}(\boldsymbol{q})$, which implies $\ell(\tilde{\tau}\boldsymbol{q}') = \ell(\boldsymbol{q})$, and so

$$\ell(\boldsymbol{q}) = \ell(\boldsymbol{q}')+1. \tag{4}$$

A argument along the same lines shows that

$$\ell(\boldsymbol{w})+1 = \ell(\boldsymbol{w}'). \tag{5}$$

Adding (4) and (5) and simplifying, we obtain (3). The proof of Case (b) is similar.

(ii) $\Leftrightarrow$ (iii). If $\mu = \tau$, it readily follows (since both $\boldsymbol{q}$ and $\boldsymbol{w}$ are concise and $S\boldsymbol{q}\tilde{\tau}\widetilde{\boldsymbol{w}}\tau = S$) that any token in $\boldsymbol{q}$ must have a reverse in $\widetilde{\boldsymbol{w}}$ and vice versa. This implies $\mathcal{C}(\boldsymbol{q}) = \mathcal{C}(\boldsymbol{w})$, which in turn imply $\ell(\boldsymbol{q}) = \ell(\boldsymbol{w})$, and so (iii) holds. As $\boldsymbol{q}\tilde{\mu}\widetilde{\boldsymbol{w}}\tau$ is vacuous, it is clear that (iii) implies (ii).

(iii) $\Rightarrow$ (iv). Since (iii) implies (ii), we have $\tau \in \widehat{Q} \setminus \widehat{N}$ by Theorem 7. But both $\boldsymbol{q}$ and $\boldsymbol{q}'$ are concise, so $\tau \in \mathcal{C}(\boldsymbol{q}') \setminus \mathcal{C}(\boldsymbol{q})$. As $\tau\boldsymbol{q}\widetilde{\boldsymbol{q}'}$ is vacuous for N, we must have $\mathcal{C}(\boldsymbol{q}) + \{\tau\} = \mathcal{C}(\boldsymbol{q}')$, yielding

$$\ell(\boldsymbol{q})+1 = \ell(\boldsymbol{q}'). \tag{6}$$

A similar argument gives $\mathcal{C}(\boldsymbol{w}) + \{\tau\} = \mathcal{C}(\boldsymbol{w}')$ and

$$\ell(\boldsymbol{w})+1 = \ell(\boldsymbol{w}'). \tag{7}$$

Adding (6) and (7) yields (iv).

(iv) $\Rightarrow$ (i). As (iv) is a special case of the first statement in (i), we only have to prove that $\mu \neq \tilde{\tau}$. Suppose that $\mu = \tilde{\tau}$. We must assign the token $\tilde{\tau}$ consistently so to ensure the vacuousness of the messages $\boldsymbol{q}\widetilde{\boldsymbol{q}}'\tau$ and $\tau\boldsymbol{w}'\widetilde{\boldsymbol{w}}$. By Theorem 7, $\mathcal{C}(\boldsymbol{q}) = \widehat{Q}\setminus\widehat{S}$. Since $\tilde{\tau} \in \widehat{Q}$ and, by Theorem 8, $\tilde{\tau} \notin \widehat{S}$, the only possibility is $\tilde{\tau} \in \mathcal{C}(\boldsymbol{q}) \setminus \mathcal{C}(\boldsymbol{q}')$. For similar reasons $\tau \in \mathcal{C}(\boldsymbol{w}) \setminus \mathcal{C}(\boldsymbol{w}')$. We obtain the two concise messages $\tilde{\tau}\boldsymbol{q}'$ and $\boldsymbol{q}$ producing Q from S, and the two concise messages $\boldsymbol{w}$ and $\tau\boldsymbol{w}'$ producing W from N. This gives $\ell(\boldsymbol{q}) = \ell(\tilde{\tau}\boldsymbol{q}')$ and $\ell(\boldsymbol{w}) = \ell(\tau\boldsymbol{w}')$. We obtain so $\ell(\boldsymbol{q}) = \ell(\boldsymbol{q}') + 1$ and $\ell(\boldsymbol{w}) = \ell(\boldsymbol{w}') + 1$, which leads to $\ell(\boldsymbol{q}) + \ell(\boldsymbol{w}) = \ell(\boldsymbol{q}') + \ell(\boldsymbol{w}') + 2$ and contradicts (iv). Thus, (iv) implies (i). We conclude that the four conditions (i)–(iv) are equivalent.

We now show that, under the hypotheses of the theorem, (ii) implies that $\boldsymbol{q}\tilde{\mu}\widetilde{\boldsymbol{w}}\tau$ is an orderly return for S with $S\boldsymbol{q}\tilde{\mu} = S\tilde{\tau}\boldsymbol{w} = W$. Both $\boldsymbol{q}$ and $\boldsymbol{w}$ are concise by hypothesis. We cannot have μ in $\mathcal{C}(\boldsymbol{q})$ because then $\tilde{\mu}$ is in $\mathcal{C}(\widetilde{\boldsymbol{q}})$ and the two concise messages $\widetilde{\boldsymbol{q}}$ and $\tau = \mu$ producing S are not jointly consistent, yielding a contradiction to Condition (vii) of Lemma 4. Similarly, we cannot have $\tilde{\mu}$ in $\mathcal{C}(\boldsymbol{q})$ since the two concise messages $\boldsymbol{q}$ and μ producing Q would not be jointly consistent. Thus, $\boldsymbol{q}\tilde{\mu}$ is a concise message producing W from S. For like reasons, with $\tau = \mu$, $\tilde{\tau}\boldsymbol{w}$ is a concise message producing W from S. We conclude that, with $\tau = \mu$, the message $\boldsymbol{q}\tilde{\mu}\widetilde{\boldsymbol{w}}\tau$ is an orderly return for S. The example of Fig. 2, in which we have

$$\mu \neq \tau, \quad \boldsymbol{q} = \alpha\tilde{\tau}, \quad \boldsymbol{w} = \tilde{\mu}\alpha, \quad \boldsymbol{w}' = \alpha\tilde{\tau}\tilde{\mu}, \quad \text{and } \boldsymbol{q}' = \alpha,$$

displays the orderly return $\alpha\tilde{\tau}\tilde{\mu}\tilde{\alpha}\mu\tau$ for S. It serves as a counterexample to the implication: if $\boldsymbol{q}\tilde{\mu}\widetilde{\boldsymbol{w}}\tau$ is an orderly return for S, then $\tau = \mu$. □

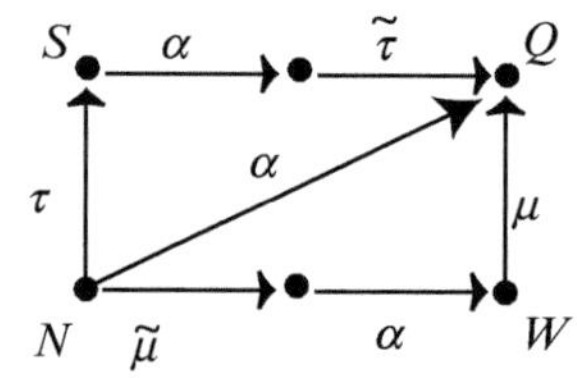

Fig. 2 Under the hypotheses of Theorem 10, the hypothesis that $\boldsymbol{q}\tilde{\mu}\widetilde{\boldsymbol{w}}\tau$ is an orderly circuit for S does not imply $\tau = \mu$, with $\boldsymbol{q} = \alpha\tilde{\tau}$, $\boldsymbol{w} = \tilde{\mu}\alpha$, $\boldsymbol{q}' = \alpha$, and $\boldsymbol{w}' = \alpha\tilde{\tau}\tilde{\mu}$

In Definition 9, the concept of an orderly circuit was specified with respect to a particular state. The next definition and theorem concern a situation in which a circuit is orderly with respect to everyone of its states. In such a case, any token occurring in the circuit must have its reverse at the exact "opposite" place in the circuit (see Theorem 12(i)).

Definition 11. Let $\tau_1 \ldots \tau_{2n}$ be an orderly return for a state S. For $1 \leq i \leq n$, the two tokens τ_i and τ_{i+n} are called *opposite*. A return $\tau_1 \ldots \tau_{2n}$ from S is *regular* if it is orderly and, for $1 \leq i \leq n$, the message $\tau_i\tau_{i+1} \ldots \tau_{i+n-1}$ is concise for $S\tau_1 \cdots \tau_{i-1}$.

Theorem 12. *Let* $\boldsymbol{m} = \tau_1 \dots \tau_{2n}$ *be an orderly return for some state S. Then the following three conditions are equivalent:*

(i) The opposite tokens of $\boldsymbol{m}$ *are mutual reverses.*
(ii) The return $\boldsymbol{m}$ *is regular.*
(iii) For $1 \le i \le 2n-1$, *the message* $\tau_i \dots \tau_{2n} \dots \tau_{i-1}$ *is an orderly return for the state* $S\tau_1 \cdots \tau_{i-1}$.

Proof. We prove (i) $\Rightarrow$ (ii) $\Rightarrow$ (iii) $\Rightarrow$ (i). In what follows $S_i = S\tau_0\tau_1 \dots \tau_i$ for $0 \le i \le 2n$, so $S_0 = S_{2n} = S$.

(i) $\Rightarrow$ (ii). Since $\boldsymbol{m}$ is an orderly return, for $1 \le j \le n$, there is only one occurrence of the pair $\{\tau_j, \tilde{\tau}_j\}$ in $\boldsymbol{m}$. Since $\tilde{\tau}_j = \tau_{j+n}$, there are no occurrences of $\{\tau_j, \tilde{\tau}_j\}$ in $\boldsymbol{p} = \tau_i \cdots \tau_{i+n-1}$, so it is a concise message for S_{i-1}.

(ii) $\Rightarrow$ (iii). Since $\boldsymbol{m}$ is a regular return, any message $\boldsymbol{p} = \tau_i \cdots \tau_{i+n-1}$ is concise, so any token of this message has a reverse in the message $\boldsymbol{q} = \tau_{i+n} \dots \tau_{2n} \dots \tau_{i-1}$. Since $\boldsymbol{p}$ is concise and $\ell(\boldsymbol{q}) = n$, the message $\boldsymbol{q}$ is concise. It follows that $\boldsymbol{pq}$ is an orderly return for the state S_{i-1}.

(iii) $\Rightarrow$ (i). Since the message $\tau_i \dots \tau_{2n} \dots \tau_{i-1}$ is an orderly return for S_{i-1}, the messages $\boldsymbol{q} = \tau_{i+1} \dots \tau_{i+n-1}$ and $\boldsymbol{q}' = \tau_i \dots \tau_{i+n-1}$ are concise for the states $S' = S_i$ and $N = S_{i-1}$, respectively, and produce the state $Q = S_{i+n-1}$. Likewise, the messages $\boldsymbol{w} = \tilde{\tau}_{i-1} \dots \tilde{\tau}_{2n} \dots \tilde{\tau}_{i+n}$ and $\boldsymbol{w}' = \tilde{\tau}_i \dots \tilde{\tau}_{2n} \dots \tilde{\tau}_{i+n}$ are concise for the states $N = S_{i-1}$ and $S' = S_i$, respectively, and produce the state $W = S_{i+n}$. It is clear that $\ell(\boldsymbol{q}) + \ell(\boldsymbol{w}) + 2 = \ell(\boldsymbol{q}') + \ell(\boldsymbol{w}')$. By Theorem 10, $\tau_{i+n} = \tilde{\tau}_i$. □

4 The Graph of a Medium

For graph-theoretical concepts and terminology, we usually follow Bondy (1995).

Definition 13. A *graph representation* of a medium $(\mathcal{S}, \mathcal{T})$ is a bijection $\gamma : \mathcal{S} \to V$, where V is a set of vertices of a graph (V, E), such that two distinct states S and T are adjacent whenever $\{\gamma(S), \gamma(T)\}$ is an edge of the graph; formally,

$$\{\gamma(S), \gamma(T)\} \in E \quad \Longleftrightarrow \quad (\exists \tau \in \mathcal{T})(S\tau = T) \quad (S, T \in \mathcal{S}, S \neq T). \tag{8}$$

We say then that the graph (V, E), which has no loops, represents the medium. A graph (V, E) representing a medium $(\mathcal{S}, \mathcal{T})$ is called the *graph of the medium* $(\mathcal{S}, \mathcal{T})$ if $V = \mathcal{S}$, the edges in E are defined as in (8), and γ is the identity mapping. Clearly, any medium has its graph. We shall prove in this paper that the converse also holds, namely: the graph of a medium defines its medium (see Theorem 35). We recall that two graphs (V, E) and (V', E') are isomorphic if there is a bijection $\varphi : V \to V'$ such that

$$\{P, Q\} \in E \Longleftrightarrow \{\varphi(P), \varphi(Q)\} \in E' \qquad (P, Q \in V, P \neq Q). \tag{9}$$

Lemma 14. *A graph isomorphic to a graph representing a medium* $\mathcal{M}$ *also represents* $\mathcal{M}$.

It is intuitively clear that shortest paths in the graph of a medium correspond to concise messages of that medium. Our next lemma states that fact precisely.

Lemma 15. *Let $\gamma: \mathcal{S} \to V$ be the representation of a medium $(\mathcal{S}, \mathcal{T})$ by a graph $G = (V, E)$. If $\boldsymbol{m} = \tau_1 \ldots \tau_m$ is a concise message producing a state T from a state S, then the sequence of vertices $(\gamma(S_i))_{0 \le i \le m}$, where $S_i = S\tau_0\tau_1 \cdots \tau_i$, for $0 \le i \le m$, forms a shortest path joining $\gamma(S)$ and $\gamma(T)$ in G. Conversely, if a sequence $(\gamma(S_i))_{0 \le i \le m}$ is a shortest path connecting $\gamma(S_0) = \gamma(S)$ and $\gamma(S_m) = \gamma(T)$, then $\boldsymbol{m} = \tau_1 \ldots \tau_m$ with $S\tau_0\tau_1 \cdots \tau_i = S_i$, for $0 \le i \le m$, is a concise message producing T from S.*

Proof. (Necessity.) Let $\gamma(P_0) = \gamma(S), \gamma(P_1), \ldots, \gamma(P_n) = \gamma(T)$ be a path in G joining $\gamma(S)$ to $\gamma(T)$. Correspondingly, there is a stepwise effective message $\boldsymbol{n} = \rho_1 \cdots \rho_n$ such that $P_i = T\rho_1 \cdots \rho_{n-i}$ for $0 \le i < n$. The message $\boldsymbol{mn}$ is a return for S. By Axiom [Mb], this message is vacuous. Since $\boldsymbol{m}$ is a concise message for S, we must have $\ell(\boldsymbol{m}) = m \le \ell(\boldsymbol{n}) = n$.

(Sufficiency.) Let $\gamma(S_0) = \gamma(S), \gamma(S_1), \ldots, \gamma(S_m) = \gamma(T)$ be a shortest path from $\gamma(S)$ to $\gamma(T)$ in G. Then, there are some tokens τ_i, $1 \le i \le m$ such that $S_i\tau_{i+1} = S_{i+1}$ for $0 \le i < m$. The message $\boldsymbol{m} = \tau_1 \ldots \tau_m$ produces the state T from the state S. An argument akin to that used in the foregoing paragraph shows that $\boldsymbol{m}$ is a concise message for S. □

We now establish a result of the same vein for the regular returns of a medium (cf. Definition 11).

Definition 16. We recall that a sequence of vertices $\boldsymbol{s}_m = (v_i)_{0 \le i \le m}$ such that $\{v_i, v_{i+1}\}$ are edges in a graph is a circuit if $v_m = v_0$ and all the vertices $v_1, \ldots, v_m$ are different. By abuse of language, we say that the edges $\{v_i, v_{i+1}\}$, for $0 \le i \le m-1$, *belong* to the circuit $\boldsymbol{s}_m$. The circuit $\boldsymbol{s}_m$ is *even* if it has an even number of edges: $m = 2n$; any two of its edges $\{v_i, v_{i+1}\}$ and $\{v_{i+n}, v_{i+n+1}\}$, $0 \le i \le n-1$ are then called *opposite*. A circuit is *minimal* if at least one shortest path between any two of its vertices is a segment of the circuit. A graph is *even* if all its circuits are even.

Lemma 17. *Let $\gamma: \mathcal{S} \to V$ be the representation of a medium $\mathcal{M} = (\mathcal{S}, \mathcal{T})$ by a graph $G = (V, E)$. If $\boldsymbol{m} = \tau_1 \ldots \tau_{2n}$ is a regular return for some state $S \in \mathcal{S}$, then the sequence of vertices $(\gamma(S_i))_{0 \le i \le 2n}$, where $S_i = S\tau_0\tau_1 \cdots \tau_i$, for $0 \le i \le 2n$, forms an even, minimal circuit of G (with $S = S_0 = S_{2n}$). Conversely, if a sequence $(\gamma(S_i))_{0 \le i \le 2n}$ is an even minimal circuit of G, then $\boldsymbol{m} = \tau_1 \ldots \tau_m$ with $S\tau_0\tau_1 \cdots \tau_i = S_i$, for $0 \le i \le 2n$ is a regular return for S in $\mathcal{M}$.*

Proof. In the notation of the lemma, let $\boldsymbol{m}$ be a regular return for state S. Thus, by definition of a regular return (cf. 11), $\tau_1 \ldots \tau_n$ and $\tilde{\tau}_{2n} \ldots \tilde{\tau}_{n+1}$ are concise messages for S. By Lemma 15, the sequence of vertices $(\gamma(S_i))_{0 \le i \le n}$, where $S_i = S\tau_0\tau_1 \cdots \tau_i$, for $0 \le i \le n$, forms a shortest path joining $\gamma(S)$ and $\gamma(T)$, with $T = S\tau_1 \cdots \tau_n$. Similarly, the sequence $\gamma(S_{2n}), \gamma(S_{2n-1}), \ldots, \gamma(S_{n+1})$ is another shortest path joining $\gamma(S)$ and $\gamma(T)$. Since γ is a 1-1 function, all the vertices $\gamma(S_i)$ are distinct, and so the sequence $(\gamma(S_i))_{0 \le i \le 2n}$ is an even circuit. This circuit is a minimal one. Indeed, by definition of a regular return, all the messages $\tau_i\tau_{i+1} \ldots \tau_{i+n-1}$ are concise for

$S\tau_1 \cdots \tau_{i-1}$. So, by Lemma 15, all the sequences $\gamma(S_i), \ldots, \gamma(S_{i+n-1})$ are shortest paths between $\gamma(S_i)$ and $\gamma(S_{i+n-1})$, which implies that at least one shortest path between any two vertices of the circuit $(\gamma(S_i))_{0 \leq i \leq 2n}$ is a segment of that circuit. We omit the proof of the converse part of this lemma. The argument is based on the converse part of Lemma 15 and is similar. □

Remark 18. A close reading of this proof shows that opposite tokens τ_i, $\tau_{i+n} = \tilde{\tau}_i$ in a regular return correspond to opposite edges $\{\gamma(S_i), \gamma(S_{i+1})\}$, $\{\gamma(S_{i+n}, \gamma(S_{i+1+n})\}$ in the even minimal circuit of the representing graph, with $S_{i+1} = S_i\tau_i$ and $S_{i+n} = S_{i+n+1}\tau_{i+n}$.

5 Media Inducing Graphs

Our next task is to characterize the graphs representing media in terms of graph concepts. Some necessary conditions are easily inferred from the axioms of a medium. For example, Axiom [Ma] forces the graph to be connected, and [Mb] demands that it is even. By convention, the graph should not have any loops. However, as shown by the two examples below, these two conditions are not sufficient to characterize the graph of a medium. To simplify the figures, only one token from each pair of mutually reverse tokens is indicated, so the graphs of media are shown as digraphs.

Two Counterexamples 19. The graphs corresponding to the digraphs A and B in Fig. 3 are connected and all their circuits are even. Moreover, they have no loops. Yet, neither A nor B can yield the graph of a medium. We leave to the reader to prove this for Fig. 3A.

Here is why in the case of B. The circuit pictured in thick lines is even and minimal. By Lemma 17, it must represent a regular return in a medium. From Remark 18, we know that the same token must be matched to opposite edges of the circuit. Accordingly, the same token ν has been assigned to the arcs JM and RW. The circuit containing the six vertices L, K, N, W, R and H is also even and minimal. Thus, the arcs LK and RW must be assigned the same token, and since RW has been assigned token ν, that token must also be assigned to LK. The argument governing the placement of the token τ are similar. The consequence, however, is that there is no concise message from L to J: any message producing J from L contains either both ν and $\tilde{\nu}$,

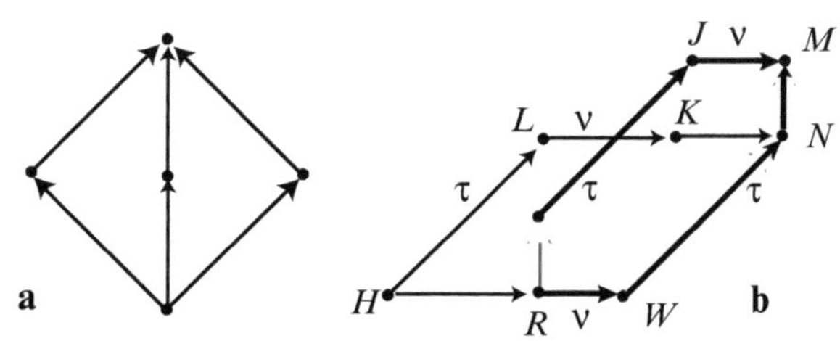

Fig. 3 Neither of these digraphs is that of a medium. The token system corresponding to Digraph (**B**) contradicts [Ma]. Which of the properties of a medium is contradicted by Digraph (**A**)?

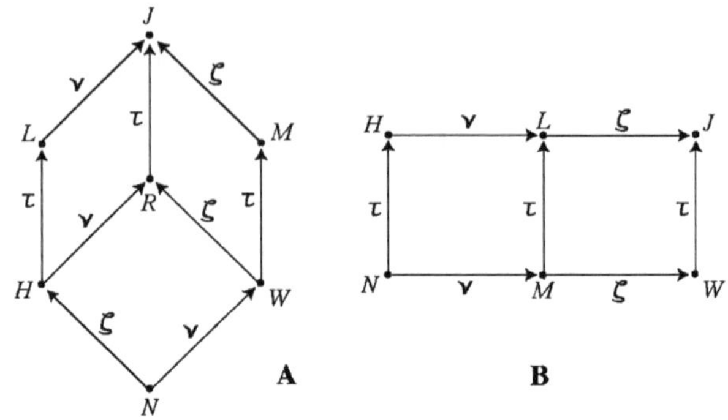

Fig. 4 Two examples of digraphs of media. In Example B, notice that different tokens are assigned to the opposite arcs HL and MW, which are opposite in the circuit N,H,L,J,W,M,N. This circuit is not minimal. Compare with the situation of the arcs LJ and NW in Example A.

or both $\tilde{\tau}$ and τ. This example will be crucial in our understanding of the appropriate axiomatization of a graph capable of representing a medium.

In our failed attempt at representing a medium in Fig. 3, we have chosen to picture the arcs representing the same token by parallel segments (forming two sides of an implicit rectangle). The intuition that the opposite arcs of even minimal circuits should be parallel is a sound one, and suggests the construction of an equivalence relation on the set of arcs of the digraph. Such a construction is delicate, however, and the two examples of media pictured below by their digraphs must be taken into account.

Examples 20 . Together with the examples of Fig. 3, Examples A and B in Fig. 4 will also guide and illustrate our choice of concepts and axioms.

Definition 21. We write $\vec{E} = \{ST \,|\, \{S,T\} \in E\}$ for the set of all the arcs of a graph $G = (V,E)$. The *like* relation of the graph G is a relation $\mathfrak{L}$ on $\vec{E}$ defined by

$$ST\,\mathfrak{L}\,PQ \Longleftrightarrow (\delta(S,P)+1=\delta(T,Q)+1=\delta(S,Q)=\delta(T,P)) \quad (\{S,T\},\{P,Q\} \in E),$$

where δ denotes the graph theoretical distance between the vertices of the graph. In Example B of Fig. 4, we have $NH\,\mathfrak{L}\,WJ$ because

$$\delta(H,J)+1 = \delta(N,W)+1 = \delta(H,W) = \delta(N,J),$$

but $HL\,\mathfrak{L}\,MW$ does not hold since

$$\delta(H,M) = \delta(L,W) = 2 \quad \text{and} \quad 3 = \delta(H,W) \neq \delta(L,M) = 1.$$

The point is that the arcs HL and MW are opposite in the circuit H,L,J,W,M,N,H, but this circuit is not minimal.

The like relation is clearly reflexive and symmetric; and moreover

$$ST\,\mathfrak{L}\,PQ \quad \Longleftrightarrow \quad TS\,\mathfrak{L}\,QP \qquad (\{S,T\},\{P,Q\} \in E). \tag{10}$$

Two binary relations on the set of edges of a graph play a central role in characterizing partial cubes. They are Djoković's relation θ (Djoković, 1973) and Winkler's relation Θ (Winkler, 1984) which are identical on bipartite graphs. These relations are germane to, but different from the like relation of this paper. Indeed, the like relation is defined on the set of arcs of a graph, whereas Djoković and Winkler's relations are defined on the set of edges. Also, the distance equations defining the like relation represent just a special instance of the distance inequality in Winkler's definition (Ovchinnikov, 2007).

We now come to the main concept of this paper. We recall that a graph is bipartite if and only if it is even (König, 1916).

Definition 22. Let $G = (V,E)$ be a graph equipped with its like relation $\mathfrak{L}$. The graph G is called *mediatic* if the following three axioms hold:

[G1] G is connected.
[G2] G is bipartite.
[G3] $\mathfrak{L}$ is transitive.

The set of vertices is not assumed to be finite. It is easily verified that any graph isomorphic to a mediatic graph is mediatic.

Axiom [G3] eliminates the counterexample of Fig. 3B. Indeed, since

$$\delta(L,J) = 4, \quad \delta(K,M) = 2, \quad \delta(L,M) = 3 = \delta(K,J)$$

we have

$$LK\,\mathfrak{L}\,RW\,\mathfrak{L}\,JM \quad \text{but not} \quad LK\,\mathfrak{L}\,JM.$$

The following result is immediate.

Lemma 23. *The like relation $\mathfrak{L}$ of a mediatic graph (V,E) is an equivalence relation on $\vec{E}$.*

Definition 24. We denote by

$$\langle ST\rangle = \{PQ \in \vec{E} \,|\, ST\,\mathfrak{L}\,PQ\}$$

the equivalence class containing the arc ST in the partition of $\vec{E}$ induced by $\mathfrak{L}$.

We will show that a graph representing a medium is mediatic (see Theorem 27). Our next lemma is the first step.

Lemma 25. *Let γ be the representation of a medium $\mathcal{M} = (\mathcal{S},\mathcal{T})$ by a graph $G = (\mathcal{S},E)$ which is equipped with its like relation $\mathfrak{L}$. Suppose that $\gamma(N)\gamma(S)\,\mathfrak{L}\,\gamma(W)\gamma(Q)$. Then $N\tau = S$ and $W\tau = Q$ for some $\tau \in \mathcal{T}$. In fact, there exists an orderly circuit $\boldsymbol{q}\tilde{\tau}\widetilde{\boldsymbol{w}}\tau$ for S in $\mathcal{M}$, with $S\boldsymbol{q}\tilde{\tau} = S\tilde{\tau}\boldsymbol{w} = W$; thus $\boldsymbol{q}$ and $\boldsymbol{w}$ are concise with $\ell(\boldsymbol{q}) = \ell(\boldsymbol{w})$. Such a circuit is not necessarily regular.*

Proof. We abbreviate our notation for this proof, and write $S^\gamma = \gamma(S)$ for all $S \in \mathcal{S}$. By definition, $N^\gamma S^\gamma \mathfrak{L} W^\gamma Q^\gamma$ implies that $\delta(S^\gamma, Q^\gamma) = \delta(N^\gamma, W^\gamma) = \delta(N^\gamma, Q^\gamma) - 1 = \delta(S^\gamma, W^\gamma) - 1$; so, there are, for some $n \in \mathbb{N}$, two shortest paths

$$S_0^\gamma = S^\gamma, S_1^\gamma, \dots, S_n^\gamma = Q^\gamma \quad \text{and} \quad N_0^\gamma = N^\gamma, N_1^\gamma, \dots, N_n^\gamma = W^\gamma$$

between S^γ and Q^γ, and N^γ and W^γ, respectively. Moreover,

$$S_0^\gamma = S^\gamma, S_1^\gamma, \dots, S_n^\gamma = Q^\gamma, W^\gamma \quad \text{and} \quad N_0^\gamma = N^\gamma, N_1^\gamma, \dots, N_n^\gamma = W^\gamma, Q^\gamma$$

are also shortest paths. Using Lemma 15, we can assert the existence of two concise messages $\boldsymbol{q}$ and $\boldsymbol{w}$ such that $S\boldsymbol{q} = Q$ and $N\boldsymbol{w} = W$, with $\ell(\boldsymbol{q}) = \ell(\boldsymbol{w}) = n$. Also, for some tokens τ and μ, we have $N\tau = S$ and $W\mu = Q$ with $\boldsymbol{q}' = \tau\boldsymbol{q}$ and $\boldsymbol{w}' = \tilde{\tau}\boldsymbol{w}$ concise for N and S, respectively, and $\ell(\boldsymbol{q}') = \ell(\boldsymbol{w}') = n + 1$. We are exactly in the situation of Theorem 10 (see Fig. 1). Using the implication (iv) $\Rightarrow$ (ii) of this theorem, we obtain $\tau = \mu$. Condition (iv) also implies that $\boldsymbol{q}\tilde{\tau}\widetilde{\boldsymbol{w}}\tau$ is an orderly circuit for S, with $S\boldsymbol{q}\tilde{\tau} = S\tilde{\tau}\boldsymbol{w} = W$. The Example B of Fig. 4 shows that, with $\boldsymbol{q} = \boldsymbol{w} = \nu\zeta$, such a circuit need not be regular. □

Convention 26. Any graph representing a medium comes implicitly equipped with its like relation $\mathfrak{L}$. When several such graphs are considered (say, for different media), their respective like relations are distinguished by diacritics, such as $\mathfrak{L}'$ or $\mathfrak{L}^*$.

Theorem 27. *Any graph representing a medium is mediatic.*

Proof. Because any graph isomorphic to a mediatic graph is mediatic, we can invoke Lemma 14 and content ourselves with proving that the graph of a medium is mediatic (which simplifies our notation). Denote the medium by $\mathcal{M} = (\mathcal{S}, \mathcal{T})$, and let $G = (\mathcal{S}, E)$ be its graph. We prove that G satisfies [G1], [G2] and [G3]:

[G1] Axiom [Ma] requires that G be connected.
[G2] Axiom [Mb] implies that G must be even. Hence, by König's Theorem, it must be bipartite.
[G3] Suppose that $NS\mathfrak{L}PR\mathfrak{L}WQ$. By Lemma 25 (applied twice), there must be some tokens τ and μ such that $N\tau = S$, $P\tau = R$, $P\mu = R$ and $W\mu = Q$, so $\tau = \mu$. Let then $\boldsymbol{q}$ and $\boldsymbol{w}'$ be two concise messages from S, and let $\boldsymbol{w}$ and bq' be two concise messages from N, such that

$$S\boldsymbol{q} = Q, \quad S\boldsymbol{w}' = W, \quad N\boldsymbol{w} = W, \quad N\boldsymbol{q}' = Q.$$

The situation is exactly as in Theorem 10, with the same notation. Because $\tau = \mu$, Condition (ii) of this theorem holds. We conclude that Conditions (iii) and (iv) also hold, which leads to

$$\delta(S, Q) + 1 = \delta(N, W) + 1 = \delta(S, W) = \delta(N, Q).$$

We have thus $NS\mathfrak{L}WQ$; so Axiom [G3] holds. □

We omit the proof of the next lemma, which is straightforward.

Lemma 28. *Let $G = (V,E)$ and $G' = (V',E')$ be two mediatic graphs, with their respective like relations $\mathfrak{L}$ and $\mathfrak{L}'$, and let φ be a bijection of V onto V'. Then φ is an isomorphism of G onto G' if and only if*

$$ST\,\mathfrak{L}PQ \iff \varphi(S)\varphi(T)\,\mathfrak{L}'\varphi(P)\varphi(Q) \qquad (S,T,P,Q \in V).$$

Remark 29. The like relation is the fundamental tool for the study of mediatic graphs. We shall see that any mediatic graph G can be used to construct a medium $\mathcal{M}$ that has G as its graph. Each of the equivalence classes $\langle ST\rangle$ of the like relation contains 'parallel' arcs of the graph, and will turn out to correspond to a particular token, say τ, of the medium under construction, with the class $\langle TS\rangle$ corresponding to the reverse token $\tilde{\tau}$. Before proceeding to such a construction, we establish in Theorem 31 a useful result which precisely links the isomorphism of media to that of their graphs.

Definition 30. Two media $(\mathcal{S},\mathcal{T})$ and $(\mathcal{S}',\mathcal{T}')$ are isomorphic if there exists a pair (α,β) of bijections $\alpha : \mathcal{S} \to \mathcal{S}'$ and $\beta : \mathcal{T} \to \mathcal{T}'$ such that

$$S\tau = V \iff \alpha(S)\beta(\tau) = \alpha(V) \qquad (S,V \in \mathcal{S},\ \tau \in \mathcal{T}). \tag{11}$$

6 Paired Isomorphisms of Media and Graphs

Isomorphic media yield isomorphic mediatic graphs, and vice versa.

Theorem 31. *Suppose that $\mathcal{M} = (\mathcal{S},\mathcal{T})$ and $\mathcal{M}' = (\mathcal{S}',\mathcal{T}')$ are two media and let $G = (\mathcal{S},E)$ and $G' = (\mathcal{S}',E')$ be their respective graphs. Then $\mathcal{M}$ and $\mathcal{M}'$ are isomorphic if and only if G and G' are isomorphic; more precisely:*

(i) *If (α,β) is an isomorphism of $\mathcal{M}$ onto $\mathcal{M}'$, then $\alpha : \mathcal{S} \to \mathcal{S}'$ is an isomorphism of G onto G' in the sense of (9).*
(ii) *If $\varphi : \mathcal{S} \to \mathcal{S}'$ is an isomorphism of G onto G' in the sense of (9), then there exists a bijection $\beta : \mathcal{T} \to \mathcal{T}'$ such that (φ,β) is an isomorphism of $\mathcal{M}$ onto $\mathcal{M}'$.*

Proof. (i) Suppose that (α,β) is an isomorphism of $\mathcal{M}$ onto $\mathcal{M}'$. For any two distinct S, T in $\mathcal{S}$, we have successively

$$\begin{aligned}
&\{S,T\} \in E \\
&\quad\iff (\exists \tau \in \mathcal{T})(S\tau = T) && (G \text{ is the graph of } \mathcal{M}) \\
&\quad\iff (\exists \tau \in \mathcal{T})(\alpha(S)\beta(\tau) = \alpha(T)) && (\mathcal{M} \text{ and } \mathcal{M}' \text{ are isomorphic}) \\
&\quad\iff \{\alpha(S),\alpha(T)\} \in E' && (G' \text{ is the graph of } \mathcal{M}'),
\end{aligned}$$

and so

$$\{S,T\} \in E \iff \{\alpha(S),\alpha(T)\} \in E' \qquad (S,T \in \mathcal{S},\ S \neq T).$$

We conclude that $\alpha : \mathcal{S} \to \mathcal{S}'$ is an isomorphism of G onto G'.

(ii) Let $\varphi : \mathcal{S} \to \mathcal{S}'$ be an isomorphism of G onto G'. Define a function $\beta : \mathcal{T} \to \mathcal{T}'$ by

$$\beta(\tau) = \tau' \iff (\forall S, T \in \mathcal{S})(S\tau = T \Leftrightarrow \varphi(S)\tau' = \varphi(T)). \tag{12}$$

We first verify that the r.h.s. of the equivalence (12) correctly defines β as a bijection of $\mathcal{T}$ onto $\mathcal{T}'$. For any $\tau \in \mathcal{T}$, there exists distinct states S and T in $\mathcal{S}$ such that $S\tau = T$ and $\{S,T\} \in E$. Fix S and T temporarily. By the isomorphism $\varphi : \mathcal{S} \to \mathcal{S}'$ of G onto G', we have $\{\varphi(S), \varphi(T)\} \in E'$, and because G' is the graph of $\mathcal{M}'$, we necessarily have $\varphi(S)\tau' = \varphi(T)$ for some $\tau' \in \mathcal{T}'$, which is unique by Lemma 4(i). The hypothesis that φ is an isomorphism of G onto G' ensures that the r.h.s. of (12) is indeed an equivalence.

Next, we show that $\beta(\tau)$ does not depend upon the choice of S and T. Let P, Q be another pair of distinct states in $\mathcal{S}$ such that $P\tau = Q$, and let $P = S\boldsymbol{m}$ and $Q = T\boldsymbol{n}$ for some concise messages $\boldsymbol{m} = \tau_1 \dots \tau_m$ and $\boldsymbol{n} = \mu_1 \cdots \mu_n$. By Condition (vii) in Lemma 4, $\tau\boldsymbol{n}$ and $\boldsymbol{m}\tau$ are concise messages, and so Theorem 10 applies. Invoking its implication (ii) $\Rightarrow$ (iii), we get $\ell(\boldsymbol{m}) = \ell(\boldsymbol{n})$ and $\mathcal{C}(\boldsymbol{m}) = \mathcal{C}(\boldsymbol{n})$, yielding $m = n$. Denote by $\mathfrak{L}$ and $\mathfrak{L}'$ the like relations of G and G' respectively. We have thus shown that $ST\,\mathfrak{L}\,PQ$. By Lemma 28, we also have

$$\varphi(S)\varphi(T)\,\mathfrak{L}'\,\varphi(P)\varphi(Q).$$

Since we have $\varphi(S)\tau' = \varphi(T)$, we can apply Lemma 25 and derive $\varphi(P)\tau' = \varphi(Q)$.

We still have to prove that β is indeed a bijection. For any $\tau' \in \mathcal{T}'$ there are some $S', T' \in \mathcal{T}'$ such that $S'\tau' = T'$. We have thus $\{S', T'\} \in E'$, and since φ is an isomorphism of G onto G', also $\{\varphi^{-1}(S'), \varphi^{-1}(T')\} \in E$, with $\varphi^{-1}(S')\tau = \varphi^{-1}(T')$ for some $\tau \in \mathcal{T}$. Thus β maps $\mathcal{T}$ onto $\mathcal{T}'$. Suppose now that $\beta(\tau) = \beta(\mu) = \tau' \in \mathcal{T}'$. This implies that for some $S, T, P, Q \in \mathcal{S}$ and $N, M \in \mathcal{S}'$, we must have

$$S\tau = T, \quad P\mu = Q, \quad \text{and} \quad N\tau' = M, \tag{13}$$

together with $\varphi(S) = \varphi(P) = N$ and $\varphi(T) = \varphi(Q) = M$ by the definition of β. As φ is a 1-1 function, we obtain $S = P$ and $T = Q$ in (13). Using Lemma 4(ii), we get $\tau = \mu$. Thus, β is a 1-1 function and so a bijection.

The fact that (φ, β) is an isomorphism of $\mathcal{M}$ onto $\mathcal{M}'$ follows from the definition of β by (12). We have

$$S\tau = T \iff \varphi(S)\beta(\tau) = \varphi(T) \qquad (S, T \in \mathcal{S})$$

whether or not $\{S,T\} \in E$. □

Having defined the graph of a medium and shown that such a graph was necessarily mediatic, we now go in the opposite direction and construct a medium from an arbitrary mediatic graph.

7 From Mediatic Graphs to Media

Definition 32. Let $G = (\mathcal{S}, E)$ be a mediatic graph and let $\mathfrak{L}$ be its like relation. For any $ST \in \vec{E}$, define a transformation $\tau_{ST} : \mathcal{S} \to \mathcal{S} : P \mapsto P\tau_{ST}$ by the formula

$$P\tau_{ST} = \begin{cases} Q & \text{if } ST \,\mathfrak{L}\, PQ, \\ P & \text{otherwise.} \end{cases} \tag{14}$$

We denote by $\mathcal{T} = \{\tau_{ST} \,|\, ST \in \vec{E}\}$ the set containing all those transformations. It is clear that the pair $(\mathcal{S}, \mathcal{T})$ is a token system. Such a token system is said to be *induced* by the mediatic graph G. The theorem below establishes that a token system $\mathcal{K}$ induced by a mediatic graph G is in fact a medium. We say that $\mathcal{K}$ is the *medium of the graph* G. Notice that, since $\mathfrak{L}$ is an equivalence relation on $\vec{E}$, we have $\tau_{ST} = \tau_{PQ}$ whenever $ST \,\mathfrak{L}\, PQ$. In such a case, we have in fact $\langle ST \rangle = \langle PQ \rangle$. The choice of a particular pair $ST \in \langle PQ \rangle$ to denote a token τ_{ST} is thus arbitrary. Notice that, as a consequence of this definition, whenever $\{S, T\} \in E$, then also $ST \,\mathfrak{L}\, ST$, and so $S\tau_{ST} = T$.

This construction is motivated by the following theorem.

Theorem 33. *The token system $(\mathcal{S}, \mathcal{T})$ induced by a mediatic graph $G = (\mathcal{S}, E)$ is a medium. In particular, the tokens τ_{ST} and τ_{TS} defined by* (14) *are mutual reverses for any $\{S, T\} \in E$.*

Proof. We verify that $(\mathcal{S}, \mathcal{T})$ satisfies Axioms [Ma] and [Mb] of a medium.

[Ma] For any $S, T \in \mathcal{S}$, there is a shortest path $S_0 = S, S_1, \ldots, S_n = T$ between S and T in G. This implies that, for $0 \leq i \leq n-1$, we have $\{S_i, S_{i+1}\} \in E$, which yields $S_i \tau_{S_i S_{i+1}} = S_{i+1}$. It follows that the message $\boldsymbol{m} = \tau_{S_0 S_1} \ldots \tau_{S_{n-1} S_n}$ produces T from S and is stepwise effective. To prove that $\boldsymbol{m}$ is concise, we must still show that it is consistent and without repetitions. The message $\boldsymbol{m}$ is consistent since otherwise we would have

$$S_h \tau_{MN} = S_{h+1} \quad \text{and} \quad S_k \tau_{NM} = S_{k+1} \tag{15}$$

for some indices h and k, with $h < k$, and some $NM \in \vec{E}$. Since τ_{MN} is the reverse of τ_{NM}, the last equality in (15) can be rewritten as $S_{k+1} \tau_{MN} = S_k$. Thus, by definition of the tokens in (14), the above statement (15) leads to $S_h S_{h+1} \,\mathfrak{L}\, MN \,\mathfrak{L}\, S_{k+1} S_k$ which, by transitivity, gives $S_k S_{k+1} \,\mathfrak{L}\, S_{h+1} S_h$. Because $h < k$, we derive by the definition of the like relation $\mathfrak{L}$

$$k + 1 - h = \delta(S_{k+1}, S_h) = \delta(S_k, S_{h+1}) = k - 1 - h$$

yielding the absurdity $1 = -1$. Thus, $\boldsymbol{m}$ is consistent. Suppose that $\boldsymbol{m}$ has repeated tokens, say $S_i \tau_{S_i S_{i+1}} = S_{i+1}$ and $S_{i+k} \tau_{S_i S_{i+1}} = S_{i+k+1}$ for some indices $0 \leq i < n$ and $0 \leq i + k < n$. This would give $S_i S_{i+1} \,\mathfrak{L}\, S_{i+k} S_{i+k+1}$, leading to

$$d(S_i, S_{i+k+1}) = k + 1 > k - 1 = d(S_{i+1}, S_{i+k}),$$

while by the definition of $\mathfrak{L}$ we should have $d(S_i, S_{i+k+1}) = d(S_{i+1}, S_{i+k})$, a contradiction. Thus, the message $\boldsymbol{m}$ is concise.

[Mb] Let $\boldsymbol{m} = \tau_{S_0 S_1} \tau_{S_1 S_2} \dots \tau_{S_{n-1} S_n}$ be a return message for some state S; we have thus $S_0 = S_n = S$. In the terminology of G, we have a closed walk $S = S_0, S_1, \dots, S_n = S$. We denote this closed walk by $\mathbf{W}$ and we write $\vec{E}_{\mathbf{W}}$ for the set of all its arcs $S_i S_{i+1}$, $0 \le i \le n-1$. By [G2] and König's Theorem, such a closed walk is even; so $n = 2q$ for some $q \in \mathbb{N}$. We prove by induction on q that $\boldsymbol{m}$ is vacuous. The case $q = 1$ (the smallest possible return) is trivial, so we suppose that [Mb] holds for any $1 \le p < q$ and prove that [Mb] also holds for $q = p$. We consider two cases.

Case 1: $\mathbf{W}$ is an isometric subgraph of G. Thus, $\mathbf{W}$ is a minimal circuit of G. Take any token $\tau_{S_i S_{i+1}}$ in $\boldsymbol{m}$. Since (with the addition modulo k in the indices), we have for $0 \le i < n$

$$\delta(S_{i+1}, S_{i+k}) = \delta(S_i, S_{i+k+1}) = k - 1,$$
$$\delta(S_i, S_{i+k}) = \delta(S_{i+1}, S_{i+k+1}) = k,$$

we obtain $S_i S_{i+1} \,\mathfrak{L}\, S_{i+k+1} S_{i+k}$. By the definition of the tokens in (14) and the transitivity and symmetry of $\mathfrak{L}$, we get for any $P, Q \in \mathcal{S}$

$$\begin{aligned} P\tau_{S_i S_{i+1}} = Q &\Longleftrightarrow S_i S_{i+1} \,\mathfrak{L}\, PQ \\ &\Longleftrightarrow S_{i+k+1} S_{i+k} \,\mathfrak{L}\, PQ \\ &\Longleftrightarrow P\tau_{S_{i+k+1} S_{i+k}} = Q \\ &\Longleftrightarrow Q\tau_{S_{i+k} S_{i+k+1}} = P. \end{aligned}$$

We conclude that $\tau_{S_{i+k} S_{i+k+1}}$ and $\tau_{S_i S_{i+1}}$ are mutual reverses, and so $\boldsymbol{m}$ is vacuous. (Note that the induction hypothesis has not been used here.)

Case 2: $\mathbf{W}$ is not an isometric subgraph of G. Then, there are two vertices S_i and S_j in $\mathbf{W}$, with $i < j$, and a shortest path $\mathbf{L}$ from S_i to S_j in G with

$$\delta_{ij} = \delta(S_i, S_j) < \min\{j - i, i + n - j\}$$

(see Fig. 5). Thus, $j - i$ and $i + n - j$ are the lengths of the two segments of $\mathbf{W}$ with endpoints S_i and S_j. For simplicity, we can assume without loss of generality that S_i and S_j are the only vertices of $\mathbf{L}$ that are also in $\mathbf{W}$. Let $\boldsymbol{p}$ the concise message producing S_j from S_i and corresponding to the shortest path $\mathbf{L}$ in the sense of Lemma 15.

We also split $\boldsymbol{m}$ into the three messages:

$$\begin{aligned} \boldsymbol{m}_{0i} &= \tau_{S_0 S_1} \dots \tau_{S_{i-1} S_i}, \\ \boldsymbol{m}_{ij} &= \tau_{S_i S_{i+1}} \dots \tau_{S_{j-1} S_j}, \\ \boldsymbol{m}_{j0} &= \tau_{S_j S_{j+1}} \dots \tau_{S_{n-1} S_0}. \end{aligned}$$

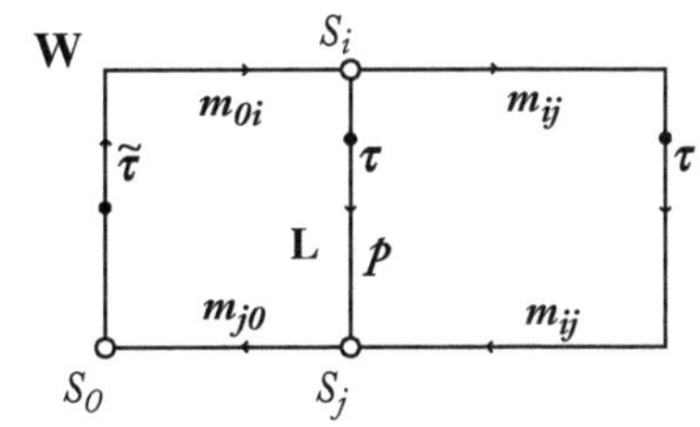

Fig. 5 Case 2 in the proof of Axiom [Mb] in Theorem 33: the closed walk **W** is not an isometric subgraph.

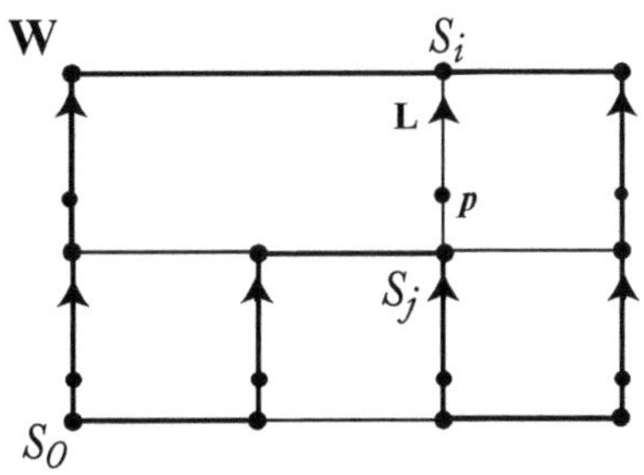

Fig. 6 The non-isometric subgraph **W** of Case 2 in the proof of [Mb] in Theorem 33 is pictured in *thick lines*. The inductive stage of the proof leads to form temporarily, in each of the two smaller closed walks delimited by the shortest path **L**, pairs $\{\mu, \tilde{\mu}\}$ and $\{\nu, \tilde{\nu}\}$ corresponding to the same pair of mutually reverse tokens in **W**.

We have thus $\boldsymbol{m} = \boldsymbol{m}_{0i}\boldsymbol{m}_{ij}\boldsymbol{m}_{j0}$. Note that the two messages $\boldsymbol{m}_{0i}\boldsymbol{p}\boldsymbol{m}_{j0}$ and $\widetilde{\boldsymbol{p}}\boldsymbol{m}_{ij}$ have a length strictly smaller that $n = 2q$. By the induction hypothesis, these two messages are vacuous. Accordingly, for any token τ of $\boldsymbol{p}$, there is an reverse token $\tilde{\tau}$ either in $\boldsymbol{m}_{0i}$ or in $\boldsymbol{m}_{j0}$. (In Fig. 5 the token $\tilde{\tau}$ is pictured as being part of $\boldsymbol{m}_{0i}$.) Considered from the viewpoint of the message $\widetilde{\boldsymbol{p}}\boldsymbol{m}_{ij}$ from S_j, the token $\tilde{\tau}$ is in $\widetilde{\boldsymbol{p}}$ with its reverse τ in $\boldsymbol{m}_{ij}$. The two reverses of the tokens in $\boldsymbol{p}$ and $\widetilde{\boldsymbol{p}}$, form a pair of mutually reverse tokens $\{\tau, \tilde{\tau}\}$ in $\boldsymbol{m}$. Such a pair can be obtained for any token τ in $\boldsymbol{p}$. Augmenting the set of all those pairs by the set of mutually reverse tokens in $\boldsymbol{m}_{0i}$, $\boldsymbol{m}_{ij}$ and $\boldsymbol{m}_{j0}$, we obtain a partition of the set $\mathcal{C}(\boldsymbol{m})$ into pairs of mutually reverse tokens, which establishes that the message $\boldsymbol{m}$ is vacuous.

We have shown that the token system $(\mathcal{S}, \mathcal{T})$ satisfies Axioms [Ma] and [Mb]. The proof is thus complete. □

Remark 34. In the above proof, the inductive argument used to establish Case 2 of [Mb] may convey the mistaken impression that the situation is always straightforward. The simple graph pictured in Fig. 5 is actually glossing over some intricacies. The non-isometric subgraph **W** is pictured by the thick lines in Fig. 6 and is not "convex". We can see how the inductive stage splitting the closed walk **W** by the shortest path **L** may lead to form, in each of the two smaller closed walks, pairs $\{\mu, \tilde{\mu}\}$ and $\{\nu, \tilde{\nu}\}$ which correspond in fact to the same pair of tokens in **W**. Since the arcs corresponding to μ and ν are in the like relation $\mathfrak{L}$, the mistaken assignment is temporary.

We finally obtain:

Theorem 35. *Let $\mathcal{S}$ an arbitrary set, with $|\mathcal{S}| \geq 2$. Denote by $\mathfrak{M}$ the set of all media on $\mathcal{S}$, and by $\mathfrak{G}$ the set of all mediatic graphs on $\mathcal{S}$. There exists a bijection $\mathfrak{f}: \mathfrak{M} \to \mathfrak{G} : \mathcal{M} \mapsto \mathfrak{f}(\mathcal{M})$ such that $G = \mathfrak{f}(\mathcal{M})$ is the graph of $\mathcal{M}$ in the sense of Definition* 13 *if and only if $\mathcal{M}$ is the medium of the mediatic graph G in the sense of Definition* 32.

Proof. Because the set $\mathcal{S}$ of states is constant in $\mathfrak{M}$ and confounded with the constant set of vertices in $\mathfrak{G}$, we could reinterpret the function $\mathfrak{f}$ as a mapping of the family $\mathfrak{T}$ of all sets of token $\mathcal{T}$ making $(\mathcal{S},\mathcal{T})$ a medium, into the family $\mathfrak{E}$ of all sets of edges E making $(\mathcal{S},E)$ a mediatic graph. However, any set of edges E of a mediatic graph on $\mathcal{S}$ is characterized by its like relation $\mathfrak{L}$, or equivalently, by the partition of $\vec{E}$ induced by $\mathfrak{L}$. We choose the latter characterization for the purpose of this proof, and denote by $\vec{\mathfrak{E}}_{|lr}$ the set of all the partitions of the sets of arcs $\vec{E}$ induced by the like relations characterizing the sets of edges in the collection $\mathfrak{E}$.

From Lemmas 27 and 33, we know that the graph of a medium is mediatic, and that the token system induced by a mediatic graph is a medium. We have to show that the functions

$$\mathfrak{f}: \mathfrak{T} \to \vec{\mathfrak{E}}_{|lr} \quad \text{and} \quad \mathfrak{g}: \vec{\mathfrak{E}}_{|lr} \to \mathfrak{T}$$

implicitly defined by (8) and (14), respectively, are mutual inverses. Note that, for any $\mathcal{T} \in \mathfrak{T}$, the partition $\mathfrak{f}(\mathcal{T})$ is defined via a function f mapping $\mathcal{T}$ into the partition $\mathfrak{f}(\mathcal{T})$. Writing as before $\langle ST \rangle$ for the equivalence class containing the arc ST, we have

$$P\tau = Q \iff f(\tau) = \langle PQ \rangle \qquad (\tau \in \mathcal{T}; P,Q \in \mathcal{S}). \tag{16}$$

Proceeding similarly, but inversely, for the function $\mathfrak{g}$, we notice that it defines, for each $\vec{E}_{|lr}$ in $\vec{\mathfrak{E}}_{|lr}$ the set of tokens $\mathfrak{g}(\vec{E}_{|lr})$ via a function g mapping $\vec{E}_{|lr}$ into the set of tokens $\mathfrak{g}(\vec{E}_{|lr})$; we obtain

$$\langle ST \rangle = \langle PQ \rangle \iff Pg(\langle ST \rangle) = Q \qquad (S,T,P,Q \in \mathcal{S}). \tag{17}$$

Combining (16) and (17) we obtain

$$P\tau = Q \iff f(\tau) = \langle PQ \rangle \iff P(g \circ f)(\tau) = Q \qquad (\tau \in \mathcal{T}; P,Q \in \mathcal{S}).$$

We have thus $g = f^{-1}$ and so $\mathfrak{g} = \mathfrak{f}^{-1}$. Conversely, we have

$$\langle ST \rangle = \langle PQ \rangle \iff Pg(\langle ST \rangle) = Q \iff (f \circ g)(\langle ST \rangle) = \langle PQ \rangle \qquad (S,T,P,Q \in \mathcal{S}),$$

yielding $f = g^{-1}$ and so $\mathfrak{f} = \mathfrak{g}^{-1}$. □

Two Examples 36. In the last paragraph of our introductory section, we announced that the collection $\mathfrak{I}$ of all the interval orders on a finite set X was representable as a mediatic graph. The argument goes as follows. Doignon and Falmagne (1997) proved that such a collection $\mathfrak{I}$ is always "well-graded", that is, for any two interval orders K and L, there exists a sequence $K_0 = K, K_1, \ldots, K_n = L$ of interval orders on

X such that $|K_i \triangle K_{i+1}| = 1$ for $0 \leq i \leq n-1$ and $|K \triangle L| = n$. It is easily shown (see Falmagne, 1997) that any well-graded family $\mathcal{F}$ can be cast as a medium $\mathcal{M}(\mathcal{F})$: the states of the medium are the sets of the family, and the tokens consist in either adding or removing an element from a set in $\mathcal{F}$. By Theorem 27, the graph of the medium $\mathcal{M}(\mathcal{J})$ is mediatic. A similar argument applies to the family of all the semiorders on X, and to some other families on X (for example, partial orders and biorders, cf. Doignon & Falmagne, 1997).

Acknowledgement We are grateful to David Eppstein for many useful exchanges pertaining to the results presented here.

References

Bondy, J. A. (1995). Basic graph theory: paths and circuits. In R. L. Graham, M. Grötschel, & L. Lovász (Eds.) *Handbook of combinatorics* (Vol. 1). Cambridge, MA: MIT.

Brightwell, G., Fishburn, P., & Winkler, P. M. (1993). Interval orders and linear extension cycles. *Ars Combinatoria, 36*, 283–288.

Djoković, D. Z. (1973). Distance preserving subgraphs of hypercubes. *Journal of Combinatorial Theory, Series B, 14*, 263–267.

Doignon, J.-P. & Falmagne, J.-Cl. (1997) Well-graded families of relations. *Discrete Mathematics, 173*, 35–44.

Eppstein, D. (2005). The lattice dimension of a graph. *European Journal of Combinatorics, 26*(6), 585–592.

Eppstein, D. & Falmagne, J.-Cl. (2008). Algorithms for media. *Discrete Applied Mathematics, 156*, 1308–1320.

Eppstein, D., Falmagne, J.-Cl., & Ovchinnikov, S. (2008). *Media theory*. Berlin: Springer.

Falmagne, J.-Cl. (1997). Stochastic token theory. *Journal of Mathematical Psychology, 41*(2), 129–143.

Falmagne, J.-Cl. & Ovchinnikov, S. (2002). Media theory. *Discrete Applied Mathematics, 121*, 83–101.

Fishburn, P. C. (1971). Betweenness, orders and interval graphs. *Journal of Pure and Applied Algebra, 1*(2), 159–178.

Fishburn, P. C. (1985). *Interval orders and interval graphs*. London: Wiley.

Fishburn, P. C. & Trotter, W. T. (1999). Split semiorders. *Discrete Mathematics, 195*, 111–126.

Graham, R. L. & Pollak, H. (1971). On addressing problem for loop switching. *Bell Systems Technical Journal, 50*, 2495–2519.

König, D. (1916). Über Graphen und ihren Anwendung auf Determinanten-theorie und Mengenlehre. *Mathematische Annalen, 77*, 453–465.

Ovchinnikov, S. (2007). Fundamentals of media theory. *International Journal of Uncertainty, Fuzziness and Knowledge-Based Systems, 15*(6), 649–680.

Ovchinnikov, S. (2008). Media theory: representations and examples. *Discrete Applied Mathematics, 156*, 1197–1219.

Ovchinnikov, S. & Dukhovny, A. (2000). Advances in media theory. *International Journal of Uncertainty, Fuzziness and Knowledge-Based Systems, 8*(1), 45–71.

Winkler, P. M. (1984). Isometric embedding in products of complete graphs. *Discrete Applied Mathematics, 7*, 221–225.

An Application of Stahl's Conjecture About the k-Tuple Chromatic Numbers of Kneser Graphs

Svata Poljak and Fred S. Roberts[†]

1 Introduction

Graph coloring is an old subject with many important applications. Variants of graph coloring are not only important in their various applications, but they have given rise to some very interesting mathematical challenges and open questions. Our purpose in this mostly expository paper is to draw attention to a conjecture of Saul Stahl's about one variant of graph coloring, k-tuple coloring. Stahl's Conjecture remains one of the long-standing, though not very widely known, conjectures in graph theory. We also apply a special case of the conjecture to answer two questions about k-tuple coloring due to N.V.R. Mahadev.

An interesting and important variant of ordinary graph coloring involves assigning a set of k colors to each vertex of a graph so that the sets of colors assigned to adjacent vertices are disjoint. Such an assignment is called a *k-tuple coloring* of the graph. k-tuple colorings were introduced by Gilbert (1972) in connection with the mobile radio frequency assignment problem (see Opsut & Roberts, 1981; Roberts, 1978, 1979; Roberts & Tesman, 2005). Other applications of multicolorings include fleet maintenance, task assignment, and traffic phasing. These are discussed in Opsut and Roberts (1981); Roberts (1979); Roberts and Tesman (2005) and elsewhere. Among the early publications on this topic are Chvátal, Garey, and Johnson (1978); Clarke and Jamison (1976); Garey and Johnson (1976); Scott (1975); Stahl (1976). Given a graph G and positive integer k, we seek the smallest number t so that there is a k-tuple coloring of G using colors from the set $\{1,2,\ldots,t\}$. This t is called the

[†]*Dedication by Fred Roberts*: This paper is dedicated to Peter Fishburn. His collaboration over the years has been a source of pleasure and inspiration to me. Because of our joint work on generalizations of ordinary graph colorings, it is especially appropriate to dedicate this paper to Peter. Not only is he a colleague, but I am pleased to call him a friend.

F.S. Roberts (✉)
Department of Mathematics and DIMACS, Rutgers University, Piscataway, NJ 08854, USA
e-mail: froberts@dimacs.rutgers.edu

S.J. Brams et al. (eds.), *The Mathematics of Preference, Choice and Order: Essays in Honor of Peter C. Fishburn*, Studies in Choice and Welfare,

k-th multichromatic number or *k-tuple chromatic number* of G and is denoted by $\chi_k(G)$. Of course, if $k=1, \chi_k(G)$ is just the ordinary chromatic number $\chi(G)$.

A *homomorphism* from graph G to graph H is a function h assigning each vertex of G to a vertex of H so that if x and y are adjacent in G, then $h(x)$ and $h(y)$ are adjacent in H. It is well known that an ordinary graph coloring of a graph G with m colors is a homomorphism from G into the complete graph K_m of m vertices. Similarly, an n-tuple coloring of a graph G with m colors is a homomorphism from G into the *Kneser graph* $K(m,n)$. This is the graph whose vertex set consists of all n-element subsets of $\{1,2,\ldots,m\}$, and which has an edge between two such subsets if they are disjoint. (We assume $m \geq 2n$, for otherwise $K(m,n)$ has no edges.) Lovász (1978) computed the ordinary chromatic number $\chi(K(m,n))$ in the process of settling the famous Kneser Conjecture:

Kneser's Conjecture: If the n-element subsets of a $2n+p$-element set are split into $p+1$ classes, then one of the classes will contain two disjoint n-element sets.

Restated, the conjecture says the following:

Kneser's Conjecture Restated: $\chi(K(2n+p,n)) \geq p+2$.

Lovász proved this conjecture by showing the following:

Theorem 1.1 (Lovász, 1978). $\chi(K(m,n)) = m-2n+2, m \geq 2n$.

This leads naturally to the question: What is $\chi_k(K(m,n))$? Stahl (1976) conjectured the following:

Stahl's Conjecture: If $k = qn-r, q \geq 1, 0 \leq r < n$, then $\chi_k(K(m,n)) = qm-2r$.

This conjecture has remained open since 1976 and very little progress has been made on it since Stahl's original paper. Section 2 summarizes what is known about Stahl's Conjecture. We make use of Lovász' Theorem and a special case of Stahl's Conjecture in Sect. 3. Our purpose is to illustrate an amusing application of these two ideas and at the same time highlight Stahl's Conjecture.

It is easy to show that $n\omega(G) \leq \chi_n(G) \leq n\chi(G)$, where $\omega(G)$ is the size of the largest clique of G. Hence, the *weakly γ-perfect* graphs, those for which $\chi = \omega$, have the property that $\chi_n(G) = n\chi(G)$. This observation led Mahadev (1990) to ask how good the lower bound $n\omega(G)$ for $\chi_n(G)$ is. In particular, he asked the following questions, which we settle in Sect. 3.

- **Question 1**: If $\chi_n(G) = n\omega(G)$, does this imply that $\chi(G) = \omega(G)$?

Question 1 suggests that if $\chi(G) \neq \omega(G)$, then $\chi_n(G) \geq n\omega(G)+1$. Mahadev conjectured that the answer to the following question is true:

- **Question 2**: Is $\chi_n(G) \geq n\omega(G) + [\chi(G) - \omega(G)]$?

In Sect. 3, we settle these questions, using Lovász' Theorem and a special case of Stahl's Conjecture.

2 Known Results Concerning Stahl's Conjecture

Here we recall some known results.

Stahl (1976) showed that the upper bound in his conjecture always holds:

Theorem 2.1 (Stahl, 1976). *If* $k = qn - r, q \geq 1, 0 \leq r < n$*, then* $\chi_k(K(m,n)) \leq qm - 2r$.

We will need the following result in the next section. It also gives a simple proof that Stahl's Conjecture holds if $1 \leq k \leq n$.

Theorem 2.2 (Stahl, 1976). *If* G *has an edge and* $n > 1$*, then* $\chi_n(G) \geq 2 + \chi_{n-1}(G)$.

Theorem 2.3 (Stahl, 1976). *If* $1 \leq k \leq n$*, then* $\chi_k(K(m,n)) = m - 2(n-k)$.

Proof. The upper bound follows by Theorem 2.1. The lower bound follows by repeated use of Theorems 2.2 and 1.1. □

Corollary 2.4 (Stahl, 1976). *Stahl's Conjecture holds if* $1 \leq k \leq n$.

Theorem 2.5 (Stahl, 1976). $\chi_{un}(K(m,n)) = um, u > 0$.

Corollary 2.6 (Stahl, 1976). *Stahl's Conjecture holds if* $k = un, u > 0$.

Theorem 2.7 (Stahl, 1976).

$$\chi_k(K(2n+1,n)) = 2k+1+\lfloor\frac{k-1}{n}\rfloor.$$

Corollary 2.8 (Stahl, 1976). *Stahl's Conjecture holds if* $m = 2n+1$.

Theorem 2.9. *Stahl's Conjecture holds for* $n = 2, 3$.†

Theorem 2.10 (Garey and Johnson, 1976). *Stahl's Conjecture holds if* $n = 3$, $k = 4$, $m \geq 6$.

By using $qn + p = (q+1)n - (n-p)$, we see that Stahl's conjecture is equivalent to

$$\text{If } k = qn + p, q \geq 0, 0 < p \leq n, \text{ then } \chi_k(K(m,n)) = qm + m - 2n + 2p.$$

By Theorem 2.1, we know that the upper bound in Stahl's conjecture holds. By Theorems 2.2 and 2.5, the lower bound follows if

$$\chi_{nq+1}(K(m,n)) \geq \chi_{nq}(K(m,n)) + m - 2n + 2,$$

† This was proven in Stahl (1998). According to Stahl (1998), it was independently and previously proven for $n = 2$ by Claude Tardif.

i.e., if the lower bound holds for $p = 1$. As Ostënyi (2007) points out, it follows from a result of Stahl (1998) that

$$\chi_{nq+1}(K(m,n)) \geq \chi_{nq}(K(m,n)) + m - n + 2 - f(n),$$

where $f(n) = n^2 - 3n + 4$. This shows that, given n and $c \in (0,1)$, we have

$$\chi_{nq+1}(K(m,n)) \geq \chi_{nq}(K(m,n)) + c[m - 2n + 2]$$

for m large enough. Note that if $m \leq n^2 - n + 4$, Stahl's results in Stahl (1998) imply that

$$\chi_{nq+1}(K(m,n)) \geq \chi_{nq}(K(m,n)) + 2.$$

Ostënyi (2007) shows that, in fact,

$$\chi_{nq+1}(K(m,n)) \geq \chi_{nq}(K(m,n)) + 3$$

for all positive integers n, m, q.[†]

There have been few other results about Stahl's conjecture over the years, though it is mentioned from time to time in the literature. Frankl and Füredi (1986) discuss extremal problems on Kneser graphs and mention the Stahl Conjecture. Tardif and Zhu (2002) show that if the conjecture is true, then only very few Kneser graphs are multiplicative. (A graph K is called multiplicative if for any two graphs G and H that are not homomorphic to K, their categorical product or tensor product is also not homomorphic to K.)

3 Answers to Mahadev's Questions

We first show that Question 2 has an affirmative answer if $\omega = 2$.

Proposition 3.1. *If* $\omega(G) = 2$, *then*

$$\chi_n(G) \geq n\omega(G) + [\chi(G) - \omega(G)].$$

Proof. Suppose $\omega(G) = 2$. By Theorem 2.2,

$$\chi_n(G) \geq 2(n-1) + \chi_1(G).$$

Thus, since $\omega(G) = 2$ and $\chi_1(G) = \chi(G)$,

$$\chi_n(G) \geq n\omega(G) + [\chi(G) - \omega(G)].$$

□

We observe next that the bound in Question 2 fails in general.

[†] The author thanks József Ostényi for sharing an early version of his paper, in which these ideas are developed.

Proposition 3.2. *There are graphs for which*

$$\chi_n(G) < n\omega(G) + [\chi(G) - \omega(G)].$$

Proof. By Theorem 2.5, we know that

$$\chi_n(K(m,n)) = m. \tag{1}$$

By Theorem 1.1,

$$\chi(K(m,n)) = m - 2n + 2. \tag{2}$$

Since a clique in $K(m,n)$ consists of a disjoint collection of n-element subsets of $\{1,2,\ldots,m\}$, we note that

$$\omega(K(m,n)) = \lfloor \frac{m}{n} \rfloor \tag{3}$$

If

$$\chi_n(G) \geq n\omega(G) + [\chi(G) - \omega(G)],$$

then by (1), (2), and (3), we have

$$m = \chi_n(K(m,n)) \geq n \times \lfloor \frac{m}{n} \rfloor + (m-2n+2) - \lfloor \frac{m}{n} \rfloor \geq n(\frac{m}{n} - 1) + (m-2n+2) - m/n,$$

so

$$3n - m + \frac{m}{n} - 2 \geq 0. \tag{4}$$

Certainly if $m = pn, p \geq 4, n \geq p$, then (4) fails. □

We next observe that the answer to Question 1 is "no".

Proposition 3.3. *There are graphs for which $\chi_n(G) = n\omega(G)$, but $\chi(G) \neq \omega(G)$.*

Proof. Consider the Kneser graph $K(m,2)$ for m even. By (1) and (3),

$$\chi_2(K(m,2)) = m = 2\omega(K(m,2)).$$

However, by (2),

$$\chi(K(m,2)) = m - 2,$$

while by (3),

$$\omega(K(m,2)) = m/2,$$

so $\chi \neq \omega$ already for $m = 6$. □

4 Closing Remarks

Several other related directions of work are of interest. Hilton, Rado, and Scott (1975) define the *multichromatic number* (sometimes called the *ultimate chromatic number*) $\chi^*(G)$ to be $\inf_k(\chi_k(G)/k)$. Clarke and Jamison (1976), Lovász (1972), and Scott (1975) independently showed that this is equal to $\chi_q(G)/q$ for some q. Of course, if G is weakly γ-perfect, then $q = 1$. Johnson et al. (1997) showed that $\chi^*(K(m,n)) = m/n$, if $n \geq 2, m > n$. They studied the relation between the multichromatic number and star chromatic number introduced by Vince (1988). Another long-standing conjecture in graph theory is the conjecture in Johnson et al. (1997) that the star chromatic number of a Kneser graph is equal to its chromatic number. Simonyi and Tardos (2004) proved this conjecture if the chromatic number is even. The star chromatic number arises by considering the set of colors $M = \{1,2,\ldots,m\}$ as residue classes modulo m. Thus, the distance $d(x,y)$ between two colors x,y in M is the distance between x and y around the circle of M points, i.e., the minimum of $(x-y)\text{mod}(m)$ and $(y-x)\text{mod}(m)$. Assume that m,D are positive integers, G has at least one edge and has chromatic number at most m. Then an *(m,D)-coloring* of G is an assignment of a color $f(a)$ to every vertex a of G using residue classes modulo m so that the minimum $d(f(a),f(b))$ is at least D. We define $\eta_m(G)$ to be the maximum D so that G has an (m,D)-coloring. The *star chromatic number* $\eta(G)$ is $\inf_m \eta_m(G)$.

Clarke and Jamison (1976), Lovász (1972), and Scott (1975) observed that the multichromatic number can be calculated by a linear program. This number and the k-tuple chromatic number are closely related to the fractional chromatic number that can also be calculated by a linear program. For an early summary of the relationships among k-tuple chromatic numbers, multiple chromatic numbers, fractional chromatic numbers, and their analogues for independence number, clique number, and clique covering number, see Hell and Roberts (1982). For a comprehensive summary of the literature of fractional chromatic number, see Scheinerman and Ullman (1997).

Klostermeyer and Zhang (2002) showed that any planar graph G with odd girth at least $10n-7, n \geq 2$, has a homomorphism to the Kneser graph $K(2n+1,n)$, i.e., $\chi_n(G) \leq 2n+1$. (The case $n = 1$ fails since that would say that every planar graph of odd girth at least 3 is 3-colorable. However, by Grötzsch's Theorem (Grötzsch, 1958/1959), every planar graph of odd girth at least 5 is 3-colorable.)

It is not hard to show that for any graph G,

$$\chi_{k+1}(G) \leq \chi_k(G) + \chi(G) \leq 2\chi_k(G). \tag{5}$$

Indeed, the first part follows from the more general result in Stahl (1976) that

$$\chi_{qp+r}(G) \leq q\chi_p(G) + \chi_r(G). \tag{6}$$

Equation (5) for $G = K(m,n)$ follows directly from Theorem 2.3 if $1 \leq k \leq n$. Lovász asked (see Erdös (n.d.)) asked whether, for every k, there are graphs G for which

$\chi_{k+1}(G) > (2-\varepsilon)\chi_k(G)$. Chvátal et al. (1978) showed that this was indeed the case.

It should be noted that $\chi_n(G)$ can be arbitrarily larger than $n\omega(G)+[\chi(G)-\omega(G)]$. Indeed, the odd cycles C_{2p+1} illustrate this point. Stahl (1976) shows that

$$\chi_n(C_{2p+1}) = 2n+1+\lfloor\frac{n-1}{p}\rfloor.$$

However,

$$n\omega(C_{2p+1})+[\chi(C_{2p+1})-\omega(C_{2p+1})] = 2n+1.$$

One can ask for a characterization of graphs for which

$$\chi_n(G) = n\omega(G)+[\chi(G)-\omega(G)]$$

and also for a characterization of graphs for which

$$\chi_n(G) > n\omega(G)+[\chi(G)-\omega(G)]$$

and one of graphs for which

$$\chi_n(G) < n\omega(G)+[\chi(G)-\omega(G)].$$

These make for intriguing open questions.

Acknowledgements Fred Roberts thanks the National Science Foundation for its support under grant CCF-0514703 to Rutgers University. He is indebted to the late Svata Poljak for his insights and ideas that form the basis for many of the results in this paper.

References

Chvátal, V., Garey, M. R., & Johnson, D. S. (1978). Two results concerning multicoloring. *Annals Discrete Mathematics, 2*, 151–154.

Clarke, F. H., & Jamison, R. E. (1976). Multicolorings, measures, and games on graphs. *Discrete Mathematics, 4*, 241–246.

Erdös, P. (n.d.). Personal communication to Chvátal, Garey, and Johnson.

Frankl, P., & Füredi, Z. (1986). Extremal problems concerning Kneser graphs. *Journal of Combinatorial Theory, B40*, 270–284.

Garey, M. R., & Johnson, D. S. (1976). The complexity of near optimal graph coloring. *Journal of the Association for Computing Machinery, 23*, 43–49.

Gilbert, E. N. (1972). Unpublished technical memorandum. Murray Hill, NJ: Bell Telephone Labs.

Grötzsch, H. (1958/1959). Ein dreifarbensatz für dreikresifreie netze auf der kugel. *Wissenschaftliche Zeitschrift der Martin Luther Universität Halle–Wittenburg, Math Naturwiss. Reihe, 8*, 109–120.

Hell, P., & Roberts, F. S. (1982). Analogues of the Shannon capacity of a graph. *Annals Discrete Mathematics, 12*, 155–168.

Hilton, A. J. W., Rado, R., & Scott, S. H. (1975). Multicolouring graphs and hypergraphs. *Nanta Mathematica, IX*, 152–155.

Johnson, A., Holroyd, F. C., & Stahl, S. (1997). Multichromatic numbers, star chromatic numbers and Kneser graphs. *Journal of Graph Theory, 26*, 137–145.

Klostermeyer, W., & Zhang, C. Q. (2002). *n*-tuple coloring of planar graphs with large odd girth. *Graphs and Combinatorics, 18*, 119–132.

Lovász, L. (1972). Minimax theorems for hypergraphs. In C. Berge, & D. Raychaudhuri (Eds.). *Hypergraph Seminar, Lecture Notes in Mathematics* (Vol. 411, pp. 111–126). Berlin: Springer.

Lovász, L. (1978). Kneser's conjecture, chromatic number and homotopy. *Journal of Combinatorial Theory, A25*, 319–324.

Mahadev, N. V. R. (1990). Personal communication, November 9, 1990.

Opsut, R. J., & Roberts, F. S. (1981). On the fleet maintenance, mobile radio frequency, task assignment, and traffic phasing problems. In G. Chartrand, et al. (Eds.) *The theory and applications of graphs* (pp. 479–492). New York: Wiley.

Ostěnyi, J. (2007). A lower bound on the multichromatic number of the Kneser graphs. Bolyai Institute, University of Szeged, 2007 (preprint).

Roberts, F. S. (1978). *Graph theory and its applications to problems of society*. CBMS-NSF Monograph No. 29. Philadelphia: Society for Industrial and Applied Mathematics.

Roberts, F. S. (1979). On the mobile radio frequency assignment problem and the traffic light phasing problem. *Annals of the New York Academy of Sciences, 319*, 466–483.

Roberts, F. S., & Tesman, B. (2005). *Applied Combinatorics* (2nd ed.). Upper Saddle River, NJ: Pearson Prentice Hall (reprint in press, Chapman & Hall/CRC).

Scheinerman, E. R., & Ullman, D. H. (1997). *Fractional graph theory: a rational approach to the theory of graphs*. New York: Wiley.

Scott, S. H. (1975). *Multiple node colorings of finite graphs*, Ph.D. Dissertation, University of Reading, England.

Simonyi, G., & Tardos, G. (2004). Local chromatic number and the Borsuk-Ulam Theorem. November 26, 2004. http://front.math.ucdavis.edu/math.CO/0407075.

Stahl, S. (1976). *n*-Tuple colorings and associated graphs. *Journal of Combinatorial Theory, B20*, 185–203.

Stahl, S. (1998). The multichromatic numbers of some Kneser graphs. *Discrete Mathematics, 185*, 287–291.

Tardif, C., & Zhu, X. (2002). The level of nonmultiplicativity of graphs. *Discrete Mathematics, 244*, 461–471.

Vince, A. (1988). Star chromatic number. *Journal of Graph Theory, 12*, 551–559.

Various Applied Mathematics Topics

Optimal Reservation Scheme Routing for Two-Rate Wide-Sense Nonblocking Three-Stage Clos Networks

Wenqing Dou and Frank K. Hwang

1 Introduction

The well-known Clos network has been widely employed for data communications and parallel computing systems, while the symmetric three-stage Clos network $C(n,m,r)$ is considered the most basic and popular multistage interconnection network. A lot of efforts have been put on the research of the three-stage Clos network. Let us first introduce some related concepts.

The *three-stage Clos network* $C(n,m,r)$ is a three-stage interconnection network symmetric with respect to the center stage. The network consists of r $(n \times m)$-crossbars (switches) in the first stage (or *input stage*), m $(r \times r)$-crossbars in the second stage (or *central stage*), r $(m \times n)$-crossbars in the third stage (or *output stage*). The n inlets (outlets) on each input (output) crossbar are the *inputs* (*outputs*) of the network. Thus the total number the *inputs* (*outputs*) of $C(n,m,r)$ is rn. There exists exactly one link between every center crossbar and every input (output) crossbar. These links are the *internal links* while the inputs and outputs are the *external links* of the network.

In the classical circuit switching, a *call* between an idle pair (input, output) is *routable* if there exists a path connecting them such that no link on the path is used by any other connection paths. A call is often referred to as a *request* before it is connected, and *connection* after it is. A network is *strictly nonblocking* (SNB) if regardless of the routing of existing connections in the network, a new request is always routable. A network is *wide-sense nonblocking* (WSNB) if a new request is always routable as long as all connections were routed according to a given routing algorithm. The problem is to estimate the minimum number m^o such that for all $m \geq m^o$, $C(n,m,r)$ is SNB (or WSNB, respectively). Clos (1953) proved that $m^o = 2n-1$ for SNB. Since SNB implies WSNB, we have $m^o \leq 2n-1$ for WSNB. Beneš (1965)

W. Dou (✉)
Department of Mathematics, Shanghai Second Polytechnic University, Shanghai 201209, People's Republic of China
e-mail: wenqingdou@yahoo.com.cn

S.J. Brams et al. (eds.), *The Mathematics of Preference, Choice and Order: Essays in Honor of Peter C. Fishburn*, Studies in Choice and Welfare,

introduced the notion of WSNB. Using the packing routing algorithm, he showed (Beneš, 1985) that for $C(n,m,2)$, $m^o = \lfloor 3n/2 \rfloor$. While m^o for WSNB is still not completely settled for $r \geq 3$, Chang, Guo, Hwang, and Lin (2004) proved that for a set of routing algorithms including almost all studied in the literature, $C(n,m,r)$ at $r \geq 3$ is WSNB if and only if $m \geq 2n-1$, the same as for SNB. Moreover, Tsai, Wang, and Hwang (2001) showed that for r sufficiently large, the latter result holds for any routing algorithm, i.e., that $m^o = 2n-1$.

Melen and Turner (1989) initiated the study on nonblocking properties in multirate interconnection networks. In the multirate environment, a *connection* is a triple (x,y,w) where x is an inlet, y an outlet, and w a weight which can be thought of as the bandwidth requirement (rate) of that connection. In the *uniform* model, each internal and external link is assumed to have the same capacity, which is normalized to be one. An external link can generate many requests, while an internal link can carry many connections, as long as the sum of rates does not exceed capacity one. In applications, the number of distinct rates is often confined to a small number k. We call this a *k-rate environment.*

Unlike the circuit switching case in which no general routing algorithm is helpful to improve m^o over the SNB result $m^o = 2n-1$, Gao and Hwang (1997) proposed a reservation-scheme routing for multirate three-stage Clos network which does the job. Suppose all rates lie in the interval $[b,B]$, where $1 \geq B \geq b > 0$. In Gao and Hwang (1997) it was proved

Theorem 1. *$C(n,m,r)$ is WSNB if $m \geq 5.75n$.*

This is achieved by employing a routing algorithm which reserves $2n-1$ central crossbars only for calls with $w > 1/2$, and another $2.25n$ central crossbars for calls with $1/2 \geq w > 1/3$. Note that currently the best SNB result was due to Melen and Turner (1989) requiring $2\lceil (n-B)/(1-B) \rceil$ central crossbars, which approaches infinity for B approaching 1.

Gao and Hwang (1997) also refined the reservation scheme to obtain

Theorem 2. *For two rates $1/2 \geq B \geq b > 0$, $C(n,m,r)$ is WSNB if $m \geq 3n$.*

Note that Theorem 2 does not cover the $B > 1/2$ case. Further, for $b > 1/2$, each link can carry only one call and hence the case is equivalent to circuit switching. In this paper, we give the best reservation-scheme routing for the $B > 1/2$ case to complete the discussion on the general two-rate case.

2 Two-Rate Routing

We first quote a recent result of Chen, Hwang, and Zhu (2004).

Theorem 3. *A multistage interconnection network is SNB for circuit switching if and only if it is so for the one-rate (any rate) environment.*

From which, we immediately obtain the following result:

Theorem 4. *$C(n,m,r)$ is two-rate WSNB for $m \geq 4n-2$.*

Proof. Consider the routing algorithm which assigns one set of $2n-1$ central crossbars to route B-calls, and another set of $2n-1$ central crossbars to route b-calls. By Theorem 3, no B-call or b-call would be blocked. □

However, we can do better than that with a reservation scheme. We first give a statement stronger than Theorem 3 when the network is the three-stage Clos network. Define $l = \lfloor 1/b \rfloor$ and $t = \lfloor (1-B)/b \rfloor$. Then l is the maximum number of b-calls a link can carry and t is the corresponding number for a link already carrying a B-call. For convenience of writing, we assume that the current request is from input i of input crossbar I to output j of output crossbar J. An I-input is an input of I, while a J-output is an output of J.

Lemma 1. *For two rates B and b satisfying $1 \geq B > 1/2 \geq b > 0$, a b-call can always be routed in a set of $2n-1$ central crossbars in $C(n,m,r)$. Furthermore, no fewer central crossbars can guarantee the routing.*

Proof. Assume that I carry exactly x B-connections. Then each of these x links can generate at most t b-connections, while the other $n-x$ links can each generate at most l b-connections. But the link which generates the current request cannot have generated a full load of connections. On the other hand, a central crossbar is not available to I if and only if either it carries l b-connections from I, or one B-connection and t b-connections. Namely, each link of I generating a full load corresponds to a central crossbar not available to I. Therefore at most $n-1$ central crossbars are not available to I. Similarly, at most $n-1$ central crossbars are not available to J. Thus when there are $2n-1$ central crossbars, at least one is available to both I and J to carry the current request.

To prove the second part of Lemma 1, suppose all calls are b-calls and every I-input and J-output generates a full load (including the current request). Then clearly $2n-1$ central crossbars are needed. □

Note that the assumption of $B > 1/2$ is necessary for otherwise the full load pattern of inputs and that of central crossbars can be different. For example, suppose $B = 0.4$ and $b = 0.3$. Then six inputs can each generate a load (B,b,b) which causes seven central crossbars to be unavailable to I with three of them having load (B,B) and four of them having load (b,b,b).

We say that a link carries a *w-saturating load* if its load exceeds $1-w$ (thus unable to carry another w-call). Let $p(w)$ denote the minimum number of w-saturated I-inputs and J-outputs to induce the full-load situation, i.e., each nonreserved crossbar has either a w-saturated I-link or a w-saturated J-link. Let $R(v)$ denote the reservation scheme which reserves v central crossbars for routing of B-calls only. We first state a general result.

Theorem 5. *Suppose $1 \geq B > 1/2 \geq b > 0$. Then $C(n,m,r)$ is two-rate WSNB under the optimal reservation scheme $R(2n-1-p(B))$ if and only if $m \geq 4n-2-p(B)$. Moreover, no other reservation scheme can improve this bound.*

Proof. Consider the routing algorithm $R(2n-1-p(B))$. By Lemma 1, a b-request can always be routed by the $2n-1$ nonreserved crossbars. Now consider a B-request from an I-input i to a J-output j. Suppose it is blocked in the $2n-1$ nonreserved crossbars. Then at least $p(B)$ I-inputs and J-outputs are B-saturated with all their loads carried by the nonreserved crossbars. Therefore the B-connections carried by the reserved crossbars must all come from the other $2n-p(B)-2$ I-inputs and J-outputs (not including i and j), while each such B-connection occupies a distinct reserved crossbar. So one extra reserved crossbar then can connect the (i,j) B-request. The total number of central crossbars needed is

$$2n-1+2n-2-p(B)+1 = 4n-2-p(B).$$

Thus we have shown that $R(2n-1-p(B))$ is WSNB.

Next we show that $R(v)$ cannot be WSNB if $v < 2n-1-p(B)$. By Lemma 1, the b-connections need $2n-1$ nonreserved crossbars. More specifically, when each I-input and each J-output carries l b-connections and the connections generated by I-inputs do not go to the J-outputs, then each of the $2n-1$ nonreserved crossbars will either has an I-link or a J-link carrying l b-connections (hence B-saturated). Delete all connections except those in the $p(B)$-pattern. Let each other I-input and J-output generates a B-connection. Then the same computation as given in the first half of proof yields a requirement of $4n-2-p(B)$ crossbars. □

We now compute $p(B)$ for the general $B > 1/2 > b$ case. Define $q = (2n-1)(t+1)/(2t+1)$.

Lemma 2. *Suppose $B > 1/2 > b$. Then*

$$p(B) = \begin{cases} \lfloor (2n-1)(t+1)/l \rfloor, & \text{if } l \geq 2t+1; \\ \min\{(2n-1)(t+1)/l - (2t+1-l)\lfloor q \rfloor/l, \lceil q \rceil\}, & \text{if } l < 2t+1. \end{cases}$$

Proof. Without loss of generality, assume among the $2n-1$ nonreserved crossbars, x of them each carries a B-call in its $I(J)$-link, and each of the others carries $t+1$ b-calls. This constitutes the minimum load which induces $2n-1$ B-saturated $I(J)$-links one for each nonreserved crossbars. We now compute $p(B)$ for the I-inputs and J-outputs. Since each B-calls must occupy a distinct $I(J)$-port, the tightest packing is to assign t b-calls to as many $I(J)$-ports (carrying B-calls) as possible. If this assignment does not exhaust the $(2n-1-x)(t+1)$ b-calls from the nonreserved crossbars, then we pack the remaining b-calls into $I(J)$-ports not carrying B-calls, each carrying l b-calls except perhaps the last one. Thus

$$p(B) = \min_{0 \leq x \leq 2n-1} f(x),$$

where x takes only integer values and

$$\begin{aligned} f(x) &= x + \max\{\lfloor ((2n-1-x)(t+1) - tx)/l \rfloor, 0\} \\ &= x + \max\{\lfloor ((2n-1)(t+1) - (2t+1)x)/l \rfloor, 0\}. \end{aligned}$$

For $x \le q$, $f(x) = \lfloor x + ((2n-1)(t+1) - (2t+1)x)/l \rfloor\}$. Since the underlying expression is linear in x, the minimum of f on $[0, \lfloor q \rfloor]$ occurs at one of the endpoints, 0 or $\lfloor q \rfloor$. More specifically, the minimum occurs at $x = 0$ if $l \ge 2t+1$ and at $x = \lfloor q \rfloor$ if otherwise (since $B > 1/2$, $l \ge 2t$; hence "otherwise" can be replaced by $l = 2t$). For $x \ge q$, $f(x) = x$ and therefore $\min_{q \le x \le 2n-1} f(x) = \lceil q \rceil$. Finally, for $l \ge 2t+1$, it is easy to see that $f(0) \le \lceil q \rceil$ since the two terms differ only in the denominator, one is l and the other $2t+1$. Thus

$$p(B) = \begin{cases} f(0), & \text{if } l \ge 2t+1; \\ \min\{f(\lfloor q \rfloor), \lceil q \rceil\}, & \text{if } l < 2t+1. \end{cases}$$

□

Corollary 1. *For $B > 1/2 \ge b$ and $B + b > 1$, a sufficient condition for $C(n,m,r)$ to be two-rate WSNB is $m \ge 4n - 2 - \lfloor (2n-1)/l \rfloor$.*

Tsai et al. (2001) proved that for r large enough, $n \ge 2$ and two rates, B and b, satisfying $B > b$ and $B + b > 1$, $C(n,m,r)$ is not WSNB if $m = \min\{l + 2n - 3, 3n - 3\} \le 3n - 3$. We show that the sufficient condition for WSNB in Corollary 1 barely crosses this $3n-3$ threshold and hence is pretty tight.

Note that $B + b > 1$ implies $t = 0$. Also, $b \le 1/2$ implies $l \ge 2$. Thus $l > 2t+1$ and by Lemma 2, $p(B) = \lfloor (2n-1)/l \rfloor$. Further, $l \ge 2 > (2n-1)/(n+1)$ implies $\lfloor (2n-1)/l \rfloor \le (2n-1)/l < n+1$. Hence $4n - 2 - \lfloor (2n-1)/l \rfloor > 3n - 3$.

3 Some Concluding Remarks

In circuit switching, we know that WSNB cannot improve over SNB for three-stage Clos network, symmetric or asymmetric, except for one special case. In fact, such a conclusion has been extended to other networks, for example, the $Log_d(N,m,p)$ network (Chang, Guo, & Hwang, 2006). We have shown that this is not the case for the two-rate model by using the reservation scheme fruitfully.

For the one-rate model, the notion of reservation scheme is not defined. So it is intriguing to ask whether WSNB can improve over SNB. By Theorem 3, one-rate and circuit switching have no difference as far as SNB is concerned. If this is also the case for WSNB, then the above question is already answered since the question in circuit switching is answered. Surprisingly, Fishbun, Hwang, Du, and Gao (1997) proved that for one-rate $C(n,m,2)$, $m^o = \lceil 3n/2 \rceil$, differing from $\lfloor 3n/2 \rfloor$ in the circuit switching case. Thus one-rate and circuit switching are not completely the same for WSNB, casting uncertainty to the question posed at the beginning of the paragraph. Our conjecture is that the answer to the question is similar to the circuit switching case, namely, $C(n,m,2)$, or its asymmetric version, is the only exception to the general rule that WSNB cannot improve over SNB.

Note that the two-rate model has a wider interpretation than its definition. Suppose a set S of rates can be classified into two groups G_1 and G_2 where each rate

r in G_i satisfies $1/k_i \geq r \geq 1/(k_i+1)$. Suppose $k_1 < k_2$. Then we can reduce S to any two rates B and b satisfying $1/k_1 \geq B > 1/(k_1+1) \geq 1/k_2 \geq b > 1/(k_2+1)$ and the routing algorithm for B, b can be used for S by substituting any G_1-call for B and any G_2-call for b. It is also easily verified that all results reported in Sect. 2 can be extended to the asymmetrical three-stage Clos networks $C(n_1, r_1, m, n_2, r_2)$ by changing $2n$ to $n_1 + n_2$.

We hope that the concept used in analyzing the two-rate model can be extended to other k-rate model for some small $k \geq 3$.

Acknowledgements We sincerely thank the referees and the editor for making many useful suggestions which we have adopted to strengthen the paper as well as to improve its readability. This research is partially supported by Specialized Research Fund for Outstanding Young Teachers in Shanghai Higher Education Institutions (RYQ307013) and Innovation Program of Shanghai Municipal Education Commission (08ZY78).

References

Beneš, V.E. (1965). *Mathematical Theory of Connecting Networks and Telephone Traffic*. New York: Academic.

Beneš, V.E. (1985). Blocking in the NAIU networks. AT&T Bell Labs Tech. Memo.

Chang, F.H., Guo, J.Y., & Hwang, F.K. (2006). Wide-sense nonblocking for multi-$Log_d N$ networks under various routing strategies. *Theoretical Computer Science A, 352*, 232–239.

Chang, F.H., Guo, J.Y., Hwang, F.K., & Lin, C.K. (2004). Wide-sense nonblocking for symmetric or asymmetric 3-stage Clos networks under various routing strategies. *Theoretical Computer Science A, 314*, 375–386.

Chen, W.R., Hwang, F.K., & Zhu, X. (2004). Equivalence of the 1-rate model to the classical model on strictly nonblocking switching networks. *SIAM Journal on Discrete Mathematics, 17*, 446-452.

Clos, C. (1953). A study of nonblocking switching networks. *Bell System Technical Journal, 32*, 406–424.

Fishburn, P., Hwang, F.K., Du, D.Z., & Gao, B. (1997). On 1-rate wide-sense nonblocking for 3-stage Clos networks. *Discrete Applied Mathematics, 78*, 75–87.

Gao, B. & Hwang, F.K. (1997). Wide-sense nonblocking for multirate 3-stage Clos networks. *Theoretical computer Science, 182*, 171–182.

Melen, R. & Turner, J.S. (1989). Nonblocking multirate networks. *SIAM Journal on Computing, 18*, 301–313.

Tsai, K.-H., Wang, D.-W., & Hwang, F.K. (2001). Lower bounds for wide-sense nonblocking clos network. *Theoret. Comput. Sci. 261*, 323–328.

Correlation Inequalities for Partially Ordered Algebras*

Siddhartha Sahi

1 Introduction

The proof of many an inequality in real analysis reduces to the observation that the square of any real number is positive. For example, the AM–GM inequality $\frac{1}{2}(a+b) \geq \sqrt{ab}$ is a restatement of the fact that $\left(\sqrt{a}-\sqrt{b}\right)^2 \geq 0$.

On the other hand, there exist useful notions of positivity in rings and algebras, for which this 'positive squares' property does not hold, viz. the square of an element is not necessarily positive. An interesting example is provided by the polynomial algebra $\mathbb{R}[x]$, where one decrees a polynomial to be positive if all its coefficients are positive. A noncommutative example is furnished by the algebra of $n \times n$ matrices, where one declares a matrix to be positive if all its entries are positive. Neither example satisfies the positive squares property, however in each case the product of two positive elements is positive.

Generalizing this, we consider the following setting:

Definition 1. A partially ordered algebra is a pair (A,P) where A is an associative algebra over $\mathbb{R}$ and P is a nonempty subset closed under addition, multiplication and multiplication by positive real numbers.

This concept goes back at least to the work of Fuchs (1963) and McShane (1953). See also Seidman and Schneider (2006) for a discussion of spectral theory in this context.

The set P is a convex cone in A, and we will call its elements *positive*. By an inequality for (A,P), we mean the assertion that some expression in A is positive. The purpose of this paper is to show that several useful inequalities continue to hold

*Dedicated to Peter Fishburn.

S. Sahi
Department of Mathematics, Rutgers University, New Brunswick, NJ 08903, USA
e-mail: sahi@math.rutgers.edu

S.J. Brams et al. (eds.), *The Mathematics of Preference, Choice and Order: Essays in Honor of Peter C. Fishburn*, Studies in Choice and Welfare,

even in this 'minimal' setting. This is somewhat surprising since traditional proofs of these inequalities tend to involve the positive squares property alluded to above.

The inequalities we consider are concerned with increasing functions on a distributive lattice Ω. For simplicity of exposition, we treat only the case in which $\Omega = 2^S$ is the lattice of all subsets of a finite set S, partially ordered by set inclusion.

We first extend the notion of increasing/decreasing functions to this setting, in the obvious manner:

Definition 2. If (A,P) is a partially ordered algebra, a function $f : 2^S \to A$ is said to be *increasing* (resp. *decreasing)* if $f(\alpha) - f(\beta)$ is positive for all $\beta \subseteq \alpha$ (resp. $\alpha \subseteq \beta$).

Our first main result is the following 'diagonal squares' theorem for partially ordered algebras:

Theorem 1. *If (A,P) is a partially ordered algebra, and $f : 2^S \to A$ is an increasing function, then the following expression is positive*

$$\Phi(S,f) := \sum_{\{\omega, S\setminus\omega\}} [f(\omega) - f(S\setminus\omega)]^2$$

(where the sum ranges over all unordered pairs of complementary subsets of S.)

The set 2^S may be naturally regarded as the vertices of an $|S|$-dimensional hypercube and the pairs $\{\omega, S\setminus\omega\}$ are precisely the endpoints of the various *principal* diagonals – *i.e.* those not contained within a proper face of the hypercube.

The theorem is nontrivial because squares are not necessarily positive in A. In fact, in the expression for $\Phi(S,f)$ only the term $[f(S) - f(\emptyset)]^2$, corresponding to the 'main' diagonal, is guaranteed to be positive.

For the statements of the remaining results we need to introduce a class of measures on 2^S.

Definition 3. A *product measure* on the set 2^S is a measure of the form

$$\mu(\omega) = \prod_{x\in\omega} a_x \prod_{y\notin\omega} (1 - a_y)$$

where each a_x satisfies $1 \geq a_x \geq 0$.

It is easy to see that such a μ is a probability measure on 2^S. Indeed $\mu(\omega)$ represents the probability of choosing the set ω if each element x is chosen independently with probability a_x. Moreover product measures are characterized by the relation

$$\mu(\alpha\cup\beta)\,\mu(\alpha\cap\beta) = \mu(\alpha)\,\mu(\beta) \text{ for all } \alpha, \beta \subseteq S.$$

If $f : 2^S \to A$ is a function and μ is a probability measure on 2^S, we define the *expectation* and *variance* of f with respect to μ as follows

$$E(f) = E(\mu,f) = \sum \mu(\omega) f(\omega),\quad V(f) = V(\mu,f) = E\left(f^2\right) - E(f)^2,$$

where by f^2 we mean the function $f^2(\omega) = f(\omega)^2$.

The diagonal squares theorem implies the following 'positive variance' result:

Theorem 2. *If* (A,P) *is a partially ordered algebra,* μ *is a product measure on* 2^S*, and* $f : 2^S \to A$ *is an increasing function, then the variance* $V(f)$ *is positive.*

Now given two functions $f,g : 2^S \to A$, we define the *covariance* of f,g to be

$$C(f,g) = C(\mu,f,g) = E(f\cdot g) - E(f)\cdot E(g).$$

Here $a\cdot b = \frac{1}{2}(ab+ba)$ is the anticommutator. Note that this expression is symmetric in f and g, even if A is noncommutative.

Our third result is the following 'positive covariance' theorem for partially ordered algebras:

Theorem 3. *If* (A,P) *is a partially ordered algebra,* μ *is a product measure on* 2^S*, and* $f,g : 2^S \to A$ *are increasing functions, then the covariance* $C(f,g)$ *is positive.*

We note that in the special case in which A is the ring of real numbers with the usual notion of positivity, the variance theorem becomes trivial. However the covariance theorem is nontrivial even in this case; in fact it is precisely the Harris inequality (Harris, 1960), which plays a key role in the study of percolation on random graphs. The beauty of our more general setting is that the covariance result becomes an immediate consequence of the variance result! This is analogous to the method of polarization in classical invariant theory.

Of course the three previous results hold equally well for decreasing functions. However one can also prove the following analog of the Cauchy–Schwartz inequality for a partially ordered commutative algebra, which involves an increasing function and a decreasing function. This proof is remarkably similar to the proof of the positive variance theorem.

Theorem 4. *If* (A,P) *is a partially ordered algebra,* A *is commutative,* μ *is a product measure on* 2^S*,* $f : 2^S \to P$ *is increasing, and* $g : 2^S \to P$ *is decreasing, then* $E(f^2)E(g^2) - E(fg)^2$ *is positive.*

We conclude with some comments on possible generalizations of the results in this paper.

First of all, the Harris inequality was generalized by Fortuin, Kasteleyen and Ginibre (1971) and Sarkar (1969) to the case in which the measure satisfies the weaker condition

$$\mu(\alpha\cup\beta)\,\mu(\alpha\cap\beta) \geq \mu(\alpha)\,\mu(\beta) \text{ for all } \alpha,\beta\subseteq S.$$

This form, called the FKG inequality, has proved extremely useful for applications (see e.g. Bricmont, Fontaine, Lebowitz, Lieb, & Spencer, 1980/81; van den Berg, Häggström, & Kahn, 2006; Fishburn, 1984, 1992; Fishburn, Doyle, & Shepp, 1988; Graham, 1983; Karlin & Rinott, 1988; Lebowitz, 1972; Percus, 1975; Shepp, 1982). Other proofs and further generalizations of the FKG inequality have been obtained in (Alon & Spencer, 1992; Den Hollander & Keane, 1986; Glimm & Jaffe,

1987; Holley, 1974; Preston, 1974) culminating in the Ahlswede–Daykin inequality (Rinott & Saks, 1993; Ahlswede & Daykin, 1978). It is natural to ask whether these more general inequalities also hold in the present setting. It turns out that the FKG inequality does hold while the Ahlswede–Daykin result does not. The proofs are quite different from the arguments in this paper and we shall report on them elsewhere (Sahi, 2006).

We also remark that higher-order analogs for the FKG inequality have been introduced in Richards (2004) and Sahi (2007). They are of the form

$$E_n(f_1,\dots,f_n) \geq 0$$

where E_n is a certain correlation functional of the n increasing positive functions $f_1,\dots,f_n : 2^S \to \mathbb{R}_{\geq 0}$, with respect to an FKG measure. For $n=2$ this is the FKG inequality, while for $n=3$ one has

$$\begin{aligned} E_3(f_1,f_2,f_3) &= 2E(f_1f_2f_3) + E(f_1)E(f_2)E(f_3) \\ &\quad - [E(f_1f_2)E(f_3) + E(f_1f_3)E(f_2) + E(f_1)E(f_2f_3)]. \end{aligned}$$

However these higher order analogs are known to hold only under additional assumptions. It seems quite likely that the ideas of this paper have some bearing on this issue and we shall consider this possibility in a future paper.

2 Diagonal Squares

In this section we give the proof of the diagonal squares theorem (Theorem 1), which asserts for increasing f the positivity of the expression

$$\Phi(S,f) := \sum_{\{\omega, S\setminus\omega\}} [f(\omega) - f(S\setminus\omega)]^2.$$

The result is trivial for $|S| = 0,1$ while the proof for $|S| = 2$ already contains the germ of the main idea. For clarity of exposition we treat this case explicitly before proceeding to the general situation. Thus suppose $S = \{1,2\}$ and write

$$a = f(S), b = f(\{1\}), c = f(\{2\}), d = f(\emptyset).$$

Then we have

$$\Phi(S,f) = (a-d)^2 + (b-c)^2$$

and the trouble, of course, is that $(b-c)^2$ need not be positive.

To get around this problem we rewrite $\Phi(S,f)$ as follows:

$$\begin{aligned} \Phi(S,f) &= \frac{1}{2}[(a-d)+(b-c)]^2 + \frac{1}{2}[(a-d)-(b-c)]^2 \\ &= \frac{1}{2}[(a-c)+(b-d)]^2 + \frac{1}{2}[(a-b)+(c-d)]^2. \end{aligned}$$

This last expression is clearly positive, since by assumption each of $a-b$, $c-d$, $a-c$, $b-d$ is positive.

Note also that the argument does not require A to be commutative or even associative. Furthermore we do not require that the product of two positive elements be positve, only that the square of a positive element is positive.

We now consider the case of general $|S|>1$, proceeding by induction on $|S|$. Fix $x \in S$, and define two functions on subsets of the set $T = S \setminus \{x\}$ as follows:

$$g(\alpha) := f(\alpha \cup \{x\}) + f(\alpha),$$
$$h(\alpha) := f(\alpha \cup \{x\}) - f(\alpha).$$

Since f is increasing on 2^S, it follows that g is increasing on 2^T and h is positive (takes values in P).

We claim the following identity holds

$$\Phi(S,f) = \frac{1}{2} \sum_{\{\alpha, T\setminus\alpha\}} [h(\alpha) + h(T\setminus\alpha)]^2 + \frac{1}{2} \sum_{\{\alpha, T\setminus\alpha\}} [g(\alpha) - g(T\setminus\alpha)]^2.$$

This suffices for the result, since the first sum is clearly positive, while the second sum, which is equal to $\frac{1}{2}\Phi(T,g)$, is positive by induction.

To prove the identity, we fix a pair $\{\alpha, T\setminus\alpha\}$ and consider all the terms on the left and right involving the quantities

$$a = f(\alpha \cup \{x\}), b = f(\alpha), c = f(T\setminus\alpha \cup \{x\}), d = f(T\setminus\alpha).$$

This reduces matters to verification of the following simple equality

$$(a-d)^2 + (b-c)^2 = \frac{1}{2}[(a-b)+(c-d)]^2 + \frac{1}{2}[(a+b)-(c+d)]^2,$$

which is the same calculation as before. □

We note again that the proof remains valid under the following weaker hypotheses:

(1) A is an arbitrary algebra over $\mathbb{R}$ (i.e. not necessarily associative)
(2) P is closed under addition, multiplication by positive scalars, and squaring (but need not be closed under multiplication).

3 Positive Variance

We now prove the positive variance theorem (Theorem 2), which asserts the positivity of the variance $V(\mu, f)$ of an increasing function f with respect to a product measure μ.

Since μ is a probability measure we can rewrite the variance in following, more homogeneous, form:

$$\begin{aligned}
V(\mu,f) &= \left[\sum_{\omega}\mu(\omega)\right]\left[\sum_{\omega}\mu(\omega)f(\omega)^2\right]-\left[\mu(\omega)\sum_{\omega}f(\omega)\right]^2\\
&= \sum_{(\alpha,\beta)}\mu(\alpha)\mu(\beta)f(\beta)^2-\sum_{(\alpha,\beta)}\mu(\alpha)\mu(\beta)f(\alpha)f(\beta)\\
&= \sum_{\{\alpha,\beta\}}\mu(\alpha)\mu(\beta)\left[f(\alpha)-f(\beta)\right]^2
\end{aligned}$$

Here the intermediate expression involves sums over ordered pair of subsets (α,β), but the final expression involves a sum over unordered pairs. In terms of the hypercube interpretation discussed in the introduction, we see that the variance is a weighted sum of squares corresponding to *all* the diagonals (and edges) of the cube, and not just the principal diagonals.

We now observe that each diagonal is a principal diagonal in a unique face of the cube. To make this precise, we note that a face of the cube is determined by a nested pair of subsets $\omega_0 \subseteq \omega_1$. The vertices of the face consist of all α such that $\omega_0 \subseteq \alpha \subseteq \omega_1$, and the principal diagonals of the face are unordered pairs $\{\alpha,\beta\}$ such that $\alpha\cap\beta=\omega_0$ and $\alpha\cup\beta=\omega_1$.

Since μ is a product measure it follows that

$$\mu(\alpha)\mu(\beta)=\mu(\alpha\cup\beta)\mu(\alpha\cap\beta)=\mu(\omega_0)\mu(\omega_1).$$

This means that in the above formula for the variance, the weight is *constant* on each face. Thus we obtain the following "facial expression" for the variance

$$V(\mu,f)=\sum_{\omega_0\subseteq\omega_1}\mu(\omega_0)\mu(\omega_1)\Phi\left(S_{\omega_0,\omega_1},f_{\omega_0,\omega_1}\right)$$

where $S_{\omega_0,\omega_1}=\omega_1\setminus\omega_0$ and f_{ω_0,ω_1} is the function on $\omega_1\setminus\omega_0$ defined by

$$f_{\omega_0,\omega_1}(\alpha):=f(\omega_0\cup\alpha).$$

Since f is an increasing function, so is f_{ω_0,ω_1}. Thus by the diagonal squares theorem, each $\Phi\left(S_{\omega_0,\omega_1},f_{\omega_0,\omega_1}\right)$ is positive, and hence so is the variance. □

This theorem too holds under the weaker hypotheses of the previous section.

4 Positive Covariance

We now prove the positive covariance result (Theorem 6), which asserts that if (A,P) is a partially ordered algebra, μ is a product measure on 2^S, and $f,g:2^S\to A$ are increasing functions, then the covariance

$$C(f,g) = E(f \cdot g) - E(f) \cdot E(g)$$

is positive.

We shall deduce this from the variance theorem by a "polarization" argument.

For this we let $A' = A[x_1, x_2]$ be the set of polynomials in two variables with coefficients in A, and let P' denote the subset of polynomials with all coefficients in P. Then A' is an algebra over $\mathbb{R}$ (with the usual definition of polynomial multiplication) and P' is a convex cone. Furthermore for polynomials p, q in A' the coefficients of pq are themselves sums of products of the coefficients of p and q. Since P is closed under addition and multiplication, it follows that (A', P') is a partially ordered algebra.

We define a function $f' : 2^S \to A'$ as follows:

$$f'(\omega) = x_1 f(\omega) + x_2 g(\omega).$$

Since f and g are increasing, it follows immediately that f' is an increasing function for the partially ordered algebra (A', P'). By the positive variance theorem it follows that $V(\mu, f')$ belongs to P', *i.e.* all its coefficients are in P.

Now we calculate as follows

$$\begin{aligned} V(\mu, f') &= E\left([x_1 f + x_2 g]^2\right) - [E(x_1 f + x_2 g)]^2 \\ &= E\left(x_1^2 f^2 + x_2^2 g^2 + 2x_1 x_2 f \cdot g\right) - [x_1 E(f) + x_2 E(g)]^2 \\ &= x_1^2 E\left(f^2\right) + x_2^2 E\left(g^2\right) + 2x_1 x_2 E(f \cdot g) \\ &\quad - x_1^2 E(f)^2 - x_2^2 E(g)^2 - 2x_1 x_2 E(f) \cdot E(g) \\ &= x_1^2 V(f) + x_2^2 V(g) + 2x_1 x_2 C(f,g). \end{aligned}$$

Considering the coefficient of $x_1 x_2$, we deduce the positivity of $C(f,g)$. □

Note that the polarization argument above (deducing positive covariance from positive variance) makes no use of the fact that μ is a product measure.

5 The Cauchy–Schwartz Inequality

We now prove the analog of the Cauchy–Schwartz inequality (Theorem 7), which asserts that if (A,P) is a partially ordered commutative algebra, μ is a product measure on 2^S, f is a P-valued increasing function, and g is a P-valued decreasing function, then $E\left(f^2\right) E\left(g^2\right) - E(fg)^2$ is positive.

We first show that if (A,P) is a partially ordered algebra, f_1, f_2 are increasing and P-valued, then the product $f_1 f_2$ is also increasing and P-valued. To see this, choose $\alpha \supseteq \beta$:and calculate as follows:

$$f_1(\alpha) f_2(\alpha) - f_1(\beta) f_2(\beta) = f_1(\alpha)[f_2(\alpha) - f_2(\beta)] + [f_1(\alpha) - f_1(\beta)] f_2(\beta).$$

Since f_1, f_2 are increasing and P-valued, it follows that the expression is positive.

Now in our situation, g is a decreasing function. It follows that $\alpha \mapsto g(S \setminus \alpha)$ is increasing, and by the above argument so is the function

$$h(\alpha) = f(\alpha)\, g(S \setminus \alpha).$$

We now compute

$$\begin{aligned}
&E(f^2)E(g^2) - E(fg)^2 \\
&= \left(\sum \mu(\alpha) f(\alpha)^2\right)\left(\sum \mu(\beta) g(\beta)^2\right) - \left(\sum \mu(\alpha) f(\alpha) g(\alpha)\right)^2 \\
&= \sum_{\alpha,\beta} \mu(\alpha)\mu(\beta) f(\alpha)^2 g(\beta)^2 - \sum_{\alpha,\beta} \mu(\alpha)\mu(\beta) f(\alpha) g(\alpha) f(\beta) g(\beta) \\
&= \sum_{\{\alpha,\beta\}} \mu(\alpha)\mu(\beta)\left[f(\alpha) g(\beta) - g(\alpha) f(\beta)\right]^2.
\end{aligned}$$

This is a weighted sum over all edges and diagonals of the hypercube 2^S and arguing as in the proof of the covariance theorem, we arrive at the following facial expression

$$E(f^2)E(g^2) - E(fg)^2 = \sum_{\omega_0 \subseteq \omega_1} \mu(\omega_0)\mu(\omega_1)\Phi\left(S_{\omega_0,\omega_1}, h_{\omega_0,\omega_1}\right)$$

where

$$h_{\omega_0,\omega_1}(\alpha) = h(\omega_0 \cup \alpha).$$

As remarked earlier in the proof, $h(\alpha) = f(\alpha)\, g(S \setminus \alpha)$ is an increasing function. Therefore so is each h_{ω_0,ω_1} and the positivity follows from the diagonal squares theorem. □

References

Ahlswede, R., & Daykin, D. (1978). An inequality for the weights of two families of sets, their unions and intersections. *Zeitschrift fr Wahrscheinlichkeitstheorie und verwandte Gebiete, 43*, 183–185.

Alon, N. & Spencer, J. (1992) *The Probabilistic Method.* New York: Wiley.

Bricmont, J., Fontaine, J.-R., Lebowitz, J., Lieb, E., & Spencer, T. (1980/81).Lattice systems with a continuous symmetry, I, II, III. *Communications in Mathematical Physics, 78*, no. 2, 281–302; no. 3, 363–371; no. 4, 545–566

van den Berg, J., Häggström, O., & Kahn, J. (2006). Some conditional correlation inequalities for percolation and related processes. *Random Struct. Algorithms 29*, 417–435

Den Hollander, W. & Keane, M. (1986). Inequalities of FKG type. *Physics A, 138*, 167–182.

Fishburn, P.C. (1984). A correlational inequality for linear extensions of a poset. *Order, 1*, 127–137.

Fishburn, P.C. (1992). Correlation in partially ordered sets. *Discrete Applied Mathematics, 39*, 173–191.

Fishburn, P.C., Doyle, P.G., & Shepp, L.A. (1988). The match set of a random permutation has the FKG property. *Annals of Probability, 16*, 1194–1214.

Fortuin, C., Kasteleyn, P., & Ginibre, J. (1971). Correlation inequalities on some partially ordered sets. *Communications in Mathematical Physics, 22*, 89–103.

Fuchs, L. (1963). *Partially Ordered Algebraic Systems*. New York: Pergamon.

Graham, R.L. (1983). Applications of the FKG inequality and its relatives. In *Mathematical Programming: The State of the Art* (pp.115-131). Berlin: Springer.

Glimm, J. & Jaffe, A. (1987). *Quantum Physics: A Functional Integral Point of View*, 2nd ed. New York: Springer.

Harris, T. (1960). A lower bound for the critical probability in a certain percolation process. *Mathematical Proceedings of the Cambridge Philosophical Society, 56*, 13–20.

Holley, R. (1974). Remarks on the FKG inequalities. *Communications in Mathematical Physics, 36*, 227–231.

Karlin, S. & Rinott, Y. (1988). A generalized Cauchy-Binet formula and applications to total positivity and majorization. *Journal of Multivariate Analysis, 27*, 284–299.

Lebowitz, J. (1972). Bounds on the correlations and analyticity properties of ferromagnetic Ising spin systems. *Communications in Mathematical Physics, 28*, 313–321.

McShane, E. (1953).*Order-Preserving Maps and Integration Processes*. New Jersey: Princeton University.

Percus, J. (1975). Correlation inequalities for Ising spin lattices. *Communications in Mathematical Physics, 40*, 283–308.

Preston, C. (1974). A generalization of the FKG inequalities. *Communications in Mathematical Physics, 36*, 233–241.

Richards, D. (2004). Algebraic method toward higher-order probability inequalities II. *Annals of Probability, 32*, 1509–1544.

Rinott, Y. & Saks, M. (1993). Correlation inequalities and a conjecture for permanents. *Combinatorica, 13*, 269–277.

Sahi, S. (2008a). Higher correlation inequalities. *Combinatorica* 28(2):209–227

Sahi, S. (2008b). The FKG Inequality for Partially Ordered Algebras, *J. Theor. Prob. 21*, 449–458

Sarkar, T. (1969). Some lower bounds of reliability. Technical Report 124, Department of Operational Research and Statistics, Stanford University.

Seidman, T. & Schneider, H. (2006). The spectral radius in partially ordered algebras. *Linear Applied Algebra, 417*, 347–369.

Shepp, L. A. (1982). The XYZ conjecture and the FKG inequality. *Annals of Probability, 10*, 824–827.

The Kruskal Count

Jeffrey C. Lagarias, Eric Rains, and Robert J. Vanderbei

1 Introduction

The *Kruskal Count* is a card trick invented by Martin D. Kruskal (who is well known for his work on solitons) which is described in Fulves and Gardner (1975) and Gardner (1978, 1988). In this card trick a magician "guesses" one card in a deck of cards which is determined by a subject using a special counting procedure that we call *Kruskal's counting procedure*. The magician has a strategy which with high probability will identify the correct card, explained below.

Kruskal's counting procedure goes as follows. The subject shuffles a deck of cards as many times as he likes. He mentally chooses a (secret) number between one and ten. The subject turns the cards of the deck face up one at a time, slowly, and places them in a pile. As he turns up each card he decreases his secret number by one and he continues to count this way till he reaches zero. The card just turned up at the point when the count reaches zero is called the *first key card* and its value is called the *first key number*. Here the value of an Ace is one, face cards are assigned the value five, and all other cards take their numerical value. The subject now starts the count over, using the first key number to determine where to stop the count at the second key card. He continues in this fashion, obtaining successive key cards until the deck is exhausted. The last key card encountered, which we call the *tapped card*, is the card to be "guessed" by the magician.

The Kruskal counting procedure for selecting the tapped card depends on the subject's secret number and the ordering of cards in the deck. The ordering is known to the magician because the cards are turned face up, but the subject's secret number is unknown. It appears impossible for the magician to know the subject's secret number. The mathematical basis of the trick is that for most orderings of the deck

R.J. Vanderbei (✉)
Department of Operations Research and Financial Engineering, Princeton University, Princeton, NJ 08544, USA
e-mail: rvdb@princeton.edu

S.J. Brams et al. (eds.), *The Mathematics of Preference, Choice and Order: Essays in Honor of Peter C. Fishburn*, Studies in Choice and Welfare,

most secret numbers produce the same tapped card. For any given deck two different secret numbers produce two different sequences of key cards, but if the two sequences ever have a key card in common, then they coincide from that point on, and arrive at the same tapped card. The magician therefore selects his own secret number and carries out the Kruskal counting procedure for it while the subject does his own count. The magician's "guess" is his own tapped card. The Kruskal Count trick succeeds with high probability, but if it fails the magician must fall back on his own wits to entertain the audience.

This paper gives a mathematical analysis of the success probability for this trick, based on two simplified mathematical models of the Kruskal counting procedure. We are concerned with the *ensemble success probability* averaged over all possible orderings of the deck (with the uniform distribution). Our objective is to estimate ensemble success probabilities for mathematical idealizations of such counting procedures. Then we numerically compare the ensemble success probabilities on a 52-card deck with that of the Kruskal Count trick itself. The success probability of the trick depends in part on the magician's strategy for choosing his own secret number. We show that the magician does best to always choose the first card in the deck as his first key card, i.e., to use secret number 1.

The general mathematical problem we consider applies the Kruskal counting procedure to a deck of N labelled cards with each card label a positive integer, in which each card has its label drawn independently from some fixed probability distribution on the positive integers $\mathbb{N}^+$. We call such distributions *i.i.d. deck distributions*; they are specified by the probabilities $\{\pi_j : j \geq 1\}$ of a fixed card having value j. We assume that the subject chooses an initial secret number from an initial probability distribution on $\mathbb{N}^+ = \{1, 2, 3, \ldots\}$, and that the magician independently does the same from a possibly different initial probability distribution, and that thereafter each follows the Kruskal counting procedure. It is convenient to view the cards of the deck as turned over at unit times, so that the card in the M-th position is turned over at time M. If the M-th card is a key card for both magician and subject and no previous card is a key card for both, then we say that M is the *coupling time* for the sequences. Let t be a random variable denoting the coupling time on the resulting probability space with $t = +\infty$ if coupling does not occur. We wish to estimate the "failure probability" $\mathbb{P}[t > N]$.

The set of permutations of a fixed deck (with uniform distribution) does not have the i.i.d. property, and is not Markovian, but it can be reasonably well approximated by such a distribution. The advantage of the simplifying assumption of an i.i.d. deck distribution is that the random variable t can be interpreted as a stopping time for a coupling method for a Markov chain, as is explained in Sect. 2.

The mathematical contents of the paper are determination of $\mathbb{P}(t > n)$ for a geometric i.i.d. deck distribution, which is carried out in Sect. 3, and estimation of $\mathbb{P}(t > n)$ for a uniform i.i.d. deck distribution, which is carried out in Sect. 4. The proofs of several results stated in Sect. 4 are given in an appendix.

In Sect. 5 we consider the actual Kruskal count trick, and compare its success probability with the approximations given by the models above. Because the Kruskal count trick using an actual deck of 52 cards involves a stochastic process that is not Markovian, we estimate the success probability by Monte Carlo simulation. We consider the effect on this success probability of varying the magician's strategy for choosing his key card, and of varying the value assigned to face cards. The magician should choose his key card value to be 1. Assuming this strategy for the magician, the success probability of the original Kruskal Count trick is just over 85%. Both the i.i.d. geometric distribution and i.i.d. uniform distribution models above give good approximations; the geometric distribution is off by less than 3%, and the uniform approximation is within 1%.

There has been previous work on mathematical models motivated by the Kruskal count. In Mallows (1975) determined that the expected value of the coupling time of two sequences $\{Z_{1,i}, Z_{2,i}\}$ given as sums $Z_{1,i} := X_1 + \cdots + X_i$, $Z_{2,i} := Y_1 + \cdots + Y_i$, of i.i.d. positive integer valued random variables X_i (resp. Y_i), possibly having different distributions, is $E[X_i]E[Y_i]$. He also observed that if the X_i were geometrically distributed then the variance of the coupling time could be determined as well. In Haga and Robins (1997) analyzed a simplified Markov chain model for the Kruskal count which is related to, but different from, the models considered here. We discuss their model at the end of Sect. 4.

2 Coupling Methods for Markov Chains

The *coupling time random variable* t is a special case of a stopping time random variable t^* associated with a coupling method for studying a Markov chain. This motivates our terminology.

To explain this connection, consider a homogeneous Markov chain $(X_n : n \geq 0)$ on a countable discrete state space S. Given two initial probability distributions $\mathbf{p}$ and $\mathbf{p}'$ on S a *coupling method* constructs a bivariate process (X_n^1, X_n^2) consisting of two copies of process X_n with X_0^1 having distribution $\mathbf{p}$, X_0^2 having distribution $\mathbf{p}'$, and the two copies evolve independently until some (random) *stopping time* t^* at which $X_{t^*}^1 = X_{t^*}^2$ and then requires them to be equal thereafter, evolving as a single process X_n. The stopping time t^* is not necessarily required to be the *first* time t at which $X_t^1 = X_t^2$ occurs, and the particular rule for choosing t^* defines the coupling method. Let $\boldsymbol{\mu}_n$, $\boldsymbol{\mu}_n'$ denote the distribution at time n of the process X_n starting from the distribution $\mathbf{p}$, $\mathbf{p}'$ respectively, at time 0, and let the *variation distance* $||\mathbf{p} - \mathbf{p}'||$ between two distributions on S be

$$||\mathbf{p} - \mathbf{p}'|| := \frac{1}{2} \sum_{s \in S} |p(s) - p'(s)| \,. \tag{1}$$

The *basic coupling inequality* is

$$||\boldsymbol{\mu}_n - \boldsymbol{\mu}_n'|| \leq \mathbb{P}[t^* > n] \,. \tag{2}$$

Such inequalities can be used to prove ergodicity of a Markov chain and to bound the speed of convergence to the equilibrium distribution, by bounding the right side of the inequality.

The first coupling method was invented by Doeblin (1938), and many other coupling methods have been proposed since, see Griffeath (1978) for a survey. Applications to card shuffling and random walks on groups are described in Aldous and Diaconis (1986) and Diaconis (1988). The basic coupling inequality (2) is also valid for non-ergodic Markov chains, e.g., null-recurrent or transient Markov chains on the state space $\mathbb{N}$, as was observed by Pitman (1976). Coupling methods are traditionally used as an auxiliary device to get information on the rate of convergence to equilibrium of an ergodic Markov chain. In this paper, we are interested in obtaining upper and lower bounds for the coupling probability itself, since it represents the failure probability of the Kruskal Count trick. We do not use the basic coupling inequality, but instead in Sect. 4 use inequalities relating coupling probabilities for various different Markov chains.

For an i.i.d. deck the Kruskal counting procedure can be viewed as moving on a Markov chain $\mathcal{M}_\pi$ on the state space $\mathbb{N}$ where a state j represents a current value of the Kruskal counting procedure, with state 0 representing being at a key card, and state j represents that the next key card be reached after exactly j more cards are turned over. Each transition of the Markov chain will correspond to turning over one card in the deck. Let the random variable X_n denote the state of the Markov chain at time n; it indicates the current Kruskal count value at location n of the deck, except that $X_n = 0$ indicates a key card at location n. The transition probability for this chain from state $j \geq 1$ is probability 1 to state $j-1$ and 0 to all other states, and from state 0 to state j is probability π_{j+1}, where $\{\pi_j : j \geq 1\}$ is the distribution π of card labels. (That is, π_1 is the probability that the key card has value 1, and the chain transitions from state 0 to state 0.) The initial distribution of secret numbers are distributions $\mathbf{p}$, $\mathbf{p}'$ on the state space $\mathbb{N}$. We define the random variable $t = t(\mathbf{p}, \mathbf{p}')$ to be the stopping time associated to the coupling method that combines the chains X_n^1 and X_n^2 at the first time that $X_n^1 = X_n^2 = 0$. (This is not necessarily the first time that $X_n^1 = X_n^2$.) The basic coupling inequality (2) for $\mathcal{M}_\pi$ and t then gives

$$||\boldsymbol{\mu}_n - \boldsymbol{\mu}_n'|| \leq \mathbb{P}[t > n] \,, \tag{3}$$

where $\boldsymbol{\mu}_n$ and $\boldsymbol{\mu}_n'$ are the n-step state probabilities for the chain $\mathcal{M}_\pi$ started with initial distributions $\mathbf{p}$ and $\mathbf{p}'$. We note that the Markov chain $\mathcal{M}_\pi$ is ergodic if $E[\pi] = \sum_{j=1}^{\infty} j\pi_j$ is finite, and is null-recurrent otherwise. In the ergodic case the stationary distribution $\tilde{\pi} = (\tilde{\pi}_0, \tilde{\pi}_1, \tilde{\pi}_2, \ldots)$ is given by

$$\tilde{\pi}_j = (1 - \pi_1 - \pi_2 - \cdots - \pi_j)(1 + E[\pi])^{-1} \tag{4}$$

for $j \geq 0$. This chain is ergodic for the deck distributions that we consider, and our object is to estimate the "failure probability" $\mathbb{P}[t > n]$.

In the remainder of the paper, rather than considering Markov chains of the type M_π, we study simplified Markov chains that jump from one key card to the next, but which retain enough information for coupling methods to apply.

3 Geometric Distribution

We consider an idealized deck consisting of cards whose labels are independently and identically distributed random variables drawn from $\mathbb{N}^+ = \{1, 2, 3, \ldots\}$ with the geometric distribution $\mathcal{G}_p$ given by $\pi_k = (1-p)p^{k-1}$, $0 < p < 1$. The geometric distribution has mean

$$E[p] \;=\; \sum_{k=1}^{\infty} k\pi_k \;=\; \frac{1}{1-p}. \tag{5}$$

Let $\mathcal{G}_N(p)$ denote the deck distribution induced on a deck of N cards.

Assume that the magician and subject both pick a secret number drawn from the same geometric distribution $\mathcal{G}_p$. Let $\mathbb{P}[t > N]$ denote the probability (choosing a deck of cards at random as above) that the magician and subject have no common key card in positions 1 through N.

For the geometric deck distribution there is a simple exact formula for all coupling probabilities.

Theorem 3.1. *For the geometric deck distribution $\mathcal{G}_N(p)$ with initial geometric value distributions $\mathcal{G}_p$,*

$$\mathbb{P}[t > N] = p^N (2-p)^N \,. \tag{6}$$

Proof. We use the memorylessness property of the geometric distribution, which is that for a $\mathcal{G}_p$-distributed variable X the conditional probability $\mathbb{P}[u = k \mid u \geq \ell]$ satisfies

$$\mathbb{P}[u = k \mid u \geq \ell] = \mathbb{P}[u = k - \ell] \,. \tag{7}$$

By direct computation

$$\mathbb{P}[t > 1] = 1 - (1-p)^2 = p(2-p) \,. \tag{8}$$

Now for $N \geq 2$,

$$\begin{aligned}\mathbb{P}[t > N] = {} & \mathbb{P}[t > N | X_1^1 \geq 2 \text{ and } X_1^2 \geq 2] \;\; \mathbb{P}[X_1^1 \geq 2 \text{ and } X_1^2 \geq 2] \\ & + \mathbb{P}[t > N | X_1^1 = 1 \text{ and } X_1^2 \geq 2] \;\; \mathbb{P}[X_1^1 = 1 \text{ and } X_1^2 \geq 2] \\ & + \mathbb{P}[t > N |\, X_1^1 \geq 2 \text{ and } X_1^2 = 1] \;\; \mathbb{P}[X_1^1 \geq 2 \text{ and } X_1^2 = 1] \\ & + \mathbb{P}[t > N |\, X_1^1 = 1 \text{ and } X_1^2 = 1] \;\; \mathbb{P}[X_1^1 = 1 \text{ and } X_1^2 = 1],\end{aligned} \tag{9}$$

in which the last condition $X_1^1 = X_1^2 = 1$ has zero probability for $N \geq 2$. Now by (7)

$$\mathbb{P}[t > N | X_1^1 \geq 2 \text{ and } X_1^2 \geq 2] = \mathbb{P}[t > N-1] \,. \tag{10}$$

In the second case $X_2^2 - 1$ is geometrically distributed, hence by (7) again

$$\mathbb{P}[t > N | X_1^1 = 1 \text{ and } X_1^2 \geq 2] = \mathbb{P}[t > N-1] \,. \tag{11}$$

The same holds for the third case, so (9) becomes

$$\begin{aligned}\mathbb{P}[t > N] &= \mathbb{P}[t > N-1]\mathbb{P}[\max(X_1^1, X_1^2) \geq 2] \\ &= p(2-p)\mathbb{P}[t > N-1] .\end{aligned}$$

The theorem follows. □

For the geometric distribution the magician can improve his chances by always selecting the first card. Let t' denote the coupling time for this process where the subject draws his secret number from $\mathcal{G}_p$. Then one finds by a similar calculation that

$$\mathbb{P}[t' > N] = p(p(2-p))^{N-1} = p^N(2-p)^{N-1} , \tag{12}$$

which is smaller than (6) by a factor $\frac{1}{2-p}$.

4 Uniform Distribution

Consider a deck of N cards having a uniform i.i.d. distribution of card values drawn from $[1,\ B]$. We estimate $\mathbb{P}[t > N]$ where t is the coupling time assuming that both the magician and the subject draw a secret value *uniformly* from $[1,\ B]$.

For our analysis we introduce two auxiliary finite state Markov chains. The first of these is a chain $\mathcal{L}_B$ that we call the *leapfrog chain.* View the subject and magician as performing the Kruskal counting procedure on two independently drawn decks. The subject will use a white pebble to mark the location of key cards and the magician will use a black pebble, according to their decks, and simultaneously each moves to their respective first key card. After this is done, the person having his pebble furthest behind in the deck moves it to his next key card. In case of a tie, where both pebbles are in the same relative position in the deck, a move consists of both persons simultaneously moving their pebbles to their next key cards, respectively. (Since the players have separate decks, the next key card values of the two players need not be the same.) The states of the chain $\mathcal{L}_B$ represent the distance the white pebble is currently ahead of or behind the black pebble in the card numbering, so there are $2B-1$ states i with $-(B-1) \leq i \leq B-1$. A transition occurs whenever a pebble is moved; a transition from state 0 corresponds to both pebbles moving (independently), while a transition from any other state corresponds to exactly one pebble being moved. A transition often involves one pebble leapfrogging over the other, hence the choice of name for $\mathcal{L}_B$. The transition probabilities p_{ij} are determined by the uniform distribution on card values. For $i \neq 0$ the transition from i to j is determined by the value v of the key card by

$$v = \text{sign}(i)(i-j) , \tag{13}$$

so that

$$p_{ij} = \begin{cases} \frac{1}{B} & \text{if } 1 \le \operatorname{sign}(i)(i-j) \le B, \\ 0 & \text{otherwise,} \end{cases} \tag{14}$$

while for $i = 0$ the transition probabilities are

$$\pi_j := p_{0j} = \frac{B - |j|}{B^2}. \tag{15}$$

This chain is ergodic, and it is easy to check that π_j in (15) gives the stationary distribution for $\mathcal{L}_B$. Table 1 gives the state transition matrix $[p_{ij}]$ for $\mathcal{L}_4$.

Now consider the case that the subject and magician perform the Kruskal counting procedure on the *same* deck. As long as their sequences of key cards remain disjoint, these key card values are independent random variables, and their relative positions of current key cards are described by transitions of the leapfrog chain. This persists until they have a key card in common, i.e., until the state 0 is reached on the leapfrog chain. Thus $\mathbb{P}[t > N]$ corresponds to the probability of those sequences of transitions in the leapfrog chain starting from 0 that avoid the 0 state until one pebble has moved to a position beyond N. We can keep track of sequences that never visit 0 by forming the *reduced leapfrog chain* $\bar{\mathcal{L}}_B$ obtained by deleting the 0 state and assigning new transition probabilities

$$\bar{P}_{ij} := (1 - p_{i0})^{-1} p_{ij}. \tag{16}$$

For $\mathcal{L}_B$ the probability of going to 0 is a constant, hence

$$\bar{P}_{ij} = \left(1 - \frac{1}{B}\right)^{-1} p_{ij}, \tag{17}$$

so that all values $\bar{P}_{ij}$ are either $\frac{1}{B-1}$ or 0. Table 2 gives the state transition probabilities $[P_{ij}]$ for $\bar{\mathcal{L}}_4$.

The *initial state distribution* on the reduced leapfrog chain $\bar{\mathcal{L}}_B$ corresponds to that obtained after one transition of the leapfrog chain from the 0 state, conditioned

Table 1 Leapfrog chain $\mathcal{L}_4$

		Exit state j						
		−3	−2	−1	0	1	2	3
	3	0	0	1/4	1/4	1/4	1/4	0
	2	0	1/4	1/4	1/4	1/4	0	0
Entering	1	1/4	1/4	1/4	1/4	0	0	0
state	0	1/16	2/16	3/16	1/4	3/16	2/16	1/16
i	−1	0	0	0	1/4	1/4	1/4	1/4
	−2	0	0	1/4	1/4	1/4	1/4	0
	−3	0	1/4	1/4	1/4	1/4	0	0

Table 2 Reduced Leapfrog chain $\bar{\mathcal{L}}_4$

		Exit state j					
		−3	−2	−1	1	2	3
	3	0	0	1/3	1/3	1/3	0
	2	0	1/3	1/3	1/3	0	0
Entering	1	1/3	1/3	1/3	0	0	0
state	−1	0	0	0	1/3	1/3	1/3
i	−2	0	0	1/3	1/3	1/3	0
	−3	0	1/3	1/3	1/3	0	0

on not staying at 0. Here let (i_b, i_w) denote the initial key card positions, yielding $j = i_b - i_w$ as the next state of the leapfrog chain, and we condition on $j \neq 0$. The resulting distribution is

$$\bar{\pi}_j := \left(1 - \frac{1}{B}\right)^{-1} \frac{B - |j|}{B^2}, \quad \text{for } 1 \leq |j| \leq B - 1. \tag{18}$$

This chain $\bar{\mathcal{L}}_B$ is ergodic and has $\bar{\pi}_j$ as its stationary distribution.

We next define a random variable $\bar{t}_{N,B}$ which counts the total number of key cards produced during the Kruskal count by the subject and magician, up to and including the first key card that occupies a position exceeding N, provided that no collision of key cards occurs until after time N. We call $\bar{t}_{N,B}$ the *travel time beyond position N*. (This random variable is undefined if a collision occurs by position N, and the trick works.) This number equals 2 plus the number of moves of the reduced leapfrog chain $\bar{\mathcal{L}}_B$ until some key card exceeds N. To determine this number from the reduced leapfrog chain sequence of states, we need to know additionally the *top key card* $i_t := \min(i_b, i_w)$, which is the position of the key cards of the magician and subject that is closest to the top of the deck. Given that initially the chain is in state $j = i_b - i_w \neq 0$, the conditional probability r_{ij} that the top key card is in position i_t is

$$r_{ij} = \frac{1}{B - |j|} \quad 1 \leq i \leq B - |j|, \tag{19}$$

and is 0 otherwise. The position of the top key card together with the sequences of successive states of the reduced $\bar{\mathcal{L}}_B$ allow the reconstruction of all moves during the Kruskal count, and the determination of the travel time $\bar{t}_{N,B}$.

Lemma 4.1. *If $N \geq B$ then*

$$\mathbb{P}[t > N] = \sum_{j=1}^{N} \left(1 - \frac{1}{B}\right)^{j-1} \mathbb{P}[\bar{t}_{N,B} = j]. \tag{20}$$

Proof. The event $[t > N]$ consists of all sequences of state transitions in the leapfrog chain $\mathcal{L}_B$ starting at state 0 that never return to 0 before some pebble moves to a position $\geq N + 1$. Such a sequence of transitions is matched (after the first move) by a corresponding sequence of state transitions in the reduced leapfrog chain $\bar{\mathcal{L}}_B$. The probabilities between $\mathcal{L}_B$ and $\bar{\mathcal{L}}_B$ differ by a multiplicative factor $\left(1-\frac{1}{B}\right)$, and these weights appear for each step of the reduced leapfrog chain. The steps of the reduced leapfrog chain do not count the first step of the leapfrog chain (which moves two pebbles), so there is one less factor of $\left(1-\frac{1}{B}\right)$ than $\bar{t}_{N,B}$ counts. □

Lemma 4.1 is useful because the distribution of the travel time $\bar{t}_{N,B}$ is strongly peaked and relatively tractable to estimate. Since no move of a pebble is larger than B, and since both pebbles are within B cards of the N-th card at the stopping time $\bar{t}_{N,B}$, one has

$$\bar{t}_{N,B} \geq \frac{2N}{B} - 1 \,. \tag{21}$$

Lemma 4.1 then yields

$$\mathbb{P}[t > N] \leq \left(1 - \frac{1}{B}\right)^{\frac{2N}{B}-2} . \tag{22}$$

This shows the (well-known) fact that $\mathbb{P}[t > N]$ decreases exponentially as a function of N.

Using large-deviation theory we obtain the following asymptotic behavior of $\mathbb{P}[t > N]$ as $N \to \infty$.

Theorem 4.1. *For fixed B there is a positive constant α_B such that*

$$\mathbb{P}[t > N] = \exp(-\alpha_B N + o(N)) \tag{23}$$

as $N \to \infty$.

We relegate the proof of this result to the appendix, where we also give a formula for α_B in (52). We easily obtain from (22) the inequality

$$\alpha_B \geq \left(\frac{2}{B}\right) \left|\log\left(1 - \frac{1}{B}\right)\right| = \frac{2}{B^2} + O\left(\frac{1}{B^3}\right) . \tag{24}$$

It is intuitively clear that the expected value of a key card is $\geq \frac{B}{2}$ in all states, hence one expects that $\mathbb{P}\left[\bar{t}_{N,B} \leq \frac{N}{B}\right] \geq \frac{1}{2}$, which with Lemma 4.1 would imply that $\alpha_B \leq \frac{4}{B^2} + O(\frac{1}{B^3})$. Theorem 4.2 below shows that $B^2\alpha_B \to 4$ as $B \to \infty$, see (33).

We next obtain upper and lower bounds for $\mathbb{P}[\bar{t}_{N,B} > k]$ by approximating the reduced leapfrog chain $\bar{\mathcal{L}}_B$ with two simpler Markov chains $\mathcal{L}_B^+$ and $\mathcal{L}_B^-$, as follows. These chains both describe the leapfrog motion of two colored pebbles at most B units apart, with the states representing the current distance the white pebble is ahead:

(1) In $\mathcal{L}_B^-$ the pebble further behind jumps v units with v drawn uniformly from the range $[1,\ B-1]$.
(2) In $\mathcal{L}_B^+$ the pebble further behind jumps v units with v drawn uniformly from the range $[2,\ B]$.

The chain $\mathcal{L}_B^-$ has $2B-1$ states labelled by $|i| \leq B-1$, while the chain $\mathcal{L}_B^+$ has $2B+1$ states labelled by $|i| \leq B$. Both these chains can occupy the state 0, i.e., they permit collisions, and from the state 0 only one pebble is moved at the next transition, namely the pebble that did not move in arriving at the 0 state. Both chains $\mathcal{L}_B^+$ and $\mathcal{L}_B^-$ have the property that the card values drawn are independent of the current state. For the chain $\mathcal{L}_B^-$ we define a *travel time* $t_{N,B}^-$ beyond position N, which is obtained by starting the chain in state 0, with both pebbles in position 0, associating a movement of pebbles on a line with each state transition, and counting the total number of state transitions up to and including the first time that a pebble is moved beyond position N. For the chain $\mathcal{L}_B^+$ we define a *travel time* $t_{N,B}^+$ beyond position N similarly. Note that these travel times are defined even when collisions occur before time N; the pebble locations may separate again after such collisions.

Lemma 4.2. *For all* N, B *and* k, *one has*

$$\mathbb{P}[t_{N,B}^+ > k] \;\geq\; \mathbb{P}[\bar{t}_{N,B} > k] \;\geq\; \mathbb{P}[t_{N,B}^- > k]\,. \tag{25}$$

We give the proof of Lemma 4.2 in the appendix. Lemmas 4.1 and 4.2 when combined yield the bounds

$$P_{N,B}^+ \geq \mathbb{P}[t > N] \geq P_{N,B}^-\,, \tag{26}$$

where

$$P_{N,B}^{\pm} := \sum_{j=1}^{N} \left(1 - \frac{1}{B}\right)^{j-1} \mathbb{P}[t_{N,B}^{\pm} = j]\,. \tag{27}$$

The simple form of the chains $\mathcal{L}_B^+$ and $\mathcal{L}_B^-$ allows the asymptotic behavior of $P_{N,B}^+$ and $P_{N,B}^-$ to be explicitly determined, as follows.

Theorem 4.2. *For fixed* B *as* $N \to \infty$ *one has*

$$P_{N,B}^{\pm} = \exp(-\alpha_B^{\pm} N + o(N)) \tag{28}$$

for explicit constants $\alpha_B^{\pm}$. *Here* $\frac{1}{2}\alpha_B^-$ *is the unique root* α *of*

$$\sum_{i=1}^{B-1} \exp(i\alpha) = B\,, \tag{29}$$

and $\frac{1}{2}\alpha_B^+$ *is the unique root of*

$$\sum_{i=1}^{B-1} \exp((i+1)\alpha) = B. \tag{30}$$

As $B \to \infty$ these quantities satisfy

$$\alpha_B^+ = \frac{4}{B^2} - \frac{20/3}{B^3} + O(B^{-4}), \tag{31}$$

$$\alpha_B^- = \frac{4}{B^2} + \frac{4/3}{B^3} + O(B^{-4}). \tag{32}$$

The proof of this result is given in the appendix. Theorem 4.1 together with the inequalities (26) shows that for large B one has

$$\mathbb{P}[t > N] = \exp\left(-\left(\frac{4}{B^2} + O\left(\frac{1}{B^3}\right)\right)(1+o(1))N\right) \tag{33}$$

as $N \to \infty$.

We relate these results to the model of Haga and Robins (1997). The Markov chain studied by Haga and Robins is obtained from the leapfrog chain by identifying states k and $-k$ for all $k \geq 1$; thus it has exactly B states. The resulting chain factors out the action of the involution sending k to $-k$ under which the chain probabilities are invariant, and this loses the "leapfrog" information which is necessary for computing exact coupling probabilities. Haga and Robins estimate instead the probability of avoiding absorption in the absorbing state 0 in the first M transitions of the resulting factor chain. This probability asymptotically decays like $O((\lambda_B)^M)$ as $M \to \infty$, where λ_B is the modulus of the second largest eigenvalue of the characteristic polynomial of their Markov chain. The characteristic polynomial of the transition matrix of the Haga–Robins Markov chain is $p_B(x) := (x+1/B)^B - (1+1/B)^B x^{B-1}$, and it can be shown that the modulus of its second largest eigenvalue satisfies

$$\lambda_B = 1 - \frac{2}{B} + O\left(\frac{1}{B^2}\right) \tag{34}$$

as $B \to \infty$. To relate λ_B to the asymptotic coupling probability decay rate $\exp(-\alpha_B)$ in Theorem 4.1, we note that the expected size of a step in the Haga–Robins chain is about $B/2$, so that after M steps the location of the chain should be around the position $N \approx MB/2$. One should therefore compare $(\lambda_B)^{2/B}$ and $\exp(-\alpha_B)$, and one finds that both of these quantities are asymptotic to $1 - \frac{4}{B^2} + O(\frac{1}{B^3})$ as $B \to \infty$, using (34) and Theorem 4.2.

5 Numerical Results: The Kruskal Count

We compare predictions obtained from the two models studied in this paper with the performance of the actual Kruskal Count trick.

For the actual Kruskal count we consider a standard deck of 52 cards, and we assume that the subject draws a key card using a uniform distribution from the set of available key card values. We study the effects of varying the magician's strategy on the success probability of the Kruskal Count trick. The magician has the freedom to choose his key card, and he also has the extra freedom to specify a rule for assigning values to the "face cards" J, Q, K. We study three possible variants:

1. Assign the values 11, 12, 13 to J, Q, K, respectively.
2. Assign the value 10 to each of J, Q, K,
3. Assign the value 5 to each of J, Q, K.

The first two of these variants are presented as "straw men" useful for comparison with the models of this paper. To obtain numerical values for the Kruskal count trick we used a Monte Carlo simulation with 10^6 trials for each data point. For simulations of the i.i.d. uniform deck distribution, an "exact" calculation was done using an enlarged Markov chain which keeps a running total of the value of the position N of the leading pebble, and enters an absorbing final state whenever a pebble jumps past the end of the deck. Since the smallest step size is 1, this chain reaches an absorbing state after a number of steps equal to the size of the deck; consequently, it suffices to compute the state of the chain after that number of steps. Simulations of the i.i.d. "semiuniform" distributions for variants 2 and 3 were done similarly to the i.i.d. uniform case.

Variant 1 corresponds to the uniform distribution on $\{1,2,\ldots,13\}$. The average key card size is 7. We therefore consider as an approximation the i.i.d. geometric deck distribution with $p = \frac{6}{7}$, which has mean key card size 7. According to Theorem 3.1, the failure probability for a magician drawing his first key card from a deck of $N = 52$ cards according to the geometric distribution is

$$FP(\mathcal{G}_a) = \left(\frac{6}{7}\right)^{52}\left(\frac{8}{7}\right)^{52} = 0.342254. \tag{35}$$

If the magician chooses the first card to be his first key card, by (3.5) his failure probability for $N = 52$ is

$$FP(\mathcal{G}'_a) = \frac{7}{8} FP(\mathcal{G}_a) = 0.299472 \tag{36}$$

Table 3 presents data for variant 1 for the Kruskal Count and the i.i.d. uniform deck distribution on $\{1,2,\ldots,13\}$. The table gives failure probabilities in which the magician's strategy is to choose as first key card the j-th card, for $1 \le j \le 13$, plus a final row that gives the failure probability when the magician draws a card uniformly in $\{1,2,\ldots,13\}$. The data in Table 3 show that the magician does best to choose $j = 1$ as his key card. The non-Markovian nature of the actual deck causes the failure probabilities to differ from the i.i.d. uniform deck distribution; the effect is a decrease of about 0.3%. We also see that the failure probability for the i.i.d. geometric distribution is an overestimate of the failure probability for the Kruskal Count when the magician picks a random card as first key card, and underestimates the failure probability when the magician picks the first card as key card.

We next consider variants 2 and 3. For variant 2 the expected key card size is $\frac{85}{13}$, so for comparison we consider the i.i.d. geometric deck distribution with $p = \frac{72}{85}$. If

Table 3 Failure probabilities for variant 1

	Kruskal	Uniform
1	0.315180	0.319486
2	0.318564	0.322994
3	0.321975	0.326492
4	0.325298	0.329981
5	0.328794	0.333461
6	0.332235	0.336929
7	0.336055	0.340385
8	0.339264	0.343827
9	0.342758	0.347251
10	0.346780	0.350655
11	0.349464	0.354034
12	0.353367	0.357385
13	0.357044	0.360703
Avg.	0.335906	0.340276

the magician chooses his first key card according to the same geometric distribution, then the failure probability is

$$FP(\mathcal{G}_b) = \left(\frac{72}{85}\right)^{52} \left(\frac{98}{85}\right)^{52} = 0.292064, \tag{37}$$

while if the magician draws the first card as his key card, then

$$FP(\mathcal{G'}_b) = \frac{85}{98} FP(\mathcal{G}_b) = 0.253320. \tag{38}$$

For variant 3 the expected key card size is $\frac{70}{13}$, so for comparison we consider the i.i.d. geometric deck distribution with $p = \frac{57}{70}$. If the magician chooses his first key card with the same geometric distribution, then the failure probability is

$$FP(\mathcal{G}_c) = \left(\frac{57}{70}\right)^{52} \left(\frac{83}{70}\right)^{52} = 0.161197, \tag{39}$$

while if the magician chooses the first card as his key card, the failure probability is

$$FP(\mathcal{G'}_c) = \frac{70}{83} FP(\mathcal{G}_c) = 0.135949. \tag{40}$$

Table 4 presents failure probability data for variants 2 and 3 for the Kruskal Count and for the i.i.d. semiuniform deck distributions which have the card values $\{1,2,\ldots,10\}$ chosen with the same probabilities as variants 2 and 3 impose on the actual deck. The non-Markovian nature of the actual deck results in the Kruskal count failure probabilities differing from the corresponding i.i.d. deck distributions; they are smaller by about 0.6%. The failure probability for the i.i.d. geometric distribution when the magician chooses the first card as first key card gives

Table 4 Failure probabilities for variants 2 and 3

	Kruskal 2	Semiuniform 2	Kruskal 3	Semiuniform 3	Uniform
1	0.277869	0.284060	0.146238	0.152658	0.150944
2	0.280756	0.287235	0.148801	0.155266	0.153684
3	0.284330	0.290447	0.151204	0.157847	0.156407
4	0.287163	0.293623	0.153736	0.160399	0.159109
5	0.290317	0.296782	0.156075	0.162918	0.161789
6	0.293557	0.299920	0.159744	0.166357	0.164444
7	0.296910	0.303034	0.162474	0.168973	0.167070
8	0.300023	0.306118	0.164977	0.171553	0.169665
9	0.303194	0.309171	0.167735	0.174094	0.172225
10	0.306383	0.312185	0.170064	0.176591	0.174747
Avg.	0.292050	0.298258	0.158105	0.164666	0.163008

an underestimate for the failure probabilities of the Kruskal Count in variants 2 and 3. The numerical results show that the magician should choose the first card as his key card. The effect of the choice of the magician's key card on the failure probability is small, at most 2.5%. In comparing variants 2 and 3 we see that the choice to have face cards take the value 5 rather than 10 has a much larger effect on the failure probability than the magician's choice of first key card position. The final column of Table 4 presents the failure probabilities for the i.i.d. uniform deck distribution on $\{1,2,\ldots,10\}$. One expects this i.i.d. uniform distribution to be comparable with rules variation 3 rather than 2, because the expected key card size is similar to case 3. (The Kruskal Count 3 mean value is slightly lower.)

To conclude: The variants that counts face cards as having value 5 rather than 10 is important to the practical success of the Kruskal Count trick; the choice of the first card as key card offers a further small improvement in success probability.

Appendix: Proofs of Theorem 4.1, Lemma 4.2 and Theorem 4.2

Proof of Theorem 4.1. Let B be fixed. In view of Lemma 4.1, one has

$$M_N \leq \mathbb{P}[t < N] \leq N M_N, \tag{41}$$

where

$$M_N := \max_{1\leq k\leq N} \left\{ \left(1 - \frac{1}{B}\right)^k \mathbb{P}\left[\bar{t}_{N,B} = k\right] \right\}. \tag{42}$$

It suffices to show that there is a positive constant α_B such that $M_N = \exp(-\alpha_B N + o(N))$ as $N \to \infty$.

We note that the travel time $\bar{t}_{N,B}$ beyond position N depends on the successive transitions of the chain $\bar{\mathcal{L}}_B$. We convert this to a problem about successive states of the *jump chain* $\bar{\mathcal{L}}_B^J$ having $2B(B-1)$ states which correspond to all possible transitions of the chain $\bar{\mathcal{L}}_B$. A jump chain state $(i,\ v)$ will mean state i of $\bar{\mathcal{L}}_B$ together with an allowable key card value v which determines the next state of $\bar{\mathcal{L}}_B$. The allowable values are $1 \le v \le B$ with $v \ne |i|$. The transition probability $P_{s,s'}$ from $s = (i,\ v)$ to $s' = (j,\ v')$ is $\frac{1}{B-1}$ when j is uniquely determined by (13) and $1 \le v' \le B$ with $v' \ne j$, and is 0 otherwise.

We let $\{(i_k,\ v_k) : k = 1,2,\ldots\}$ denote a sequence of states of $\bar{\mathcal{L}}_B^J$, and introduce the *modified travel time*

$$\tilde{t}_{N,B} := \min\{k : v_1 + \ldots + v_k \ge 2N\}. \tag{43}$$

It is easy to show that

$$\bar{t}_{N,B} \le \tilde{t}_{N,B} \le \bar{t}_{N,B} + B, \tag{44}$$

and this yields

$$\mathbb{P}[\tilde{t}_{N,B} \le k] \le \mathbb{P}[\bar{t}_{N,B} \le k] \le \mathbb{P}[\tilde{t}_{N,B} \le k - B]\ . \tag{45}$$

Now, for a fixed $0 < \gamma \le 1$, the quantity $\mathbb{P}[\tilde{t}_{N,B} \le \gamma N]$ can be estimated using large deviation theory, as a special case of Theorem 1 of Donsker and Varadhan (1975), see also Varadhan (1984) for the general theory.

Theorem A.1. *For fixed B and fixed γ with $0 < \gamma < 1$, one has as $N \to \infty$,*

$$\mathbb{P}[\tilde{t}_{N,B} \le \gamma N] = \exp(-f_B(\gamma)N + o(N)), \tag{46}$$

where the function $f_B(\gamma) := \gamma I^*(\gamma)$ where

$$I^*(\gamma) := \inf\{I(\mu) : \text{weight}(\mu) \ge \frac{2}{\gamma}\}\ . \tag{47}$$

Here μ runs over the set of probability measures on the state space S of the chain $\bar{\mathcal{L}}_B^J$, and

$$\text{weight}(\mu) := \sum_{s=(i,\ v)\in S} v\mu((i,\ v)), \tag{48}$$

gives the expected value of a card drawn using the measure μ, and

$$I(\mu) := -\inf_u \left\{ \sum_{s\in S} \log\left(\frac{\pi_u(s)}{u(s)}\right) \mu(s) : u : S \to \mathbb{R}^+ \right\}, \tag{49}$$

where $\pi_u(s) := \sum_{s'\in S} p_{s,s'} u(s')$.

This theorem computes a large deviations rate for an additive functional determined from pairs of successive states of an (irreducible, aperiodic) finite Markov chain, a situation already treated by Miller (1961), prior to the development of the general theory of large deviations. An important feature in the (irreducible, aperiodic) finite Markov chain case is that the function $f_B(\gamma)$ is a continuous (real-analytic), strictly convex function, in the interval where it is finite (Miller, 1961, Theorem 2). In Theorem A.1 the function $f_B(\gamma) = +\infty$ for $0 \le \gamma \le \frac{2}{B}$ by virtue of

(21), $f_B(\gamma)$ is positive and finite on $\frac{2}{B} < \gamma < 1$ and one can show $f_B(1) = 0$. The constant given in the $o(N)$ error term in (46) depends on γ, and can be made uniform on $[\frac{2}{B}, 1]$. Sharper bounds on the error are available in the literature, see the treatment of large deviations for additive Markov processes in Iscoe, Ney, and Nummelin (1985, Example 7(ii)) and Ney and Nummelin (1987a 1987b); strict convexity of the rate function in more generality appears in Ney and Nummelin (1987b, Lemma 3.3).

Combining Theorem A.1 and (45) with $k = \gamma N$ we see that for fixed γ the quantities $\bar{t}_{N,B}$ and $\tilde{t}_{N,B}$ have the same asymptotic behavior, with

$$\mathbb{P}[\bar{t}_{N,B} \leq \gamma N] = \exp(-f_B(\gamma)N + o(N)) \tag{50}$$

as $N \to \infty$. Now we have from (42) that

$$\begin{aligned} \log M_N &= \max_{1 \leq k \leq N} \left\{ k \log\left(1 - \frac{1}{B}\right) + \log \mathbb{P}[\bar{t}_{N,B} = k] \right\} \\ &= \max_{0 < \gamma < 1} \left\{ \gamma N \log\left(1 - \frac{1}{B}\right) - f_B(\gamma)N + o_\gamma(N) \right\}. \end{aligned} \tag{51}$$

We now define the desired constant by the formula

$$-\alpha_B := \sup_{0 < \gamma < 1} \left[\gamma \log\left(1 - \frac{1}{B}\right) - f_B(\gamma) \right]. \tag{52}$$

Here strict convexity of $f_B(\gamma)$ guarantees a unique point γ_B attaining the supremum. We derive from (51) and (52) the estimate

$$M_N = \exp(-\alpha_B N + o(N)) \tag{53}$$

as $N \to \infty$, provided one shows that the error term $o_\gamma(N)$ in (51) is bounded uniformly in $0 < \gamma < 1$. This can be done; we omit details. The bound of Theorem 4.1 follows. □

Proof of Lemma 4.2. We first compare the reduced leapfrog chain $\bar{\mathcal{L}}_B$ with the chain $\mathcal{L}_B^-$, to prove the right side inequality in (25). Let $q_{ij}^-(k)$ and $q_{ij}(k)$ denote the probability distributions of the locations (i, j) of the white and black pebbles after k pebble moves, of $\mathcal{L}_B^-$ and $\bar{\mathcal{L}}_B$, respectively. We use the convention that the initial state $\bar{\mathcal{L}}_B$ (plus a choice of top key card i_t) counts as 2 pebble moves. Note that

$$\mathbb{P}[\bar{t}_{N,B} > k] = 1 - \sum_{\substack{i \leq N \\ j \leq N}} q_{ij}(k) \tag{54}$$

and

$$\mathbb{P}[t_{N,B}^- > k] = 1 - \sum_{\substack{i \leq N \\ j \leq N}} q_{ij}^-(k). \tag{55}$$

Therefore it suffices to prove that the majorization inequalities

$$\sum_{\substack{i \le i_0 \\ j \le j_0}} q_{ij}^-(k) \ge \sum_{\substack{i \le i_0 \\ j \le j_0}} q_{ij}(k)\,, \quad \text{all} \quad i_0 \ge 1,\ j_0 \ge 1, \tag{56}$$

hold for all $k \ge 2$. Comparing these formulas (56) for $i_0 = j_0 = N$ with (55) and (54), yields

$$\mathbb{P}[\bar{t}_{N,B} > k] \ge \mathbb{P}[t_{N,B}^- > k] \tag{57}$$

for all $k \ge 2$, which is the desired right side inequality of (25). We now establish (56) by induction on k, using a stochastic dominance argument. The following is the base case $k = 2$ of the induction.

Claim. *There is a mapping ϕ^- of the probability mass $q_{ij}(2)$ on $(i,\ j)$ for $\bar{\mathcal{L}}_B$ to various (i', j') having $i' \le i$ and $j' \le j$ whose image is the distribution $q_{ij}^-(2)$.*

To prove the claim, we have by definition

$$q_{ij}^-(2) = \begin{cases} \frac{1}{(B-1)^2} & 1 \le i, j \le B-1, \\ 0 & \text{otherwise;} \end{cases} \tag{58}$$

$$q_{ij}(2) = \begin{cases} \frac{1}{B(B-1)} & 1 \le i, j \le B \text{ with } i \ne j, \\ 0 & \text{otherwise.} \end{cases} \tag{59}$$

One can find by hand a direct shifting of mass of (58)–(59) to establish the claim; we omit details. It is well known that a coordinate-monotone probability rearrangement ϕ^- as in the claim is equivalent to a two-dimensional majorization inequality, see Marshall and Olkin (1979), which here are exactly the majorization formulas (56) for $k = 2$, and all (i_0, j_0).

Now let $(i,\ j) \succ (i',\ j')$ mean $\min(i,j) \ge \min(i',j')$ and $\max(i,j) \ge \max(i',j')$, i.e., the two pebbles $(i,\ j)$ are both moved further along the line than $(i',\ j')$, ignoring their colors. The claim establishes for $k = 2$ a (stochastic) pairing of pebble positions $(i,\ j) \xrightarrow{\phi} (i',\ j')$ such that $(i,\ j) \succ (i',\ j')$ between $\bar{\mathcal{L}}_B$ and $\mathcal{L}_B^-$. For each subsequent move, both chains have $B-1$ possible transitions with probabilities each $\frac{1}{B-1}$. For $\bar{\mathcal{L}}_B$ in state $k+i-j$ the set of admissible values of the next move is $\{1,\ 2, \ldots, B\} - \{|k|\}$. We map these transitions to transitions of $\mathcal{L}_B^-$ in linear order, with a mapping $\psi_{|k|}$ having $\psi_{|k|}(i) = i$ for $i \le |k|$ and $\phi_j(i) = i-1$ for $i \ge |k|+1$. One easily sees that if pebbles in $\bar{\mathcal{L}}_B$ are at $(i_1,\ j_1)$ and the corresponding ones are at $(i_2,\ j_2)$ with $(i_1,\ j_1) \succ (i_2,\ j_2)$, and if the pebble closer to the origin is moved i resp. $\phi_{|k|}(i)$ for the two chains resulting in positions $(i_1^*,\ j_1^*)$, $(i_2^*,\ j_2^*)$ then $(i_1^*,\ j_1^*) \succ (i_2^*,\ j_2^*)$. This gives a stochastic pairing of pebble positions at all subsequent moves, with both pebbles of $\mathcal{L}_B^-$ always being behind those of $\bar{\mathcal{L}}_B$ in the ordering $\succ$. Consequently we deduce by induction on k that the majorization inequalities (56) hold for all $k \ge 2$.

We next compare $\bar{\mathcal{L}}_B$ and $\mathcal{L}_B^+$, to deduce the left side inequality in (25). This is proved in similar fashion. Here we have

$$q_{ij}^-(2) = \begin{cases} \frac{1}{(B-1)^2} & 2 \le i, j \le B, \\ 0 & \text{otherwise.} \end{cases} \tag{60}$$

If $q_{ij}^+(t)$ is the probability that the pebbles are at $(i,\ j)$ after t steps, then we obtain similarly the majorization inequalities

$$\sum_{\substack{i \le i_0 \\ j \le j_0}} q_{ij}(t) \ge \sum_{\substack{i \le i_0 \\ j \le j_0}} q_{ij}^+(t)\,, \quad \text{all } i_0 \ge 1\,,\ j_0 \ge 1, \tag{61}$$

for all $t \ge 2$. □

Proof of Theorem 4.2. We let $A(N) \approx B(N)$ mean $A(N) = B(N)^{1+o(1)}$ as $N \to \infty$.

Consider first $P_{N,B}^-$. Let $\tilde{t}_B^-(M)$ denote the travel time for the chain $\mathcal{L}_B^-$, which counts the number of transitions up to and including the transition at which the sum of the jumps of the chain exceeds M. Then for any fixed sequence of transitions

$$t_B^-(2N) \ge t_{N,B}^- \ge t_B^-(2N - N)\,. \tag{62}$$

Hence

$$P_N^- \le N \max_{1 \le j \le N} \left\{ \left(1 - \frac{1}{B}\right)^{j-1} \mathbb{P}[\tilde{t}_B^-(2N) = j] \right\} \tag{63}$$

and

$$P_N^- \ge N \max_{1 \le j \le N} \left\{ \left(1 - \frac{1}{B}\right)^{j-1} \mathbb{P}[\tilde{t}_B^-(2N - B) = j] \right\}. \tag{64}$$

It is easy to check that

$$\begin{aligned} Q_N^- &:= \max_{1 \le j \le N} \left\{ \left(1 - \frac{1}{B}\right)^{j-1} \mathbb{P}[\tilde{t}_B^-(2N) = j] \right\} \\ &\approx \max_{1 \le j \le N} \left\{ \left(1 - \frac{1}{B}\right)^{j-1} \mathbb{P}[\tilde{t}_B^-(2N - B) = j] \right\} \end{aligned} \tag{65}$$

using $\mathbb{P}[t_B^-(2N-B) \ge j] \ge \mathbb{P}[t_B^-(2N) \ge j + B]$. This shows that it suffices to asymptotically estimate Q_N^-. Since the sizes of the steps of the chain $\mathcal{L}_B^-$ are identically distributed independent of which state we are in the chain, one obtains

$$Q_N^- \approx \sup_{0 \le \gamma \le 1} \left\{ B^{-\gamma N} \binom{\gamma N}{\gamma_1 N, \gamma_2 N, \ldots, \gamma_{B-1} N} : \gamma_i \ge 0\,,\ \sum_{i=1}^{B-1} \gamma_i = \gamma\,,\ \sum_{i=1}^{B-1} i\gamma_i = 2 \right\}. \tag{66}$$

(Here $j \approx \gamma N$ and $B^{-\gamma N}$ arises as $\left(1-\frac{1}{B}\right)^{\gamma N}(B-1)^{-\gamma N}$.) Using Stirling's formula to estimate the multinomial coefficient, one obtains $Q_N^- \approx \exp(Z^- N)$ where Z^- is the optimal value of the constrained maximization problem (M^-) given by:

$$\text{maximize } Z = -\gamma \log B + \gamma \log \gamma - \sum_{i=1}^{B-1} \gamma_i \log \gamma_i \tag{67}$$

subject to

$$\sum_{i=1}^{B-1} i\gamma_i = 2, \tag{68}$$

$$\sum_{i=1}^{B-1} \gamma_i = \gamma, \tag{69}$$

$$\gamma_i \geq 0 \quad \text{for } 1 \leq i \leq B-1. \tag{70}$$

To determine Z^-, introduce Lagrange multipliers λ_1, λ_2 for the two equality constraints and let

$$G = -\gamma \log B + \gamma \log \gamma - \sum_{i=1}^{B-1} \gamma_i \log \gamma_i + \lambda_1 \left(\sum_{i=1}^{B-1} i\gamma_i - 2 \right) + \lambda_2 \left(\sum_{i=1}^{B-1} \gamma_i - \gamma \right) \tag{71}$$

denote the Lagrangian. Necessary conditions for an interior extremal are

$$\frac{\partial G}{\partial \gamma} = -\log B + 1 + \log \gamma - \lambda_2 = 0, \tag{72}$$

$$\frac{\partial G}{\partial \gamma_i} = -1 - \log \gamma_i + i\lambda_1 + \lambda_2 = 0. \tag{73}$$

These yield

$$\gamma = B \exp(\lambda_2 - 1), \tag{74}$$

$$\gamma_i = \exp(\lambda_2 - 1)\exp(i\lambda_1), \quad 1 \leq i \leq B-1. \tag{75}$$

Substituting these expressions into (69) and cancelling $\exp(\lambda_2 - 1)$ from both sides yields

$$\sum_{i=1}^{B-1} \exp(i\lambda_1) = B, \tag{76}$$

which shows that $\lambda_1 = \frac{1}{2}\alpha_B^-$, as defined in the theorem statement as the unique real root of (29). (Uniqueness follows from strict convexity of the left side of (76) as a function of λ_1.) Substituting the same expressions into (68) and simplifying yields

$$\exp(\lambda_2 - 1) = \frac{2}{\sum_{i=1}^{B-1} i \exp(i\lambda_1)}. \tag{77}$$

Using this formula, (74) and (75) become

$$\gamma = \frac{2B}{\sum_{i=1}^{B-1} i\exp(i\lambda_1)}, \tag{78}$$

$$\gamma_i = \frac{2\exp(i\lambda_1)}{\sum_{i=1}^{B-1} i\exp(i\lambda_1)}. \tag{79}$$

Using these formulas the objective function value Z is evaluated at the maximum (with $F = \sum_{i=1}^{B-1} i\exp(i\lambda_1)$) as

$$\begin{aligned} Z^- &= \frac{2B}{F}\left(\log\frac{2B}{F} - \log B\right) - \sum_{i=1}^{B-1} \frac{2\exp(i\lambda_1)}{F}\log\left(\frac{2\exp(i\lambda_1)}{F}\right) \\ &= \frac{2B}{F}\log\frac{2}{F} - \frac{2}{F}\left(\sum_{i=1}^{B-1}\exp(i\lambda_1)\left[\log\frac{2}{F} + i\lambda_1\right]\right) \\ &= -\frac{2\lambda_1}{F}\left(\sum_{i=1}^{B-1} i\exp(i\lambda_1)\right) = -2\lambda_1 = -\alpha_B^-. \end{aligned} \tag{80}$$

Thus $Z^- = -\alpha_B^-$, and this gives $P_{N,B}^- \approx Q_N^- = \exp(-\alpha_B^- N(1+o(1)))$, which is (28) provided one establishes that the maximum of (M^-) occurs at an interior point where all $\gamma_i > 0$. We omit the details of checking that boundary extremals having some $\gamma_i = 0$ do not give the absolute maximum in (M^-).

The case of $P_{N,B}^+$ is handled by analogous arguments. One reduces it to finding the optimal value Z^+ of the constrained maximization problem (M^+) given by

$$\text{maximize } Z = -\gamma\log B + \gamma\log\gamma - \sum_{i=1}^{B-1}\gamma_i\log\gamma_i \tag{81}$$

subject to

$$\sum_{i=1}^{B-1}(i+1)\gamma_i = 2, \tag{82}$$

$$\sum_{i=1}^{B-1}\gamma_i = \gamma, \tag{83}$$

$$\gamma_i \geq 0 \quad \text{for } 1 \leq i \leq B-1. \tag{84}$$

Again $Z^+ = -2\lambda_1$ at the extremal point, and λ_1 is determined as the unique real root of

$$\sum_{i=1}^{B-1}\exp((i+1)\lambda_1) = B, \tag{85}$$

so that $\lambda_1 = \frac{1}{2}\alpha_B^+$, cf. (30).

Finally, the asymptotic formulae (31) and (32) as $B \to \infty$ are obtained from the formulas (29) and (30), by setting $\exp(\frac{1}{2}\alpha) = 1 + \frac{2}{B^2} + \frac{\delta}{B^3} + O(\frac{1}{B^4})$, in which δ is an unknown to be determined. We find it by noting that

$$\exp(i\alpha) = 1 + 2i\left(\frac{2}{B^2} + \frac{\delta}{B^3}\right) + \frac{4i^2}{B^4} + O\left(\frac{i}{B^4}\right), \tag{86}$$

substituting these formulas (29) and (30) and asking for the $O(\frac{1}{B})$ term to vanish. □

References

Aldous, D., & Diaconis P. (1986). Shuffling cards and stopping times. *American Mathematical Monthly, 93*, 333–348.

Diaconis, P. (1988). *Group representations in probability and statistics, IMS Lecture Notes – Monograph Series No. 11*. Hayward, CA: Institute of Mathematical Statistics.

Doeblin, W. (1938). Exposé de la theorie des chaines simple constantes de Markov á un nombre fini d'etats. *Revue Mathematique de l'Union Interbalkanique, 2*, 77–105.

Donsker, M. D., & Varadhan, S. R. S. (1975). Asymptotic evaluation of certain Markov process expectations for large time I. *Communications on Pure and Applied Mathematics, 28*, 1–47.

Fulves, C., & Gardner M. (1975). The Kruskal principle. *The Pallbearer's Review*, Vol. 10 No. 8 (June), 967–976.

Gardner, M. (1978). Mathematical games. *Scientific American, 238*(2), 19–32.

Gardner, M. (1988). *From Penrose tiles to Trapdoor ciphers*. New York: W. H. Freeman (Chap. 19).

Griffeath, D. (1978). Coupling methods for Markov processes. In G. C. Rota (Ed.) *Studies in probability and ergodic theory* (pp. 1–43). New York: Academic.

Haga, W., & Robins, S. (1997). On Kruskal's principle. In J. Borwein, P. Borwein, L. Jorgenson, & R. Corless (Eds.) *Organic mathematics, Canadian Mathematical Society Conference Proceedings* (Vol. 20, pp. 407–412). Providence, RI: AMS.

Iscoe, I., Ney, P., & Nummelin E. (1985) Large devations of uniformly recurrent Markov additive processes. *Advances in Applied Mathematics, 6*, 373–412.

Mallows, C. L. (1975). On a probability problem suggested by M. D. Kruskal's card trick. Bell Laboratories memorandum, April 18, 1975 (unpublished)

Marshall, A. W., & Olkin, I. (1979). *Inequalities: Theory of majorization and its applications*. New York: Academic.

Miller, H. D. (1961). A convexity property in the theory of random variables defined on a finite Markov chain. *Annals of Mathematical Statistics, 32*, 1260–1270.

Ney, P., & Nummelin, E. (1987a). Markov Additive processes I. Eigenvalue properties and Limit theorems. *Annals of Probability, 15*, 561–592.

Ney, P., & Nummelin, E. (1987b). Markov Additive processes II. Large deviations *Annals of Probability, 15*, 593–609.

Pitman, J. (1976). On coupling of Markov Chains. *Zeitschrift fur Wahrscheinlichkeitheorie, 35*, 315–322.

Varadhan, S. R. S. (1984). Large deviations and applications. In *CBMS-NSF Regional Conference No. 46*. SIAM.

Descending Dungeons and Iterated Base-Changing

David Applegate, Marc LeBrun, and N.J.A. Sloane

TO OUR FRIEND AND FORMER COLLEAGUE PETER FISHBURN, ON THE OCCASION OF HIS 70TH BIRTHDAY.

1 Introduction

The starting point for this paper was the question: what is the asymptotic behavior of the sequences

$$
\begin{aligned}
&10,\ 10_{11},\ 10_{11_{12}},\ 10_{11_{12_{13}}},\ \ldots,\\
&10,\ 11_{10},\ 12_{11_{10}},\ 13_{12_{11_{10}}},\ \ldots,
\end{aligned}
\tag{1}
$$

where, for real numbers $a, b > 1$, a_b (or, more conveniently although less graphically, a_b) denotes the result of interpreting a in base b instead of base 10? That is, if a is a real number > 1, with decimal expansion

$$a = \sum_{i=-\infty}^{k} c_i 10^i, \quad \text{for some } k \geq 0, \text{ all } c_i \in \{0, 1, \ldots, 9\}, \text{ and } c_k \neq 0, \tag{2}$$

and b is a real number > 1, then

$$a_b := a\_b := \sum_{i=-\infty}^{k} c_i b^i. \tag{3}$$

N.J.A. Sloane (✉)
AT&T Shannon Labs, 180 Park Ave., Florham Park, NJ 07932-0971, USA
e-mail: njas@research.att.com

S.J. Brams et al. (eds.), *The Mathematics of Preference, Choice and Order: Essays in Honor of Peter C. Fishburn*, Studies in Choice and Welfare,

We use text-sized subscripts in expressions like a_{b} to help distinguish them from symbols with ordinary subscripts. The sum in (3) converges, since

$$1 < a_{\text{b}} < 9b^{k+1}/(b-1), \tag{4}$$

and a_{b} is well-defined if we agree to avoid decimal expansions ending with infinitely many 9's. This restriction is needed, since (for example) $3_{\text{b}} = 3$ for any $b > 1$, whereas

$$2.999\ldots_{\text{b}} = 2 + \frac{9}{b} + \frac{9}{b^2} + \frac{9}{b^3} + \cdots = 2 + \frac{9}{b-1} \neq 3$$

unless $b = 10$. Equation (3) is meaningful for some values of a and $b \leq 1$, but to avoid exceptions we only consider $a, b > 1$. In this range a_b is a binary operation for which 10 is both a left and right unit.

In fact, since the iterated subscripts can be grouped either from the bottom upwards or from the top downwards, there are really four sequences to be considered (it is convenient to index these sequences starting at 10):

$$\begin{aligned}
(\alpha) = (\alpha_{10}, \alpha_{11}, \alpha_{12}, \ldots) &:= 10,\ 10_11,\ 10_(11_12),\ 10_(11_(12_13)), \ldots,\\
(\beta) = (\beta_{10}, \beta_{11}, \beta_{12}, \ldots) &:= 10,\ 10_11,\ (10_11)_12,\ ((10_11)_12)_13, \ldots,\\
(\gamma) = (\gamma_{10}, \gamma_{11}, \gamma_{12}, \ldots) &:= 10,\ 11_10,\ 12_(11_10),\ 13_(12_(11_10)), \ldots,\\
(\delta) = (\delta_{10}, \delta_{11}, \delta_{12}, \ldots) &:= 10,\ 11_10,\ (12_11)_10,\ ((13_12)_11)_10, \ldots.
\end{aligned}$$

Sequence (α), for example, begins

$$10,\ \ 10\_11 = 11,\ \ 10\_(11\_12) = 10\_13 = 13,$$

$$10\_(11\_(12\_13)) = 10\_(11\_15) = 10\_16 = 16,$$

$$10\_(11\_(12\_(13\_14))) = 10\_(11\_(12\_17)) = 10\_(11\_19) = 10\_20 = 20, \ldots.$$

The terms grow quite rapidly – see Table 1. These are now sequences A121263, A121265, A121295 and A121296 in Sloane (2008).

In Theorem 1 we will show that, if s_n is the nth term in any of the four sequences (α), (β), (γ) or (δ), indexed by $n = 10, 11, \ldots$, then

$$\log\log s_n \sim n \log\log n \quad \text{as} \quad n \to \infty \tag{5}$$

(in this paper all logarithms are to the base 10).

Since expressions like

$$10^{11^{12^{13}}}$$

are called *towers*, we will call expressions like those in (1) and (α), (β), (γ) or (δ), *dungeons*. For reasons that will be given in Sect. 2, we believe that the standard parenthesizing of dungeons should be from the bottom upwards, and we will take this as the default meaning if the parentheses are omitted. For towers of exponents,

Table 1 Initial terms of sequence (α), (β), (γ), (δ)

n	(α)	(β)	(γ)	(δ)
10	10	10	10	10
11	11	11	11	11
12	13	13	13	13
13	16	16	16	16
14	20	20	20	20
15	25	30	25	28
16	31	48	31	45
17	38	76	38	73
18	46	132	46	133
19	55	420	55	348
20	65	1640	110	4943
21	87	11991	221	22779
22	135	249459	444	537226
23	239	14103793	891	11662285
24	463	5358891675	1786	46524257772
25	943	19563802363305	3577	1092759075796059
...	...	...	...	...
30	38959	$3.6053\ldots \times 10^{80}$	171999	$2.5841\ldots \times 10^{89}$
...	...	...	...	...
35	9153583	$8.6168\ldots \times 10^{643}$	41795936	$1.2327\ldots \times 10^{898}$
...	...	...	...	...
100	$4.0033\ldots \times 10^{57}$	...	$4.9144\ldots \times 10^{114}$	...
...	...	...	...	...
...	$6.8365\ldots \times 10^{1098}$ at $n = 109$	...	$3.4024\ldots \times 10^{917}$ at $n = 103$	...

parenthesizing from the top downwards is clearly better (for otherwise the tower collapses). The tower with nth term

$$t_n := 10 \uparrow (11 \uparrow (12 \uparrow \cdots ((n-1) \uparrow n) \cdots)), \quad n = 10, 11, \ldots,$$

(where $a \uparrow b$ denotes a^b) has the property that the iterated logarithm $\log^{(n)} t_n \to \infty$ (note that $\log^{(n)} t_n$ is well-defined for n sufficiently large). When parenthesized from the bottom upwards, the tower with nth term

$$u_n := (\cdots((10 \uparrow 11) \uparrow 12) \cdots (n-1)) \uparrow n = 10^{11 \cdot 12 \cdot \ldots \cdot n}, \quad n = 10, 11, \ldots,$$

has the property that $\log \log u_n \sim n \log n$. Equation (5) shows that the dungeon sequences have a slower growth rate than either version of the tower.

In Sects. 3 and 4 we prove Theorem 1 and give some other properties of these sequences, such as the fact that sequence (α) converges 10-adically – for example, from a certain point on, the last 10 digits are always $\ldots 9163204655$.

In Sect. 5 we investigate the behavior as n increases of the sequence with nth term $(n = 1, 2, \ldots)$

$$a(n) := a_{\_}(a_{\_}(a_{\_}(a_{\_} \cdots a))) \quad \text{(with } n \text{ copies of } a\text{)} \tag{6}$$

for a fixed real number $a > 1$. If the parameter a exceeds 10 this sequence certainly diverges, and for $a = 10$ we have $a(n) = 10$ for all $n \geq 1$. Somewhat surprisingly, it seems hard to say precisely what happens for $1 < a < 10$. The mapping from $a(n)$ to $a(n+1) = a_{a(n)}$ is a discrete dynamical system, which converges either to a single number (e.g., to the golden ratio if the parameter $a = 1.1$), to a two-term limit cycle (e.g., if $a = 1.05$) or diverges (e.g., if $a = \frac{100}{99}$). But we do not have a simple characterization of the parameters a that fall into the different classes.

Section 2 contains some general properties of the subscript notation.

The following definition will be used throughout. If $a > 1$ is a fixed real number with decimal expansion given by (2) and x is any real number, we define the Laurent series

$$L^{\langle a \rangle}(x) := \sum_{i=-\infty}^{k} c_i x^i, \tag{7}$$

so that $a\_b = L^{\langle a \rangle}(b)$. We use angle brackets to show the dependence on the parameter a. Note also that $L^{\langle a \rangle}(10) = a_{10} = a$ for all a.

Remark 1. The choice of base 10 in this paper was a matter of personal preference.

Remark 2. To answer a question raised by some readers of an early draft of this paper, as far as we know there is no connection between this work and the base-changing sequences studied by Goodstein (1944).

2 Properties of the Subscript Notation

In this and the following section we will be concerned with the numbers a_b defined in (3) when a and b are integers ≥ 10.

Lemma 1. *Let $N = \sum_{i=0}^{k} \nu_i 10^i$, where the ν_i are nonnegative integers (not necessarily in the range 0–9), and suppose b is an integer ≥ 10. Then*

$$N_b \geq \sum_{i=0}^{k} \nu_i b^i . \tag{8}$$

Proof. If the ν_i are all in the range $\{0,\ldots,9\}$ then the two sides of (8) are equal. Any $\nu_i \geq 10$, say $\nu_i = 10q + r$, $q \geq 1$, $r \in \{0,\ldots,9\}$, causes the term $\nu_i b^i$ on the right-hand side of (8) to be replaced by $qb^{i+1} + rb^i \geq (10q+r)b^i = \nu_i b^i$ on the left-hand side, and so the difference between the two sides can only increase. □

Corollary 1. *If $f(x)$ is a polynomial with nonnegative integer coefficients, and b is an integer ≥ 10, then $f(10)_b \geq f(b)$.*

Lemma 2. *Assume a, b, a', b' are integers ≥ 10. Then:*

(1) *$a' \geq a$ if and only if $a'_b \geq a_b$.*
(2) *$b' \geq b$ if and only if $a_{b'} \geq a_b$.*

(3) $(a+a')_b \geq a_b + a'_b$.
(4) $a_{(b+b')} \geq a_b + a_{b'}$.
(5) $a_b \geq \max\{a,b\}$.

Proof. (1) Suppose $a' = \sum_{i=0}^{r'} c'_i 10^i > a = \sum_{i=0}^{r} c_i 10^i$, with all c'_i, $c_i \in \{0,\ldots,9\}$, and let k be the largest i such that $c'_i \neq c_i$. Then $a'_b - a_b = \sum_{i=0}^{k}(c'_i - c_i)b^i \geq b^k - \sum_{i=0}^{k-1} 9b^i > 0$. The converse has a similar proof. Claims (2), (4) and (5) are immediate, and (3) follows from Lemma 1. □

Note that all parts of Lemma 2 may fail if we allow a and b to be less than 10 (e.g., $12_2 = 4 < 7_2 = 7$; $6_3 = 6 \geq 6_4 = 6$, but $3 < 4$).

Lemma 3. *Assume a,b,c are integers ≥ 10. Then*

$$(a_b)_c \geq a_{(b_c)}\,. \tag{9}$$

Proof. The left-hand side of (9) is (in the notation of (7)) $L^{\langle a\rangle}(L^{\langle b\rangle}(10))_c = (L^{\langle a\rangle} \circ L^{\langle b\rangle})(10)_c$, where $\circ$ denotes composition. The right-hand side is $L^{\langle a\rangle}(L^{\langle b\rangle}(c)) = (L^{\langle a\rangle} \circ L^{\langle b\rangle})(c)$, and the result now follows from Corollary 1. □

We can now explain why we prefer the "bottom-up" parenthesizing of dungeons. The reason can be stated in two essentially equivalent ways. First, $a_{(b_{(c_d)})}$, say, is simply

$$L^{\langle a\rangle} \circ L^{\langle b\rangle} \circ L^{\langle c\rangle}(d)\,,$$

whereas no such simple expression holds for $((a_b)_c)_d$. To put this another way, consider evaluating the nth term of sequence (α) of Sect. 1. To do this, we must repeatedly calculate values of r_s where r is $\leq n$ and s is huge. But to find the nth term of (β), we must repeatedly calculate values of r_s where r is huge and $s \leq n$. The latter is a more difficult task, since it requires finding the decimal expansion of r. Again, when computing the sequence $a(1), a(2), a(3), \ldots$ for a given values of a (see (6)), as long as the terms are parenthesized from the bottom upwards, only one decimal expansion (of a itself) is ever needed.

In Sect. 3 we will also need numerical estimates of a_b. If $a, b \geq 10$ then a_b is roughly $10^{\log a \log b}$ (remember that all logarithms are to the base 10). More precisely, we have

Lemma 4. *Assume a,b are integers ≥ 10. Then*

$$10^{\lfloor \log a \rfloor \lfloor \log b \rfloor} \leq 10^{\lfloor \log a \rfloor \log b} \leq a_b \leq 10^{\log a \log b}\,. \tag{10}$$

Proof. Suppose $a = \sum_{i=0}^{k} c_i 10^i$ where $k := \lfloor \log a \rfloor, c_i \in \{0,1,\ldots,9\}$ for $i = 0,1,\ldots,k$, $c_k \neq 0$. The left-hand inequalities in (10) are immediate. For the right-hand inequality we must show that

$$\sum_{i=0}^{k} c_i b^i \leq b^{\log a}\,,$$

or equivalently that

$$\log\left\{c_k b^k(1+\sum_{i=0}^{k-1}\frac{c_i}{c_k b^{k-i}})\right\} \le (\log b)\left(\log\{c_k 10^k(1+\sum_{i=0}^{k-1}\frac{c_i}{c_k 10^{k-i}})\}\right),$$

and this is easily checked to be true using $b \ge 10$. □

3 Growth Rate of the Sequences (α), (β), (γ), (δ)

Theorem 1. *If s_n $(n \ge 10)$ denotes the nth term in any of the sequences (α), (β), (γ), (δ) then*

$$\log\log s_n \sim n\log\log n \quad \textit{as } n \to \infty.$$

Proof. From Lemma 4 it follows that

$$\prod_{i=10}^{n} \lfloor \log i \rfloor \le \log s_n \le \prod_{i=10}^{n} \log i.$$

For the upper bound, we have

$$\log\log s_n \le \sum_{i=10}^{n} \log\log i \le n\log\log n.$$

For the lower bound,

$$\log s_n \ge \prod_{i=10}^{n} \lfloor \log i \rfloor \ge \prod_{i=10}^{n} \log i(1-\frac{1}{\log i}),$$

$$\log\log s_n \ge \sum_{i=10}^{n} \log\log i - \sum_{i=10}^{n}\frac{1}{\log i},$$

and the right-hand side is $\sim n\log\log n + O(n)$. □

A slight tightening of this argument shows that there are positive constants c_1, c_2 such that

$$n\log\log n - c_1\frac{n}{\log n} < \log\log s_n < n\log\log n - c_2\frac{n}{\log n}$$

for all sufficiently large n.

Table 1 suggests that sequences (β) and (δ) grow faster than (α) and (γ). We can prove three of these four relationships.

Theorem 2. *For $n \ge 10$, $\beta_n \ge \alpha_n$ and $\delta_n \ge \gamma_n$.*

Proof. This follows by repeated application of Lemma 3. □

Lemma 5. *If for some real number $k > 10$ we have $a \ge kb$ and $\log c \ge \log k/(\log k - 1)$, then $a\_c \ge k(c\_b)$.*

Proof. From Lemma 4 and the assumed bounds, we have

$$\begin{aligned} a\_c &\geq 10^{\lfloor \log a \rfloor \log c} \\ &\geq 10^{(\log a - 1)\log c} \\ &\geq 10^{(\log b + \log k - 1)\log c} \\ &= 10^{(\log c)(\log k - 1)} 10^{\log b \log c} \\ &\geq k(c\_b)\,. \end{aligned}$$

□

Theorem 3. *For $n \geq 10$, $\beta_n \geq \gamma_n$.*

Proof. From Table 1, this is true for $n \leq 23$. For $n > 23$, since $\beta_{n+1} = (\beta_n)\_(n+1)$ and $\gamma_{n+1} = (n+1)\_\gamma_n$, the previous lemma (with $k = 10^4$) gives us the result by induction. □

4 p-Adic Convergence of the Sequence (α)

For the next theorem we need a further lemma. Let us say that a polynomial $f(x) \in \mathbb{Z}[x]$ is *m-stable*, for a positive integer m, if all its coefficients except the constant term are divisible by m. In particular, if $f(x)$ is m-stable, $f(x) \equiv f(0) \pmod{m}$.

Lemma 6. *If the polynomial $f(x) \in \mathbb{Z}[x]$ is m-stable and the polynomial $g(x) \in \mathbb{Z}[x]$ is n-stable, then the polynomial $h(x) := f \circ g(x)$ is mn-stable.*

Proof. If $f(x) := \sum_i f_i x^i$, $g(x) := \sum_j g_j x^j$, then $h(x) = \sum_i f_i \left(\sum_j g_j x^j\right)^i = \sum_k h_k x^k$ (say). When the expression for h_k $(k > 0)$ is expanded as a sum of monomials, each term contains both a factor f_i for some $i > 0$ and a factor g_j for some $j > 0$. □

Theorem 4. *The sequence $\alpha_{10}, \alpha_{11}, \alpha_{12}, \ldots$ converges 10-adically.*

Proof. We know from the above discussions that, for any $10 \leq k < n$,

$$\alpha_n = \Phi^{[k]}((k+1)\_(k+2)\_(k+3)\_\ldots\_n)\,,$$

where $\Phi^{[k]}(x)$ is the polynomial

$$\Phi^{[k]}(x) := L^{\langle 10\rangle} \circ L^{\langle 11\rangle} \circ L^{\langle 12\rangle} \circ \cdots \circ L^{\langle k\rangle}(x)\,.$$

(We would normally write $\Phi_k(x)$, but since there are already two different kinds of subscripts in this paper, we will use the temporary notation $\Phi^{[k]}(x)$ in this proof instead.) Now $L^{\langle 20\rangle}(x)$, $L^{\langle 21\rangle}(x), \ldots, L^{\langle 29\rangle}(x)$ are 2-stable and $L^{\langle 50\rangle}(x), \ldots, L^{\langle 59\rangle}(x)$ are 5-stable, so by Lemma 6, $\Phi^{[59]}(x)$ is 10^{10}-stable. This means that for $n \geq 60$, $\alpha_n \equiv \Phi^{[59]}(0) \pmod{10^{10}}$, and so is a constant (in fact 5564023619) mod 10^{10}. Similarly, $L^{\langle 500\rangle}(x)$, $L^{\langle 501\rangle}(x), \ldots, L^{\langle 509\rangle}(x)$ are 5-stable, so α_n is a constant mod 10^{20} for $n \geq 510$; and so on. □

Remark 3. The same proof shows that $\alpha_{10}, \alpha_{11}, \alpha_{12}, \ldots$ converges l-adically, for any l all of whose prime factors are less than 10.

5 The Limiting Value of $a \_ a \_ a \_ a \_ \ldots$

In this section we consider the behavior of the sequence $a(1), a(2), a(3), \ldots$ (see (6)) as n increases, for a fixed real number a in the range $1 < a < 10$. For example, we have the amusing identity

$$1.1^{1.1^{1.1^{1.1^{1.1^{1.1^{1.1^{\cdots}}}}}}} = \frac{1+\sqrt{5}}{2}. \tag{11}$$

The sequence (6) is the trajectory of the discrete dynamical system $x \mapsto L^{\langle a \rangle}(x)$ when started at $x = a$. (Since $L^{\langle a \rangle}(10) = a$, we could also start all trajectories at 10.)

Suppose $a = \sum_{i=0}^{\infty} c_i 10^{-i}$ with all $c_i \in \{0, 1, \ldots, 9\}$ and $c_0 \neq 0$. The graph of $y = L^{\langle a \rangle}(x)$ is a convex curve, illustrated[1] for $a = 1.1$ in Fig. 1, which decreases monotonically from its value at $x = 1$ (which may be infinite) and approaches c_0 as $x \to \infty$. This curve therefore meets the line $y = x$ at a unique point $x = \omega$ (say) in the range $x > 1$. The point ω is the unique fixed point for the dynamical system in the range of interest.

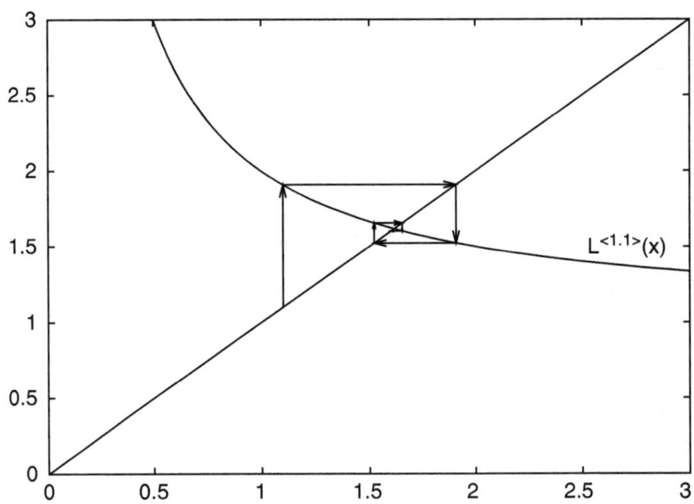

Fig. 1 Trajectory of $L^{<1.1>}(x)$ starting at $x = 1.1$

[1] This is a "cobweb" picture – compare Fig. 1.4 of Devaney (1989).

The general theory of dynamical systems (Devaney, 1989; Lauwerier, 1986) tells us that the fixed point ω is respectively an attractor, a neutral point or a repelling point, according to whether the value of the derivative $L^{\langle a\rangle\prime}(\omega)$ is between 0 and -1, equal to -1, or less than -1. For our problem this does not tell the whole story, since we are constrained to start at a. However, since $L^{\langle a\rangle}(x)$ is a monotonically decreasing function, there are only a few possibilities. Cycles of length three or more cannot occur.

Theorem 5. *For a fixed real number a in the range* $1 < a < 10$, *and an initial real starting value* $x > 1$, *consider the trajectory* x, $L^{\langle a\rangle}(x)$, $L^{\langle a\rangle} \circ L^{\langle a\rangle}(x)$, $L^{\langle a\rangle} \circ L^{\langle a\rangle} \circ L^{\langle a\rangle}(x)$, ... *Then one of the following holds:*

(*1*) $x = \omega$ *is the fixed point, and the trajectory is simply* $\omega, \omega, \omega, \ldots$.
(*2*) *The trajectory converges to* ω .
(*3*) *x is in a two-term cycle, and the trajectory simply repeats that cycle.*
(*4*) *The trajectory converges to a two-term limit cycle.*
(*5*) *The trajectory diverges, alternately approaching* 1 *and* ∞.

Proof. If a is an integer, then the trajectory is simply $x, a, a, a, \ldots$, and either case (1) or (2) holds. Suppose then that a is not an integer. Since a is fixed, we abbreviate $L^{\langle a\rangle}$ by L in this discussion, and write $L^{(k)}$ to indicate the k-fold composition of L, for $k = 1, 2, \ldots$. Because $L(x)$ is strictly decreasing, if $L^{(2)}(x) > x$, then $L^{(3)}(x) < L(x)$, $L^{(4)}(x) > L^{(2)}(x) > x$; if $L^{(2)}(x) < x$, then $L^{(3)}(x) > L(x)$, $L^{(4)}(x) < L^{(2)}(x) < x$; and if $L^{(2)}(x) = x$, then $L^{(3)}(x) = L(x)$, $L^{(4)}(x) = L^{(2)}(x) = x$. Hence if $x < L^{(2)}(x)$, then $x < L^{(2)}(x) < L^{(4)}(x) < \cdots$ and if $x > L^{(2)}(x)$, then $x > L^{(2)}(x) > L^{(4)}(x) > \cdots$. This means the even-indexed iterates form a monotonic sequence, so either converge or are unbounded, and similarly for the odd-indexed iterates. Equation (4) implies that if the trajectory diverges then the lower limit must be 1. □

Note also that if $x < y < L^{(2)}(x)$, then $L^{(2k)}(x) < L^{(2k)}(y) < L^{(2k+2)}(x)$, and if $x > y > L^{(2)}(x)$, then $L^{(2k)}(x) > L^{(2k)}(y) > L^{(2k+2)}(x)$. So every y between x and $L^{(2k)}(x)$ converges to the same limiting two-cycle as x does, or diverges as x does.

The following examples illustrate the five cases in the situation which most interests us, the trajectory $a, a\_a, a\_(a\_a), \ldots$ of (6), that is, when we set $x = a$ in the theorem:

1. This case holds if and only if a is one of $\{2, 3, \ldots, 9\}$.
2. Examples are $a = 1 + \frac{m}{10}$, for $m \in \{1, \ldots, 9\}$, when $\omega = (1 + \sqrt{4m+1})/2$ is an attractor (see (11)); $a = 1 + \frac{m}{100}$ for $m \in \{1, 2, 3\}$, when ω, the real root 1.465..., 1.695... or 1.863... of $x^3 - x^2 - m = 0$ is an attractor; and $a = 1 + \frac{4}{100}$, when $\omega = 2$ is neutral, but the trajectory still converges to ω.
3. Examples are $a = 1 + \frac{m}{9}$, $m \in \{1, \ldots, 8\}$, ω is a neutral point, and the two-term cycle is $\{a, 10\}$. (The trajectory does not include ω.)
4. Examples are $a = 1 + \frac{m}{100}$, $m \in \{5, \ldots, 9\}$, ω is a repelling point, and the trajectory approaches a two-term limit cycle consisting of a pair of solutions to $L^{\langle a\rangle} \circ L^{\langle a\rangle}(x) = x$; also $a = 1.1110000099$, ω is an attractor, but again the trajectory approaches a two-term cycle given by $L^{\langle a\rangle} \circ L^{\langle a\rangle}(x) = x$.

5. Examples are $a = 1 + \frac{1}{10^r - 1}, r \in \{2, 3, \ldots\}$, ω is a repelling point, and the trajectory alternately approaches 1 or ∞.

We do not know which values of a fall into classes (2) through (5). The distribution of the five classes for $1 < a < 10$ seems complicated.

References

Devaney, R. L. (1989). Dynamics of simple maps. In R. L. Devaney and L. Keen (Eds.), *Chaos and fractals*, Proceedings of Symposia in Applied Mathematics (Vol. 39). Providence, RI: American Mathematical Society.

Goodstein R. L. (1944). On the restricted ordinal theorem. *Journal of Symbolic Logic, 9*, 33–41.

Lauwerier H. A. (1986). One-dimensional iterative maps. In A. V. Holden (Ed.), *Chaos* (pp. 39–57). Princeton: Princeton University Press.

Sloane N. J. A. (2008) *The on-line encyclopedia of integer sequences*. www.research.att.com/ ~njas/ sequences/.

Updating Hardy, Littlewood and Pólya with Linear Programming

Larry Shepp

1 Proving Two Simple Inequalities with Convexity Methods

Some of the standard inequalities that mathematicians use can be proven with convexity arguments or linear programming.[1] Perhaps others cannot, so we might say that an inequality is "simple" if there is a convexity based proof. The Cauchy–Schwarz inequality, which may be the most famous and useful inequality ever found is simple in this sense Steele (2004), but there are so many proofs of it that it seems that almost any method will give one, so it may be that it is simple in any sense. The Schwarz inequality can be stated for a general measure space but it easily reduces to the statement that

$$EX^2EY^2 \geq (EXY)^2,$$

where X and Y are any r.v.'s on a common probability space, Ω. Equality holds if and only if X and Y are proportional.

To give a convexity based proof, one thinks of the probability measure, μ, on the space, $(\Omega, \mathcal{F}, P)$, on which X and Y are defined as an element of the convex set of all probability measures. Then one finds a linear functional, $L(\mu)$, of μ, which is based on a function, $f_L(x,y)$ which must be everywhere nonnegative and whose integral, $L(\mu) = \int_\Omega f_L(X,Y)d\mu$, would therefore be nonnegative. This would then give an inequality which must then be the same as the Schwarz inequality. At this stage, this is not possible because the Schwarz inequality by its very nature is *non-linear* in μ – indeed both terms, $(EXY)^2$ and EX^2EY^2 are quadratic when looked upon as

[1] The methods of linear programming came along only in 1944 with the appearance of the book of von Neumann and Morganstern (1944). The book of Hardy, Littlewood, and Polya (1934), which systematically developed the theory of inequalities, appeared 10 years earlier. It is not surprising that the methods of the earlier book did not include linear programming. This paper is an attempt to call attention to the need for a systematic update of the theory of inequalities in which the method of linear programming is used.

L. Shepp
Rutgers University, Statistics Department, Piscataway, NJ 08854, USA
e-mail: shepp@stat.rutgers.edu

S.J. Brams et al. (eds.), *The Mathematics of Preference, Choice and Order: Essays in Honor of Peter C. Fishburn*, Studies in Choice and Welfare,

functionals of μ. Instead we must *encode* the quadraticity in the statement in order to linearize the problem and we do this by going to the product probability space, $\Omega\times\Omega$, with the product measure, so that X_1, Y_1 and X_2, Y_2 are two *independent* pairs of r.v.'s on the product space each with the joint distribution of X, Y. Note that for any four numbers, x_1, y_1, x_2, y_2, the homogeneous polynomial

$$f(x_1,y_1,x_2,y_2) = x_1^2y_2^2 + x_2^2y_1^2 - 2x_1y_1x_2y_2 = (x_1y_2 - y_1x_2)^2 \geq 0,$$

is indeed nonnegative. If we now substitute X_i, Y_i for $x_i, y_i, i = 1, 2$ and take expectations, using the independence of r.v.'s with different subscripts we obtain that

$$2EX^2EY^2 \geq 2(EXY)^2$$

and the proof is complete after dividing by 2. The only case of equality is when $X_1Y_2 - X_2Y_1 \equiv 0$, that is when the ratios $\frac{X_i}{Y_i}$ are constant since they are independent for $i = 1, 2$ and it is easy to see that two independent r.v.'s which are equal must each be constant. We have linearized the problem and we have encoded all the conditions we need to make the proof work. This is what one does in linear programming, and so this is a "linear programming" or "convexity" proof. This proof appeared in the paper with Olkin and Shepp (2006), which uses the same technique to prove a more difficult inequality due to M. Brown, among others which I now discuss. I reproduce this material here for clarity and convenience.

M. Brown's inequality states that for *positive and independent* r.v.'s on a common probability space

$$\frac{E\frac{1}{X+Y}}{E\frac{1}{(X+Y)^2}} \geq \frac{E\frac{1}{X}}{E\frac{1}{X^2}} + \frac{E\frac{1}{Y}}{E\frac{1}{Y^2}}.$$

Equality holds if and only if both X and Y are constants.

The Brown inequality is proved in Olkin and Shepp (2006) by the same method, and the encoding of all the conditions is similar. I refer to Olkin and Shepp (2006) for some details but the idea of the proof is completely analogous. We first "clear of fractions" by multiplying by $EX^{-2}EY^{-2}E(X+Y)^{-2}$, and so it is equivalent to show that

$$E\frac{1}{X+Y}EX^{-2}EY^{-2} - E\frac{1}{(X+Y)^2}(EX^{-1}EY^{-2} + EX^{-2}EY^{-1} \geq 0.$$

To use the linear functional or convexity method as above for the Schwarz inequality, we again construct the product probability space on which two independent pairs of independent r.v.'s X_i, Y_i are defined. Then *if* we could show that for any four numbers x_1, y_1, x_2, y_2 the function

$$f(x_1,y_1,x_2,y_2) \equiv \frac{1}{x_1+y_1}\frac{1}{x_2^2}\frac{1}{y_2^2} - \frac{1}{(x_1+y_1)^2}(x_2^{-1}y_2^{-2} + x_1^{-2}y_2^{-1})$$

is everywhere nonnegative, then it would easily follow that

$$Ef(X_1,Y_1,X_2,Y_2) \geq 0$$

which would then prove the Brown inequality, but (alas) f takes negative values. Alternatively, *if* we could show that $f(x_1,y_1,x_2,y_2)+f(x_2,y_2,x_1,y_1)$ is everywhere nonnegative, then the same proof would give the Brown inequality because upon substituting r.v.'s X_i,Y_i for x_i,y_i we would get the desired inequality after dividing by two. Again (alas), there are numbers $x_i,y_i,i=1,2$ for which this form is also negative. Fortunately, we have one last chance. If we can show that the doubly mixed (symmetric in x_1,x_2 and in y_1,y_2) form

$$f(x_1,y_1,x_2,y_2)+f(x_1,y_2,x_2,y_1)+f(x_2,y_1,x_1,y_2)+f(x_2,y_2,x_1,y_1)\geq 0$$

for all positive values of $x_i,y_i,i=1,2$, then substituting X_i,Y_i for x_i,y_i, taking expectations and using the independence of X_1,X_2,Y_1,Y_2 we get the Brown inequality after dividing by 4. The last inequality is true and the proof is easy provided one does it in the right way. I refer to Olkin and Shepp (2006) for the details. Note in the Brown inequality we have to encode the independence and all the symmetry of the problem, but it is all quite natural.

2 An Inequality with Both a Convexity Proof and an Alternative Proof

Another example of how linear programming methodology can provide new inequalities is taken from a paper of Reeds, Shepp, and Win (in press). Some recent work in wireless communications engineering Conti, Win, and Chiani (2003) raised the problem of determining the best constants L and U such that

$$\prod_{k=1}^{n}\frac{1}{L+a_k}\leq\int_0^\infty\prod_{k=1}^{n}\frac{1}{x+a_k}m(dx)\leq\prod_{k=1}^{n}\frac{1}{U+a_k}$$

hold uniformly for all values of $a_k>0$, $1\leq k\leq n$, where

$$m(dx)=\frac{1}{\pi\sqrt{x(1-x)}}\,dx \text{ on } [0,1].$$

If we can prove the following general result for a given probability measure μ, and if it holds for $\mu(dx)=m(dx)$, then we can easily find the best values of L and U.

Given a probability measure μ on $[0,\infty)$, and positive a_k, define $c(\mathbf{a},\mu)=c(a_1,\ldots,a_n,\mu)$ to be that positive real value of c such that

$$\int_0^\infty\prod_{k=1}^{n}\frac{c+a_k}{x+a_k}\mu(dx)=1.$$

Theorem 1. *For any probability measure μ, $c(\mathbf{a},\mu)$ is monotone increasing in each a_k. More precisely, $c(a_1,\ldots,a_n,\mu)$ is defined by the implicit equation*

$$H(a_1,\ldots,a_n,\mu,c(a_1,\ldots,a_n,\mu))=1,$$

where H is defined by

$$H(a_1,\ldots,a_n,\mu,c)=\int_0^\infty \prod_{k=1}^{n}\frac{c+a_k}{x+a_k}\mu(dx).$$

One can interpret $c(a_1,\ldots,a_n,\mu)$ as a *generalized mean*, $\mathcal{M}_\phi[\mu]$, in the sense of Hardy, Littlewood, and Pólya (1934), Chap. VI, for a suitable function $\phi(x)=\phi(x,a_1,\ldots,a_n)$. A generalized mean wrt. a strictly monotonic function $\phi(x)$, denoted $\mathcal{M}_\phi[\mu]$, is defined by the equation

$$\phi(\mathcal{M}_\phi[\mu])=\int\phi(x)\mu(dx).$$

In our case, $\phi(x,a_1,\ldots,a_n)=\prod_{k=1}^n\frac{1}{x+a_k}$ for $x\geq 0$ gives $\mathcal{M}_\phi[\mu]=c(a_1,\ldots,a_n,\mu)$.

We prove by linear programming arguments that the desired monotonicity follows for any probability measure μ. It can alternatively be proved by applying a criterion of Hardy et al. (1934, Chap. III), which gives necessary and sufficient conditions on pairs of monotonic functions (ϕ,ψ) for $\mathcal{M}_\phi[\mu]\leq\mathcal{M}_\psi[\mu]$ for all μ. For details of the alternative proof see Reeds et al. (in press) [Appendix]. Our proof seems simpler but it is also indirect.

We reformulation the problem by first implicitly differentiating H with respect to a_i which gives

$$\frac{\partial H(\mathbf{a},\mu,\mathbf{c})}{\partial a_i}\Big|_{c=c(\mathbf{a},\mu)}+\left[\frac{\partial c(\mathbf{a},\mu)}{\partial a_i}\right]\frac{\partial H(\mathbf{a},\mu,\mathbf{c})}{\partial c}=0.$$

Clearly $\frac{\partial H}{\partial c}(\mathbf{a},\mu,\mathbf{c})>\mathbf{0}$, so to show that $c(\mathbf{a},\mu)$ is increasing in a_i it suffices to show that

$$H_i\equiv\frac{\partial H(\mathbf{a},\mu,\mathbf{c})}{\partial a_i}\Big|_{c=c(\mathbf{a},\mu)}\leq 0,$$

whenever $H(\mathbf{a},\mu,\mathbf{c})=\mathbf{1}$.

This leads to consideration of the set $\mathcal{C}(\mathbf{a},\mathbf{c})\subset\mathcal{R}^{\mathbf{2}}$ of possible values of the pair $(H(\mathrm{a},\mu,\mathrm{c}),\ \mathrm{H_i}(\mathrm{a},\mu,\mathrm{c}))$ as μ ranges over all probability measures, for fixed c and $\mathbf{a}$. Indeed, the closure of $\mathcal{C}(\mathbf{a},\mathbf{c})$ is the convex hull of the union of $(0,0)\in\mathcal{R}^2$ and the curve

$$x\mapsto(h(x),\ h_i(x))$$

in $\mathcal{R}^2$ traced out by $x\in[0,\infty)$, where

$$h(x)\ =\ \prod_{k=1}^{n}\frac{c+a_k}{x+a_k} \tag{1}$$

and

$$h_i(x)\ =\ \frac{\partial}{\partial a_i}\left(\prod_{k=1}^{n}\frac{c+a_k}{x+a_k}\right)\Bigg|_{c=c(\mathbf{a},\mu)}.$$

Then

$$H = \int_0^\infty h(x)\mu(dx)$$

and

$$H_i = \int_0^\infty h_i(x)\mu(dx)\,.$$

The main fact that we use about $\mathcal{C}(\mathbf{a},\mathbf{c})$ is a linear inequality.

Lemma 1. *There exists a* $\lambda > 0$ *such that for all* $(s,t) \in \mathcal{C}(\mathbf{a},\mathbf{c})$,

$$s + \lambda t \leq 1\,.$$

The theorem follows directly from the lemma since from

$$H(\mathbf{a},\mu,\mathbf{c}) + \lambda\,\mathbf{H_i}(\mathbf{a},\mu,\mathbf{c}) \leq \mathbf{1}$$

and $H(\mathbf{a},\mu,\mathbf{c}) = \mathbf{1}$ it follows that $H_i(\mathbf{a},\mu,\mathbf{c}) \leq \mathbf{0}$, as desired.

Proof. We will exhibit a $\lambda > 0$ such that for all $x \geq 0$,

$$h(x) + \lambda h_i(x) \leq 1 \tag{2}$$

and

$$h_i(x) = \left[\frac{1}{c+a_i} - \frac{1}{x+a_i}\right] h(x)\,. \tag{3}$$

Integrating (2) against μ will then finish the proof. □

To see that such a λ exists, we rearrange (2) into the equivalent form

$$x + a_i + \lambda\frac{x-c}{c+a_i} \leq (x+a_i)\prod_{k=1}^{n}\frac{x+a_k}{c+a_k}\,. \tag{4}$$

The right hand side of (4), which we are to prove, is a polynomial function in x with positive coefficients, and hence convex on $[0,\infty)$. The left hand side is an affine function of x, agreeing with the right hand side when $x = c$. With appropriate choice of λ the left hand side's derivative matches the right hand side's derivative at $x = c$, too. For that choice of λ, (4) will hold for all $x \geq 0$.

It remains to check the positivity of the chosen λ, namely, of the solution of

$$\left.\frac{d}{dx}\right|_{x=c} x + a_i + \lambda\frac{x-c}{c+a_i} = \left.\frac{d}{dx}\right|_{x=c} (x+a_i)\prod_{k=1}^{n}\frac{x+a_k}{c+a_k}\,,$$

that is, of

$$1 + \frac{\lambda}{c+a_i} = 1 + (c+a_i)\sum_{k=1}^{n}\frac{1}{c+a_k}\,.$$

But clearly

$$\lambda = (c+a_i)^2 \sum_{k=1}^{n} \frac{1}{c+a_k} > 0,$$

which finishes the proof. □

3 Discussion

The argument given here follows the pattern of a typical application of the "weak duality theorem" of finite dimensional linear programming. Finite dimensional linear programming deals with the problem of maximizing a linear form such as (c,x) with respect to $x \in \mathcal{R}^n$ subject to constraints of the form

$$\sum_{j=1}^{n} a_{ij}x_j = b_i \quad i = 1,\dots,mx_j \geq 0 \quad j = 1,\dots,n.$$

Associated with each such problem Vanderbei (2001, pp. 73,74), is its dual problem, that of minimizing (y,b) with respect to $y \in \mathcal{R}^m$, subject to the constraints

$$\sum_{i=1}^{m} y_i a_{ij} \geq c_j \quad j = 1,\dots,n.$$

This in fact is not the "standard form" for presenting primal and dual linear programming problems, but an equivalent one which matches this application more exactly.

The weak duality theorem Vanderbei (2001, p. 58), asserts that if $x \in \mathcal{R}^n$ and $y \in \mathcal{R}^m$ satisfy the constraints of the original problem and of its dual, respectively, then $(c,x) \leq (y,b)$.

In our case, working formally and ignoring all differences between finitely many dimensions and uncountably many, consider the problem of finding a finite signed measure μ on $[0,\infty)$ which maximizes the linear form $H_i(\mu) = \int_0^\infty h_i(x)\mu(dx)$ subject to the constraints

$$\int_0^\infty 1\mu(dx) = 1; \int_0^\infty h(x)\mu(dx) = 1; \mu(dx) \geq 0 \text{ for all } x \geq 0.$$

The dual problem would be that of minimizing $u+v$ over $\mathcal{R}^2$, subject to the uncountably many constraints

$$u \cdot 1 + vh(x) \geq h_i(x) \quad for \quad all \quad x \geq 0. \tag{5}$$

The weak duality theorem would then say that if μ and (u,v) satisfied their respective constraints, then

$$\int_0^\infty h_i(x)\mu(dx) \le u+v. \tag{6}$$

But the $\lambda > 0$ of the lemma obeys (2), namely $h(x) + \lambda h_i(x) \le 1$ for all $x \ge 0$, which means $(u,v) = (1/\lambda, -1/\lambda)$ satisfies the constraint (5). So (6) would then imply

$$H_i = \int_0^\infty h_i(x)\mu(dx) \le 0,$$

which of course gives us the theorem.

Although our proof of the theorem and lemma would have been perfectly comprehensible to mathematicians such as Caratheodory and Markov working in the early 1900s, the formalism of linear programming duality – which seems to have originated half a century later Vanderbei (2001), p.87, – would not have been available to them.

4 Examples of Known Inequalities Where There May or May Not be a Convexity Proof

My next two examples are incomplete and suggestions for further work; I suggest trying to prove each of the Schur and FKG inequalities via convexity, which may or may not be possible.

Schur: It would be nice to see a proof of Schur's inequality via convexity. Schur's inequality (Morehead's inequality in Hardy et al. (1934) states that if $f(x_1,\dots,x_n)$ is permutation symmetric in its arguments $x_j \ge 0$, and differentiable, and if

$$\frac{\partial f(x)}{\partial x_1} \ge \frac{\partial f(x)}{\partial x_2}$$

whenever $x_1 \ge x_2$, then $f(x) \ge f(y)$ whenever $y = Ax$ where A is a matrix with nonnegative entries and row sums one. An example is $f(x) = \sum_{i=1}^n x_i^2$. It's clear that the set, C, of all such f is convex and so there has to be a linear functional,

$$F_{x,y}(g) = \int g(x,y,z)\mu_{x,y}(dz)$$

for which the hyperplane $f(x) - f(y) \ge 0$ for all $f \in C$. Indeed the usual proof as in Hardy et al. (1934) gives such a μ, if written from this point of view, see also Olkin and Shepp (2006). Is μ unique?

FKG: It would be nice to see a proof of the FKG (1971), inequality via convexity. This extremely useful inequality came along after Hardy et al. (1934). Ahlswede and Daykin (1978) gave a later and sometimes more useful version Fishburn and Shepp (2000), but I will discuss only the original version. Here Λ is a partially ordered set which is also a distributive lattice, and $\mu \ge 0$ is a function on $x \in \Lambda$ satisfying the property $\mathcal{P}$: $\mu(x \vee y)\mu(x \wedge y) \ge \mu(x)\mu(y)$, where $x \vee y$ is the unique largest element of Λ smaller than both x and y and $x \wedge y$ is the unique smallest element of Λ larger than both. If now f,g are a pair of monotinic functions on Λ, $f(x) \leq f(y)$ and $g(x) \leq g(y)$ when $x < y$, then the FKG inequality asserts that

$$\sum_{x \in \Lambda} f(x)g(x) \sum_{y \in \Lambda} \mu(y) \geq \sum_{x \in \Lambda} f(x)\mu(x) \sum_{y \in \Lambda} g(y)\mu(y).$$

A new proof has been given by Siddharti Sahi in this volume, but I would like to see a convexity proof. There are at least two approaches to a convexity proof: one can look at the convex set of all f satisfying the given inequalities with g,μ held fixed; or one can fix f,g and try to imitate the proof of the Schwarz or Brown inequality. It is well known that property $\mathcal{P}$ is not "sharp", but how to make a better general condition on μ for the same conclusion is unclear. It's a long shot, but maybe convexity can help.

5 An Example Where We Can Show that There is No Convexity Proof

Finally, I discuss a conjectured inequality that I will show *cannot* be proved by using convexity. I believe it to be true but I have no proof for it. This example has not been published before and I owe the formulation of the problem to J. Denny.

Let X,Y denote a pair of symmetric (about zero) independent r.v.'s with common variance σ^2, given, and with cumulative distribution functions, F,G, respectively. Define the functional $\mathcal{F}(F,G) = P(X+Y \geq 1)$, and let $\phi(\sigma)$ denote the supremum of $\mathcal{F}(F,G)$ over all such choices of F,G, so that,

$$\phi(\sigma) = \sup_{X,Y} P(X+Y \geq 1)$$

under the constraints that each of X and Y are symmetric about zero, each has variance σ^2, and X,Y are independent. We want to explicitly find $\phi(\sigma)$. First note that

$$\mathcal{F}(F,G) = P(X+Y \geq 1) = \int_{-\infty}^{\infty} \int_{-\infty}^{\infty} \chi(x+y \geq 1) F(dx) G(dy)$$

is bi-linear in the pair F,G but is not convex. This causes the problem and it becomes even worse (quadratic) if we impose $F = G$ as is the case if $X \sim Y$ are assumed to have the same distribution. In case F and G can be chosen separately, the for each fixed F, $\mathcal{F}(F,G)$ is linear in G and for each fixed G, $\mathcal{F}(F,G)$ is linear in F. Since the class of symmetric distributions with given variance is convex, a theorem of Lester Dubins (1962), says that for the extreme "points" at which the supremum of $\mathcal{F}(F,G)$ must occur, each of X and Y have at most 4 values. Thus the answer can be *numerically* determined via a search through a 4 dimensional parameter space. This leads me to believe that I know the exact formula for $\phi(\sigma)$ for every σ, but a rigorous proof that the formula is correct escapes me. For $\sigma < c$ where $c \sim 0.724$, I indicate next that a rigorous proof might be devised based on linear programming, but for $\sigma > c$, it can be *proven* that the linear programming argument breaks down and some other method of proof has to be found. I do not know how to move ahead to find an honest proof that the search gives the correct maximum.

If we relax the independence condition on X,Y, which is a non-linear condition, and replace it by the weaker condition that

$$E[X^2 - \sigma^2 | Y] \equiv 0$$

and

$$E[Y^2 - \sigma^2 | X] \equiv 0$$

which of course holds if X,Y are independent, then the problem becomes a linear programming problem (infinite dimensional, but linear). We have linearized the problem and in the wider class of X,Y pairs satisfying the last two conditions if it turns out that the maximum value is just the one obtained by the search then this in principle would give a rigorous proof that the search found the optimum. For $\sigma < \sim 0.724$ the upper bound seems to coincide with the value for $\phi(\sigma)$ obtained from the search (at least numerically) which would then give a rigorous proof that there is no "duality gap" between the linear and nonlinear problems. But we show that a duality gap does exist for larger σ. The best pair under the relaxed conditions on conditional expectations above are no longer independent for $\sigma > 0.724$. Some method other than linear programming will be required to give a rigorous proof for these values of σ.

We find convincingly that the maximum value of $\phi(\sigma) = P(X+Y \geq 1)$ taken over all pairs of independent symmetric r.v.'s X,Y each with variance $\leq \sigma^2$ strictly increases in σ for $0 \leq \sigma \leq 1$, and thereafter, i.e., for $\sigma \geq 1$, $\phi(\sigma) \equiv \frac{1}{2}$, although there is no i.i.d. pair that achieves the value $\frac{1}{2}$ for any value of σ. The range $[0,1]$ breaks up into 5 intervals, $(\sigma_i, \sigma_{i+1}), i = 0, \ldots, 4$, with $\sigma_0 = 0, \sigma_1 = \frac{1}{2}, \sigma_4 = 1$. In each interval the optimal X,Y has a different form and in the first, third, and fourth interval the optimal pair are i.i.d., but this is not so for the second and fifth intervals, which seems surprising. More precisely, we have that

I. for $0 \leq \sigma \leq \sigma_0$ an optimal X,Y pair (not necessarily unique) is *identically distributed* and each of X,Y is equal to ± 1 w.p. $\frac{\sigma^2}{2}$ and equal to zero w.p. $1-\sigma^2$. It is easy to verify that this makes $\phi(\sigma) = \sigma^2 - \frac{3}{4}\sigma^4$ for $0 \leq \sigma \leq \sigma_0$.

II. for $\sigma_0 < \sigma \leq \frac{1}{2} = \sigma_1$, surprisingly, the (apparently unique up to an interchange of X,Y) optimal X,Y pair is *not* identically distributed (except at $\sigma = \frac{1}{2}$). One of the variables, say X has the distribution $X = \pm\sigma$ w.p. $\frac{1}{2}$, while Y has the unequal distribution $Y = \pm(1-\sigma)$ w.p. $\frac{\sigma^2}{2(1-\sigma)^2}$ and $Y = 0$ w.p. $1 - (\frac{\sigma}{1-\sigma})^2$. Of course X and Y could be interchanged here so there are at least two different optimal pairs now. It is easy to check that for σ in this range this gives $\phi(\sigma) = \frac{1}{4}(\frac{\sigma}{1-\sigma})^2$. Here $\sigma_0 \sim 0.46$ is the root of the quartic equation

$$\sigma^2 - \frac{3}{4}\sigma^4 = \frac{1}{4}\left(\frac{\sigma}{1-\sigma}\right)^2,$$

required to make ϕ continuous at σ_0 (smooth-fitting).

III. for the next range, $\frac{1}{2} \le \sigma \le \sigma_1 \sim 0.65$, the optimal X, Y pair are again identically distributed and

$$X \sim Y = \pm\frac{3}{2}\text{w.p.}\frac{1}{4}\left(\sigma^2 - \frac{1}{4}\right)$$

and

$$X \sim Y = \pm\frac{1}{2}\text{w.p.}\frac{1}{4}\left(\frac{9}{4} - \sigma^2\right).$$

This gives the value $\phi(\sigma) = \frac{23}{128} + \frac{5}{16}\sigma^2 - \frac{1}{8}\sigma^4$.

IV. for the next range, $\sigma_1 \le \sigma \le \sigma_2 \sim 0.78$, the optimal pair X, Y is again identically distributed and has the same distribution as the earlier identical pair for $0 \le \sigma \le \sigma_0$ and the same formula for ϕ, $\phi(\sigma) = \sigma^2 - \frac{3}{4}\sigma^4$.

V. for $\sigma_2 \le \sigma \le 1$, the optimal pair is not identically distributed and have the distributions

$$X = \pm 1\text{w.p. } \frac{1}{2}\sigma^2, X = 0 \text{ w.p. } 1 - \sigma^2,$$

$$Y = \pm 2\text{w.p. } \frac{\sigma^2}{8}, Y = 0 \text{ w.p. } 1 - \frac{\sigma^2}{4}.$$

This gives the value for this range of σ as

$$\phi(\sigma) = \frac{5\sigma^2 - \sigma^4}{8}, \sigma_2 \le \sigma \le 1.$$

The above particular choices of the pair X, Y give a lower bound on $\phi(\sigma)$. In the next section we will use linear programming, or duality, to obtain an upper bound which is tight for some σ, but alas, not for all σ. The upper bound will be a "linear programming bound" despite the fact that we are maximizing a non-linear (actually bilinear) functional:

$$P(X + Y \ge 1) = \int_{-\infty}^{\infty}\int_{-\infty}^{\infty} \chi(x + y \ge 1)F(dx)G(dy)$$

which is linear in the d.f. F of X for each fixed d.f. G of Y, and is linear in G for each fixed F, but is not linear in the pair F, G since it involves products. It is even "less" linear if $X \sim Y$ is imposed as an extra condition – more on this later. It is well-known that a linear functional on a compact convex set C is maximized at an extreme point of C. Here C might be the set of pairs F, G with G fixed and F the d.f. of any symmetric X with variance σ. It's a consequence of the Dubins–Caratheodory theorem that an extreme point of C is a distribution with at most 4 points in its support which is a set of the form $\{-b, -a, a, b\}$ or $\{-a, 0, a\}$, or $\{-a, a\}$ or (for $\sigma = 0$, just $\{0\}$). Similarly for each fixed F the best G has at most four points in its support. It follows rigorously that the best F, G are each "four-pointers". A search of this set of distributions for each of F, G led to the conjecture above for $\phi(\sigma)$. Each of the above 4 particular choices for X, Y gives a lower bound for $\phi(\sigma)$, and the maximum for each σ of the 4 lower bounds is also a lower bound for $\phi(\sigma)$ which we call $\phi_0(\sigma)$. Note that pair I (or IV) can be used for any value of $\sigma \le 1$ but II is allowed for $\sigma \le \frac{1}{2}$, III for $\sigma \in [\frac{1}{2}, \frac{3}{2}]$, and V for $\sigma \in [\frac{1}{2}, 1]$.

We summarize this as follows, $\phi(\sigma) \ge \phi_0(\sigma)$, where:

$$\phi_0(\sigma) = \sigma^2 - \frac{3}{4}\sigma^4, 0 \le \sigma \le \sigma_0,$$

$$\phi_0(\sigma) = \frac{1}{4}\frac{\sigma^2}{(1-\sigma)^2}, \sigma_0 \le \sigma \le \sigma_1 = \frac{1}{2},$$

$$\phi_0(\sigma) = \frac{23}{128} + \frac{5}{16}\sigma^2 - \frac{1}{8}\sigma^4, \frac{1}{2} \le \sigma \le \sigma_2,$$

$$\phi_0(\sigma) = \sigma^2 - \frac{3}{4}\sigma^4, \sigma_2 \le \sigma \le \sigma_3,$$

$$\phi_0(\sigma) = \frac{5\sigma^2 - \sigma^4}{8}, \sigma_3 \le \sigma \le 1,$$

where the values of $\sigma_i, i = 0,1,2,3$ are defined to make $\phi_0(\sigma)$ continuous, and each is the solution of an algebraic equation of degree at most four. Note also the value of $\phi_0(\sigma) \equiv \frac{1}{2}$ for $\sigma \ge 1$ and since $P(X+Y \ge 1) \le \frac{1}{2}$ because $X+Y$ is symmetric, we see that $\phi(\sigma) \equiv \frac{1}{2}$ for $\sigma \ge 1$.

Remark: Under the additional constraint that $X \sim Y$ are equi-distributed it is impossible to make $P(X+Y \ge 1) = \frac{1}{2}$. Maximizing $P(X+Y \ge 1)$ under this additional constraint is a harder problem because this functional of the d.f. of X is non-linear (quadratic). The optimal X is not necessarily a four-pointer.

A graph of $\phi_0(\sigma)$ is shown in Fig. 1.

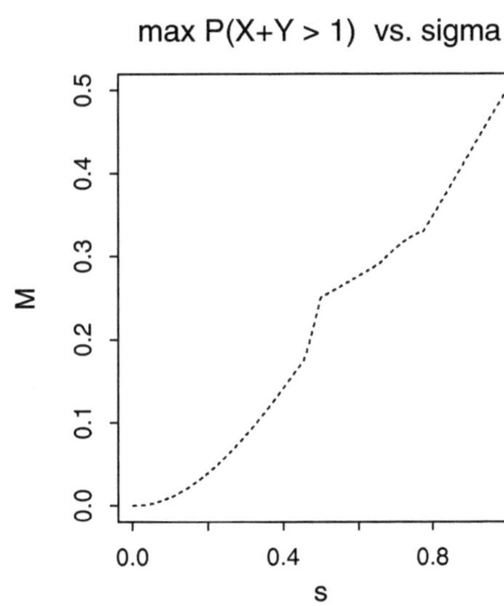

Fig. 1 The maximum value of $P(X+Y \ge 1)$ as a function of σ for independent and symmetric r.v.'s X, Y with variance σ^2

At this point we only know that $\phi(\sigma) \geq \phi_0(\sigma)$. We wanted to show that the two functions are the same, but we now believe that this is actually false in general although it seems to hold for $\sigma <\sim 0.724$. In the course of conducting the search our "best" guess kept getting better especially as we also studied upper bounds given in the next section. Searches are only as good as the searcher, and it's better to have a rigorous proof that you have found the least upper bound.

5.1 Upper Bounds

Even though the above problem is not linear, as remarked above, we may use a Chebyshev method, or a "duality" method to obtain upper bounds for $\phi(\sigma)$. Suppose for a given value of θ there is a function $f = f(x), 0 \leq x < \infty$ for which $|f(x)| \leq A + Bx^2$ for some fixed constants A, B, and for which for all values of $x, y \in (-\infty, \infty)$,

$$f(x)(y^2 - \sigma^2) + f(y)(x^2 - \sigma^2) + \theta$$

$$\geq \frac{1}{4}(\chi\{x+y \geq 1\} + \chi\{x-y \geq 1\} + \chi\{-x+y \geq 1\} + \chi\{-x-y \geq 1\}).$$

Then θ is an upper bound on $\phi(\sigma)$. The proof is to observe that if $X(\omega), Y(\omega)$ are defined on some probability space Ω, and X, Y are independent, symmetric, and have variance σ^2, then we may set $x = X(\omega), y = Y(\omega)$ and take expectations. By the symmetry, each of the four expectations on the right is the same, i.e. $P(X+Y \geq 1$, and so this immediately gives that for any such X, Y pair, $P(X+Y >= 1) \leq \theta$. The infimum of all such θ is called $\phi_1(\sigma)$ and we therefore have

$$\phi_0(\sigma) \leq \phi(\sigma) \leq \phi_1(\sigma).$$

We can make the problem slightly simpler if we take into account the symmetry. All we really need is to find $f(x)$ for a given θ such that for all $x, y \in [0, \infty)$, rather than on the whole line, we have

$$f(x)(y^2 - \sigma^2) + f(y)(x^2 - \sigma^2) + \theta \geq \frac{1}{4}(\chi\{x+y \geq 1\} + \chi\{|x-y| \geq 1\}). \quad (7)$$

Then we can merely set $f(x) = f(-x)$ to define f for negative values of x, and we can easily verify that the first set of inequalities hold for all x, y. Note that we do not expect that f will be unique. We remark that, of course, for $\theta < \phi_0(\sigma)$ such an f cannot exist.

If for some σ, the *strict* inequality $\phi_0(\sigma) < \phi_1(\sigma)$ then there is a "duality gap". The problem of minimizing θ subject to the linear constraints above *is* a (continuous) linear programming problem in the infinitely many unknowns, $f(x), 0 \leq x < \infty$ and θ. Minimizing θ, which is a linear form in the unknowns, subject to the inequalities (7) above for all x, y is thus a linear problem. The dual problem is equivalent to maximizing $P(X' + Y' \geq 1)$ over all jointly distributed symmetric r.v.'s X', Y' which now are not necessarily independent but which satisfy $E[X^2 - \sigma^2 | Y = y] \equiv 0$ and

$E[Y^2 - \sigma^2 | X = x] \equiv 0$ for all $x, y \in [0, \infty)$. To see this, observe that if we multiply in (7) by $t(x,y) \geq 0$ for x, y in a discrete finite set $\mathcal{S}$ and add we get the latter problem provided that

$$\sum_{x,y} t_{x,y} = 1,$$

$$\sum_{x} t_{x,y}(x^2 - \sigma^2) \equiv 0, \text{ and}$$

$$\sum_{y} t_{x,y}(y^2 - \sigma^2) \equiv 0,$$

which becomes the latter problem if we interpret $t_{x,y}$ as $P(X = x, Y = y)$. This conditional expectation condition is of course weaker than independence, if X, Y are actually independent then the conditional expectations vanish because the variances are σ^2. Thus the dual version of the upper bound problem drops the independence assumption. Does this increase the value of $\phi(\sigma)$ and leave a "duality gap"? Alas there is gap for some values of σ as we will see below, but for small $\sigma <\sim 0.724$ there seems to be no gap.

We believe one can produce $f(\cdot, \sigma)$ for $\sigma <\sim 0.724$, which satisfies (7), *at least numerically* on a fine lattice x, y consisting of all multiples of a small spacing. This seems to leave little doubt that there is no duality gap for these cases. A rigorous proof that the inequalities (7) hold for the given f's for all x, y and these σ's still needs to be supplied.

For the first range, $0 \leq \sigma \leq \sigma_0$, the f, which was found by discretizing the problem to a finite set of values, $x_1, \ldots, x_n$, and then solving the linear programming problem of finding the least θ for which the inequalities (7) hold for some values $f_i = f(x_i), i = 1, \ldots, N$. We then guessed the solution to the continuous version by finding the set of i, j for which equality holds in the inequalities with $x = x_i, y = x_j$.

This led to the guess that the inequalities (7) hold with equality when $0 \leq \sigma \leq \sigma_0$ when $y = y(x) = x, x \geq \frac{1}{2}$, which indicates that, for

$$\theta_0 = \phi_0(\sigma) = \sigma^2 - \frac{3}{4}\sigma^4,$$

$$f(x) = \frac{\frac{1}{2} - \theta_0}{2(x^2 - \sigma^2)}, x \geq \frac{1}{2}.$$

The linear programming solution also indicated that there is a function, $a = a(\sigma)$ for $0 \leq \sigma \leq \sigma_0$ such that for $0 \leq x \leq a$, equality holds in (7) when $y = y(x) = x + 1$. This indicates that

$$f(x) = \frac{\frac{1}{2} - \theta_0 - \frac{(\frac{1}{4} - \theta_0)(x^2 - \sigma^2)}{((1+x)^2 - \sigma^2)}}{(1+x)^2 - \sigma^2}, 0 \leq x \leq a,$$

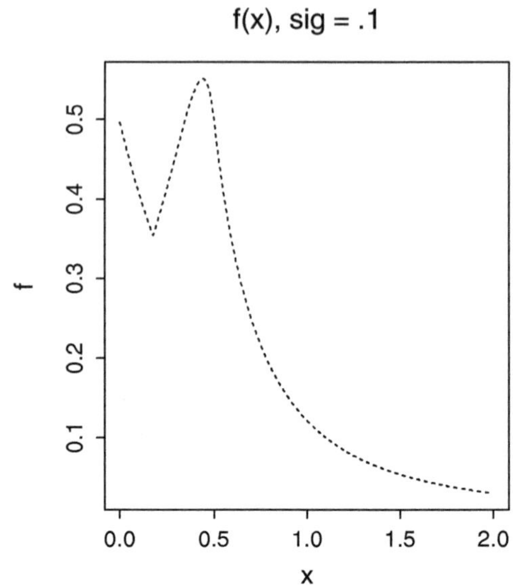

Fig. 2 The function $f(x, \mathit{sigma} = .1)$ in (7) giving the least upper bound

and also when $a \le x \le \frac{1}{2}$, that equality holds when $y = 1 - x$ which, in turn, indicates that

$$f(x) = (\frac{1}{4} - \theta_0)\frac{1-(x^2-\sigma^2)}{((1-x)^2-\sigma^2)}(1-x)^2 - \sigma^2, a \le x \le \frac{1}{2},$$

and $f(x)$ is defined for all $x \ge 0$. There is only one value of $a(\sigma)$ that makes f continuous and consistently defined. Thus for example for $\sigma = 0.1$ the value of $a(\sigma) \sim 0.1777$, found only numerically. A graph of $f(x, \sigma = 0.1)$ is given in Fig. 2.

I tried to use the same technique to guess a function $f(x) = f(x, \sigma)$ for σ in the second range $\sigma_0 < \sigma \le \frac{1}{2}$ but I could not verify all the inequalities and I am not sure whether or not this case has a duality gap or not.

The linear program now gives different $y = y(x)$ where equality holds. It now indicates that for $x \ge 1 - \sigma$, equality holds when $y = x$, which indicates that with the new optimal value for θ,

$$\theta_1 = \phi_0(\sigma) = \frac{1}{4}\frac{\sigma^2}{(1-\sigma)^2},$$

$$f(x) = \frac{\frac{1}{2} - \theta_1}{2(x^2-\sigma^2)}, x \ge 1 - \sigma.$$

For this second range of σ, $\sigma_0 \le \sigma \le \frac{1}{2}$, there is a threshold $b = b(\sigma) \le \sigma$, analogous to the threshold $a(\sigma$ for the range $0 \le \sigma \le \sigma_0$ and now, for $0 \le x \le b$, equality holds when $y = 1 + x$, whereas for $b \le x \le \sigma$, the l.p. indicates that $y = 1 - x$. This leads to the guessed f obeying the equations

$$f(x)((1+x)^2-\sigma^2)+f(x+1)(x^2-\sigma^2)+\theta_1=\frac{1}{2}, \text{ for } 0\leq x\leq b$$

which determines $f(x)$ on this range since we know $f(x+1)$, *except that we don't yet know* b, in particular we now know $f(0)$. Next,

$$f(x)(0^2-\sigma^2)+f(0)(x^2-\sigma^2)+\theta_1=0, \text{ for } \sigma\leq x\leq\frac{1}{2}$$

which determines $f(x)$ in this range. Next.

$$f(x)((1-x)^2-\sigma^2)+f(1-x)(x^2-\sigma^2)+\theta_1=\frac{1}{4}, \text{ for } b\leq x\leq\frac{1}{2}, \text{ for } b\leq x\leq\frac{1}{2},$$

which determines $f(x)$ in the range $b\leq x\leq\sigma$ (noting that this range reflects into $[1-\sigma,1-b]$ where f is known). There is a unique b which will make $f(x)$ continuous on $[0,\frac{1}{2}]$. Supposing b is so defined (there is an equation for b and we have found numerically, for example, for $\sigma=0.48>\sigma_0$, that $b(\sigma)\sim 0.311784$. The rest of the range $(b,\frac{1}{2})$ reflects into $(\frac{1}{2},1-b)$ and since $1-b>1-\sigma$ this allows us to determine $f(x)$ on $(\frac{1}{2},1-\sigma)$. This makes f well-defined everywhere. A graph of $f(x,\sigma=0.48)$ is given in Fig. 3.

It remains to determine $f(x,\sigma)$. It is an indication of trouble that the f in Fig. 2 indicates a discontinuity near $x=0.5$, and that (7) does not seem to hold as cleanly as in case I. Though I believe there is no duality gap for $\sigma<\sim 0.724$, the above method to find $f(x,\sigma)$ for this range is not a good one because it produces an extreme point of the set of functions f satisfying (7). It would be better to look for a smooth f or one in the "center" of the set of solutions of (7).

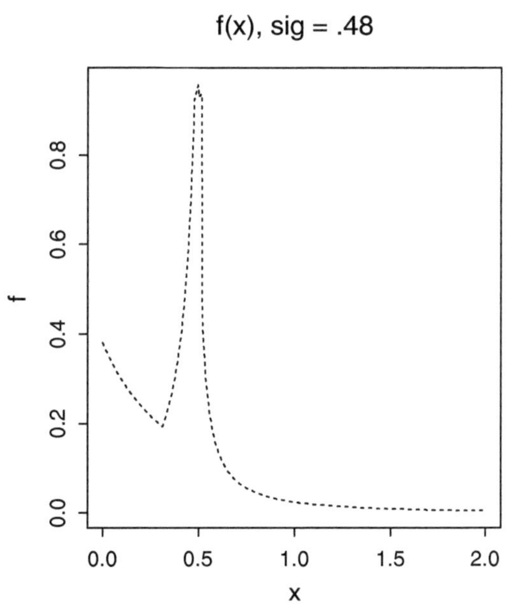

Fig. 3 The function $f\,x,\sigma=$ .48) in (7) giving the least upper bound

5.2 A Duality Gap Seems to Exist for $\sigma >\sim 0.724$

Since the problem of finding the smallest value of θ for which there exists a function $f = f(x)$ satisfying the inequalities of (7) for all nonnegative x, y is a linear programming problem (even finite if we restrict x, y to a discrete truncated lattice of values, it can be solved numerically for any fixed value of σ. We did this for various values of σ and found that the maximal value of θ is $\phi_0(\sigma)$ for $\sigma <$ around 0.7 or so, but for σ around 0.8 or so a strictly larger θ was found, and it was further found that the dual linear program was solved by a distribution of X, Y which satisfies

$$E[X^2 - \sigma^2|Y] = E[Y^2 - \sigma^2|X] \equiv 0 \tag{8}$$

but one for which X, Y fail to be independent. The actual example found for the case $\sigma = 0.8$ led to the following distribution:

$$P(X = Y = 0) = \alpha = \frac{4}{S}\frac{(1-\sigma^2)(4-\sigma^2)}{\sigma^4},$$

$$P(X = \pm\frac{1}{2}, Y = \pm\frac{1}{2}) = \beta = \frac{1}{S}\frac{(4-\sigma^2)(1-\sigma^2)}{(\sigma^2 - \frac{1}{4})^2},$$

$$P(X = 0, Y = \pm 1) = P(X = \pm 1, Y = 0) = \gamma = \frac{1}{S}\frac{4-\sigma^2}{\sigma^2},$$

$$P(X = \pm\frac{1}{2}, Y = \pm 2) = P(X = \pm 2, Y = \pm\frac{1}{2} = \delta = \frac{1}{S}\frac{4-\sigma^2}{\sigma^2 - \frac{1}{4}},$$

$$P(X = \pm 1, Y = \pm 2) = P(X = \pm 2, Y = \pm 1) = \varepsilon = \frac{1}{S},$$

where S is chosen so that all these probabilities add to unity. It's easy to verify that this distribution of X, Y satisfies (8) and that

$$P(X + Y \geq 1) = \beta + 4(\gamma + \delta + \varepsilon).$$

We will call the right side of the last equation $\phi_1(\sigma)$; a graph is given in Fig. 4.

We have verified numerically that $\phi_0(\sigma) > \phi_1(\sigma)$ for $\sigma < \sim 0.724$ but thereafter $\phi(\sigma) \ll \phi_1(\sigma)$. There is little doubt that this is right because we have run a linear program to maximize $P(X + Y \geq 1)$ over X, Y satisfying (8) and taking values in a large finite set S for various values of σ. We found that for σ slightly less than 0.724 the best X, Y pair are independent but this fails for σ slightly greater than 0.724. The value 0.724 is approximately the σ where $P(X + Y \geq 1)$ in the above example coincides with $\phi(\sigma)$ and occurs in the range (σ_2, σ_3). This of course just means that the upper bound given by linear programming is no longer tight. We still believe that for all σ,

$$\phi(\sigma) = \phi_0(\sigma),$$

though we see no way to prove this. It seems that the only proof rests on a maximization of $P(X + Y \geq 1)$ over all 4 point symmetric independent r.v.'s.

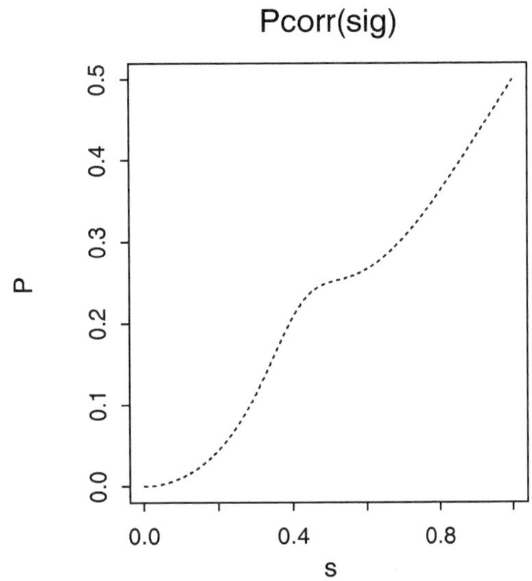

Fig. 4 The function $\phi_1(\sigma) = P(X+Y \geq 1)$ for the nonindependent pair satisfying (8)

5.3 *More General Versions of the Problem*

Consider the problem of maximizing $P(X+Y \geq 1)$ under the restriction that X, Y are independent and symmetric and have variances less than or equal to σ^2. This will have the same answer as if the variance are equal to σ^2 because we can always place ε probability of X or Y far out and increase the variance without changing the value, so any value that can be achieved can also be achieved with variance equal to σ^2.

What if we impose the additional condition that $X \sim Y$ are equi-distributed? It seems likely that the optimal X will be discrete and the number of points in its support will go to infinity with σ. If we denote by $\psi(\sigma)$ the maximum value of $P(X+Y \geq 1)$ over all r.v.'s X, Y, symmetric, identically distributed, and independent then it seems very hard to determine ψ except that when the maximum of the former problem is attained for iid X, Y then of course $\phi(\sigma) = \psi(\sigma)$. In particular this happens for $\sigma \leq \sigma_0$, and for $\frac{1}{2} \leq \sigma \leq \sigma_2$. But for $\sigma \geq 1$ it can be seen that

$$\phi(\sigma) = \frac{1}{2} > \psi(\sigma), \sigma \geq 1$$

because no iid pair can ever give $P(X+Y \geq 1) = \frac{1}{2}$.

Acknowledgements I am grateful to each of my collaborators, Jack Denny, Peter Fishburn, Ingram Olkin, Jim Reeds, and Moe Win for permission to republish joint work under my own name and even moreso if they will remain my collaborators after they learn that I have written this paper without actually asking for this permission! I am also grateful to Jack Denny for posing the last problem as a general way to bound the probability that two different methods of measuring the same quantity, each with errors respectively, X, Y, modelled as symmetric and having variances σ^2, given, would differ by at least one, using normalized units, that is the probability that $|X-Y|>1$. Because X,Y are assumed symmetric this problem reduces to the form given in the last section.

This paper is prepared for a Festschrift in honor of Peter Fishburn and the problems were chosen in the hopes that he will resolve them as he has so frequently done in the past.

The author was supported in part by the NSF grant 0203086 – Statistical Methods in Fast Functional MRI.

References

Ahlswede, R., Daykin, D. E. (1978). An inequality for the weights of two families of sets, their unions and intersections. *Zeitschrift für Wahrscheinlichkeitstheorie und verwandte Gebiete, 43*, 183–185.

Brown, M. (2006). Exploiting the waiting time paradox: Applications of the size biasing transformations. *Probability in the Engineering and Informational Sciences, 20*, 195–230.

Conti, A., Win, M. Z., Chiani, M. (2003). On the inverse symbol error probability for diversity reception. *IEEE Transactions on Communications 51*, 753–756.

Dubins, L. E. (1962). On extreme points of convex sets. *Journal of Mathematical Analysis and Applications, 5*, 237–244.

Fishburn, P. C., Shepp, L. (2000). The Ahlswede–Daykin theorem. In Althofer et al. (Eds.), *Numbers, information, and complexity* (pp. 501–516). New York: Kluwer.

Fortuin, C. M., Ginibre, J., Kastelyn, P. N. (1971). Correlational inequalities for partially ordered sets. *Communications in Mathematical Physics, 22*, 89–103.

Hardy, G. H., Littlewood, J. E., Pólya, G. (1934). *Inequalities*. Cambridge: Cambridge University Press.

Olkin, I., Shepp, L. (2006). Several colorful inequalities. *The American Mathematical Monthly, 113*, 817–822.

Reeds, J., Shepp, L., Win, M. (in press). Monotone parametric dependence of a generalized mean. *The American Mathematical Monthly*.

Siddhartha Sahi, this volume.

Steele, J. M. (2004). *The Cauchy–Schwarz master class, an introduction to the art of mathematical inequalities*. Cambridge: Cambridge University Press.

Vanderbei, R. J. (2001). *Linear programming: Foundations and extensions* (2nd ed.). Norwell MA: Kluwer.

von Neumann, J., Morganstern, O. (1944). *Theory of games and economic behavior*. Princeton: Princeton University Press.

FSC
www.fsc.org
MIX
Papier aus verantwortungsvollen Quellen
Paper from responsible sources
FSC® C105338

If you have any concerns about our products,
you can contact us on
ProductSafety@springernature.com

In case Publisher is established outside the EU,
the EU authorized representative is:
Springer Nature Customer Service Center GmbH
Europaplatz 3, 69115 Heidelberg, Germany

Printed by Libri Plureos GmbH
in Hamburg, Germany